WORLD CHECKLIST AND BIBLIOGRAPHY OF

Euphorbiaceae 3

(with Pandaceae)

WORLD CHECKLIST AND BIBLIOGRAPHY OF

Euphorbiaceae 3

(with Pandaceae)

Euphorbiaceae:
Fahrenheitia – Oxydectes

Rafaël Govaerts, David G. Frodin

and

Alan Radcliffe-Smith

(assisted by Susan Carter, Mike Gilbert and Victor Steinmann for Euphorbiinae, Hans-Jürgen Esser for Hippomaneae, and Petra Hoffman for *Antidesma*)

First published March 2000

World Checklists and Bibliographies, 4.
[The first three in this series were published respectively as Magnoliaceae (October 1996), Fagales (May 1998) and Coniferae (August 1998)].

Address of the principal authors:
Herbarium, Royal Botanic Gardens, Kew, Richmond, Surrey TW9 3AE, United Kingdom

ISBN 1 900347 85 7

Cover photograph: *Jatropha integerrima* Jacq.
Royal Botanic Gardens, Kew, 1996, *Andrew McRobb*

Cover design by Jeff Eden for Media Resources, Information Services Department, Royal Botanic Gardens, Kew

Page make-up by Media Resources, Information Services Department, Royal Botanic Gardens, Kew, from text generated by David G. Frodin using Microsoft Access 2.0® and Microsoft Word 6.0®

Printed in The European Union by Redwood Books Limited, Trowbridge, Wiltshire, UK.

Published by
The Royal Botanic Gardens, Kew
2000

Summary

Volume 1

Preface 1
Introduction 2
Euphorbiaceae: family and regional references 13
Euphorbiaceae: Aalius–Crossophora 43-415

Volume 2

Euphorbiaceae: Croton–Excoecariopsis 417-921

Volume 3

Euphorbiaceae: Fahrenheitia–Oxydectes 923-1232

Volume 4

Euphorbiaceae: Pachystemon–Zygospermum 1233-1594
Pandaceae: family and regional references 1595
Pandaceae: genera 1597
Summary of unplaced taxa and names 1603
Summary of excluded taxa 1615
Addendum 1620

Contents, Volume 3

List of Colour Plates xi
Euphorbiaceae: Checklist:
Fahrenheitia–Oxydectes 923-1232

Euphorbiaceae: Fahrenheitia – Oxydectes

Fahrenheitia 923
Falconeria 923
Flueggea 925
Fluggea 930
Fluggeopsis 930
Foersteria 930
Fontainea 931
Forsteria 932
Fourneaua 932
Fragariopsis 932
Frankia 932
Freireodendron 932
Friesia 932
Furcaria 932

Gaedawakka 933
Galarhoeus 933
Galorhoeus 933
Galurus 933
Garcia 933
Garumbium 934
Gatnaia 934
Gavarretia 934
Geblera 935
Geiseleria 935
Gelonium 935
Geminaria 936
Genesiphyla 936
Gentilia 936
Gitara 936
Givotia 936
Glochidion 937
Glochidinopsis 974
Glochidionopsis 974
Glochisandra 974
Glycydendron 974
Glyphostylus 975
Godefroya 975
Gonatogyne 975
Grimmeodendron 975
Grossera 976
Guarania 977
Gussonia 977
Gymnalypha 977
Gymnanthes 977
Gymnanthus 981
Gymnobothrys 981
Gymnocarpus 981
Gymnostillingia 981
Gynamblosis 982
Gynoon 982

Haematospermum 982
Haematostemon 982
Hamilcoa 982
Halecus 983
Halliophytum 983
Hancea 983
Hasskarlia 983
Hebecocca 983
Hecatea 985
Hecaterium 985
Hedraiostylus 985
Hedycarpus 985
Hemecyclia 985
Hemicicca 985
Hemicyclia 985
Hemiglochidion 986
Hendecandra 986
Henribaillonia 986
Heptallon 986
Hermesia 986
Heterocalymnantha 987
Heterocalyx 987
Heterochlamys 987
Heterocroton 987
Hevea 987
Hexadena 995
Hexadenia 995
Hexakestra 995
Hexakistra 995
Hexaspermum 995
Heywoodia 995
Hieronima 997
Hieronyma 997
Hippocrepandra 1001
Hippomane 1001
Holstia 1002
Homalanthus 1002
Homonoia 1008
Hotnima 1010
Humblotia 1010
Hura 1010
Hyaenanche 1011
Hyeronima 1011
Hylandia 1011

Hylococcus 1013
Hymenocardia 1013
Hypocoton 1016

Ibina 1016

Jablonskia 1016
Janipha 1017
Jatropa 1017
Joannesia 1045
Julocroton 1046
Junghuhnia 1048
Jussieuia 1048

Kairothamnus 1048
Kaluhaburunghos 1048
Kanopikon 1048
Keayodendron 1050
Keraselma 1050
Kirganelia 1051
Klaineanthus 1051
Kleinodendron 1052
Klotzschiphytum 1052
Kobiosis 1052
Koelera 1052
Koilodepas 1052
Kunstlerodendron 1054
Kurkas 1054
Kurziodendron 1054

Lacanthis 1054
Lachnostylis 1054
Lambertya 1055
Laneasagum 1055
Lasiococca 1055
Lasiocroton 1057
Lasiogyne 1058
Lasiostyles 1058
Lasipana 1058
Lassia 1058
Lathyris 1059
Lautembergia 1059
Lebidiera 1059
Lebidieropsis 1059
Leeuwenbergia 1059
Leichhardtia 1060
Leidesia 1060
Leiocarpus 1060
Leiopyxis 1061
Leontia 1061
Lepadena 1061
Lepidanthus 1061
Lepidococea 1061
Lepidocroton 1061
Lepidostachys 1061
Lepidoturus 1062
Leptemon 1062
Leptobotrys 1062
Leptomeria 1062
Leptonema 1062
Leptopus 1063
Leptorhachis 1064
Leucadenia 1064
Leucandra 1064
Leucocroton 1064
Lingelsheimia 1066
Linostachys 1068
Liodendron 1068
Liparena 1068
Liparene 1068
Lithoxylon 1068
Lobanilia 1068
Lobocarpus 1069
Loerzingia 1069
Lomanthes 1070
Longetia 1070
Lophobios 1070
Lortia 1070
Loureira 1071
Luntia 1071
Lyciopsis 1071

Mabea 1071
Maborea 1077
Macaranga 1077
Macraea 1106
Macrocroton 1107
Maesobotrya 1107
Mallotus 1111
Mancanilla 1130
Mancinella 1131
Mandioca 1131
Manihot 1131
Manihotoides 1150
Manniophyton 1151
Mappa 1151
Maprounea 1153
Mareya 1156
Mareyopsis 1158
Margaritaria 1158
Martretia 1163
Maschalanthus 1164
Mazinna 1164
Meborea 1164

Mecostylis 1164
Medea 1164
Medusea 1165
Megabaria 1165
Megalostylis 1165
Megistostigma 1165
Meialisa 1166
Meineckia 1166
Melanolepis 1170
Melanthes 1171
Melanthesa 1171
Melanthesopsis 1172
Menarda 1172
Mercadoa 1172
Mercurialis 1172
Mercuriastrum 1177
Merleta 1177
Mesandrinia 1177
Meterana 1177
Mettenia 1177
Micrandra 1177
Micrandropsis 1179
Micranthea 1180
Micrantheum 1180
Micrococca 1181
Microelus 1183
Micropetalum 1183
Microsepala 1183
Microstachys 1183
Middelbergia 1193
Mildbraedia 1193
Minutalia 1195
Mirabellia 1195
Mischodon 1195
Moacroton 1195
Moeroris 1196
Molina 1196
Monadenium 1197
Monguia 1204
Monotaxis 1204
Moultonianthus 1207
Mozinna 1207
Muricococcum 1207
Murtekias 1208
Myladenia 1208
Myricanthe 1208
Myriogomphus 1209

Nageia 1209
Nanopetalum 1209
Nealchornea 1209
Necepsia 1210
Nellica 1211
Neoboutonia 1211
Neochevaliera 1213
Neogoetzia 1213
Neoguillauminia 1213
Neoholstia 1214
Neojatropha 1214
Neomanniophyton 1214
Neomphalea 1214
Neopalissya 1214
Neopeltandra 1214
Neopycnocoma 1215
Neoroepera 1215
Neoscortechinia 1215
Neoshirakia 1217
Neotrewia 1217
Neotrigonostemon 1218
Neowawraea 1218
Nepenthandra 1218
Nephrostylus 1218
Niedenzua 1218
Niruri 1218
Niruris 1218
Nisomenes 1218
Nymania 1218
Nymphanthus 1219

Ocalia 1219
Octospermum 1219
Odonteilema 1219
Odotalon 1219
Oldfieldia 1221
Oligoceras 1221
Omalanthus 1223
Omphalandria 1223
Omphalea 1223
Omphellantha 1228
Ophthalmoblapton 1228
Orbicularia 1229
Oreoporanthera 1229
Orfilea 1230
Ostodes 1230
Owataria 1232
Oxalistylis 1232
Oxydectes 1232

List of Colour Plates

Plate 1

Hippomane mancinella L. Galápagos Is., 1996-97, *S. C. Darwin* — Private Collection

Monadenium torrei L. C. Leach. Masasi, Tanzania, 1973, *Bally 16936* (photograph from cultivated material, 1976) — Kew Slide Collection

Plate 2

Glochidion multiloculare (Rottler ex Willd.) Voigt Icones Roxburghianae, no. 1699 (as *Bradleia multiloculare* (Rottler ex Willd.) Spreng.) — Kew Illustrations Collection

Plate 3

Homalanthus populneus (Geiseler) Pax Horsfield Collection, no. 158 (as 'Carumbium populifolium'; later determined as *Homalanthus leschenaultianus* A. Juss.) — Kew Illustrations Collection

Plate 4

Hura crepitans L. *G. D. Ehret/J. J. Haid*; in C. J. Trew, *Plantae selectae*, pl. 34 (1754) — Kew Illustrations Collection

Plate 5

Joannesia princeps Vell. Brazil; in Martius, *Auswahl merkwürdiger Pflanzen*, pl. 1 (1829, as *Anda brasiliensis* Raddi) — Kew Illustrations Collection

Plate 6

Manihot esculenta Crantz (*M. utilissima* Pohl). *T. Nicholson*; *Bot. Mag.* **58**: pl. 3071 (1831) — Kew Illustrations Collection

Plate 7

Microstachys bidentata (Mart. & Zucc.) Esser. Brazil; in Martius, *Nova genera et species plantarum* 1: pl. 43 (1824, as *Cnemidostachys bidentata* Mart. & Zucc.) — Kew Illustrations Collection

Plate 8

Neoroepera banksii Benth. Kew (raised from seed sent from Queensland by A. Cunningham), 1825, *T. Duncanson* (as 'Phyllanthus banksii') — Kew Illustrations Collection

Fahrenheitia

Reduced to *Paracroton* by Balakrishnan & Chakrabarty (1993; see that genus).

Synonyms:

Fahrenheitia Rchb. & Zoll. ex Müll.Arg. === **Paracroton** Miq.
Fahrenheitia collina Rchb. & Zoll. ex Müll.Arg. === **Paracroton pendulus** (Hassk.) Miq. subsp. **pendulus**
Fahrenheitia integrifolia (Airy Shaw) Airy Shaw === **Paracroton integrifolius** (Airy Shaw) N.P.Balakr. & Chakrab.
Fahrenheitia minor (Thwaites) Airy Shaw === **Paracroton zeylanicus** (Müll.Arg.) N.P.Balakr. & Chakrab.
Fahrenheitia pendula (Hassk.) Airy Shaw === **Paracroton pendulus** (Hassk.) Miq.
Fahrenheitia sterrhopoda Airy Shaw === **Paracroton sterrhopodus** (Airy Shaw) Radcl.-Sm. & Govaerts
Fahrenheitia zeylanica (Thwaites) Airy Shaw === **Paracroton pendulus** subsp. **zeylanicus** (Thwaites) N.P.Balakr. & Chakrab.

Falconeria

1 species, S. China, S. & SE. Asia, and Malesia (but rare in the last-named); recently again segregated from *Sapium* s.l. (formerly its sect. *Falconeria*). Glabrous laticiferous small to large trees with grid-cracked, papery bark and terminal spicate inflorescences to 50 cm above the last 'whorl' of leaves. In Sri Lanka it prefers hot dry areas and in Peninsular Malaysia, limestone hills. For arguments regarding its distinctness, see Esser (1994) and Esser et al. (1997; under **Euphorbiaceae (except Euphorbieae)**). The latter authors' cladogram suggests an affinity with *Stillingia lineata*. (Euphorbioideae (except Euphorbieae))

Pax, F. (with K. Hoffmann) (1912). *Sapium*. In A. Engler (ed.), Das Pflanzenreich, IV 147 V (Euphorbiaceae-Hippomaneae): 199-258. Berlin. (Heft 52.) La/Ge. — No. 76 (pp. 241-243, with figure), *S. insigne* (in sect. *Falconeria*), is *Falconeria insignis*.

Esser, H.-J. (1999). A partial revision of the Hippomaneae (Euphorbiaceae) in Malesia. Blumea 44: 149-215, illus., maps. En. — *Falconeria*, pp. 160-165; protologue of genus and treatment of 1 species with description, synonymy, references, types, indication of distribution and habitat, illustration (*F. insignis*), map and commentary; all general references, identification list and index to botanical names at end of paper. [*F. insignis* reaches 40 m. in height.]

Falconeria Royle, Ill. Bot. Himal. Mts.: 354 (Febr. 1839).
S. China, Trop. Asia. 36 40 41 42.

Falconeria insignis Royle, Ill. Bot. Himal. Mts.: 354 (1839). *Excoecaria insignis* (Royle) Müll.Arg. in A.P.de Candolle, Prodr. 15(2): 1212 (1866). *Sapium insigne* (Royle) Benth. in G.Bentham & J.D.Hooker, Gen. Pl. 3: 335 (1880). *Sapium insigne* var. *genuinum* Pax in H.G.A.Engler, Pflanzenr., IV, 147, V: 242 (1912), nom. inval. – FIGURE, p. 924.
WC. Himalaya to SW. China and Pen. Malaysia. 36 CHC 40 ASS EHM IND NEP WHM 41 BMA CBD THA VIE 42 MLY. Nanophan. or phan.
Falconeria wallichiana Royle, Ill. Bot. Himal. Mts.: 354 (1839).
Falconeria malabarica Wight, Icon. Pl. Ind. Orient. 5: t. 1866 (1852). *Sapium insigne* var. *malabaricum* (Wight) Hook.f., Fl. Brit. India 5: 472 (1888).

Synonyms:

Falconeria malabarica Wight === **Falconeria insignis** Royle
Falconeria wallichiana Royle === **Falconeria insignis** Royle

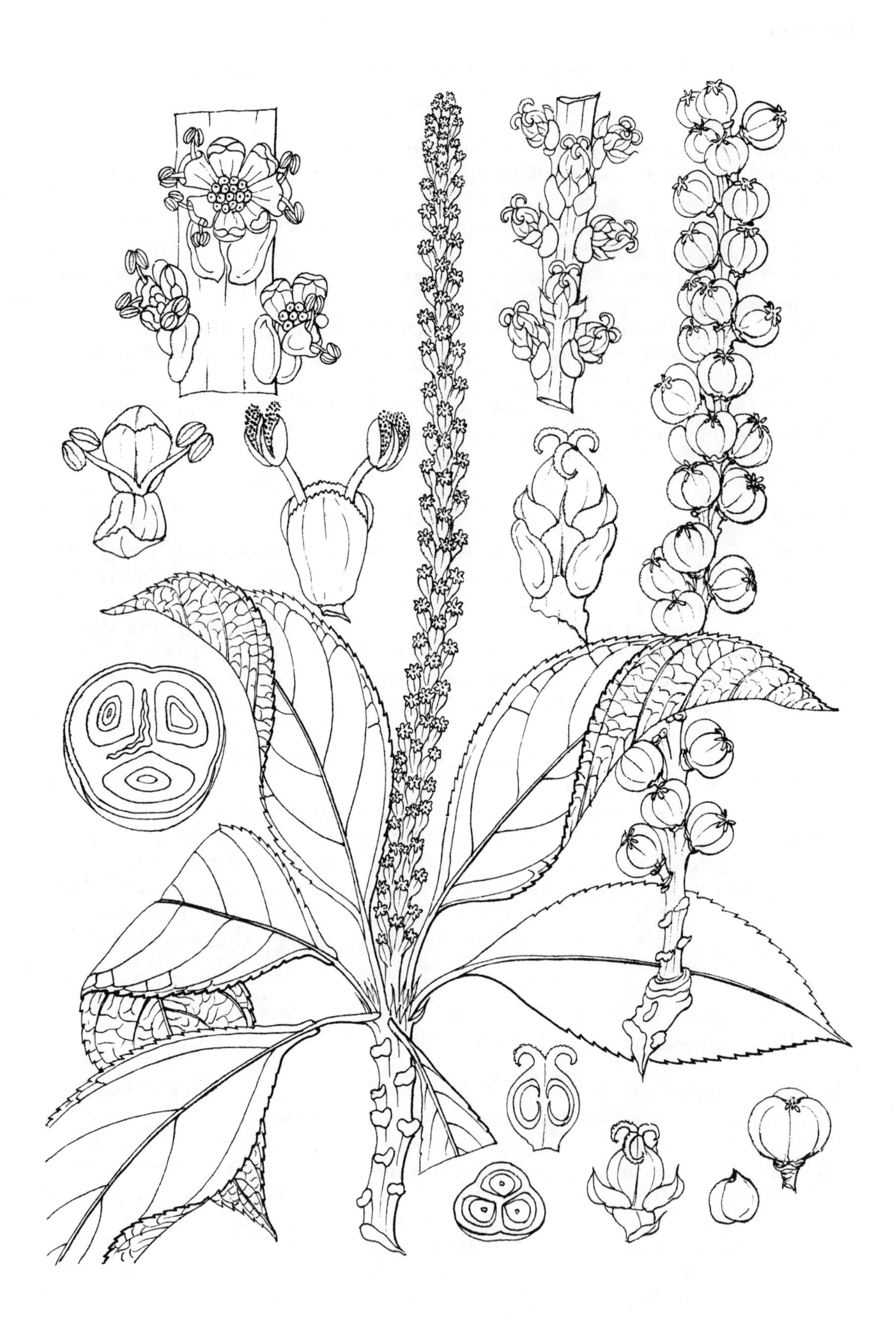

Falconeria insignis Royle (as *F. malabarica* Wight)
Artist: Govindoo
Wight, Icon. Pl. Ind. Orient. 5: pl. 1866 (1852)
KEW ILLUSTRATIONS COLLECTION

Flueggea

15 species, widespread though scattered in tropics as well as in temperate E Asia and SW Europe (for distribution map, see Webster 1984: 272); shrubs or trees to 20 m (*F. neowawrea* up to 30 m × 2 m diam.) with distichously arranged leaves and axillary flowers. The genus was never revised for *Pflanzenreich* but later Pax & Hoffmann (1931) included *Flueggea* (along with *Meineckia*) in *Securinega* and maintained *Pleiostemon* as distinct. *Flueggea* was finally revived by Webster (1984) who recognised 2 sections based primarily on the absence or presence of some kind of indumentum or scabrid surfaces. The latter, the geographically rather scattered sect. *Pleiostemon*, includes the former genus *Pleiostemon*. Since publication of Webster's revision two species have been added to the genus, one involving the former Hawaiian genus *Neowawrea* (Hayden 1987). Suggestions have been made that *Flueggea* has to an extent a 'relict' distribution, with no notable secondary centres of development. On the other hand, the very widely distributed *F. virosa* and its subspecies are successful components of more or less open seasonal vegetation from Asia through Malesia to Australia. Related genera include *Securinega* and *Savia*. [Some authors have retained *Pleiostemon* or included it in *Phyllanthus*. Any return to generic rank of Webster's section would, however, call for the name *Colmeiroa* which has priority.] (Phyllanthoideae)

Pax, F. & K. Hoffmann (1931). *Pleiostemon*. In A. Engler (ed.), Die natürlichen Pflanzenfamilien, 2. Aufl., 19c: 60. Leipzig. Ge. — Synopsis with description of genus. [1 species, *P. verrucosa*; S Africa. Not revised for Pflanzenreich. United with *Flueggea* by Webster (1984) and Webster (1994; see General).]

Pax, F. & K. Hoffmann (1931). *Securinega*. In A. Engler (ed.), Die natürlichen Pflanzenfamilien, 2. Aufl., 19c: 60. Leipzig. Ge. — Synopsis with description of genus. [*Flueggea* included here as sect. *Flueggea* and incorporating *Meineckia*. Never revised for Pflanzenreich; for revision of *Flueggea* alone, see Webster (1984).]

Leandri, J. (1958). *Flueggea*. Fl. Madag. Comores 111 (Euphorbiacées), I: 105-106. Paris. Fr. — Flora treatment (1 species, *F. obovata*).

• Webster, G. (1984). A revision of *Flueggea* (Euphorbiaceae). Allertonia 3: 271-312. En. — 13 species, widespread in tropics, temperate E Asia, and SW Europe; distribution, however, relict. Merged with *Securinega* by several workers. [Now includes the Hawaiian *Neowawraea*; see Hayden (1987).]

Brunel, J. F. (1987). *Pleiostemon*. In *idem*, Sur le genre *Phyllanthus* L. et quelques genres voisins de la tribu des Phyllantheae Dumort. Strasbourg. [See also **Phyllanthus**.] Fr. — Description of genus, with illustrations of key characters. 1 species only (S Africa).

Hayden, W. J. (1987). The identity of the genus *Neowawraea* (Euphorbiaceae). Brittonia 39: 268-277, illus., maps En. — Reduction of *Neowawraea phyllanthoides* J.F. Rock (and the genus) of the Hawaiian Is. to *Flueggea*, the species being renamed as *F. neowawraea*. Specimens seen, with localities, appear at the end of the paper. [*F. neowawraea* is characteristic of drier western parts of the islands and with low population densities is threatened, with insect (borer) predation also a problem. It appears most closely related to the Malesian and Oceanian *F. flexuosa* in sect. *Flueggea* (cf. Webster, 1984).]

Flueggea Willd., Enum. Pl.: 1013 (1809).

Trop. & Subtrop. 12 20 22 23 24 25 26 27 29 30 31 34 35 36 37 38 40 41 42 50 60 63 81 83 84. Nanophan. or phan.

Acidoton P.Browne, Civ. Nat. Hist. Jamaica: 355 (1756), nom. rejic.

Fluggea Willd., Sp. Pl. 4: 757 (1806).

Bessera Spreng., Pl. Min. Cogn. Pug. 2: 90 (1815), nom. illeg.

Geblera Fisch. & C.A.Mey., Index Seminum (LE) 1: 28 (1835).

Colmeiroa Reut. in P.E.Boissier & G.F.Reuter, Diagn. Pl. Nov. Hisp.: 23 (1842).

Coilmeroa Endl., Gen. Pl., Suppl. 3: 99 (1843).

Pleiostemon Sond., Linnaea 23: 135 (1850).

Villanova Pourr. ex Cutanda, Fl. Comp. Madrid: 595 (1861).

Neowawraea Rock, Indig. Trees Haw. Isl.: 243 (1913).

Flueggea acicularis (Croizat) G.L.Webster, Allertonia 3: 304 (1984).
China (W. Hubei, E. Sichuan, Yunnan). 36 CHC. Nanophan.
* *Securinega acicularis* Croizat, J. Arnold Arbor. 21: 491 (1940).

Flueggea acidoton (L.) G.L.Webster, Allertonia 3: 299 (1984).
Caribbean. 81 BAH CUB DOM HAI JAM LEE PUE. Nanophan. or phan.
* *Adelia acidoton* L., Syst. Nat. ed. 10: 1298 (1759). *Securinega acidoton* (L.) Fawc., J. Bot. 57: 68 (1919).
Flueggea acidothamnus Griseb., Nachr. Königl. Ges. Wiss. Georg-Augusts-Univ. 1865: 164 (1865). *Securinega acidothamnus* (Griseb.) Müll.Arg. in A.P.de Candolle, Prodr. 15(2): 451 (1866). *Acidoton acidothamnus* (Griseb.) Kuntze, Revis. Gen. Pl. 2: 592 (1891).

Flueggea anatolica Gemici, Edinburgh J. Bot. 50: 75 (1993).
S. Turkey. 34 TUR. Nanophan.

Flueggea elliptica (Spreng.) Baill., Étude Euphorb.: 592 (1858).
Ecuador. 83 ECU. Nanophan. or phan.
Phyllanthus ellipticus Kunth in F.W.H.von Humboldt, A.J.A.Bonpland & C.S.Kunth, Nov. Gen. Sp. 2: 114 (1817), nom. illeg. *Kirganelia elliptica* Spreng., Syst. Veg. 3: 48 (1826). *Securinega elliptica* (Spreng.) Müll.Arg.in A.P.de Candolle, Prodr. 15(2): 448 (1866). *Acidoton ellipticus* (Spreng.) Kuntze, Revis. Gen. Pl. 2: 592 (1891).
Phyllanthus rubellus Müll.Arg. in A.P.de Candolle, Prodr. 15(2): 379 (1866).

Flueggea flexuosa Müll.Arg., Linnaea 34: 76 (1865). *Securinega flexuosa* (Müll.Arg.) Müll.Arg. in A.P.de Candolle, Prodr. 15(2): 450 (1866). *Acidoton flexuosus* (Müll.Arg.) Kuntze, Revis. Gen. Pl. 2: 592 (1891).
Philippines to Samoa. 42 BIS MOL NWG PHI 60 FIJ SAM SOL TON WAL. Phan.
Phyllanthus acuminatissimus C.B.Rob., Philipp. J. Sci., C 3: 200 (1908). *Securinega acuminatissima* (C.B.Rob.) C.B.Rob., Philipp. J. Sci., C 4: 73 (1909).
Securinega samoana Croizat, Bernice P. Bishop Mus. Bull. 184: 45 (1945).

Flueggea jullienii (Beille) G.L.Webster, Allertonia 3: 308 (1984).
Cambodia, Laos, Vietnam. 41 CBD LAO VIE. Nanophan.
* *Phyllanthus jullienii* Beille in H.Lecomte, Fl. Indo-Chine 5: 576 (1927).

Flueggea leucopyrus Willd., Sp. Pl. 4: 757 (1806). *Phyllanthus leucopyrus* (Willd.) D.Koenig ex Roxb., Fl. Ind. ed. 1832, 3: 658 (1832). *Securinega leucopyrus* (Willd.) Müll.Arg. in A.P.de Candolle, Prodr. 15(2): 451 (1866). *Cicca leucopyrus* (Willd.) Kurz, Forest Fl. Burma 2: 353 (1877). *Acidoton leucopyrus* (Willd.) Kuntze, Revis. Gen. Pl. 2: 592 (1891).
S. India, Sri Lanka, China (Sichuan, Yunnan). 36 CHC 40 IND SRL. Nanophan. or phan.
Phyllanthus lucena B.Heyne ex Roth, Nov. Pl. Sp.: 185 (1821).
Xylophylla lucena Roth, Nov. Pl. Sp.: 185 (1821).
Flueggea xerocarpa A.Juss., Euphorb. Gen.: 106 (1824).
Phyllanthus albicans Wall., Numer. List: 7937 (1847), nom. inval.
Flueggea wallichiana Baill., Étude Euphorb.: 592 (1858).

Flueggea monticola G.L.Webster, Allertonia 3: 283 (1984).
China (Sichuan, Yunnan). 36 CHC. Nanophan. or phan.

Flueggea neowawraea W.J.Hayden, Brittonia 39: 268 (1987).
Hawaiian Is. 63 HAW. Phan.
* *Neowawraea phyllanthoides* Rock, Indig. Trees Haw. Isl.: 245 (1913). *Drypetes phyllanthoides* (Rock) Sherff, Publ. Field Mus. Nat. Hist., Bot. Ser. 17: 562 (1939).

Flueggea schuechiana (Müll.Arg.) G.L.Webster, Allertonia 3: 277 (1984).
Brazil (Pernambuco). 84 BZE. Nanophan.

** Securinega schuechiana* Müll.Arg. in C.F.P.von Martius, Fl. Bras. 11(2): 78 (1873).
Acidoton schuechianus (Müll.Arg.) Kuntze, Revis. Gen. Pl. 2: 592 (1891).

Flueggea spirei Beille in H.Lecomte, Fl. Indo-Chine 5: 529 (1927). *Securinega spirei* (Beille) Croizat, J. Arnold Arbor. 23: 32 (1942).
Laos. 41 LAO. Nanophan.

Flueggea suffruticosa (Pall.) Baill., Étude Euphorb.: 502 (1858).
Temp. Asia. 30 CTA 31 AMU PRM 36 CHC CHH CHI CHM CHN CHQ CHS CHT 37 MON 38 JAP KOR TAI. Nanophan.
** Pharnaceum suffruticosum* Pall., Reise Russ. Reich. 3: 716 (1776). *Geblera suffruticosa* (Pall.) Fisch. & C.A.Mey., Index Seminum (LE) 1: 28 (1835). *Securinega suffruticosa* (Pall.) Rehder, J. Arnold Arbor. 13: 338 (1932).
Xylophylla ramiflora Aiton, Hort. Kew. 1: 376 (1789). *Phyllanthus ramiflorus* (Aiton) Pers., Syn. Pl. 2: 591 (1807). *Securinega ramiflora* (Aiton) Müll.Arg. in A.P.de Candolle, Prodr. 15(2): 449 (1866), nom. illeg. *Acidoton ramiflorus* (Aiton) Kuntze, Revis. Gen. Pl. 2: 592 (1891).
Xylophylla parviflora Bellardi ex Colla, Herb. Pedem. 5: 106 (1836).
Geblera chinensis Rupr., Bull. Cl. Phys.-Math. Acad. Imp. Sci. Saint-Pétersbourg 15: (1857).
Geblera sungariensis Rupr., Bull. Cl. Phys.-Math. Acad. Imp. Sci. Saint-Pétersbourg 15: 357 (1857).
Phyllanthus fluggeoides Müll.Arg., Linnaea 32: 16 (1863). *Securinega fluggeoides* (Müll.Arg.) Müll.Arg. in A.P.de Candolle, Prodr. 15(2): 450 (1866). *Acidoton flueggeoides* (Müll.Arg.) Kuntze, Revis. Gen. Pl. 2: 592 (1891). *Flueggea flueggeoides* (Müll.Arg.) G.L.Webster, Brittonia 18: 373 (1967).
Securinega japonica Miq., Ann. Mus. Bot. Lugduno-Batavi 3: 128 (1867). *Flueggea japonica* (Miq.) Pax in H.G.A.Engler & K.A.E.Prantl, Nat. Pflanzenfam. 3(5): 18 (1890).
Phyllanthus trigonocladus Ohwi, Bull. Natl. Sci. Mus. Tokyo, n.s., 1(34): 7 (1954).
Securinega multiflora S.B.Liang, Bull. Bot. Res., Harbin 8(4): 89 (1988).

Flueggea tinctoria (L.) G.L.Webster, Allertonia 3: 302 (1984).
Portugal, Spain. 12 POR SPA. Nanophan.
** Rhamnus tinctoria* L. in P.Loefling, Iter Hispan.: 302 (1758). *Securinega tinctoria* (L.) Rothm., Repert. Spec. Nov. Regni Veg. 49: 276 (1940).
Adelia gracilis Salisb., Prodr. Stirp. Chap. Allerton: 388 (1796).
Adelia virgata Poir. in J.B.A.M.de Lamarck, Encycl., Suppl. 1: 132 (1810). *Securinega virgata* (Poir.) Maire, Bull. Soc. Hist. Nat. Afrique N. 27: 259 (1936).
Colmeiroa buxifolia Reut. in P.E.Boissier & G.F.Reuter, Diagn. Pl. Nov. Hisp.: 23 (1842). *Securinega buxifolia* (Reut.) Müll.Arg. in A.P.de Candolle, Prodr. 15(2): 452 (1866). *Acidoton buxifolius* (Reut.) Kuntze, Revis. Gen. Pl. 2: 592 (1891).

Flueggea verrucosa (Thunb.) G.L.Webster, Allertonia 3: 305 (1984).
SE. Cape Prov., S. KwaZulu-Natal. 27 CPP NAT. Nanophan.
** Phyllanthus verrucosus* Thunb., Prodr. Pl. Cap.: 24 (1794). *Pleiostemon verrucosum* (Thunb.) Sond., Linnaea 23: 136 (1850). *Securinega verrucosa* (Thunb.) Benth. ex Pax in H.G.A.Engler & K.A.E.Prantl, Nat. Pflanzenfam. 3(5): 18 (1890).

Flueggea virosa (Roxb. ex Willd.) Voigt, Hort. Suburb. Calcutt.: 152 (1845).
– FIGURE, p. 928.
Africa to Australia. 20 EGY 22 BKN GAM GHA GUI IVO MLI NGA SIE SEN TOG 23 BUR CAF CMN RWA ZAI 24 CHA ETH SOM SUD 25 ALL 26 ALL 27 BOT NAM NAT SWZ TVL 29 REU 35 YEM 36 CHC CHH CHS 38 TAI 40 ASS EHM NEP 41 BMA NCB 42 BOR LSI MOL NWG PHI SUM 50 NTA QLD WAU. Nanophan. or phan.
** Phyllanthus virosus* Roxb. ex Willd., Sp. Pl. 4: 578 (1805). *Securinega virosa* (Roxb. ex Willd.) Baill., Adansonia 6: 334 (1866). *Acidoton virosus* (Roxb. ex Willd.) Kuntze, Revis. Gen. Pl. 2: 592 (1891).

Flueggea virosa (Roxb. ex Willd.) Voigt subsp. *virosa* (as *F. virosa* Baill.)
Artist: William E. Trevithick
Fl. West Tropical Africa 1: 292, fig. 119 (1928)
KEW ILLUSTRATIONS COLLECTION

subsp. **himalaica** D.G.Long, Notes Roy. Bot. Gard. Edinburgh 44: 167 (1986).
C. Himalaya to Burma. 40 ASS EHM NEP 41 BMA. Nanophan.

subsp. **melanthesoides** (F.Muell.) G.L.Webster, Allertonia 3: 294 (1984).
New Guinea, N. Australia. 42 NWG 50 NTA QLD WAU. Nanophan. or phan.
* *Leptonema melanthesoides* F.Muell., Hooker's J. Bot. Kew Gard. Misc. 9: 17 (1857).
Flueggea melanthesoides (F.Muell.) F.Muell., Trans. Bot. Soc. Edinburgh 7: 490 (1863).
Securinega melanthesoides (F.Muell.) Airy Shaw, Kew Bull. 31: 352 (1976).
Securinega virosa var. *australiana* Baill., Adansonia 6: 334 (1866).
Securinega keyensis Warb., Bot. Jahrb. Syst. 13: 358 (1891). *Flueggea keyensis* (Warb.) Boerl., Handl. Fl. Ned. Ind. 3: 216 (1900).
Flueggea novoguineensis Valeton ex Hallier f., Meded. Rijks Herb. 36: 5 (1918).
Flueggea virosa f. *reticulata* Domin, Biblioth. Bot. 22(89): 878 (1927).
Flueggea virosa var. *aridicola* Domin, Biblioth. Bot. 22(89): 878 (1927). *Securinega melanthesoides* var. *aridicola* (Domin) Airy Shaw, Muelleria 4: 213 (1980).

subsp. **virosa**
Africa, Trop. & Subtrop. Asia. 20 EGY 22 BKN GAM GHA GUI IVO MLI NGA SIE SEN TOG 23 BUR CAF CMN RWA ZAI 24 CHA ETH SOM SUD 25 ALL 26 ALL 27 BOT NAM NAT SWZ TVL 29 REU 35 YEM 36 CHC CHH CHS 38 TAI 40 ASS 41 BMA NCB 42 BOR LSI MOL PHI SUM. Nanophan. or phan.
Phyllanthus hamrur Forssk., Fl. Aegypt.-Arab.: 159 (1775), nom rejic.
Xylophylla obovata Willd., Enum. Pl.: 329 (1809). *Securinega obovata* (Willd.) Müll.Arg. in A.P.de Candolle, Prodr. 15(2): 449 (1866). *Cicca obovata* (Willd.) Kurz, Forest Fl. Burma 2: 354 (1877). *Flueggea obovata* (Willd.) Wall. ex Fern.-Vill. in F.M.Blanco, Fl. Filip., ed. 3, 4(13A): 189 (1880). *Acidoton obovatus* (Willd.) Kuntze, Revis. Gen. Pl. 2: 592 (1891).
Bessera inermis Spreng., Pl. Min. Cogn. Pug. 2: 90 (1815).
Drypetes bengalensis Spreng., Syst. Veg. 3: 902 (1826).
Flueggea microcarpa Blume, Bijdr.: 580 (1826). *Securinega microcarpa* (Blume) Müll.Arg. in A.P.de Candolle, Prodr. 15(2): 434 (1866).
Phyllanthus angulatus Schumach. & Thonn. in C.F.Schumacher, Beskr. Guin. Pl.: 415 (1827). *Flueggea angulata* (Schumach. & Thonn.) Schrank, Fl. Syll. Ratisb. 2: 65 (1828).
Phyllanthus dioicus Schumach. & Thonn. in C.F.Schumacher, Beskr. Guin. Pl.: 416 (1827). *Bradleia dioica* (Schumach. & Thonn.) Gaertn. ex Vahl, Adansonia 1: 80 (1860).
Phyllanthus obtusus Schrank, Syll. Pl. Nov. 2: 65 (1828).
Cicca pentandra Blanco, Fl. Filip.: 701 (1837).
Phyllanthus lucidus Steud., Nomencl. Bot., ed. 2, 2: 327 (1841).
Phyllanthus polygamus Hochst. ex A.Rich., Tent. Fl. Abyss. 2: 256 (1850), pro syn.
Securinega abyssinica A.Rich., Tent. Fl. Abyss. 2: 256 (1850). *Flueggea abyssinica* (A.Rich.) Baill., Étude Euphorb.: 593 (1858).
lueggea phyllanthoides Baill., Étude Euphorb.: 592 (1858). *Acidoton phyllanthoides* (Baill.) Kuntze, Revis. Gen. Pl. 2: 592 (1891).
Phyllanthus leucophyllus Strachey & Winterb. ex Baill., Étude Euphorb.: 593 (1858).
Phyllanthus reichenbachianus Sieber ex Baill., Adansonia 1: 80 (1860).
Flueggea senensis Klotzsch in W.C.H.Peters, Naturw. Reise Mossambique: 106 (1861).
Securinega grisea Müll.Arg. in A.P.de Candolle, Prodr. 15(2): 451 (1866). *Acidoton griseus* (Müll.Arg.) Kuntze, Revis. Gen. Pl. 2: 592 (1891).
Securinega leucopyrus Brandis, Forest Fl. N.W. India: 456 (1874).
Diasperus portoricensis Kuntze, Revis. Gen. Pl. 2: 602 (1891). *Phyllanthus portoricensis* (Kuntze) Urb., Symb. Antill. 4: 338 (1905). *Conami portoricensis* (Kuntze) Britton, Bot. Porto Rico 5: 475 (1924).
Flueggea obovata var. *luxurians* A.Chev. ex Beille, Bull. Soc. Bot. France 55(8): 55 (1908).

Synonyms:
Flueggea abyssinica (A.Rich.) Baill. === **Flueggea virosa** (Roxb. ex Willd.) Voigt subsp. **virosa**
Flueggea acidothamnus Griseb. === **Flueggea acidoton** (L.) G.L.Webster
Flueggea angulata (Schumach. & Thonn.) Schrank === **Flueggea virosa** (Roxb. ex Willd.) Voigt subsp. **virosa**
Flueggea bailloniana (Müll.Arg.) Pax === **Margaritaria discoidea** var. **triplosphaera** Radcl.-Sm.
Flueggea capillipes Pax === **Andrachne chinensis** Bunge
Flueggea eglandulosa Baill. === **Margaritaria anomala** (Baill.) Fosberg
Flueggea fagifolia Pax === **Margaritaria discoidea** var. **fagifolia** (Pax) Radcl.-Sm.
Flueggea flueggeoides (Müll.Arg.) G.L.Webster === **Flueggea suffruticosa** (Pall.) Baill.
Flueggea hilariana Baill. === **Meineckia neogranatensis** subsp. **hilariana** (Baill.) G.L.Webster
Flueggea japonica (Miq.) Pax === **Flueggea suffruticosa** (Pall.) Baill.
Flueggea javanica Blume === **Streblus spinosus** (Blume) Corner (Moraceae)
Flueggea keyensis (Warb.) Boerl. === **Flueggea virosa** subsp. **melanthesoides** (F.Muell.) G.L.Webster
Flueggea major Baill. === **Margaritaria anomala** (Baill.) Fosberg
Flueggea meineckia Müll.Arg. === **Meineckia phyllanthoides** Baill.
Flueggea melanthesoides (F.Muell.) F.Muell. === **Flueggea virosa** subsp. **melanthesoides** (F.Muell.) G.L.Webster
Flueggea microcarpa Blume === **Flueggea virosa** (Roxb. ex Willd.) Voigt subsp. **virosa**
Flueggea nitida Pax === **Margaritaria discoidea** var. **nitida** (Pax) Radcl.-Sm.
Flueggea novoguineensis Valeton ex Hallier f. === **Flueggea virosa** subsp. **melanthesoides** (F.Muell.) G.L.Webster
Flueggea obovata Baill. === **Margaritaria discoidea** var. **triplosphaera** Radcl.-Sm.
Flueggea obovata (Willd.) Wall. ex Fern.-Vill. === **Flueggea virosa** (Roxb. ex Willd.) Voigt subsp. **virosa**
Flueggea obovata var. *luxurians* A.Chev. ex Beille === **Flueggea virosa** (Roxb. ex Willd.) Voigt subsp. **virosa**
Flueggea phyllanthoides Baill. === **Flueggea virosa** (Roxb. ex Willd.) Voigt subsp. **virosa**
Flueggea senensis Klotzsch === **Flueggea virosa** (Roxb. ex Willd.) Voigt subsp. **virosa**
Flueggea serrata Miq. === **Celastrus hindsii** Benth. (Celastraceae)
Flueggea trichogynis Baill. === **Meineckia trichogynis** (Baill.) G.L.Webster
Flueggea virosa var. *aridicola* Domin === **Flueggea virosa** subsp. **melanthesoides** (F.Muell.) G.L.Webster
Flueggea virosa f. *reticulata* Domin === **Flueggea virosa** subsp. **melanthesoides** (F.Muell.) G.L.Webster
Flueggea wallichiana Baill. === **Flueggea leucopyrus** Willd.
Flueggea xerocarpa A.Juss. === **Flueggea leucopyrus** Willd.

Fluggea

An orthographic variant of *Flueggea*.

Synonyms:
Fluggea Willd. === **Flueggea** Willd.

Fluggeopsis

of K. Schumann = *Phyllanthus*

Foersteria

Synonyms:
Foersteria Scop. === **Breynia** J.R.Forst. & G.Forst.

Fontainea

9 species, E Australia, New Guinea, Vanuatu (Aneityum) and New Caledonia (cf. distribution map in Jessup and Guymer, 1985); shrubs or small to medium forest trees related to *Codiaeum*. The genus is absent from the Solomon Islands. The revision by Jessup and Guymer is in effect limited to Australia with for the New Caledonian *F. pancheri* a note to the effect that 'three taxa [were identified] amongst the specimens seen from New Caledonia and Vanuatu'. New Guinean records of *F. pancheri* were subsequently described as two new species by Forster (1997). (Crotonoideae)

Heckel, E. M. (1870). Étude au point de vue botanique et théraputique sur le *Fontainea pancheri* (nobis). Montpellier. (Thése, Faculté de Médecine, Univ. Montpellier.) — Includes the protologue of the genus; from New Caledonia.

Pax, F. (with K. Hoffmann) (1911). *Fontainea*. In A. Engler (ed.), Das Pflanzenreich, IV 147 III (Euphorbiaceae-Cluytieae): 30-31. Berlin. (Heft 47.) La/Ge. — 1 species, New Caledonia and E Australia. [Treatment now well out of date.]

Airy-Shaw, H. K. (1974). Notes on Malesian and other Asiatic Euphorbiaceae, CLXXXIV. The genus *Fontainea* Heckel in New Guinea. Kew Bull. 29: 326-328. En. — Extension of range of *F. pancheri* to widely scattered parts of Papua New Guinea; the species had been known from Queensland, New Caledonia and Vanuatu. [The three records concerned have now been referred to other species; see Forster (1997).]

Jessop, L. W. & G. Guymer (1985). A revision of *Fontainea* Heckel (Euphorbiaceae-Cluytieae). Austrobaileya 2: 112-125, illus., maps. En. — Descriptive revision (6 species, 5 in Australia) with key, synonymy, references, types, localities with exsiccatae, indication of distribution and habitat, and commentary. [See also L. W. Jessop in Austral. Pl. 16: 182-183, 185 (1991) for a sketch with illustrations.]

McPherson, G. & C. Tirel (1987). *Fontainea*. Fl. Nouvelle-Calédonie, 14 (Euphorbiacées, I): 74-78. Paris. Fr. — Flora treatment (1 endemic species out of the 6 known for the genus).

Forster, P. I. (1997). Three new species of *Fontainea* Heckel (Euphorbiaceae) from Australia and Papua New Guinea. Austrobaileya 5: 29-37, illus. En. — Descriptions of 3 new species, 2 from Papua New Guinea (accounting for all records referred to *F. pancheri* by Airy-Shaw), 1 from southeastern Queensland. In New Guinea *F. borealis* is from Eastern Highlands Province and *F. subpapuana* from lowland Central Province.

Forster, P. I. & P. C. van Welzen (1999). The Malesian species of *Choriceras*, *Fontainea* and *Petalostigma* (Euphorbiaceae). Blumea 44: 99-107, illus. En. — *Fontainea*, pp. 101-104; treatment of 2 species (in New Guinea) with genus and species descriptions, key, synonymy, references, types, indication of distribution and ecology, and general notes; general references and lists of specimens seen at end of paper.

Fontainea Heckel, Étude Fontainea (Thèse Inaug. Montpellier): 9 (1870).
New Guinea, Australia, Vanuatu, New Caledonia. 42 50 60.

Fontainea australis Jessup & Guymer, Austrobaileya 2: 116 (1985).
New South Wales (NW. of Tyalgum). 50 NSW. Nanophan. or phan.

Fontainea borealis P.I.Forst., Austrobaileya 5: 29 (1997).
Papua New Guinea. 42 NWG. Nanophan. or phan.

Fontainea fugax P.I.Forst., Austrobaileya 5: 34 (1997).
Queensland (Burnett Distr.). 50 QLD. Nanophan. or phan.

Fontainea oraria Jessup & Guymer, Austrobaileya 2: 119 (1985).
New South Wales (S. of Lennox Head). 50 NSW. Nanophan. or phan.

Fontainea pancheri (Baill.) Heckel, Étude Fontainea (Thése Inaug. Montpellier): 11 (1870).
Vanuatu ?, New Caledonia (incl. Loyalty Is.). 60 NWC VAN? Phan.
* *Baloghia pancheri* Baill., Adansonia 2: 214 (1862). *Codiaeum pancheri* (Baill.) Müll.Arg. in A.P.de Candolle, Prodr. 15(2): 1117 (1866).

Fontainea picrosperma C.T.White, Contr. Arnold Arbor. 4: 55 (1933).
NE. Queensland (Atherton Tableland). 50 QLD. Phan.

Fontainea rostrata Jessup & Guymer, Austrobaileya 2: 114 (1985).
SE. Queensland (Gympie). 50 QLD. Nanophan. or phan.

Fontainea subpapuana P.I.Forst., Austrobaileya 5: 32 (1997).
Papua New Guinea. 42 NWG. Nanophan. or phan.

Fontainea venosa Jessup & Guymer, Austrobaileya 2: 122 (1985).
Queensland (SW. of Beenleigh). 50 QLD. Phan.

Forsteria

An orthographic variant of *Foersteria.*

Synonyms:
Forsteria Steud. === **Breynia** J.R.Forst. & G.Forst.

Fourneaua

Synonyms:
Fourneaua Pierre ex Prain === **Grossera** Pax

Fragariopsis

Reduced to *Plukenetia* by Gillespie (1993; see that genus).

Synonyms:
Fragariopsis A.St.-Hil. === **Plukenetia** L.
Fragariopsis paxii Pittier === **Plukenetia volubilis** L.
Fragariopsis scandens A.St.-Hil. === **Plukenetia serrata** (Vell.) L.J.Gillespie
Fragariopsis warmingii Müll.Arg. === **Plukenetia serrata** (Vell.) L.J.Gillespie

Frankia

Synonyms:
Frankia Steud. === **Phyllanthus** L.

Freireodendron

Synonyms:
Freireodendron Müll.Arg. === **Drypetes** Vahl
Freireodendron sessiliflorum (Allemão) Müll.Arg. === **Drypetes sessiliflora** Allemão

Friesia

A synonym of *Crotonopsis*; that genus is now in turn part of *Croton.*

Synonyms:
Friesia Spreng. === **Croton** L.

Furcaria

Synonyms:
Furcaria Boivin ex Baill. === **Croton** L.
Furcaria boiviniana Baill. === **Croton boivinianus** (Baill.) Baill.

Gaedawakka

Synonyms:
Gaedawakka L. ex Kuntze === **Chaetocarpus** Thwaites
Gaedawakka schomburgkiana Kuntze === **Chaetocarpus schomburgkianus** (Kuntze) Pax & K.Hoffm.

Galarhoeus

A segregate of *Euphorbia* and, in particular, its subgen. *Esula*.

Synonyms:
Galarhoeus Haw. === **Euphorbia** L.
Galarhoeus androsaemifolius Haw. === **Euphorbia lucida** Waldst. & Kit.
Galarhoeus croizatii Hurus. === **Euphorbia leoncroizatii** Oudejans
Galarhoeus densiusculiformis Pazij === **Euphorbia densiusculiformis** (Pazij) Botsch.
Galarhoeus ebracteolatus var. *anwheiensis* Hurus. === **Euphorbia ebracteolata** Hayata
Galarhoeus helioscopius f. *litoralis* Hurus. === **Euphorbia helioscopia** L. subsp. **helioscopia**
Galarhoeus jolkinii f. *insularis* Hurus. === **Euphorbia jolkinii** Boiss.
Galarhoeus kudrjaschevii Pazij === **Euphorbia kudrjaschevii** (Pazij) Prokh.
Galarhoeus lasiocaulus f. *maritima* Hurus. === **Euphorbia pekinensis** Rupr. subsp. **pekinensis**
Galarhoeus lasiocaulus var. *pseudolucorum* Hurus. === **Euphorbia pekinensis** Rupr. subsp. **pekinensis**
Galarhoeus sieboldianus f. *grandifolius* Franch. & Sav. ex Hurus. === **Euphorbia sieboldiana** Morris & Decne.
Galarhoeus sieboldianus var. *ohsumiensis* Hurus. === **Euphorbia sieboldiana** Morris & Decne.
Galarhoeus spathulifolius Haw. === ?
Galarhoeus subtilis Prokh. === **Euphorbia esula** L. subsp. **esula**
Galarhoeus togakusensis f. *intermedius* Hurus. === **Euphorbia togakusensis** Hayata
Galarhoeus zhiguliensis Prokh. === **Euphorbia esula** L. subsp. **esula**

Galorhoeus

An orthographic variant of *Galarhoeus*.

Synonyms:
Galorhoeus Endl. === **Euphorbia** L.

Galurus

Synonyms:
Galurus Spreng. === **Acalypha** L.

Garcia

2 species, Mexico (with one species, *G. nutans*, possibly native elsewhere); trees representing the closest American relatives of the Asian Aleuritinae. *G. nutans* was thought by Lundell (1945: 1-2) to be native only to San Luis Potosí, Mexico with records elsewhere reflecting cultivation or naturalisation. The other species is limited to Tabasco, Mexico. The seeds of *G. nutans* yield a hard, quick-drying oil similar to tung oil (*Vernicia fordii*); studies towards possible commercial development were made during World War II (Lundell 1945). (Crotonoideae)

Pax, F. (1910). *Garcia*. In A. Engler (ed.), Das Pflanzenreich, IV 147 [I] (Euphorbiaceae-Jatropheae): 14-15. Berlin. (Heft 42.) La/Ge. — 1 species, Middle America (Mexico and West Indies) and NW S America.

Lundell, C. L. (1945). The genus *Garcia* Vahl, a potential source of superior hard quick-drying oil. Wrightia 1: 1-12. En. — Introduction; revision (2 species, one new and very local in Tabasco, Mexico) with key, descriptions, synonymy, vernacular names, localities with exsiccatae, and commentary; notes on garcia oil and its commercial exploitation (pp. 7-12). [The long-known *G. nutans* is here shown naturally to be less widely distributed than was supposed, being endemic to San Luis Potosí in Mexico.]

Garcia Vahl ex Rohr, Skr. Naturhist.-Selsk. 2: 217 (1792).
Mexico to Brazil. 79 80 81 83 84.
Carcia Raeusch., Nomencl. Bot.: 275 (1797).

Garcia nutans Vahl ex Rohr, Skr. Naturhist.-Selsk. 2: 217 (1792).
Mexico (E. San Luis Potosí, NW. Veracruz) to NW. Costa Rica. 79 MXE MXG mxt 80 COS pan (83) clm. Nanophan. or phan.
Garcia mayana Britton, Bot. Porto Rico 6: 357 (1926).

Garcia parviflora Lundell, Wrightia 1: 6 (1945).
Mexico (Tabasco). 79 MXT. Nanophan. or phan.

Synonyms:
Garcia mayana Britton === **Garcia nutans** Vahl ex Rohr

Garumbium

Synonyms:
Garumbium Blume === **Homalanthus** A.Juss.

Gatnaia

Synonyms:
Gatnaia Gagnep. === **Baccaurea** Lour.
Gatnaia annamica Gagnep. === **Baccaurea ramiflora** Lour.

Gavarretia

1 species, N South America (Guayana, Amazonia); shrubs. (Acalyphoideae)

Pax, F. (with K. Hoffmann) (1914). *Gavarretia*. In A. Engler (ed.), Das Pflanzenreich, IV 147 VII (Euphorbiaceae-Acalypheae-Mercurialinae): 213. Berlin. (Heft 63.) La/Ge. — Monotypic, S America (Amazon Basin).
Jablonski, E. (1967). *Gavarretia*. Euphorbiaceae, Guayana Highland (Mem. New York Bot. Gard. 17(1)): 130-131. New York. En. — 1 species, *G. terminalis*.

Gavarretia Baill., Adansonia 1: 185 (1861).
S. Trop. America. 82 83 84.

Gavarretia terminalis Baill., Adansonia 1: 186 (1861). *Conceveiba terminalis* (Baill.) Müll.Arg., Linnaea 34: 167 (1865).
S. Venezuela, Guyana, Brazil (Amazonas), Peru. 82 GUY VEN 83 PER 84 BZN. Nanophan.

Geblera

Synonyms:
Geblera Fisch. & C.A.Mey. === **Flueggea** Willd.
Geblera chinensis Rupr. === **Flueggea suffruticosa** (Pall.) Baill.
Geblera suffruticosa (Pall.) Fisch. & C.A.Mey. === **Flueggea suffruticosa** (Pall.) Baill.
Geblera sungariensis Rupr. === **Flueggea suffruticosa** (Pall.) Baill.

Geiseleria

Synonyms:
Geiseleria Klotzsch === **Croton** L.
Geiseleria chamaedrifolia Klotzsch === **Croton trinitatis** Millsp.
Geiseleria corchorifolia Klotzsch === **Croton trinitatis** Millsp.

Gelonium

A formerly widely used alternative to *Suregada.*

Synonyms:
Gelonium Roxb. ex Willd. === **Suregada** Roxb. ex Rottl.
Gelonium aequoreum Hance === **Suregada aequoreum** (Hance) Seem.
Gelonium affine S.Moore === **Suregada multiflora** (A.Juss.) Baill.
Gelonium angolense Prain === **Tetrorchidium didymostemon** (Baill.) Pax & K.Hoffm.
Gelonium angustifolium Müll.Arg. === **Suregada lanceolata** (Willd.) Kuntze
Gelonium arboreum (J.F.Gmel.) Kuntze === **Molinaea arborea** J.F.Gmel. (Sapindaceae)
Gelonium baronii S.Moore === **Suregada boiviniana** Baill. var. **boiviniana**
Gelonium bifarium Roxb. ex Willd. === **Suregada bifaria** (Roxb. ex Willd.) Baill.
Gelonium borbonicum Pax & K.Hoffm. === **Suregada borbonica** (Pax & K.Hoffm.) Croizat
Gelonium borneense Pax & K.Hoffm. === **Suregada glomerulata** (Blume) Baill.
Gelonium brevipes (Radlk.) Kuntze === **Molinaea brevipes** Radlk. (Sapindaceae)
Gelonium cicerospermum Gagnep. === **Suregada cicerosperma** (Gagnep.) Croizat
Gelonium comorense S.Moore === **Suregada comorensis** Baill.
Gelonium congoense S.Moore === **Suregada gossweileri** (S.Moore) Croizat
Gelonium glandulosum Elmer === **Rinorea bengalensis** (Wall.) Kuntze (Violaceae)
Gelonium gossweileri S.Moore === **Suregada gossweileri** (S.Moore) Croizat
Gelonium humbertii Leandri === **Suregada humbertii** (Leandri) Radcl.-Sm.
Gelonium ivorense Aubrév. & Pellegr. === **Suregada ivorensis** (Aubrév. & Pellegr.) J.Léonard
Gelonium lanceolatum Willd. === **Suregada lanceolata** (Willd.) Kuntze
Gelonium lithoxylon Pax & K.Hoffm. === **Suregada lithoxyla** (Pax & K.Hoffm.) Croizat
Gelonium meliocarpum Elmer === **Suregada glomerulata** (Blume) Baill.
Gelonium microcarpum Pax & K.Hoffm. === **Suregada glomerulata** (Blume) Baill.
Gelonium mindanaense Elmer === **Suregada glomerulata** (Blume) Baill.
Gelonium multiflorum A.Juss. === **Suregada multiflora** (A.Juss.) Baill.
Gelonium occidentale Hoyle === **Suregada occidentalis** (Hoyle) Croizat
Gelonium oxyphyllum Miq. === **Suregada multiflora** (A.Juss.) Baill.
Gelonium papuanum Pax === **Suregada glomerulata** (Blume) Baill.
Gelonium perrieri Leandri === **Suregada perrieri** (Leandri) Radcl.-Sm.
Gelonium philippinense Pax & K.Hoffm. === **Suregada glomerulata** (Blume) Baill.
Gelonium pinatubense Elmer === **Casearia trivalvis** (Blanco) Merr. (Flacourtiaceae)
Gelonium procerum Prain === **Suregada procera** (Prain) Croizat
Gelonium pulgarense Elmer === **Suregada glomerulata** (Blume) Baill.
Gelonium pycnantherum Pax & K.Hoffm. === **Suregada boiviniana** Baill. var. **boiviniana**
Gelonium racemulosum Merr. === **Suregada racemulosa** (Merr.) Croizat
Gelonium rubrum Ridl. === **Suregada glomerulata** (Blume) Baill.

Gelonium scaligerianum A.Massal. === ?
Gelonium stenophyllum Merr. === **Suregada stenophylla** (Merr.) Croizat
Gelonium subglomerulatum Elmer === **Suregada glomerulata** (Blume) Baill.
Gelonium sumatranum S.Moore === **Suregada multiflora** (A.Juss.) Baill.
Gelonium tenuifolium Ridl. === **Suregada multiflora** (A.Juss.) Baill.
Gelonium trifidum Elmer === **Rinorea bengalensis** (Wall.) Kuntze (Violaceae)

Geminaria

Synonyms:
Geminaria Raf. === **Phyllanthus** L.

Genesiphyla

An orthographic variant of *Genesiphylla*.

Synonyms:
Genesiphyla Raf. === **Phyllanthus** L.

Genesiphylla

Synonyms:
Genesiphylla L'Hér. === **Phyllanthus** L.
Genesiphylla asplenifolia L'Hér. === **Phyllanthus latifolius** (L.) Sw.
Genesiphylla speciosa (Jacq.) Raf. === **Phyllanthus arbuscula** (Sw.) J.F.Gmel.

Gentilia

Synonyms:
Gentilia Beille === **Bridelia** Willd.
Gentilia hygrophila Beille === **Bridelia ndellensis** Beille

Gitara

Synonyms:
Gitara Pax & K.Hoffm. === **Acidoton** Sw.
Gitara venezolana Pax & K.Hoffm. === **Acidoton nicaraguensis** (Hemsl.) G.L.Webster

Givotia

4 species, E. and NE. tropical Africa (1), C. & W. Madagascar (2), S. Asia and Sri Lanka (1); deciduous shrubs or trees to 30 m in seasonally dry areas with a soft, light, weak wood and more or less palmately lobed leaves. Of the two sections recognised *Afrogivotia* is in Africa and *Givotia* extends from Madagascar to India. Revised by Radcliffe-Smith (1968). Allies include *Ricinodendron* and *Schinziophyton*, both African; in turn, they may be allied to *Jatropheae*. [The modern distribution of *Givotia* is intriguing and possibly a relict of a former closer association of the land masses; it could also represent a remnant of a former wider distribution. Sect. *Givotia* is, howe notably, shared only between Madagascar and Asia.] (Crotonoideae)

Pax, F. (with K. Hoffmann) (1911). *Givotia*. In A. Engler (ed.), Das Pflanzenreich, IV 147 III (Euphorbiaceae-Cluytieae): 44-45. Berlin. (Heft 47.) La/Ge. — 2 species, 1 in S India and Sri Lanka, the other in Madagascar.

Radcliffe-Smith, A. (1968). An account of the genus *Givotia* Griff. Kew Bull. 22: 493-505, illus., maps. En. — 4 species (2 new), NE Tropical Africa, Madagascar (2) and India and Sri Lanka, in 2 sections; key, descriptions of novelties, references and citations, localities with exsiccatae, and commentary. Notes on the uses of *G. gosai* are also included.

Sreemadhavan, C. P. (1975). Reinstatement of an alleged *nomen confusum* and typification of *Givotia* Griff. (Euphorbiaceae). Taxon 24: 695-696. En. — Nomenclatural.

Givotia Griff., Calcutta J. Nat. Hist. 4: 388 (1843).
E. Africa, Madagascar, India, Sri Lanka. 24 25 29 40.

Givotia gosai Radkl.-Sm., Kew Bull. 22: 499 (1968). – FIGURE, p. 938.
Ethiopia, Somalia, Kenya. 24 ETH SOM 25 KEN. Phan.

Givotia madagascariensis Baill., Bull. Mens. Soc. Linn. Paris 1: 810 (1889).
W. Madagascar. 29 MDG. Phan.

Givotia moluccana (L.) Sreem., Taxon 24: 696 (1975).
India, Sri Lanka. 40 IND SRL. Phan.
* *Croton moluccanus* L., Sp. Pl.: 1005 (1753).
Givotia rottleriformis Griff. ex Wight, Icon. Pl. Ind. Orient. 5: 24 (1852).

Givotia stipularis Radcl.-Sm., Kew Bull. 22: 503 (1968).
W. Madagascar. 29 MDG. Phan.

Synonyms:
Givotia rottleriformis Griff. ex Wight === **Givotia moluccana** (L.) Sreem.

Glochidion

318 species, Indo-Pacific, east to SE Polynesia and south into Australia; also in Madagascar (with 7 species, although Brunel has proposed transfer of these to *Phyllanthus*). The name is conserved against *Agyneia* L. *Glochidion* comprises shrubs or small to medium trees with distichously arranged foliage but without 'phyllanthoid branching' and furthermore featuring a distinctive ribbed fruit. The genus is here and there very well represented, notably New Guinea where some 65 or more species are currently recognised. In Malesia as a whole, it is at some 150 species the largest genus of the family. It is also well-developed in the South Pacific Islands. There appears to be a particular preference for montane (e.g. in Madagascar and New Guinea) or 'oceanic' habitats. Until the late nineteenth century it was sometimes included with *Phyllanthus*, e.g. by Mueller in 1866 and Bentham in 1880. According to Webster (Synopsis, 1994), it represents a specialised line of development from 'within' that genus. Like it, *Glochidion* was never revised for *Das Pflanzenreich*. The infrageneric arrangement largely remains that of Mueller with alterations by Pax for the first edition of *Die natürlichen Pflanzenfamilien* and by Pax & Hoffmann for the second edition (1931). Sections 5-17 in the last-named work have since, however, mostly been transferred back to *Phyllanthus* (e.g. *Pentaglochidion*, *Physoglochidion* and *Eleutherogynium*, all from New Caledonia, and the more widespread *Gomphidium*) while sect. 4, *Synostemon*, is now part of *Sauropus* (Airy-Shaw 1980; see **Australasia**). Very many additions and partial revisions were essayed by Airy-Shaw from the 1960s until 1981, and the species in several Pacific island groups have also been revised in recent years (the latest by Jacques Florence for SE. Polynesia). S. Asian species were revised by Chakrabarty and Gangopadhyay in 1995. A new study of the Malesian species has now been undertaken by Peter van Welzen (Leiden) in preparation for a *Flora Malesiana* account, with utilisation of a wider range of herbaria than assayed by Airy-Shaw. He has indicated (personal communication, 1998) that the legacy of the Kew author was with respect to West Malesia more firmly founded than in Papuasia, where 'everything must be new'; in any case a goodly number of reductions were seen as in prospect. (Phyllanthoideae)

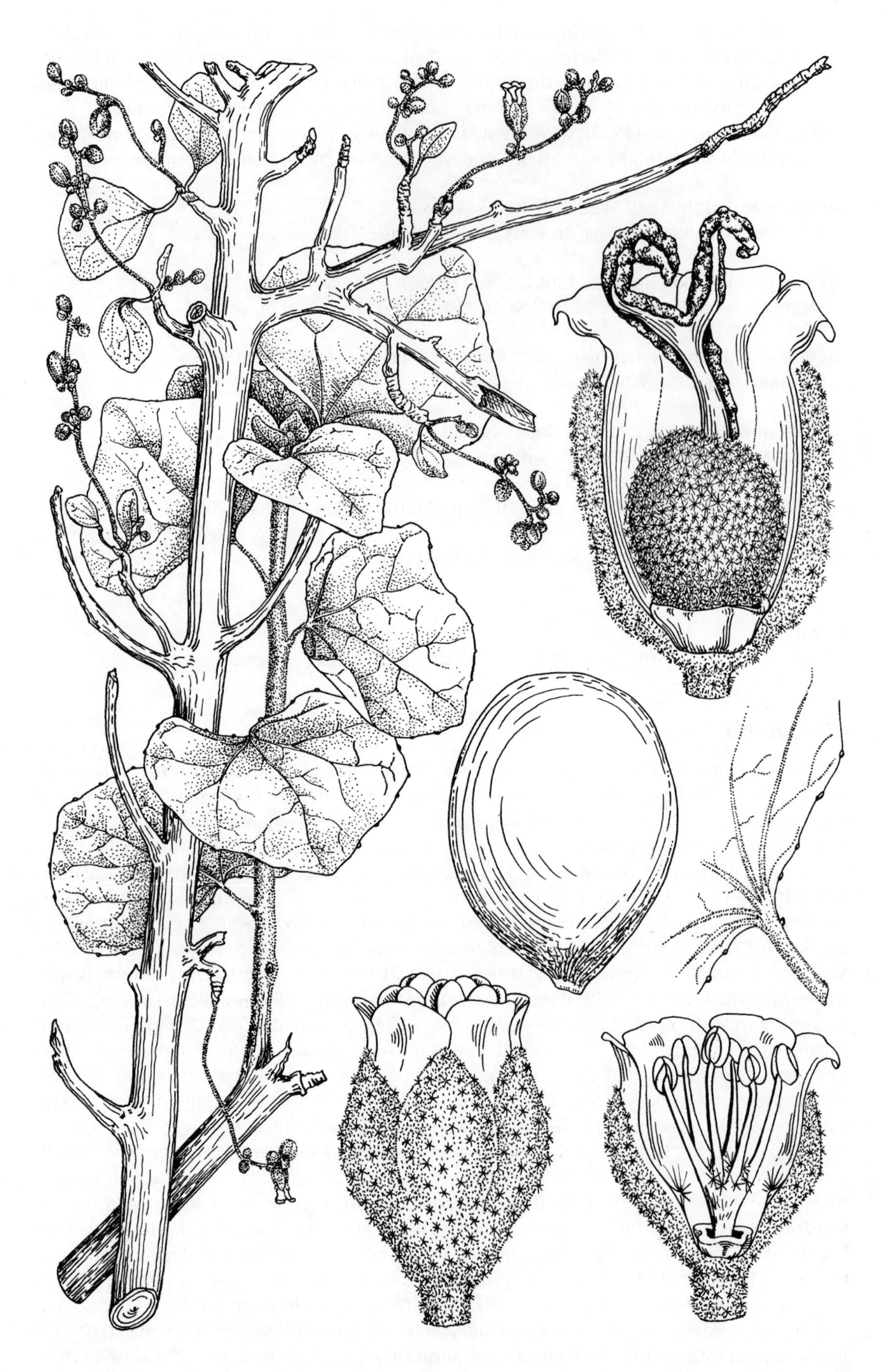

Givotia gosai Radcl. Sm.

Artist: Mary Grierson

Kew Bull. 22: 500, fig. 3 (1968); also in Fl. Trop. East Africa, Euphorbiaceae 1: 330, fig. 62 (1987)

KEW ILLUSTRATIONS COLLECTION

Alston, A. H. G. (1928). The Ceylon species of *Glochidion*. Ann. Roy. Bot. Gard. Peradeniya (Ceylon J. Sci. A, Bot.) 11: 1-10, pl. 1. En. — General notes with analytical key; treatment of 10 species (two new) with synonymy, references, citations, localities with exsiccatae, and commentary; references and plate at end. [Superseded by the treatment of the genus in the *Revised Handbook*, vol. 11: 242-258 (1997).]

Pax, F. & K. Hoffmann (1931). *Glochidion*. In A. Engler (ed.), Die natürlichen Pflanzenfamilien, 2. Aufl., 19c: 56-59, illus. Leipzig. Ge. — Description of genus; synoptic précis with 17 sections accepted. [An estimate of 280 species was offered.]

Leandri, J. (1958). *Glochidion*. Fl. Madag. Comores 111 (Euphorbiacées), I: 21-30. Paris. Fr. — Flora treatment (7 species); key. [Proposals have been made to transfer these to *Phyllanthus*; see Brunel, 1987.]

Airy-Shaw, H. K. (1969). Notes on Malesian and other Asiatic Euphorbiaceae, XCVIII. New or noteworthy species of *Glochidion* J. R. et G. Forst. Kew Bull. 23: 6-26. En. — 23 species, many from Papuasia; several new. No sections are indicated.

Airy-Shaw, H. K. (1971). Notes on Malesian and other Asiatic Euphorbiaceae, CXXI. New or noteworthy species of *Glochidion* Forst. Kew Bull. 25: 481-488. En. — 7 species, 6 new or redescribed.

Airy-Shaw, H. K. (1972). Notes on Malesian and other Asiatic Euphorbiaceae, CLI: New or noteworthy species of *Glochidion* J. R. and G. Forst. Kew Bull. 27: 6-63. En. — A substantial contribution, with many novelties and other notes from across the region and Australia; key to varieties of *G. mindoroense*.

Airy-Shaw, H. K. (1972). Notes on Malesian and other Asiatic Euphorbiaceae, CLII: The genus *Glochidion* J. R. and G. Forst. in New Guinea. Kew Bull. 27: 63-74. En. — Very concise synopsis of some 60 species (of which nearly two-thirds are endemic) with key and indication of distributions; arrangement of species alphabetical. Some citations and notes are also given. [About a quarter of the species feature large leaves and large, multilocular capsules.]

Airy-Shaw, H. K. (1974). Notes on Malesian and other Asiatic Euphorbiaceae, CLXXI. New or noteworthy species of *Glochidion* Forst. Kew Bull. 29: 287-294. En. — Novelties and notes.

Airy-Shaw, H. K. (1978). Notes on Malesian and other Asiatic Euphorbiaceae, CCVIII. *Glochidion* J. R. et G. Forst. Kew Bull. 33: 27-35. En. — Novelties and notes; 10 species, several from Papuasia.

Airy-Shaw, H. K. (1978). Notes on Malesian and other Asiatic Euphorbiaceae, CXC. New or noteworthy species of *Glochidion* Forst. Kew Bull. 32: 370-379. En. — Seven novelties, a new combination and notes; the greater part from New Guinea and the western Pacific. No sections indicated.

Airy-Shaw, H. K. (1980). New or noteworthy Australian Euphorbiaceae, II. Muelleria 4: 207-241. En. —*Glochidion*, pp. 208-213; notes and novelties.

Airy-Shaw, H. K. (1980). Notes on Euphorbiaceae from Indomalesia, Australia and the Pacific, CCXXXI. *Glochidion* Forst. Kew Bull. 35: 385-386. En. — New name for *G. cataractum* and description of *G. phellocarpum* from Peninsular Malaysia.

Airy-Shaw, H. K. (1981). Notes on Asiatic, Malesian and Melanesian Euphorbiaceae, CCXLII. *Glochidion* Forst. Kew Bull. 36: 599-603. En. — Descriptions of 5 new species and a new variety of *G. granulare*.

Brunel, J. F. (1987). *Glochidion*. In *idem*, Sur le genre *Phyllanthus* L. et quelques genres voisins de la tribu des Phyllantheae Dumort. Strasbourg. [See also **Phyllanthus**.] Fr. — Description of the genus; synopsis (the species in Madagascar formerly credited here are transferred to *Phyllanthus*, with *Glochidion* thus excluded west of Pakistan).

- Chakrabarty, T. & M. Gangopadhyay (1995). The genus *Glochidion* (Euphorbiaceae) in the Indian subcontinent. J. Econ. Taxon. Bot. 19: 173-234, illus. En. — Detailed regional revision (32 species) with key, synonymy, references and citations, types, descriptions, phenology, vernacular names, indication of distribution and habitat, localities with exsiccatae, and commentary; list of Malesian types of *Glochidion* in CAL and references at end. [Largely based on materials in CAL. Covers Sri Lanka and Burma (Myanmar) as well as India, Bangladesh, Bhutan, Nepal and Pakistan.]

Glochidion J.R.Forst. & G.Forst., Char. Gen. Pl.: 57 (1775), nom. cons.
Madagascar, Trop. & Subtrop. Asia, Pacific. 29 36 38 40 41 42 50 60 61 62. Nanophan. or phan.
Agyneia L., Mant. Pl. 2: 161 (1771), nom. rejic.
Bradleia Banks ex Gaertn., Fruct. Sem. Pl. 2: 127 (1791).
Gynoon A.Juss., Mém. Mus. Hist. Nat. 10: 335 (1823).
Glochidionopsis Blume, Bijdr.: 588 (1826).
Lobocarpus Wight & Arn., Prodr. Fl. Ind. Orient.: 7 (1834).
Episteira Raf., Sylva Tellur.: 20 (1838).
Glochidionopsis Steud., Nomencl. Bot., ed. 2, 1: 689 (1840).
Glochisandra Wight, Icon. Pl. Ind. Orient. 5(2): 28 (1852).
Zarcoa Llanos, Bot. Zeitung (Berlin) 15: 423 (1857).
Coccoglochidion K.Schum. in K.M.Schumann & C.A.G.Lauterbach, Fl. Schutzgeb. Südsee, Nachtr.: 292 (1905).
Tetraglochidion K.Schum. in K.M.Schumann & C.A.G.Lauterbach, Fl. Schutzgeb. Südsee, Nachtr.: 291 (1905).

Glochidion acuminatissimum Airy Shaw, Kew Bull. 27: 55 (1972).
Irian Jaya. 42 NWG. Phan.

Glochidion acuminatum Müll.Arg., Linnaea 32: 68 (1863).
Nepal to Thailand and China (S. Yunnan). 36 CHC 40 ASS BAN? EHM NEP 41 BMA THA. Nanophan. or phan.

var. **acuminatum**
Nepal, Sikkim, Bhutan, Assam, China (S. Yunnan). 36 CHC 40 ASS BAN? EHM NEP. Nanophan. or phan.

var. **siamense** Airy Shaw, Kew Bull. 26: 273 (1972). *Glochidion triandrum* var. *siamense* (Airy Shaw) P.T.Li, Acta Phytotax. Sin. 26: 62 (1988).
Burma, Thailand, China (Yunnan). 36 CHC 41 BMA THA. Nanophan. or phan.

Glochidion acustylum Airy Shaw, Kew Bull. 34: 591 (1980).
Papua New Guinea. 42 NWG. Phan.

Glochidion album (Blanco) Boerl., Handl. Fl. Ned. Ind. 3: 275 (1900).
Philippines, Sulawesi. 42 PHI SUL.
* *Kirganelia alba* Blanco, Fl. Filip.: 713 (1837). *Phyllanthus albus* (Blanco) Müll.Arg., Flora 48: 387 (1865), nom. illeg.
Zarcoa phillipica Llanos, Bot. Zeitung (Berlin) 15: 423 (1857).
Glochidion cumingii Müll.Arg., Linnaea 32: 61 (1863). *Phyllanthus cumingii* (Müll.Arg.) Müll.Arg., Flora 48: 371 (1865).
Glochidion leytense Elmer, Leafl. Philipp. Bot. 1: 303 (1908).
Glochidion mindanaense Elmer, Leafl. Philipp. Bot. 7: 2642 (1915).
Glochidion brachystylum Merr., Philipp. J. Sci. 16: 543 (1920).

Glochidion alstonii Airy Shaw, Kew Bull. 36: 600 (1981).
Sulawesi. 42 SUL. Phan.

Glochidion alticola Airy Shaw, Kew Bull. 27: 8 (1972).
W. Sumatera. 42 SUM. Phan.

Glochidion aluminescens Airy Shaw, Kew Bull. 27: 16 (1972).
Borneo (SW. Sarawak). 42 BOR. Nanophan. or phan.

Glochidion ambiguum Airy Shaw, Kew Bull. 27: 62 (1972).
Papua New Guinea, Bismarck Archip., Solomon Is. 42 BIS NWG 60 SOL. Phan.

Glochidion amentuligerum (Müll.Arg.) Croizat, Sargentia 1: 46 (1942).
Fiji. 60 FIJ. Nanophan. or phan.
* *Phyllanthus amentuliger* Müll.Arg., Flora 48: 390 (1865).

Glochidion andamanicum Kurz, J. Asiat. Soc. Bengal, Pt. 2, Nat. Hist. 42(2): 238 (1873).
Andaman Is., Burma. 41 AND BMA. Phan.

var. **andamanicum**
S. Andaman Is. 41 AND. Phan.

var. **desmogyne** (Hook.f.) Chakrab. & M.G.Gangop., J. Econ. Taxon. Bot. 19: 186 (1995).
Andaman Is., Burma. 41 AND BMA. Phan.
* *Glochidion desmogyne* Hook.f., Fl. Brit. India 5: 310 (1887).
Glochidion airyshawii N.P.Balakr. & Chakrab., Bull. Bot. Surv. India 25: 220 (1983 publ. 1985).

Glochidion andersonii Airy Shaw, Kew Bull. 29: 287 (1974).
Borneo (SW. Sarawak). 42 BOR. Nanophan. or phan.

Glochidion anfractuosum Gibbs, J. Linn. Soc., Bot. 39: 168 (1909).
Fiji. 60 FIJ. Nanophan. or phan.

Glochidion angulatum C.B.Rob., Philipp. J. Sci., C 4: 91 (1909).
Solomon Is., Papua New Guinea, Philippines, Borneo (Sabah), Sulawesi, Maluku. 42 BOR MOL NWG PHI SUL 60 SOL. Phan.

Glochidion ankaratrae Leandri, Mém. Inst. Sci. Madagascar, Sér. B, Biol. Vég. 8: 214 (1957).
C. Madagascar (Ambositra). 29 MDG.

Glochidion apodogynum Airy Shaw, Kew Bull. 27: 44 (1972).
Papua New Guinea, Northern Territory, Queensland. 42 NWG 50 NTA QLD. Phan. – Close to *G. fluvirameum*.

Glochidion atalotrichum A.C.Sm., Contr. U. S. Natl. Herb. 37: 74 (1967).
Fiji (Viti Levu). 60 FIJ. Phan.

Glochidion atrovirens A.C.Sm., Fl. Vit. Nova 2: 491 481 (1981).
Fiji (S. Viti Levu). 60 FIJ. Nanophan. or phan.

Glochidion auii Airy Shaw, Kew Bull. 27: 19 (1972).
Borneo (NE. Sarawak). 42 BOR. Phan.

Glochidion azaleon Airy Shaw, Kew Bull. 27: 15 (1972).
Borneo (E. Kalimantan). 42 BOR. Nanophan. or phan.

Glochidion bachmaense Thin, Euphorb. Vietnam: 49 (1995).
Vietnam. 41 VIE. Phan.

Glochidion balansae Beille in H.Lecomte, Fl. Indo-Chine 5: 628 (1927).
Vietnam. 41 VIE.

Glochidion barronense Airy Shaw, Kew Bull. 31: 343 (1976).
Queensland (Cook). 50 QLD. Nanophan. or phan.

Glochidion beccarii Airy Shaw, Kew Bull. 23: 10 (1969).
W. Sumatera. 42 SUM. Phan.

Glochidion beguinii Airy Shaw, Kew Bull. 27: 27 (1972).
Maluku (NW. Halmahera). 42 MOL. Nanophan. or phan.

Glochidion benguetense Elmer, Leafl. Philipp. Bot. 1: 304 (1908).
Philippines. 42 PHI.
Glochidion sablanense Elmer, Leafl. Philipp. Bot. 1: 306 (1908).

Glochidion benthamianum Domin, Biblioth. Bot. 89: 318 (1927).
New Guinea, Queensland. 42 NWG 50 QLD. Nanophan. or phan.
Glochidion obscurum var. *papuanum* J.J.Sm., Nova Guinea 8: 223 (1910).
Glochidion capitis-york Airy Shaw, Kew Bull. 23: 25 (1969).

Glochidion billardieri Baill., Adansonia 2: 241 (1862). *Phyllanthus billardieri* (Baill.) Müll.Arg., Flora 48: 375 (1865). – FIGURE, p. 943.
New Caledonia (incl. Loyalty Is.). 60 NWC. Nanophan. or phan.
Bradleia glauca Labill., Sert. Austro-Caledon.: 76 (1825). *Glochidion glaucum* (Labill.) Müll.Arg., Linnaea 32: 62 (1863), nom. illeg.
Glochidion kanalense Baill., Adansonia 2: 241 (1862).
Glochidion heterolobum Müll.Arg., Linnaea 32: 63 (1863).
Phyllanthus kanalophilus Müll.Arg., Flora 48: 375 (1865).
Phyllanthus wagapensis Müll.Arg., Flora 48: 372 (1865). *Glochidion wagapense* (Müll.Arg.) Briq., Annuaire Conserv. Jard. Bot. Genève: 226 (1900).

Glochidion borgmannii Airy Shaw, Kew Bull. 27: 49 (1972).
New Guinea. 42 NWG. Nanophan. or phan.

Glochidion borneense (Müll.Arg.) Boerl., Handl. Fl. Ned. Ind. 3: 276 (1900).
Pen. Malaysia (incl. Singapore), Borneo, Sumatera (Bangka), Jawa. 42 BOR JAW MLY SUM. Phan.
* *Phyllanthus borneensis* Müll.Arg., Flora 48: 377 (1865).
Phyllanthus polycarpus Müll.Arg., Flora 48: 387 (1865).
Glochidion microbotrys Hook.f., Fl. Brit. India 5: 319 (1887).
Glochidion trilobum Ridl., Bull. Misc. Inform. Kew 1923: 364 (1923).

Glochidion bourdillonii Gamble, Bull. Misc. Inform. Kew 1925: 330 (1925).
Bhutan, S. India. 40 EHM IND. Phan.

var. **bhutanicum** (D.G.Long) Chakrab. & M.G.Gangop., J. Econ. Taxon. Bot. 19: 190 (1995).
Bhutan. 40 EHM. Phan.
* *Glochidion bhutanicum* D.G.Long, Notes Roy. Bot. Gard. Edinburgh 44: 169 (1986).

var. **bourdillonii**
S. India. 40 IND. Phan.

Glochidion bracteatum Gillespie, Bernice P. Bishop Mus. Bull. 91: 15 (1932).
Fiji (Viti Levu). 60 FIJ. Nanophan. or phan.

Glochidion brideliifolium Airy Shaw, Kew Bull. 23: 15 (1969).
Irian Jaya. 42 NWG. Phan.

Glochidion brooksii Ridl., Bull. Misc. Inform. Kew 1925: 88 (1925).
Pen. Malaysia, Sumatera, Borneo (Sarawak). 42 BOR MLY SUM. Phan.
Glochidion laevigatum var. *cuspidatum* Ridl., Fl. Malay Penins. 3: 215 (1924).

Glochidion brothersonii Florence, Fl. Polynésie Fr. 1: 68 (1997).
Society Is. (Raiatea). 61 SCI. Nanophan.

Glochidion brunnescens A.C.Sm., Fl. Vit. Nova 2: 491 (1981).
Fiji. 60 FIJ. Nanophan. or phan.

Glochidion bullatissimum Airy Shaw, Kew Bull. 32: 376 (1978).
NE. Papua New Guinea. 42 NWG. Phan.

Glochidion billiardieri Baill. (as *Bradleia glauca* Labill.)
Artist: P.J.F. Turpin
Labilliardière, Sert. Austro-Caledon.: pl. 77 (1825)
KEW ILLUSTRATIONS COLLECTION

Glochidion butonicum Airy Shaw, Kew Bull. 29: 291 (1974).
Sulawesi (Butung I.). 42 SUL. Nanophan. or phan.

Glochidion cacuminum Müll.Arg., Linnaea 32: 60 (1863).
Jawa. 42 JAW.

Glochidion cagayanense C.B.Rob., Philipp. J. Sci., C 6: 328 (1911).
Philippines (Luzon). 42 PHI.

Glochidion calciphilum Croizat, Sargentia 1: 46 (1942).
Fiji. 60 FIJ. Nanophan.

Glochidion caledonicum Müll.Arg., Linnaea 32: 62 (1863). *Phyllanthus caledonicus* (Müll.Arg.) Müll.Arg., Flora 48: 376 (1865).
New Caledonia (incl. I. des Pins, Loyalty Is.). 60 NWC. Nanophan. or phan.
Bradleia zeylanica Labill., Sert. Austro-Caledon.:76 (1825), nom. illeg.
Glochidion diospyroides Schltr., Bot. Jahrb. Syst. 39: 147 (1906).

Glochidion calocarpum Kurz, J. Bot. 13: 330 (1875).
Andaman Is., Nicobar Is. 41 AND NCB. Phan.

Glochidion caloneurum Airy Shaw, Kew Bull. 23: 16 (1969).
Irian Jaya. 42 NWG. Phan.

Glochidion calospermum Airy Shaw, Kew Bull. 27: 13 (1972).
Borneo. 42 BOR. Nanophan. or phan.

Glochidion camiguinense Merr., Philipp. J. Sci., C 3: 414 (1909).
Philippines (Babuyan). 42 PHI.

Glochidion candolleanum (Wight & Arn.) Chakrab. & M.G.Gangop., J. Econ. Taxon. Bot. 19: 191 (1995).
S. India, Sri Lanka. 40 IND SRL. Phan.
Glochidion arboreum var. *pauciflorum* Hook.f. in ?, . *Glochidion pauciflorum* (Hook.f.) Gamble, Fl. Madras: 1307 (1925).
* *Lobocarpus candolleanus* Wight & Arn., Prodr. Fl. Ind. Orient.: 7 (1834).
Glochidion arboreum Wight, Icon. Pl. Ind. Orient. 5: t. 1907 (1852).
Glochidion neilgherrense Wight, Icon. Pl. Ind. Orient. 5: t. 1907 (1852). *Phyllanthus neilgherrensis* (Wight) Müll.Arg., Flora 48: 385 (1865).
Phyllanthus arboreus Müll.Arg., Flora 48: 385 (1865).
Phyllanthus perrottetianus Müll.Arg., Flora 48: 387 (1865).
Phyllanthus pycnocarpus Müll.Arg., Flora 48: 386 (1865). *Glochidion pycnocarpum* (Müll.Arg.) Bedd., Fl. Sylv. S. India: 194 (1872).
Glochidion perrottetianum Bedd., Fl. Sylv. S. India: 194 (1872).
Glochidion pycnocarpum var. *ellipticum* Hook.f., Fl. Brit. India 5: 316 (1887).
Glochidion sisparense Gamble, Fl. Madras: 1307 (1925).
Glochidion pachycarpum Alston, Ann. Roy. Bot. Gard. (Peradeniya) 11: 5 (1928).

Glochidion canescens Elmer, Leafl. Philipp. Bot. 8: 3082 (1919).
Philippines (Luzon), Maluku (Ceram), Timor. 42 LSI MOL PHI. Phan.

Glochidion carrickii Airy Shaw, Kew Bull. 25: 482 (1971).
Pen. Malaysia. 42 MLY. Phan.

Glochidion carrii Airy Shaw, Kew Bull. 23: 20 (1969).
Papua New Guinea. 42 NWG. Phan.

Glochidion castaneum Airy Shaw, Kew Bull. 27: 36 (1972).
Irian Jaya. 42 NWG. Nanophan. or phan.

Glochidion cauliflorum Merr., Philipp. J. Sci. 16: 545 (1920).
Philippines (Mindanao, Leyte). 42 PHI.

B

PLATE 1

A. *Hippomane mancinella* **B.** *Monadenium torrei*

PLATE 2

Glochidion multiloculare

PLATE 3

Homalanthus populneus

PLATE 4

Hura crepitans

Anda brasiliensis.

PLATE 5

Joannesia princeps

PLATE 6

Manihot esculenta

Tab. 43. Mart. Nov. Gen.

PLATE 7

Microstacys bidentata

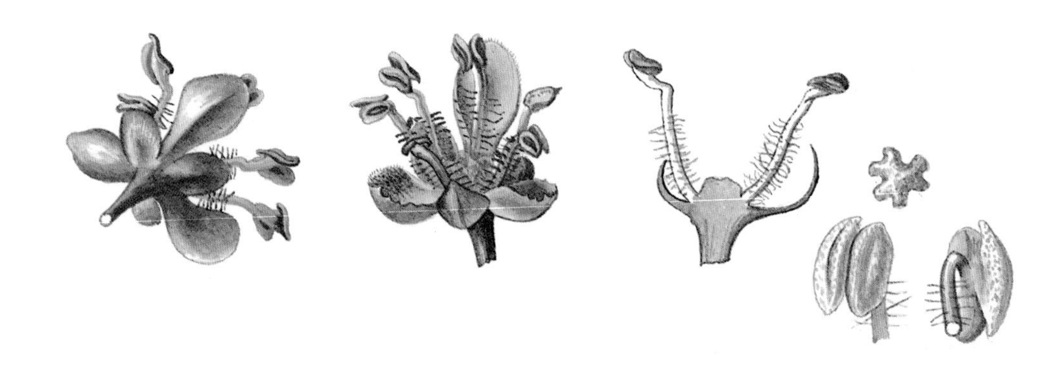

PLATE 8

Neoroepera banksii

Glochidion cavaleriei H.Lév., Repert. Spec. Nov. Regni Veg. 12: 183 (1913).
S. China. 36 CHS.

Glochidion celastroides (Müll.Arg.) Pax in H.G.A.Engler & K.A.E.Prantl, Nat. Pflanzenfam. 3(5): 24 (1890).
Borneo (SE. Kalimantan), Sumatera (Bangka, Billiton). 42 BOR SUM. Phan.
* *Phyllanthus celastroides* Müll.Arg., Flora 48: 390 (1865).

Glochidion cenabrei Merr., Philipp. J. Sci. 26: 463 (1925).
Philippines (Palawan). 42 PHI.

Glochidion chademenosocarpum Hayata, Icon. Pl. Formosan. 9: 94 (1920).
Taiwan. 38 TAI. Nanophan.

Glochidion chevalieri Beille in H.Lecomte, Fl. Indo-Chine 5: 615 (1927).
Laos. 41 LAO.

Glochidion chlamydogyne Airy Shaw, Kew Bull. 23: 14 (1969).
Papua New Guinea. 42 NWG. Phan.

Glochidion chodoense C.S.Lee & H.T.Im, Korean J. Pl. Taxon. 24: 13 (1994).
S. Korea. 38 KOR.

Glochidion chondrocarpum Airy Shaw, Kew Bull. 32: 372 (1978).
Papua New Guinea. 42 NWG. Nanophan. or phan.

Glochidion christopheraenii Croizat, Occas. Pap. Bernice Pauahi Bishop Mus. 17: 213 (1943).
Samoa. 60 SAM.

Glochidion cleistanthoides Fosberg, Willdenowia 20: 263 (1991).
Caroline Is. 62 CRL.

Glochidion coccineum (Buch.-Ham.) Müll.Arg., Linnaea 32: 60 (1863).
Assam, Indo-China, S. China. 36 CHC CHH CHS 40 ASS 41 BMA CBD LAO THA VIE. Nanophan. or phan.
* *Agyneia coccinea* Buch.-Ham. in M.Symes, Embassy Ava, ed. 2, 3: 317 (1809). *Bradleia coccinea* (Buch.-Ham.) Wall., Numer. List: 7868 (1847). *Phyllanthus coccineus* (Buch.-Ham.) Müll.Arg., Flora 48: 370 (1865).

Glochidion collectorum Airy Shaw, Kew Bull. 33: 33 (1978).
Solomon Is. 60 SOL. Phan. – Close to *G. ambiguum*.

Glochidion collinum A.C.Sm., Fl. Vit. Nova 2: 494 (1981).
Fiji (Viti Levu). 60 FIJ. Phan.

Glochidion comitum Florence, Novon 7: 29 (1997).
Pitcairn Is. 61 PIT. Nanophan. or phan.

Glochidion concolor Müll.Arg., Linnaea 32: 62 (1863). *Phyllanthus concolor* (Müll.Arg.) Müll.Arg., Flora 48: 374 (1865).
Fiji, Tonga. 60 FIJ TON. Nanophan. or phan.
Glochidion concolor var. *obovatum* Müll.Arg., Linnaea 32: 62 (1863). *Phyllanthus concolor* var. *obovatus* (Müll.Arg.) Müll.Arg. in A.P.de Candolle, Prodr. 15(2): 290 (1866).
Glochidion ramiflorum var. *lanceolatum* Müll.Arg., Linnaea 32: 63 (1863).
Phyllanthus ramiflorus var. *lanceolatus* Müll.Arg., Flora 48: 374 (1865).
Phyllanthus concolor var. *ellipticus* Müll.Arg. in A.P.de Candolle, Prodr. 15(2): 290 (1866).

Glochidion conostylum Airy Shaw, Kew Bull. 23: 23 (1969).
Bismarck Archip. 42 BIS. Phan.

Glochidion cordatum Seem. ex Müll.Arg., Linnaea 32: 64 (1863). *Phyllanthus cordatus* (Seem. ex Müll.Arg.) Müll.Arg., Flora 48: 376 (1865).
Fiji. 60 FIJ. Nanophan. or phan.

Glochidion coriaceum Thwaites, Enum. Pl. Zeyl.: 285 (1861). *Phyllanthus coriaceus* (Thwaites) Müll.Arg., Flora 48: 386 (1865).
Sri Lanka. 40 SRL. Nanophan. or phan.
Bradleia coriacea Wall., Numer. List: 7872 (1847), nom. nud.
Glochidion brachylobum Müll.Arg., Linnaea 32: 62 (1863). *Phyllanthus brachylobus* (Müll.Arg.) Müll.Arg., Flora 48: 373 (1865).

Glochidion coronulatum C.B.Rob., Philipp. J. Sci., C 4: 94 (1909).
Philippines. 42 PHI.
Glochidion balsahanense Elmer, Leafl. Philipp. Bot. 4: 1289 (1911).

Glochidion cupreum Airy Shaw, Kew Bull. 27: 15 (1972).
Borneo (E. Kalimantan). 42 BOR. Nanophan. or phan.

Glochidion curranii C.B.Rob., Philipp. J. Sci., C 4: 102 (1909).
Philippines (Culion ,Palawan). 42 PHI.

Glochidion cyrtophyllum Miq., Fl. Ned. Ind. 1(2): 378 (1859).
Sumatera. 42 SUM.

Glochidion cyrtostylum Miq., Fl. Ned. Ind. 1(2): 378 (1859). *Phyllanthus cyrtostylus* (Miq.) Müll.Arg., Flora 48: 387 (1865). *Phyllanthus cyrtophyllus* (Miq.) Müll.Arg., Flora 48: 390 (1865).
Jawa. 42 JAW.

Glochidion daltonii (Müll.Arg.) Kurz, Forest Fl. Burma 2: 344 (1877).
Sikkim, Assam, Burma, N. Thailand, Pen. Malaysia, Vietnam, S. China. 36 CHC CHS 40 ASS EHM 41 BMA THA VIE 42 MLY. Nanophan. or phan.
* *Phyllanthus daltonii* Müll.Arg., Flora 48: 388 (1865).
Glochidion gamblei Hook.f., Fl. Brit. India 5: 310 (1887).

Glochidion dasyanthum Kurz ex Teijsm. & Binn., Tijdschr. Ned.-Indië 27: 47 (1864). *Phyllanthus dasyanthus* (Kurz ex Teijsm. & Binn.) Müll.Arg., Flora 48: 390 (1865).
Jawa. 42 JAW.

Glochidion dasystylum Kurz, J. Asiat. Soc. Bengal, Pt. 2, Nat. Hist. 42(2): 237 (1873).
Thailand, Burma, China (Yunnan). 36 CHC 41 BMA THA. Nanophan. or phan.

var. **dasystylum**
Burma, China (Yunnan). 36 CHC 41 BMA. Nanophan. or phan.

var. **kerrii** (Craib) Chakrab. & M.G.Gangop., J. Econ. Taxon. Bot. 13: 710 (1989).
Thailand. 41 THA. Nanophan. or phan.
* *Glochidion kerrii* Craib, Bull. Misc. Inform. Kew 1911: 458 (1911).

Glochidion decorum J.J.Sm., Nova Guinea 8: 222 (1910).
Irian Jaya. 42 NWG. Phan.

Glochidion delticola Airy Shaw, Kew Bull. 29: 292 (1974).
Papua New Guinea. 42 NWG. Phan.

Glochidion dichromum Airy Shaw, Kew Bull. 27: 23 (1972).
N. Sulawesi. 42 SUL. Nanophan.

Glochidion discogyne Airy Shaw, Kew Bull. 32: 374 (1978).
NE. Papua New Guinea. 42 NWG. Phan.

Glochidion disparilaterum S.Moore, J. Bot. 63(Suppl.): 95 (1925).
Sumatera. 42 SUM.

Glochidion disparipes Airy Shaw, Kew Bull. 27: 43 (1972).
Papua New Guinea, N. Australia. 42 NWG 50 NTA QLD WAU. Phan.

Glochidion dodecapterum Airy Shaw, Kew Bull. 36: 602 (1981).
C. Sulawesi. 42 SUL. Nanophan. or phan.

Glochidion dolichostylum Merr., Philipp. J. Sci., C 9: 483 (1914 publ. 1915).
Philippines (Palawan). 42 PHI.

Glochidion drypetifolium Airy Shaw, Kew Bull. 27: 34 (1972).
NE. Papua New Guinea. 42 NWG. Phan.

Glochidion dumicola Airy Shaw, Kew Bull. 27: 47 (1972).
Irian Jaya. 42 NWG. Phan.

Glochidion elaphrocarpum Airy Shaw, Kew Bull. 27: 35 (1972).
NE. Papua New Guinea. 42 NWG. Phan.

Glochidion ellipticum Wight, Icon. Pl. Ind. Orient. 5: t. 1906 (1852).
India to Taiwan. 36 CHC CHH 38 TAI 40 ASS BAN EHM INDNEP 41 BMA THA VIE. Nanophan. or phan.
Glochidion assamicum var. *brevipedicellatum* Hurus. & Yu.Tanaka in ?
Bradleia wightiana Wall., Numer. List: 7862 (1847), nom. nud.
Phyllanthus assamicus Müll.Arg., Flora 48: 378 (1865). *Glochidion assamicum* (Müll.Arg.) Hook.f., Fl. Brit. India 5: 319 (1887).
Phyllanthus diversifolius var. *longifolius* Müll.Arg., Flora 48: 378 (1865).
Phyllanthus diversifolius var. *wightiana* Müll.Arg., Flora 48: 378 (1865). *Glochidion diversifolium* var. *wightianum* (Müll.Arg.) Bedd., Fl. Sylv. S. India: 193 (1872). *Glochidion ellipticum* var. *wightiana* (Müll.Arg.) Hook., Fl. Brit. India 5: 321 (1887).
Phyllanthus malabaricus Müll.Arg., Linnaea 34: 69 (1865). *Glochidion malabaricum* (Müll.Arg.) Bedd., Fl. Sylv. S. India: 194 (1872).
Glochidion diversifolium Bedd., Fl. Sylv. S. India: 193 (1872).
Phyllanthus andersonii Müll.Arg., Flora 55: 3 (1872), nom. illeg.

Glochidion elmeri Merr., Univ. Calif. Publ. Bot. 15: 139 (1929).
Borneo (Sabah). 42 BOR. Nanophan. or phan.

Glochidion emarginatum J.W.Moore, Bernice P. Bishop Mus. Bull. 102: 30 (1933).
Society Is. (Raiatea). 61 SCI. Nanophan. or phan.
Glochidion raiateense J.W.Moore, Bernice P. Bishop Mus. Bull. 102: 30 (1933).

Glochidion eriocarpum Champ. ex Benth., Hooker's J. Bot. Kew Gard. Misc. 6: 6 (1854).
Phyllanthus eriocarpus (Champ. ex Benth.) Müll.Arg., Flora 48: 387 (1865).
S. China, Taiwan, Indo-China, Pen. Malaysia. 36 CHC CHH CHS 38 TAI 41 BMA THA VIE 42 MLY. Nanophan. or phan.
Glochidion villicaule Hook.f., Fl. Brit. India 5: 326 (1887).
Glochidion anamiticum Kuntze, Revis. Gen. Pl. 2: 601 (1891).

Glochidion esquirolii H.Lév., Repert. Spec. Nov. Regni Veg. 12: 186 (1913).
Glochidion annamense Beille in H.Lecomte, Fl. Indo-Chine 5: 627 (1927).

Glochidion eucleoides S.Moore, J. Bot. 61(Suppl.): 44 (1923).
New Guinea (incl. D'Entrecasteaux Is., Louisiade Archip.). 42 NWG. Phan.

Glochidion euryodes A.C.Sm., J. Arnold Arbor. 33: 373 (1952).
Fiji (Viti Levu: Mt. Koromba). 60 FIJ. Phan.

Glochidion excorticans Fosberg, Willdenowia 20: 260 (1991).
Caroline Is. 62 CRL.

var. **calvum** Fosberg, Willdenowia 20: 261 (1991).
Caroline Is. 62 CRL.

var. **excorticans**
Caroline Is. 62 CRL.

Glochidion falcatilimbum Merr., Philipp. J. Sci. 16: 542 (1920).
Philippines (Luzon). 42 PHI.

Glochidion ferdinandii (Müll.Arg.) F.M.Bailey, Queensl. Fl. 5: 1423 (1902).
Northern Territory, Queensland, New South Wales. 50 NSW NTA QLD. Phan.
* *Phyllanthus ferdinandii* Müll.Arg., Flora 48: 379 (1865).
Glochidion ferdinandii var. *minor* Benth., Fl. Austral. 6: 96 (1873).
Glochidion ferdinandii var. *pubens* Maiden ex Airy Shaw, Kew Bull. 31: 346 (1976).

Glochidion flavidum Kurz ex Teijsm. & Binn., Tijdschr. Ned.-Indië 27: 46 (1864).
Phyllanthus flavidus (Kurz ex Teijsm. & Binn.) Müll.Arg., Flora 48: 390 (1865).
Jawa. 42 JAW.

Glochidion formanii Airy Shaw, Kew Bull. 27: 25 (1972).
NE. Sulawesi. 42 SUL. Phan.

Glochidion frodinii Airy Shaw, Kew Bull. 27: 51 (1972).
NE. Papua New Guinea. 42 NWG. Nanophan. or phan.

Glochidion fulvirameum Miq., Fl. Ned. Ind. 1(2): 376 (1859). *Phyllanthus fulvirameus* (Miq.) Müll.Arg., Flora 48: 385 (1865).
Jawa, New Guinea (incl. Aru I.), Solomon Is. 42 JAW NWG 50 SOL. Nanophan. or phan.
Glochidion salomonis Briq., Annuaire Conserv. Jard. Bot. Genève: 226 (1900).

Glochidion galorii Airy Shaw, Kew Bull. 25: 484 (1971).
Irian Jaya ?, NE. Papua New Guinea. 42 NWG. Nanophan. or phan. – The specimens from Irian Jaya could represent a different taxon.

Glochidion gaudichaudii (Müll.Arg.) Boerl., Handl. Fl. Ned. Ind. 3: 276 (1900).
Maluku, New Guinea. 42 MOL NWG.
* *Phyllanthus gaudichaudii* Müll.Arg., Flora 48: 379 (1865).

Glochidion geoffrayi Beille in H.Lecomte, Fl. Indo-Chine 5: 619 (1927).
Cambodia. 41 CBD.

Glochidion gigantifolium (Vidal) J.J.Sm., Nova Guinea 8: 223 (1910).
Philippines. 42 PHI. Nanophan. or phan.
* *Phyllanthus gigantifolius* Vidal, Revis. Pl. Vasc. Filip.: 237 (1886).

Glochidion gillespiei Croizat, Sargentia 1: 46 (1942).
Fiji (Viti Levu). 60 FIJ. Nanophan. or phan.

Glochidion gimi (K.Schum.) Pax & K.Hoffm. in H.G.A.Engler, Nat. Pflanzenfam. ed. 2, 19c: 58 (1931).
NE. Papua New Guinea. 42 NWG. Phan.
* *Tetraglochidion gimi* K.Schum. in K.M.Schumann & C.A.G.Lauterbach, Fl. Schutzgeb. Südsee, Nachtr.: 291 (1905).

Glochidion glabrum J.J.Sm., Nova Guinea 8: 224 (1910). *Glochidion mindorense* subsp. *glabrum* (J.J.Sm.) Airy Shaw, Kew Bull. 27: 22 (1972).
Irian Jaya. 42 NWG. Nanophan.

Glochidion glaucescens Merr., Philipp. J. Sci., C 8: 381 (1913).
Philippines (Leyte). 42 PHI.

Glochidion glaucops Airy Shaw, Kew Bull. 27: 10 (1972).
Sumatera. 42 SUM. Nanophan. or phan.

Glochidion glomerulatum (Miq.) Boerl., Handl. Fl. Ned. Ind. 3: 276 (1900).
India, Vietnam, Thailand, Pen. Malaysia, Sumatera (incl. Bangka), Jawa, Borneo. 40 IND 41 THA VIE 42 BOR JAW MLY SUM. Phan.
Bridelia heterantha Wall., Numer. List: 7878 (1847), nom. inval.
* *Agyneia glomerulata* Miq., Fl. Ned. Ind., Eerste Bijv.: 447 (1861). *Phyllanthus glomerulatus* (Miq.) Müll.Arg., Flora 48: 375 (1865).
Glochidion wallichianum Müll.Arg., Linnaea 32: 67 (1863). *Phyllanthus wallichianus* (Müll.Arg.) Müll.Arg., Flora 48: 387 (1865). *Glochidion glomerulatum* var. *wallichianum* (Müll.Arg.) Chakrab. & M.G.Gangop., J. Econ. Taxon. Bot. 19: 204 (1995).
Phyllanthus nanogynus Müll.Arg., Flora 48: 376 (1865). *Glochidion nanogynum* (Müll.Arg.) Hook.f., Fl. Brit. India 5: 318 (1887).
Glochidion curtisii Hook.f., Fl. Brit. India 5: 327 (1887).
Glochidion desmocarpum Hook.f., Fl. Brit. India 5: 318 (1887).
Glochidion palustre Koord., Bull. Jard. Bot. Buitenzorg, III, 1: 145 (1918).

Glochidion goniocarpum Hook.f., Fl. Brit. India 5: 309 (1887).
Pen. Malaysia (incl. Singapore). 42 MLY. Phan.

Glochidion goniocladum Airy Shaw, Kew Bull. 34: 592 (1980).
Papua New Guinea. 42 NWG. Nanophan.

Glochidion gracile Airy Shaw, Kew Bull. 36: 601 (1981).
C. Sulawesi. 42 SUL. Phan.

Glochidion gracilentum (Müll.Arg.) Boerl., Handl. Fl. Ned. Ind. 3: 276 (1900).
Jawa. 42 JAW.
* *Phyllanthus gracilentus* Müll.Arg., Flora 48: 380 (1865).

Glochidion grantii Florence, Bull. Mus. Natl. Hist. Nat., B, Adansonia 18: 250 (1996).
Society Is. (Raiatea, Tahaa). 61 SCI. Nanophan.

Glochidion granulare Airy Shaw, Kew Bull. 27: 42 (1972).
NE. Papua New Guinea. 42 NWG. Phan.
Glochidion granulare var. *heterogyne* Airy Shaw, Kew Bull. 36: 602 (1981).

Glochidion grayanum (Müll.Arg.) Florence, Bull. Mus. Natl. Hist. Nat., B, Adansonia 18: 253 (1996).
Society Is. (Tahiti). 61 SCI. Nanophan. or phan.
* *Phyllanthus grayanus* Müll.Arg., Flora 48: 380 (1865).

Glochidion grossum Airy Shaw, Kew Bull. 25: 483 (1971).
NE. Papua New Guinea. 42 NWG. Phan. – Close to *G. nobile.*

Glochidion harveyanum Domin, Biblioth. Bot. 89: 319 (1927). *Glochidion mindorense* subsp. *harveyanum* (Domin) Airy Shaw, Kew Bull. 27: 23 (1972).
Queensland. 50 QLD.
Glochidion harveyanum var. *pubescens* Airy Shaw, Kew Bull. 31: 347 (1976).

Glochidion helferi (Müll.Arg.) Hook.f., Fl. Brit. India 5: 311 (1887).
S. Burma. 41 BMA.
* *Phyllanthus helferi* Müll.Arg., Flora 48: 372 (1865).
Phyllanthus andamanicus Kurz, Rep. Veg. Andaman Isl.: 47 (1870), pro syn.
Glochidion subscandens Kurz, Forest Fl. Burma 2: 344 (1877), nom. illeg.

Glochidion heterocalyx Airy Shaw, Kew Bull. 27: 57 (1972).
Irian Jaya. 42 NWG. Phan. – Close to *G. acuminatissimum.*

Glochidion heterodoxum (Müll.Arg.) Pax & K.Hoffm. in H.G.A.Engler, Nat. Pflanzenfam. ed. 2, 19c: 58 (1931).
Fiji. 60 FIJ. – Provisionally accepted.
* *Phyllanthus heterodoxus* Müll.Arg. in A.P.de Candolle, Prodr. 15(2): 321 (1866).

Glochidion heyneanum (Wight & Arn.) Wight, Icon. Pl. Ind. Orient. 5: t. 1908 (1852).
Phyllanthus heyneanus (Wight) Müll.Arg., Flora 48: 389 (1865), nom. illeg.
Pakistan to China (Yunnan). 36 CHC 40 ASS BAN EHM IND NEP PAK WHM 41 BMA CBD LAO THA VIE. Nanophan. or phan.
Bradleia ovata Wall., Numer. List: 7052 (1832), nom. nud.
* *Gynoon heyneanum* Wight & Arn., Edinburgh New Philos. J. 14: 300 (1833).
Glochidion velutinum Wight, Icon. Pl. Ind. Orient. 5: t. 1907 (1852). *Phyllanthus velutinus* (Wight) Müll.Arg., Flora 48: 387 (1865).
Eriococcus glaucescens Zoll., Tijdschr. Ned.-Indië 14: 173 (1857).
Phyllanthus asperus Müll.Arg., Flora 48: 377 (1865). *Glochidion asperum* (Müll.Arg.) Bedd., Fl. Sylv. S. India: 193 (1872).
Phyllanthus nepalensis Müll.Arg., Flora 48: 375 (1865).
Glochidion nepalense Kurz, Forest Fl. Burma 2: 344 (1877).

Glochidion hivaoaense Florence, Fl. Polynésie Fr. 1: 74 (1997).
Marquesas (Hiva Oa). 61 MRQ. Nanophan. or phan.

Glochidion hohenackeri (Müll.Arg.) Bedd., Fl. Sylv. S. India: 193 (1872).
S. India. 40 IND. Nanophan. or phan.
* *Phyllanthus hohenackeri* Müll.Arg., Flora 48: 373 (1865).

var. **hohenackeri**
S. India. 40 IND. Nanophan. or phan.
Bridelia sinica J.Graham, Cat. Pl. Bombay: 179 (1839).
Phyllanthus fagifolius Müll.Arg., Flora 48: 373 (1865). *Glochidion fagifolium* (Müll.Arg.) Miq. ex Bedd., Fl. Sylv. S. India: 193 (1872).
Glochidion ralphii Hook.f., Fl. Brit. India 5: 314 (1887).

var. **johnstonei** (Hook.f.) Chakrab. & M.G.Gangop., J. Econ. Taxon. Bot. 19: 210 (1995).
S. India. 40 IND. Phan.
* *Glochidion johnstonei* Hook.f., Fl. Brit. India 5: 314 (1887).

Glochidion hosokawae Fosberg, Willdenowia 20: 261 (1991).
Caroline Is. 62 CRL.

Glochidion huahineense Florence, Fl. Polynésie Fr. 1: 75 (1997).
Society Is. (Huahine). 61 SCI. Nanophan. or phan.

Glochidion humbertii Leandri, Mém. Inst. Sci. Madagascar, Sér. B, Biol. Vég. 8: 215 (1957).
EC. Madagascar. 29 MDG.

Glochidion humile Merr., Philipp. J. Sci. 16: 544 (1920).
Philippines (Mindanao). 42 PHI.

Glochidion huntii Airy Shaw, Kew Bull. 27: 61 (1971).
olomon Is. 60 SOL. Nanophan. or phan.

Glochidion hylandii Airy Shaw, Kew Bull. 31: 347 (1976).
Queensland (Cook). 50 QLD. Nanophan. or phan.

Glochidion impuber (Roxb.) Govaerts in R.Govaerts, D.G.Frodin & A.Radcliffe-Smith, World Checklist Bibliogr. Euphorbiaceae: 951 (2000).
Maluku. 42 MOL.
 * *Bradleia impuber* Roxb., Fl. Ind. ed. 1832, 3: 698 (1832). *Agyneia impuber* (Roxb.) Miq., Fl. Ned. Ind. 1(2): 367 (1859).
 Glochidion lambertianum Müll.Arg., Linnaea 32: 60 (1863). *Phyllanthus lambertianus* (Müll.Arg.) Müll.Arg., Flora 48: 372 (1865).

Glochidion insectum Airy Shaw, Kew Bull. 27: 54, 68 (1972).
New Guinea, Bismarck Archip. 42 BIS NWG. Phan.

Glochidion insigne (Müll.Arg.) J.J.Sm. in S.H.Koorders & T.Valeton, Bijdr. Boomsoort. Java 12: 160 (1910).
Thailand ?, Sumatera, Jawa. 41 THA? 42 JAW SUM. Phan.
 * *Phyllanthus insignis* Müll.Arg. in A.P.de Candolle, Prodr. 15(2): 1271 (1866).

Glochidion insulanum Müll.Arg., Linnaea 32: 67 (1863). *Phyllanthus insulanus* (Müll.Arg.) Müll.Arg., Flora 48: 387 (1865).
Maluku. 42 MOL.

Glochidion intercastellanum Airy Shaw, Kew Bull. 23: 24 (1969).
New Guinea (D'Entrecasteaux Is.). 42 NWG. Phan.

Glochidion inusitatum A.C.Sm., Fl. Vit. Nova 2: 493 (1981).
Fiji (Vanua Levu: Mt. Ndelanathau). 60 FIJ. Nanophan.

Glochidion kanehirae Hosok., Trans. Nat. Hist. Soc. Taiwan 25: 22 (1935).
Caroline Is. (Palau). 62 CRL.

Glochidion karnaticum Chakrab. & M.G.Gangop., J. Econ. Taxon. Bot. 19: 211 (1995).
India (Karnataka). 40 IND.

Glochidion katikii Airy Shaw, Kew Bull. 32: 373 (1978).
Papua New Guinea. 42 NWG. Phan.

Glochidion kerangae Airy Shaw, Kew Bull. 27: 18 (1972).
Borneo (C. Sarawak, Brunei, Sabah). 42 BOR. Phan.

Glochidion khasicum (Müll.Arg.) Hook.f., Fl. Brit. India 5: 324 (1887).
Bhutan, Sikkim, Assam, N. Thailand, Andaman Is., China (Yunnan, Guangxi). 36 CHC CHS 40 ASS EHM 41 AND THA. Nanophan. or phan.
* *Phyllanthus khasicus* Müll.Arg., Flora 48: 389 (1865).

var. **bilobulatum** (Vasud.Nair & Chakrab.) Chakrab. & M.G.Gangop., J. Econ. Taxon. Bot. 19: 213 (1995).
Andaman Is. 41 AND. Nanophan. or phan.
* *Glochidion bilobulatum* Vasud.Nair & Chakrab., J. Econ. Taxon. Bot. 5: 936 (1984).

var. **khasicum**
Bhutan, Sikkim, Assam, N. Thailand, S. China. 36 CHC CHS 40 ASS EHM 41 THA. Nanophan. or phan.

Glochidion kopiaginis Airy Shaw, Kew Bull. 33: 30 (1978).
NE. Papua New Guinea. 42 NWG. Phan.

Glochidion korthalsii (Müll.Arg.) Boerl., Handl. Fl. Ned. Ind. 3: 276 (1900).
Borneo (SE. Kalimantan). 42 BOR. Phan.
* *Phyllanthus korthalsii* Müll.Arg., Flora 48: 377 (1865).

Glochidion kostermansii Airy Shaw, Kew Bull. 27: 29 (1972).
Lesser Sunda Is. 42 LSI. Phan.

Glochidion kunstlerianum Gage, Rec. Bot. Surv. India 9: 220 (1922).
Pen. Malaysia, Sumatera, Borneo (E. Kalimantan). 42 BOR MLY SUM. Phan.

Glochidion kusuktsense Hayata, Icon. Pl. Formosan. 9: 96 (1920).
Taiwan. 38 TAI. Nanophan.

Glochidion lalae Airy Shaw, Kew Bull. 23: 21 (1969).
Papua New Guinea. 42 NWG. Phan.

Glochidion lambiricum Airy Shaw, Kew Bull. 36: 599 (1981).
Borneo (Sarawak). 42 BOR. Phan.

Glochidion lamprophyllum (Müll.Arg.) Pax & K.Hoffm. in H.G.A.Engler, Nat. Pflanzenfam. ed. 2, 19c: 58 (1931).
Jawa (Madura I.). 42 JAW.

Glochidion lanceilimbum Merr., Philipp. J. Sci. 26: 462 (1925). *Phyllanthus lanceilimbus* (Merr.) Merr., Philipp. J. Sci. 30: 402 (1926).
S. Philippines to New Guinea (incl. D'Entrecasteaux Is.), Bismarck Archip., Solomon Is., Queensland. 42 BIS BOR MOL NWG PHI SUL ? 50 QLD 60 SOL. Nanophan. or phan. – Related to *G. perakense*.

Glochidion lanceisepalum Merr., J. Straits Branch Roy. Asiat. Soc. 86: 319 (1922).
Burma, Borneo. 41 BMA 42 BOR. Nanophan. or phan.

Glochidion lanceolarium (Roxb.) Voigt, Hort. Suburb. Calcutt.: 153 (1845).
N. India to China. 36 CHC CHH CHS 40 ASS BAN EHM IND NEP 41 BMA CBD LAO THA VIE. Nanophan. or phan.
Phyllanthus fraxinifolius Lodd., Bot. Cab.: t. 839 (1824). – Provisional synonym.
* *Bradleia lanceolaria* Roxb., Fl. Ind. ed. 1832, 3: 697 (1832). *Phyllanthus lanceolarius* (Roxb.) Müll.Arg., Flora 48: 371 (1865).
Glochidion macrophyllum Benth., London J. Bot. 1: 491 (1842).
Glochisandra acuminata Wight, Icon. Pl. Ind. Orient. 5(2): 28 (1852).

Phyllanthus benthamianus Müll.Arg., Flora 48: 371 (1865), nom. illeg.
Glochidion cantoniense Hance, Ann. Sci. Nat., Bot., V, 5: 241 (1866).
Glochidion subsessile var. *birmanicum* Chakrab. & M.G.Gangop., J. Econ. Taxon. Bot. 13: 716 (1989).

Glochidion lanceolatum Hayata, J. Coll. Sci. Imp. Univ. Tokyo 20(3): 16 (1904). *Glochidion zeylanicum* var. *lanceolatum* (Hayata) M.J.Deng & J.C.Wang, Fl. Taiwan, ed. 2, 3: 480 (1993).
Taiwan. 38 TAI. Nanophan. or phan.
Glochidion kotoense Hayata, Icon. Pl. Formosan. 9: 96 (1920).

Glochidion latistylum C.B.Rob., Philipp. J. Sci., C 4: 93 (1909).
Philippines (Mindanao). 42 PHI.

Glochidion leptostylum Airy Shaw, Kew Bull. 27: 37 (1972).
Irian Jaya. 42 NWG. Phan.

Glochidion leucocarpum Airy Shaw, Kew Bull. 36: 303 (1981).
Sumatera. 42 SUM.

Glochidion leucogynum Miq., Fl. Ned. Ind. 1(2): 376 (1859).
Jawa. 42 JAW.
Phyllanthus leucogynus Müll.Arg., Flora 48: 75 (1865).

Glochidion lichenisilvae Leandri ex Humbert, Notul. Syst. (Paris) 6: 29 (1937).
C. Madagascar (Tsaratanana). 29 MDG.

Glochidion ligulatum C.B.Rob., Philipp. J. Sci., C 6: 329 (1911).
Philippines (Luzon: Ilocos Norte). 42 PHI.

Glochidion littorale Blume, Bijdr.: 585 (1826). *Bradleia littorea* (Blume) Steud., Nomencl. Bot., ed. 2, 1: 222 (1840). *Phyllanthus littoralis* (Blume) Müll.Arg., Flora 48: 370 (1865). – FIGURE, p. 955.
S. India, Sri Lanka, Thailand, Vietnam, Pen. Malaysia, Sumatera, Jawa, Borneo, Philippines. 40 IND SRL 41 THA VIE 42 BOR JAW MLY PHI SUM. Nanophan. or phan.
Agyneia impuber Wall., Numer. List: 7869 (1847), pro syn.
Bradleia obtusa Wall., Numer. List: 7869 (1847), nom. nud.
Glochidion littorale var. *caudatum* Airy Shaw, Kew Bull. 29: 290 (1974).
Glochidion littorale var. *culminicola* Airy Shaw, Kew Bull. 29: 290 (1974).

Glochidion lobocarpum (Benth.) F.M.Bailey, Queensl. Fl. 5: 1424 (1902).
Papua New Guinea, CE. Queensland. 42 NWG 50 QLD. Phan.
* *Phyllanthus lobocarpus* Benth., Fl. Austral. 6: 97 (1873).

Glochidion loerzingii Airy Shaw, Kew Bull. 23: 9 (1969). – FIGURE, p. 955.
NW. Sumatera. 42 SUM. Phan.

Glochidion longfieldiae (Ridl.) F.Br., Bernice P. Bishop Mus. Bull. 130: 141 (1935).
Tubuai Is. (Rapa I.). 61 TUB. Nanophan. or phan.
* *Phyllanthus longfieldiae* Ridl., Bull. Misc. Inform. Kew 1926: 55 (1926).

Glochidion longipedicellatum Yamam., J. Soc. Trop. Agric. 5: 178 (1933).
Taiwan. 38 TAI.

Glochidion longistylum C.B.Rob., Philipp. J. Sci., C 4: 96 (1909).
Philippines (Luzon). 42 PHI.

Glochidion lucidum Blume, Bijdr.: 584 (1826). *Bradleia lucida* (Blume) Steud., Nomencl. Bot., ed. 2, 1: 222 (1840). *Phyllanthus lucidus* (Blume) Müll.Arg., Flora 48: 370 (1865).
Pen. Malaysia, Sumatera (incl. Bangka), Jawa, Maluku (Morotai), Philippines, Irian Jaya. 42 JAW MLY MOL NWG PHI SUM. Phan.

Glochidion lutescens Blume, Bijdr.: 585 (1826). *Bradleia lutescens* (Blume) Steud., Nomencl. Bot., ed. 2, 1: 222 (1840). *Phyllanthus lutescens* (Blume) Müll.Arg., Flora 48: 377 (1865). – FIGURE, p. 955.
S. China, Indo-China, Malesia. 36 CHC CHS 40 ASS? 41 BMA MLY THA VIE 42 BIS BOR JAW MOL NWG PHI SUL SUM. Nanophan. or phan.
Bradleia laevigata Wall., Numer. List: 7853 (1847), nom. nud.
Anisonema hypoleucum Miq., Fl. Ned. Ind., Eerste Bijv.: 449 (1861). *Phyllanthus hypoleucus* (Miq.) Müll.Arg., Flora 48: 374 (1865), nom. illeg. *Glochidion hypoleucum* (Miq.) Boerl., Handl. Fl. Ned. Ind. 3: 275 (1900).
Glochidion glaucifolium Müll.Arg., Linnaea 32: 65 (1863). *Phyllanthus glaucifolius* (Müll.Arg.) Ridl., J. Straits Branch Roy. Asiat. Soc. 61: 59 (1912).
Phyllanthus kollmannianus Müll.Arg., Flora 48: 378 (1865). *Glochidion kollmannianum* (Müll.Arg.) J.J.Sm. in S.H.Koorders & T.Valeton, Bijdr. Boomsoort. Java 12: 166 (1910).
Phyllanthus laevigatus Müll.Arg., Flora 48: 374 (1865). *Glochidion laevigatum* (Müll.Arg.) Hook.f., Fl. Brit. India 5: 319 (1887).
Glochidion breynioides C.B.Rob., Philipp. J. Sci., C 4: 95 (1909).
Glochidion hollandianum J.J.Sm., Nova Guinea 12: 544 (1917).
Glochidion pusilliflorum S.Moore, J. Bot. 63(Suppl.): 93 (1925).

Glochidion luzonense Elmer, Leafl. Philipp. Bot. 1: 301 (1908).
Philippines (Luzon). 42 PHI.

Glochidion macphersonii Govaerts & Radcl.-Sm., Kew Bull. 51: 175 (1996).
New Caledonia. 60 NWC.
* *Bradleia macrophylla* Labill., Sert. Austro-Caledon.: 77 (1825). *Glochidion macrophyllum* (Labill.) Müll.Arg., Linnaea 32: 61 (1863), nom. illeg. *Phyllanthus macrophyllus* (Labill.) Müll.Arg., Flora 48: 370 (1865).

Glochidion macrocarpum Blume, Bijdr.: 584 (1826). *Bradleia macrocarpa* (Blume) Steud., Nomencl. Bot., ed. 2, 1: 222 (1840). *Phyllanthus macrocarpus* (Blume) Müll.Arg., Flora 48: 370 (1865). *Cicca macrocarpa* (Blume) Kurz, J. Asiat. Soc. Bengal, Pt. 2, Nat. Hist. 42(2): 239 (1873).
Sumatera to New Guinea. 42 JAW LSI NWG SUM. Phan.

subsp. **macrocarpum**
Jawa, Sumatera. 42 JAW SUM. Phan.

subsp. **orientale** Airy Shaw, Kew Bull. 23: 12 (1969).
Irian Jaya. 42 NWG. Phan.

var. **sumbawanum** Airy Shaw, Kew Bull. 33: 28 (1978).
Lesser Sunda Is. (W. Sumbawa). 42 LSI. Phan.

Glochidion macrosepalum Hosok., Trans. Nat. Hist. Soc. Taiwan 25: 21 (1935).
Caroline Is. (Palau). 62 CRL.

Glochidion macrostigma Hook.f., Fl. Brit. India 5: 313 (1887).
Pen. Malaysia, Jawa, Borneo. 42 BOR JAW MLY. Phan.
Glochidion capitatum J.J.Sm. in S.H.Koorders & T.Valeton, Bijdr. Boomsoort. Java 12: 133 (1910).

Glochidion littorale Blume (A), *G. loerzingii* Airy-Shaw (B), *G. rubrum* Bl. (C), *G. lutescens* Bl. (D)

Artist: Mary Millar Watt
Kew Bull. 36: 305, fig. 6 (1981)
KEW ILLUSTRATIONS COLLECTION

Glochidion maingayi Gage, Rec. Bot. Surv. India 9: 221 (1922).
Pen. Malaysia. 42 MLY.

Glochidion malindangense C.B.Rob., Philipp. J. Sci., C 4: 101 (1909).
Philippines (Mindanao). 42 PHI.

Glochidion manono Baill. ex Müll.Arg., Linnaea 32: 65 (1863). *Phyllanthus manono* (Baill. ex Müll.Arg.) Müll.Arg., Flora 48: 377 (1865).
Society Is. (Moorea Tahiti). 61 SCI. Nanophan. or phan.

Glochidion marchionicum F.Br., Bernice P. Bishop Mus. Bull. 130: 142 (1935).
Marquesas. 61 MRQ. Nanophan. or phan.

Glochidion marianum Müll.Arg., Linnaea 32: 65 (1863).
Marianas. 62 MRN.

Glochidion marojejiense Leandri, Mém. Inst. Sci. Madagascar, Sér. B, Biol. Vég. 8: 215 (1957).
C. Madagascar (Marojejy Mts.). 29 MDG.

Glochidion × marquesanum (F.Br.) Croizat, Sargentia 1: 48 (1942). G. marchionicum × C. tooviianum.
Marquesas (Nuku Hiva). 61 MRQ. Nanophan. or phan.
* *Glochidion ramiflorum* var. *marquesanum* F.Br., Bernice P. Bishop Mus. Bull. 130: 144 (1935).

Glochidion medogense T.L.Chin, Acta Phytotax. Sin. 18: 251 (1980).
SE. Tibet. 36 CHT. Phan.

Glochidion mehipitense Pax & K.Hoffm., Mitt. Inst. Allg. Bot. Hamburg 7: 225 (1931).
Borneo (SW. Kalimantan). 42 BOR. Phan. – Perhaps identical with *G. calospermum*.

Glochidion meijeri Airy Shaw, Kew Bull. 27: 11 (1972).
Sumatera (Mt. Sago). 42 SUM. Nanophan.
Glochidion physocarpum Airy Shaw, Kew Bull. 32: 370 (1978).

Glochidion melvilliorum Airy Shaw, Kew Bull. 25: 487 (1971).
Fiji (Viti Levu). 60 FIJ. Phan.

Glochidion merrillii C.B.Rob., Philipp. J. Sci., C 4: 100 (1909).
Philippines (Luzon). 42 PHI.

Glochidion mindorense C.B.Rob., Philipp. J. Sci., C 4: 98 (1909).
Borneo (Sabah), Philippines, Jawa, Lesser Sunda Is., Maluku, Sulawesi. 42 BOR JAW LSI MOL PHI SUL. Nanophan. or phan.

Glochidion mitrastylum Airy Shaw, Kew Bull. 33: 31 (1978).
NE. Papua New Guinea, Bismarck Archip. 42 BIS NWG. Phan.

Glochidion molle Blume, Bijdr.: 586 (1826). *Bradleia mollis* (Blume) Steud., Nomencl. Bot., ed. 2, 1: 222 (1840). *Phyllanthus mollis* (Blume) Müll.Arg., Flora 48: 387 (1865).
Borneo (Sabah), Philippines, Sulawesi, Moluccas, Tenimber Is., Jawa. 42 BOR JAW MOL PHI SUL. Phan.
Phyllanthus tomentosus Noronha, Verh. Batav. Genootsch. Kunsten 5(4): 22 (1790).
Bradleia blumei Steud., Nomencl. Bot., ed. 2, 1: 222 (1840).
Glochidion villosum Miq., Fl. Ned. Ind. 1(2): 376 (1859).

Glochidion moluccanum Blume, Bijdr.: 587 (1826). *Bradleia moluccana* (Blume) Steud., Nomencl. Bot., ed. 2, 1: 222 (1840). *Phyllanthus moluccanus* (Blume) Müll.Arg., Flora 48: 380 (1865).
Sulawesi. 42 SUL. Phan.

var. **glabrescens** Airy Shaw, Kew Bull. 27: 25 (1972).
WC. & SE. Sulawesi. 42 SUL. Phan.

var. **moluccanum**
NE. Sulawesi. 42 SUL. Phan.

Glochidion monostylum Airy Shaw, Kew Bull. 29: 288 (1974).
Borneo (Sabah: Mt. Kinabalu). 42 BOR. Nanophan. or phan.

Glochidion moonii Thwaites, Enum. Pl. Zeyl.: 286 (1861). *Phyllanthus moonii* (Thwaites) Müll.Arg. in A.P.de Candolle, Prodr. 15(2): 312 (1866).
Sri Lanka. 40 SRL.

var. **moonii**
Sri Lanka. 40 SRL.
* *Phyllanthus pubescens* Moon, Cat. Pl. Ceylon: 65 (1824), nom. nud.
Glochidion montanum Thwaites, Enum. Pl. Zeyl.: 286 (1861).
Phyllanthus glaucogynus Müll.Arg., Flora 48: 389 (1865). *Glochidion glaucogynum* (Müll.Arg.) Bedd., Fl. Sylv. S. India: 195 (1872). *Glochidion moonii* var. *glaucogynum* (Müll.Arg.) Trimen, Handb. Fl. Ceylon 4: 52 (1898). *Glochidion moonii* f. *glaucogynum* (Müll.Arg.) Alston, Ann. Roy. Bot. Gard. (Peradeniya) 11: 6 (1928).
Phyllanthus symplocoides Müll.Arg., Flora 48: 389 (1865). *Glochidion symplocoides* (Müll.Arg.) Bedd., Fl. Sylv. S. India: 195 (1872).

var. **subglabrum** Trimen, Handb. Fl. Ceylon 4: 31 (1898). *Glochidion montanum* f. *subglabrum* (Trimen) Alston, Ann. Roy. Bot. Gard. (Peradeniya) 11: 6 (1928).
Sri Lanka. 40 SRL. Nanophan. or phan.

Glochidion moorei P.T.Li, Acta Phytotax. Sin. 20: 117 (1982).
Society Is. (Raiatea). 61 SCI. Nanophan.
* *Glochidion salicifolium* J.W.Moore, Bernice P. Bishop Mus. Bull. 226: 13 (1963), nom. illeg.

Glochidion mop Airy Shaw, Kew Bull. 27: 50 (1972).
Irian Jaya. 42 NWG. Phan.

Glochidion muelleri Briq., Annuaire Conserv. Jard. Bot. Genève: 225 (1900).
Jawa. 42 JAW.

Glochidion multilobum A.C.Sm., Fl. Vit. Nova 2: 493 (1981).
Fiji (Vanua Levu). 60 FIJ. Nanophan. or phan.

Glochidion multiloculare (Rottler ex Willd.) Voigt, Hort. Suburb. Calcutt.: 152 (1845).
India, Assam, Sikkim, Nepal, Bangladesh, N. Burma. 40 ASS BAN EHM IND NEP 41 BMA. Nanophan. or phan.
* *Agyneia multilocularis* Rottler ex Willd., Ges. Naturf. Freunde Berlin Neue Schriften 4: 206 (1803). *Bradleia multiloculare* (Rottler ex Willd.) Spreng., Syst. Veg. 3: 19 (1826). *Phyllanthus multilocularis* (Rottler ex Willd.) Müll.Arg., Flora 48: 370 (1865).

var. **multiloculare**
India, Assam, Sikkim, Nepal, Bangladesh, N. Burma. 40 ASS BAN EHM IND NEP 41 BMA. Nanophan. or phan.
Agyneia pubera Wall., Numer. List: 7870 (1847), nom. nud.

var. **pubescens** Chakrab. & M.G.Gangop., J. Econ. Taxon. Bot. 14: 720 (1990).
Assam, Sikkim. 40 ASS EHM.

Glochidion muscisilvae Airy Shaw, Kew Bull. 33: 29 (1978).
Irian Jaya. 42 NWG. Phan.

Glochidion myrianthum (Müll.Arg.) Pax & K.Hoffm. in H.G.A.Engler, Nat. Pflanzenfam. ed. 2, 19c: 58 (1931).
Vanuatu. 60 VAN.
* *Phyllanthus myrianthus* Müll.Arg. in A.P.de Candolle, Prodr. 15(2): 317 (1866).

Glochidion myrtifolium J.W.Moore, Bernice P. Bishop Mus. Bull. 226: 10 (1963).
Society Is. 61 SCI. Nanophan. or phan.
Glochidion longipedicellatum J.W.Moore, Bernice P. Bishop Mus. Bull. 226: 9 (1963), nom. illeg. *Glochidion longipes* P.T.Li, Acta Phytotax. Sin. 20: 117 (1982).

Glochidion nadeaudii Florence, Bull. Mus. Natl. Hist. Nat., B, Adansonia 18: 253 (1996).
Society Is. (Moorea). 61 SCI. Phan.

Glochidion namilo Guillaumin, Bull. Mus. Natl. Hist. Nat., II, 9: 300 (1937).
Vanuatu. 60 VAN.

Glochidion nemorale Thwaites, Enum. Pl. Zeyl.: 286 (1861). *Phyllanthus nemoralis* (Thwaites) Müll.Arg., Flora 48: 389 (1865).
Sri Lanka. 40 SRL. Nanophan. or phan.
Glochidion gardneri Thwaites, Enum. Pl. Zeyl.: 286 (1861).
Phyllanthus leptogynus Müll.Arg., Flora 48: 389 (1865). *Glochidion leptogynum* (Müll.Arg.) Bedd., Fl. Sylv. S. India: 195 (1872).
Glochidion acutifolium Alston, Ann. Roy. Bot. Gard. (Peradeniya) 11: 7 (1928).

Glochidion nervosum Alston in H.Trimen, Handb. Fl. Ceylon 6(Suppl.): 260 (1931).
Sri Lanka. 40 SRL. – Provisionally accepted.

Glochidion nesophilum Airy Shaw, Kew Bull. 27: 45 (1972).
Papua New Guinea (incl. D'Entrecasteaux Is., Louisiade Archip.), Bismarck Archip. 42 BIS NWG. Phan.

Glochidion nobile Airy Shaw, Kew Bull. 32: 371 (1978).
Papua New Guinea. 42 NWG. Phan. – Close to *G. grossum*.

Glochidion nothofageticum Airy Shaw, Kew Bull. 34: 593 (1980).
Bismarck Archip. 42 BIS. Phan.

Glochidion novaegeorgiae Airy Shaw, Kew Bull. 23: 23 (1969).
Solomon Is. 60 SOL. Phan.

Glochidion novoguineense K.Schum. in K.M.Schumann & C.A.G.Lauterbach, Fl. Schutzgeb. Südsee, Nachtr.: 287 (1905).
Maluku, New Guinea, Bosmarck Archip. 42 BIS MOL NWG. Nanophan. or phan.
Phyllanthus submollis var. *glabra* Lauterb. & K.Schum., Fl. Schutzgeb. Südsee: 391 (1900).
Glochidion globosum J.J.Sm., Nova Guinea 8: 783 (1912).
Glochidion platycladum Airy Shaw, Kew Bull. 23: 18 (1969).
Glochidion vinkiorum Airy Shaw, Kew Bull. 23: 19 (1969).

Glochidion nubigenum Hook.f., Fl. Brit. India 5: 315 (1887). *Glochidion velutinum* var. *nubigenum* (Hook.f.) Chakrab. & M.G.Gangop., J. Econ. Taxon. Bot. 14: 720 (1990).

Nepal to N. Thailand. 36 CHT 40 ASS EHM NEP 41 BMA THA. Phan.
Glochidion metanubigenum Hurus. & Yu.Tanaka, Fl. E. Himal.: 177 (1966).

Glochidion oblatum Hook.f., Fl. Brit. India 5: 312 (1887).
Sikkim, Assam, Bangladesh, Burma, N. Thailand, China (Yunnan). 40 ASS BAN EHM 41 BMA THA. Nanophan. or phan.

Glochidion oblongifolium Airy Shaw, Kew Bull. 33: 28 (1978).
New Guinea. 42 NWG. Nanophan. or phan.

Glochidion obovatum Siebold & Zucc., Abh. Math.-Phys. Cl. Königl. Bayer. Akad. Wiss. 4(2): 143 (1846). *Phyllanthus obovatus* (Siebold & Zucc.) Müll.Arg., Flora 48: 387 (1865).
Japan (S. Honshu, Shikoku, Kyushu) to Taiwan, China (Fujian, Zhejiang). 36 CHS 38 JAP NNS TAI. Nanophan. or phan.
Bradleia sinensis Siebold ex Miq., Ann. Mus. Bot. Lugduno-Batavi 3: 128 (1867).

Glochidion obscurum (Roxb. ex Willd.) Blume, Bijdr.: 585 (1826).
Indo-China, W. & C. Malesia. 41 CBD THA VIE 42 BOR MLY MOL SUL SUM. Nanophan. or phan.
* *Phyllanthus obscurus* Roxb. ex Willd., Sp. Pl. 4: 581 (1805).
Glochidion glaucum Blume, Bijdr.: 587 (1826). *Glochidion blumeanum* Müll.Arg., Linnaea 32: 65 (1863).
Bradleia pinnata Roxb., Fl. Ind. ed. 1832, 3: 700 (1832).
Bradleia kipareh Steud., Nomencl. Bot., ed. 2, 1: 222 (1840).
Bradleia glaucophylla Hassk., Cat. Hort. Bot. Bogor.: 242 (1844).
Glochidion pinnatum Voigt, Hort. Suburb. Calcutt.: 153 (1845).
Glochidion roxburghianum Müll.Arg., Linnaea 32: 61 (1863).
Phyllanthus kipareh Müll.Arg., Flora 48: 373 (1865).

Glochidion oligotrichum (Müll.Arg.) Boerl., Handl. Fl. Ned. Ind. 3: 276 (1900).
Jawa. 42 JAW.
* *Phyllanthus oligotrichus* Müll.Arg., Flora 48: 379 (1865).

Glochidion oogynum Airy Shaw, Kew Bull. 27: 46 (1972).
NE. Papua New Guinea. 42 NWG. Nanophan.

Glochidion oreichtitum (Leandri) Leandri, Bull. Soc. Bot. France 84: 65 (1937).
C. Madagascar. 29 MDG.
* *Phyllanthus monticola* Leandri, Bull. Soc. Bot. France 81: 450 (1934), nom. illeg.
Phyllanthus oreichtitus Leandri, Cat. Pl. Madag. Euphorbiac.: 24 (1935).

subsp. **oreichtitum**
C. Madagascar (Andringitra). 29 MDG.

subsp. **tsaratananense** Leandri, Mém. Inst. Sci. Madagascar, Sér. B, Biol. Vég. 8: 217 (1957).
C. Madagascar (Mt. Tsaratanana). 29 MDG.

Glochidion ornatum Kurz ex Teijsm. & Binn., Tijdschr. Ned.-Indië 27: 46 (1864).
Phyllanthus ornatus (Kurz ex Teijsm. & Binn.) Müll.Arg., Flora 48: 390 (1865).
Sumatera. 42 SUM.

Glochidion orohenense J.W.Moore, Occas. Pap. Bernice Pauahi Bishop Mus. 16: 6 (1940).
Society Is. (Tahiti). 61 SCI. Phan.

Glochidion oxygonum Airy Shaw, Kew Bull. 23: 12 (1969).
Maluku (Halmahera). 42 MOL. Phan.

Glochidion pachyconum Airy Shaw, Kew Bull. 27: 52 (1972).
Papua New Guinea. 42 NWG. Phan.

Glochidion palauense Hosok., Trans. Nat. Hist. Soc. Taiwan 25: 22 (1935).
Caroline Is. (Palau). 62 CRL.

Glochidion palawanense Elmer, Leafl. Philipp. Bot. 4: 1290 (1911).
Philippines (Palawan). 42 PHI.

Glochidion paludicola (Airy Shaw) Airy Shaw, Kew Bull., Addit. Ser. 8: 111 (1980).
NE. Papua New Guinea. 42 NWG. Phan.
* *Glochidion mindorense* subsp. *paludicola* Airy Shaw, Kew Bull. 27: 22 (1972).

Glochidion papenooense Florence, Bull. Mus. Natl. Hist. Nat., B, Adansonia 18: 254 (1996).
Society Is. (Tahiti). 61 SCI. Nanophan.

Glochidion peltiferum S.Moore, J. Bot. 63(Suppl.): 94 (1925).
Sumatera. 42 SUM.

Glochidion perrieri Leandri, Bull. Soc. Bot. France 81: 606 (1934).
C. Madagascar. 29 MDG.

Glochidion phellocarpum Airy Shaw, Kew Bull. 35: 385 (1980).
Pen. Malaysia, Sumatera, Jawa ? 42 JAW MLY SUM.

Glochidion philippicum (Cav.) C.B.Rob., Philipp. J. Sci., C 4: 103 (1909).
S. China, Hainan, Taiwan, Trop. Asia, N. Australia, Solomon Is. 36 CHC CHH CHS 38 TAI 42 BIS BOR JAW MOL NWG PHI SUL SUM (Enggano isl.) 50 NTA QLD 60 SOL. Phan.
* *Bradleia philippica* Cav., Icon. 4: 48 (1798).
Bradleia philippensis Willd., Sp. Pl. 4: 592 (1805).
Glochidion philippinense Benth., Fl. Hongk.: 314 (1861), nom. illeg. *Phyllanthus philippinensis* Müll.Arg., Flora 48: 376 (1865).
Glochidion compressicaule Kurz ex Teijsm. & Binn., Tijdschr. Ned.-Indië 27: 45 (1864). *Phyllanthus compressicaulis* (Kurz ex Teijsm. & Binn.) Müll.Arg., Flora 48: 376 (1865).
Phyllanthus quercinus Müll.Arg., Flora 48: 386 (1865). *Glochidion quercinum* (Müll.Arg.) Boerl., Handl. Fl. Ned. Ind. 3: 276 (1900).
Phyllanthus kurzianus Müll.Arg. in A.P.de Candolle, Prodr. 15(2): 1272 (1866).
Phyllanthus ferdinandii var. *mollis* Benth., Fl. Austral. 6: 97 (1873).
Glochidion formosanum Hayata, J. Coll. Sci. Imp. Univ. Tokyo 20(3): 20 (1904).
Coccoglochidion erythrococcus K.Schum. in K.M.Schumann & C.A.G.Lauterbach, Fl. Schutzgeb. Südsee, Nachtr.: 293 (1905).

Glochidion phyllanthoides Merr., Philipp. J. Sci. 16: 541 (1920).
Philippines (Luzon). 42 PHI.

Glochidion phyllochlamys Airy Shaw, Kew Bull. 33: 30 (1978).
SE. Irian Jaya. 42 NWG. Phan.

Glochidion pilosum (Lour.) Merr., Trans. Amer. Philos. Soc., n.s., 24: 232 (1935).
Vietnam. 41 VIE.
* *Nymphanthus pilosus* Lour., Fl. Cochinch.: 544 (1790). *Emblica pilosa* (Lour.) Spreng., Syst. Veg. 3: 20 (1826). *Phyllanthus pilosus* (Lour.) Müll.Arg. in A.P.de Candolle, Prodr. 15(2): 432 (1866).

Glochidion pitcairnense (F.Br.) H.St.John, Trans. Roy. Soc. New Zealand, Bot. 1: 187 (1962).
Pitcairn Is. (incl. Henderson I.). 61 PIT.
* *Glochidion taitense* var. *pitcairnense* F.Br., Bernice P. Bishop Mus. Bull. 130: 142 (1935).

Glochidion plagiophyllum Airy Shaw, Kew Bull. 37: 23 (1982).
Lesser Sunda Is. 42 LSI.

Glochidion pleiosepalum Airy Shaw, Kew Bull. 27: 56 (1972).
Irian Jaya. 42 NWG. Phan.

Glochidion podocarpum (Müll.Arg.) C.B.Rob., Philipp. J. Sci., C 6: 330 (1911).
Fiji. 60 FIJ. Nanophan.
* *Phyllanthus podocarpus* Müll.Arg., Flora 48: 388 (1865).

Glochidion pomiferum Airy Shaw, Kew Bull. 23: 13 (1969).
New Guinea. 42 NWG. Phan.

Glochidion ponapense Hosok., Trans. Nat. Hist. Soc. Taiwan 25: 24 (1935).
Caroline Is. (Ponape). 62 CRL.

Glochidion praeclarum Airy Shaw, Kew Bull. 33: 27 (1978).
Borneo (Sarawak). 42 BOR. Phan.

Glochidion prinoides S.Moore, J. Bot. 63(Suppl.): 95 (1925).
Jawa. 42 JAW.

Glochidion pruinosum Airy Shaw, Muelleria 4: 211 (1980).
Queensland (Cook). 50 QLD. Nanophan. or phan.

Glochidion psidioides C.B.Rob., Philipp. J. Sci., C 4: 92 (1909).
Philippines (Luzon). 42 PHI.

Glochidion puberum (L.) Hutch. in C.S.Sargent, Pl. Wilson. 2: 518 (1916).
China, Taiwan. 36 CHC CHH CHN CHS CHT 38 TAI. Nanophan. or phan.
Agyneia impubes L., Mant. Pl. 2: 296 (1771).
* *Agyneia pubera* L., Mant. Pl. 2: 296 (1771). *Bradleia pubera* (L.) Roxb., Fl. Ind. ed. 1832, 3: 698 (1832). *Phyllanthus puberus* (L.) Müll.Arg., Flora 48: 387 (1865).
Bradleia sinica Gaertn., Fruct. Sem. Pl. 2: 127 (1791). *Glochidion sinicum* (Gaertn.) Hook. & Arn., Bot. Beechey Voy.: 210 (1837).
Agyneia pinnata Miq., Fl. Ned. Ind. 1(2): 368 (1859).
Agyneia sinica Miq., Fl. Ned. Ind. 1(2): 368 (1859).
Glochidion distichum Hance, Ann. Sci. Nat., Bot., IV, 18: 228 (1862).
Glochidion fortunei Hance, Ann. Sci. Nat., Bot., IV, 18: 228 (1862).
Glochidion pseudo-obscurum Pamp., Nuovo Giorn. Bot. Ital. 17: 411 (1910).
Glochidion bodinieri H.Lév., Repert. Spec. Nov. Regni Veg. 12: 183 (1913).

Glochidion pubescens Zipp. ex Span., Linnaea 15: 346 (1841).
Lesser Sunda Is. (Timor). 42 LSI.

Glochidion pubicapsa Airy Shaw, Kew Bull., Addit. Ser. 4: 130 (1975).
Borneo, Sumatera, Jawa, Philippines. 42 BOR JAW PHI SUM. Phan.

var. **brunneiforme** Airy Shaw, Kew Bull., Addit. Ser. 4: 131 (1975).
Borneo (Sarawak). 42 BOR. Phan.

var. **pubicapsa**
Borneo, Sumatera, Jawa, Philippines. 42 BOR JAW PHI SUM. Phan.
Glochidion pubicarpum Elmer, Leafl. Philipp. Bot. 10: 3733 (1939), no latin descr.

Glochidion pulgarense Elmer, Leafl. Philipp. Bot. 4: 1291 (1911).
Philippines (Palawan: Mt. Pulgar). 42 PHI.

Glochidion punctatum Pax & K.Hoffm., Mitt. Inst. Allg. Bot. Hamburg 7: 226 (1931).
Borneo (SW. Kalimantan). 42 BOR. Phan.

Glochidion pungens Airy Shaw, Kew Bull. 31: 349 (1976).
Queensland (Cook). 50 QLD. Phan.

Glochidion pyriforme Airy Shaw, Kew Bull. 27: 33 (1972).
NE. Papua New Guinea. 42 NWG. Phan.

Glochidion raivavense F.Br., Bernice P. Bishop Mus. Bull. 130: 142 (1935).
Tubuai Is. (Raivavae, Rurutu, Tubuai). 61 TUB. Nanophan. or phan.

Glochidion ramiflorum J.R.Forst. & G.Forst., Char. Gen. Pl.: 57 (1775). *Bradleia glochidion* Gaertn., Fruct. Sem. Pl. 2: 128 (1791). *Phyllanthus ramiflorus* (J.R.Forst. & G.Forst.) Müll.Arg., Flora 48: 374 (1865).
Vanuatu, Samoa, Niue. (36) chs 60 fij NUE SAM VAN. Nanophan. or phan.
Glochidion uniflorum J.F.Gmel., Syst. Nat.: 1555 (1792).
Glochidion tannaense Guillaumin, J. Arnold Arbor. 13: 90 (1932).

Glochidion rapaense Florence, Bull. Mus. Natl. Hist. Nat., B, Adansonia 18: 258 (1996).
Tubuai Is. (Rapa I.). 61 TUB. Nanophan.

Glochidion reinwardtii (Müll.Arg.) Boerl., Handl. Fl. Ned. Ind. 3: 276 (1900).
Jawa. 42 JAW.
* *Phyllanthus reinwardtii* Müll.Arg., Flora 48: 379 (1865).

Glochidion reticulatum Elmer, Leafl. Philipp. Bot. 1: 302 (1908).
Philippines (Luzon). 42 PHI.

Glochidion retinerve Airy Shaw, Kew Bull. 27: 39, 67 (1972).
NE. Papua New Guinea. 42 NWG. Phan.

Glochidion robinsonii Elmer, Leafl. Philipp. Bot. 7: 2643 (1915).
Philippines. 42 PHI.
Glochidion sorsogonense Elmer ex Merr., Enum. Philipp. Fl. Pl. 2: 402 (1923).

Glochidion rubrum Blume, Bijdr.: 586 (1826). *Bradleia rubra* (Blume) Steud., Nomencl. Bot., ed. 2, 1: 222 (1840). – FIGURE, p. 955.
China (Fujian), Taiwan, Japan, Trop. Asia. 36 CHS 38 JAP TAI 41 BMA CBD THA VIE 42 BOR JAW LSI MLY MOL PHI SUL SUM. Nanophan. or phan.

var. **brevistylum** (Chakrab. & Gang.) Chakrab. & M.G.Gangop., J. Econ. Taxon. Bot. 19: 223 (1995).
Burma. 41 BMA. Nanophan. or phan.
* *Glochidion insulare* var. *brevistylum* Chakrab. & M.G.Gangop., J. Econ. Taxon. Bot. 13: 715 (1989).

var. **rubrum**
Trop. & Subtrop. Asia. 36 CHS 38 JAP TAI 41 BMA CBD THA VIE 42 BOR JAW LSI MLY MOL PHI SUL SUM. Nanophan. or phan.
Bradleia coronata Wall., Numer. List: 7857 (1847), nom. nud.
Bridelia glauca Wall., Numer. List: 7875 (1847), nom. inval.
Phyllanthus diversifolius Miq., Fl. Ned. Ind., Eerste Bijv.: 448 (1861). *Glochidion diversifolium* (Miq.) Merr., Forest. Bur. Philipp. Bull. 1: 29 (1903).
Phyllanthus penangensis Müll.Arg., Flora 48: 388 (1865). *Glochidion penangense* (Müll.Arg.) Airy Shaw, Kew Bull. 23: 6 (1969).
Glochidion leiostylum Kurz, J. Asiat. Soc. Bengal, Pt. 2, Nat. Hist. 42(2): 237 (1873).

Glochidion coronatum Hook.f., Fl. Brit. India 5: 326 (1887).
Glochidion insulare Hook.f., Fl. Brit. India 5: 310 (1887).
Glochidion rubrum f. *longistylis* J.J.Sm. in ?, : 152 (1910).
Glochidion foliosum S.Moore, J. Bot. 63(Suppl.): 96 (1925).
Glochidion grave S.Moore, J. Bot. 63(Suppl.): 96 (1925).
Glochidion versicolor S.Moore, J. Bot. 63(Suppl.): 94 (1925).
Glochidion thorelii Beille in H.Lecomte, Fl. Indo-Chine 5: 622 (1927).

Glochidion rufoglaucum (Müll.Arg.) Boerl., Handl. Fl. Ned. Ind. 3: 275 (1900).
Jawa. 42 JAW.
* *Phyllanthus rufoglaucus* Müll.Arg., Flora 48: 372 (1865).

Glochidion rugulosum Airy Shaw, Kew Bull. 23: 17 (1969).
Irian Jaya. 42 NWG. Nanophan. or phan.

Glochidion runikerae Airy Shaw, Kew Bull. 33: 33 (1978).
Solomon Is. (San Cristobal). 60 SOL. Nanophan. or phan.

Glochidion saccocarpum Airy Shaw, Kew Bull. 25: 487 (1971).
Solomon Is. 60 SOL. Phan.

Glochidion sambiranense (Leandri) Leandri, Notul. Syst. (Paris) 6: 29 (1937).
Madagascar. 29 MDG.
* *hyllanthus sambiranensis* Leandri, Bull. Soc. Bot. France 81: 451 (1934).

var. **sambiranense**
C. Madagascar (Upper Sambirano). 29 MDG.

var. **trapezophyllum** Leandri, in Fl. Madag. 111: 29 (1958).
NC. Madagascar (Mt. Tsaratanana). 29 MDG.

Glochidion santisukii Airy Shaw, Kew Bull. 35: 385 (1980).
Thailand. 41 THA. Phan.
* *Glochidion cataractarum* Airy Shaw, Kew Bull. 32: 76 (1977), nom. illeg.

Glochidion seemannii Müll.Arg., Linnaea 32: 63 (1863). *Phyllanthus seemannianus* (Müll.Arg.) Müll.Arg., Flora 48: 374 (1865).
Fiji. 60 FIJ. Nanophan. or phan.

Glochidion semicordatum (Müll.Arg.) Boerl., Handl. Fl. Ned. Ind. 3: 276 (1900).
Jawa. 42 JAW.
* *Phyllanthus semicordatus* Müll.Arg., Flora 48: 376 (1865).

Glochidion senyavinianum Glassman, Bernice P. Bishop Mus. Bull. 209: 71 (1952).
Caroline Is. (Ponape). 62 CRL.

Glochidion sericeum (Blume) Zoll. & Moritzi, Natuur.-Geneesk. Arch. Ned.-Indië 2: 585 (1845).
Pen. Malaysia, Sumatera, Jawa, Borneo. 42 BOR JAW MLY SUM. Nanophan. or phan.
* *Glochidionopsis sericea* Blume, Bijdr.: 588 (1826). *Phyllanthus sericeus* (Blume) Müll.Arg., Flora 48: 390 (1865).

Glochidion sessiliflorum Airy Shaw, Kew Bull. 31: 350 (1976).
Queensland. 50 QLD. Nanophan. or phan.
Glochidion sessiliflorum var. *pedicellatum* Airy Shaw, Kew Bull. 35: 640 (1980).
Glochidion sessiliflorum var. *stylosum* Airy Shaw, Kew Bull. 35: 641 (1980).

Glochidion singaporense Gage, Rec. Bot. Surv. India 9: 221 (1922).
Pen. Malaysia (incl. Singapore), Sumatera, Borneo (Sabah, NE. Kalimantan). 42 BOR MLY SUM. Nanophan. or phan.

Glochidion societatis Florence, Fl. Polynésie Fr. 1: 90 (1997).
Society Is., Tubuai Is. (Rimatara). 61 SCI TUB. Nanophan. or phan.

Glochidion sphaerogynum (Müll.Arg.) Kurz, Forest Fl. Burma 2: 346 (1877).
Sikkim, Bhutan, Assam, Burma, Thailand, Vietnam, China. 36 CHC CHH CHS 40 ASS EHM 41 BMA THA VIE. Nanophan. or phan.
* *Phyllanthus sphaerogynus* Müll.Arg., Flora 48: 375 (1865).

Glochidion stellatum (Retz.) Bedd., Fl. Sylv. S. India: 194 (1872).
Sri Lanka. 40 SRL. Nanophan. or phan.
* *Phyllanthus stellatus* Retz., Observ. Bot. 5: 29 (1788).
Gynoon rigidum A.Juss., Euphorb. Gen.: 107 (1824). *Glochidion rigidum* (A.Juss.) Müll.Arg., Linnaea 32: 67 (1863).
Gynoon jussieuianum Wight, Icon. Pl. Ind. Orient. 5: 29 (1852). *Glochidion jussieuianum* (Wight) Thwaites, Enum. Pl. Zeyl.: 285 (1861). *Phyllanthus jussieuianus* (Wight) Müll.Arg., Flora 48: 386 (1865).
Gynoon tetrandrum D.Dietr., Syn. Pl. 5: 388 (1852).
Glochidion thwaitesii Müll.Arg., Linnaea 32: 66 (1863).
Gynoon triandrum Wight & Arn., Edinburgh New Philos. J. 14: 300 (Apr. 1833).

Glochidion stenophyllum Airy Shaw, Kew Bull. 27: 31 (1972).
Maluku, Irian Jaya. 42 MOL NWG.

Glochidion stilpnophyllum Govaerts & Radcl.-Sm., Kew Bull. 51: 175 (1996).
Borneo (Sabah), Philippines (Luzon, Polillo, Samar). 42 BOR PHI.
* *Glochidion nitidum* Merr., Philipp. J. Sci., C 9: 483 (1914 publ. 1915), nom. illeg.

Glochidion stipulare Airy Shaw, Kew Bull. 32: 377 (1978).
Vanuatu. 60 VAN. Nanophan. or phan.

Glochidion striatum J.J.Sm., Nova Guinea 8: 784 (1912).
Irian Jaya. 42 NWG. Nanophan. or phan.

Glochidion styliferum J.J.Sm., Bull. Jard. Bot. Buitenzorg, III, 1: 391 (1920).
Borneo. 42 BOR. Nanophan. or phan.

Glochidion stylosum Ridl., Bull. Misc. Inform. Kew 1923: 364 (1923).
Pen. Malaysia. 42 MLY.

Glochidion subfalcatum Elmer, Leafl. Philipp. Bot. 1: 305 (1908).
N. Borneo, Philippines. 42 BOR PHI.

var. **nitidum** Chakrab. & M.G.Gangop., J. Econ. Taxon. Bot. 19: 232 (1995).
Borneo (Sabah), Philippines. 42 BOR PHI.

var. **subfalcatum**
Philippines. 42 PHI.

Glochidion submolle (K.Schum. & Lauterb.) Airy Shaw, Kew Bull. 32: 377 (1978).
New Guinea. 42 NWG. Phan.
* *Phyllanthus submollis* K.Schum. & Lauterb., Fl. Schutzgeb. Südsee: 390 (1900).
Glochidion magnificum K.Schum. in K.M.Schumann & C.A.G.Lauterbach, Fl. Schutzgeb. Südsee, Nachtr.: 288 (1905).

Glochidion subsessile N.P.Balakr. & Chakrab., Proc. Indian Acad. Sci., Pl. Sci. 92: 359 (1983).
Andaman Is. 41 AND. Phan.

Glochidion subterblancum C.E.C.Fisch., Bull. Misc. Inform. Kew 1927: 211 (1927). *Breynia subterblanca* (C.E.C.Fisch.) C.E.C.Fisch., Bull. Misc. Inform. Kew 1939: 98 (1938).
S. Burma. 41 BMA.

Glochidion suishaense Hayata, Icon. Pl. Formosan. 9: 97 (1920).
Taiwan (Shuishe). 38 TAI. Nanophan.

Glochidion superbum Baill. ex Müll.Arg., Linnaea 32: 64 (1863). *Phyllanthus superbus* (Baill. ex Müll.Arg.) Müll.Arg., Flora 48: 375 (1865).
S. Thailand, Pen. Malaysia, Sumatera (incl. Bangka), Borneo. 41 THA 42 BOR MLY SUM. Phan.
Bradleia finlaysoniana Wall., Numer. List: 7860 (1847), nom. nud.
Glochidion dasyphyllum Miq., Fl. Ned. Ind., Eerste Bijv.: 451 (1861), nom. illeg.

Glochidion symingtonii Airy Shaw, Kew Bull. 27: 6 (1972).
Pen. Malaysia. 42 MLY. Nanophan. or phan.

Glochidion taitense Baill. ex Müll.Arg., Linnaea 32: 66 (1863). *Phyllanthus taitensis* (Baill. ex Müll.Arg.) Müll.Arg., Flora 48: 380 (1865).
Society Is. (Moorea, Tahiti). 61 SCI. Nanophan. or phan.
Glochidion majus Baill., Étude Euphorb.: 637 (1858), nom. nud.

Glochidion talmyanum Beille in H.Lecomte, Fl. Indo-Chine 5: 614 (1927).
Vietnam. 41 VIE.

Glochidion temehaniense J.W.Moore, Bernice P. Bishop Mus. Bull. 226: 15 (1963).
Society Is. (Huahine, Raiatea, Tahaa). 61 SCI. Nanophan.

Glochidion tenuistylum Stapf, Trans. Linn. Soc. London, Bot. 4: 223 (1894). *Glochidion penangense* var. *tenuistylum* (Stapf) Airy Shaw, Kew Bull. 23: 7 (1969).
Borneo (Sabah: Mt. Kinabalu). 42 BOR. Nanophan.

Glochidion ternateum Airy Shaw, Kew Bull. 27: 28 (1972).
Maluku (Ternate). 42 MOL. Nanophan.

Glochidion tetrapteron Gage, Rec. Bot. Surv. India 9: 223 (1922).
Pen. Malaysia. 42 MLY. Phan.

Glochidion thomsonii (Müll.Arg.) Hook.f., Fl. Brit. India 5: 318 (1887).
Assam, Bangladesh, SE. Tibet. 36 CHT 40 ASS BAN. Nanophan. or phan.
* *Phyllanthus thomsonii* Müll.Arg., Flora 48: 375 (1865).
Glochidion rubidulum T.L.Chin, Acta Phytotax. Sin. 18: 251 (1980).

Glochidion timorense Airy Shaw, Kew Bull. 27: 31 (1972).
Lesser Sunda Is. (Timor). 42 LSI. Nanophan. or phan.

Glochidion tooviianum Florence, Bull. Mus. Natl. Hist. Nat., B, Adansonia 18: 260 (1996).
Marquesas (Nuku Hiva). 61 MRQ. Nanophan. or phan.

Glochidion triandrum (Blanco) C.B.Rob., Philipp. J. Sci., C 4: 92 (1909).
Nepal, Sikkim, Assam, S. China, Cambodia, S. Japan, Nansei-shoto, Taiwan, Philippines. 36 CHC CHS 38 JAP NNS TAI 40 ASS EHM NEP 41 CBD 42 PHI. Nanophan.
* *Kirganelia triandra* Blanco, Fl. Filip.: 711 (1837). *Phyllanthus triandrus* (Blanco) Müll.Arg., Flora 48: 379 (1865).

Bradleia acuminata Wall., Numer. List: 7855 (1847), nom. nud.
Bridelia acuminata Wall., Numer. List: 7885 (1847), nom. inval.
Glochidion eleutherostylum Müll.Arg., Linnaea 32: 69 (1863).
Phyllanthus bicolor Müll.Arg., Flora 48: 389 (1865). *Glochidion bicolor* (Müll.Arg.) Hayata, J. Coll. Sci. Imp. Univ. Tokyo 20(3): 18 (1904).
Glochidion quinquestylum Elmer, Leafl. Philipp. Bot. 1: 303 (1908).
Glochidion hypoleucum Hayata, Icon. Pl. Formosan. 9: 95 (1920), nom. illeg. *Glochidion hayatae* Croizat & Hara, J. Jap. Bot. 16: 316 (1940).

Glochidion trichogynum Müll.Arg., Linnaea 32: 66 (1863). *Phyllanthus trichogynus* (Müll.Arg.) Müll.Arg., Flora 48: 380 (1865).
Philippines. 42 PHI.

Glochidion trichophorum Merr., Philipp. J. Sci., C 9: 484 (1914 publ. 1915).
Philippines (Luzon). 42 PHI.

Glochidion trusanicum Airy Shaw, Kew Bull. 27: 20 (1972).
Borneo (NE. Sarawak). 42 BOR. Phan.

Glochidion tuamotuense Florence, Fl. Polynésie Fr. 1: 98 (1997).
Tuamotu Is. (Niau, Taravai). 61 TUA. Nanophan.

Glochidion ultrabasicola Airy Shaw, Kew Bull. 36: 601 (1981).
Sulawesi. 42 SUL. Nanophan.

Glochidion urceolare Airy Shaw, Kew Bull. 27: 40 (1972).
NE. Papua New Guinea. 42 NWG. Nanophan.

Glochidion urophylloides Elmer, Leafl. Philipp. Bot. 1: 300 (1908).
Philippines. 42 PHI.
Glochidion fenicis Merr., Philipp. J. Sci., C 3: 414 (1909).
Glochidion lanceifolium C.B.Rob., Philipp. J. Sci., C 4: 90 (1909).
Glochidion subangulatum Elmer, Leafl. Philipp. Bot. 7: 2644 (1915).

Glochidion vaniotii H.Lév., Fl. Kouy-Tchéou: 164 (1914).
China (Guizhou). 36 CHS.

Glochidion varians Miq., Fl. Ned. Ind., Eerste Bijv.: 450 (1861).
Sumatera. 42 SUM.
Phyllanthus fuscus Müll.Arg., Flora 48: 380 (1865). *Glochidion fuscum* (Müll.Arg.) Boerl., Handl. Fl. Ned. Ind. 3: 276 (1900).
Phyllanthus varians Müll.Arg., Flora 48: 380 (1865).

Glochidion venulosum (Müll.Arg.) P.T.Li, Guihaia 14: 131 (1994).
Fiji. 60 FIJ.
* *Phyllanthus venulosus* Müll.Arg., Flora 48: 374 (1865).

Glochidion vitiense (Müll.Arg.) Gillespie, Bernice P. Bishop Mus. Bull. 91: 17 (1932).
Fiji. 60 FIJ. Nanophan. or phan.
* *Phyllanthus vitiensis* Müll.Arg., Flora 48: 374 (1865).

Glochidion weberi C.B.Rob., Philipp. J. Sci., C 6: 330 (1911).
Philippines (Mindanao). 42 PHI.

Glochidion websteri Fosberg, Willdenowia 20: 262 (1991).
Caroline Is. 62 CRL.

Glochidion wilderi Florence, Fl. Polynésie Fr. 1: 99 (1997).
Tuamotu Is. (Mangarave, Makatea). 61 TUA. Nanophan.

Glochidion williamsii C.B.Rob., Philipp. J. Sci., C 3: 199 (1908).
Philippines, Borneo (Sabah: Mt. Kinabalu). 42 BOR PHI. Phan.

Glochidion wilsonii Hutch. in C.S.Sargent, Pl. Wilson. 2: 518 (1916).
S. China. 36 CHC CHS. Nanophan.

Glochidion wisselense Airy Shaw, Kew Bull. 27: 38 (1972).
Irian Jaya. 42 NWG. Nanophan. or phan.
Glochidion soegengii Airy Shaw, Kew Bull. 29: 293 (1974).

Glochidion wonenggau Airy Shaw, Kew Bull. 27: 26 (1972).
NE. Sulawesi. 42 SUL. Phan.

Glochidion woodii Merr., Philipp. J. Sci. 26: 464 (1925).
Philippines (Mindanao). 42 PHI.

Glochidion wrightii Benth., Fl. Hongk.: 313 (1861). *Phyllanthus wrightii* (Benth.) Müll.Arg., Flora 48: 378 (1865).
S. China, Hainan. 36 CHC CHH CHS. Nanophan. or phan.

Glochidion xerocarpum (O.Schwartz) Airy Shaw, Kew Bull., Addit. Ser. 8: 115 (28 Apr. 1980).
Borneo (Sabah), S. Philippines, Maluku, Lesser Sunda Is., NE. Papua New Guinea, Northern Territory. 42 BOR JAW LSI MOL NWG SUL PHI 50 NTA. Phan.
* *Phyllanthus xerocarpus* O.Schwarz, Repert. Spec. Nov. Regni Veg. 24: 87 (1927).

Glochidion xestophyllum Airy Shaw, Kew Bull. 23: 11 (1969).
Pen. Malaysia ?, Borneo (E. Kalimantan). 42 BOR MLY. Phan.

Glochidion zeylanicum (Gaertn.) A.Juss., Euphorb. Gen.: 107 (1824).
India to Solomon Is. 36 CHC CHH CHS 38 JAP NNS TAI 40 ASS BAN EHM IND SRL 41 AND BMA NCB THA VIE 42 BIS BOR JAW MLY NWG SUM 50 NTA QLD 60 SOL. Nanophan. or phan.
* *Bradleia zeylanica* Gaertn., Fruct. Sem. Pl. 2: 128 (1791). *Phyllanthus zeylanicus* (Gaertn.) Müll.Arg. in A.P.de Candolle, Prodr. 15(2): 281 (1866), nom. illeg.

var. **arborescens** (Blume) Chakrab. & M.G.Gangop., J. Econ. Taxon. Bot. 19: 228 (1995).
China (Yunnan), Assam to Borneo. 36 CHC 40 ASS BAN IND 41 AND BMA NCB THA 42 BOR JAW MLY SUM. Nanophan. or phan.
* *Glochidion arborescens* Blume, Bijdr.: 584 (1826). *Bradleia arborescens* (Blume) Steud., Nomencl. Bot., ed. 2, 1: 222 (1840). *Phyllanthus arborescens* (Blume) Müll.Arg., Flora 48: 370 (1865).
Glochidion bancanum Miq., Fl. Ned. Ind., Eerste Bijv.: 450 (1861).
Phyllanthus silheticus Müll.Arg., Flora 48: 378 (1865). *Glochidion silheticum* (Müll.Arg.) Croizat, J. Arnold Arbor. 21: 492 (1940).
Phyllanthus teysmannii Müll.Arg. in A.P.de Candolle, Prodr. 15(2): 1270 (1866).
Glochidion sclerophyllum Hook.f., Fl. Brit. India 5: 313 (1887).

var. **arunachalense** Chakrab. & M.G.Gangop., J. Econ. Taxon. Bot. 19: 229 (1995).
Arunachal Pradesh. 40 EHM. Nanophan. or phan.

var. **talbotii** (Hook.f.) Haines, Bot. Bihar Orissa 2: 132 (1921).
India to Taiwan. 36 CHC CHH CHS 38 TAI 40 ASS BAN EHM IND SRL 41 THA. Nanophan. or phan.
Bradleia hirsuta Roxb., Fl. Ind. ed. 1832, 3: 699 (1832). *Glochidion hirsutum* (Roxb.) Voigt, Hort. Suburb. Calcutt.: 153 (1845). *Phyllanthus hirsutus* (Roxb.) Müll.Arg., Flora 48: 371 (1865).

Glochidion molle Hook. & Arn., Bot. Beechey Voy.: 210 (1837), nom. illeg.
Glochidion tomentosum Dalzell, Hooker's J. Bot. Kew Gard. Misc. 2: 38 (1851). *Phyllanthus tomentosus* (Dalzell) Müll.Arg., Flora 48: 371 (1865). *Glochidion zeylanicum* var. *tomentosum* (Dalzell) Chakrab. & M.G.Gangop., J. Econ. Taxon. Bot. 19: 229 (1995), nom. illeg.
Glochidion dasyphyllum K.Koch, Hort. Dendrol.: 85, 3 (1853).
Agyneia hirsuta Miq., Fl. Ned. Ind. 1(2): 368 (1859).
Glochidion arnottianum Müll.Arg., Linnaea 32: 60 (1863).
Phyllanthus arnottianus Müll.Arg., Flora 48: 370 (1865).
Glochidion mishmiense Hook.f., Fl. Brit. India 5: 327 (1887).
* *Glochidion tomentosum* var. *talbotii* Hook.f., Fl. Brit. India 5: 311 (1887).
Glochidion hongkongense var. *puberulum* Chakrab. & M.G.Gangop., J. Econ. Taxon. Bot. 13: 712 (1989).

var. **zeylanicum**

India to Solomon Is. 36 CHC CHS 38 JAP NNS TAI 40 ASS BAN IND SRL 41 BMA THA VIE 42 BIS BOR JAW MLY NWG SUM 50 NTA QLD 60 SOL. Nanophan. or phan.
Agyneia obliqua Willd., Sp. Pl. 4: 568 (1805). *Bradleia obliqua* (Willd.) Spreng., Syst. Veg. 3: 19 (1826). *Glochidion obliquum* (Willd.) Decne., Nouv. Ann. Mus. Hist. Nat. 3: 481 (1834). *Phyllanthus obliquus* (Willd.) Müll.Arg., Flora 48: 371 (1865).
Bradleia nitida Roxb., Fl. Ind. ed. 1832, 3: 699 (1832). *Glochidion nitidum* (Roxb.) Voigt, Hort. Suburb. Calcutt.: 153 (1845). *Phyllanthus nitidus* (Roxb.) Reinw. ex Blume, Mus. Bot. 2: 110 (1856). *Glochidion zeylanicum* var. *nitidum* (Roxb.) Haines, Bot. Bihar Orissa 2: 132 (1921).
Bradleia timoriensis Steud., Nomencl. Bot., ed. 2, 1: 222 (1840).
Glochidion subscandens Zoll. & Moritzi, Natuur.-Geneesk. Arch. Ned.-Indië 2: 584 (1845). *Phyllanthus subscandens* (Zoll. & Moritzi) Müll.Arg., Flora 48: 372 (1865).
Agyneia flexuosa B.Heyne ex Wall., Numer. List: 7863 (1847).
Glochidion littorale Benth., Fl. Hongk.: 314 (1861).
Glochidion sumatranum Miq., Fl. Ned. Ind., Eerste Bijv.: 450 (1861).
Glochidion hongkongense Müll.Arg., Linnaea 32: 60 (1863). *Phyllanthus hongkongensis* (Müll.Arg.) Müll.Arg., Flora 48: 371 (1865).
Phyllanthus canaranus Müll.Arg., Flora 48: 871 (1865). *Glochidion canaranum* (Müll.Arg.) Bedd., Fl. Sylv. S. India: 192 (1872).
Phyllanthus ferdinandii var. *supra-axillaris* Benth., Fl. Austral. 6: 96 (1873). *Glochidion ferdinandii* var. *supra-axillare* (Benth.) F.M.Bailey, Queensl. Fl. 5: 1424 (1902). *Glochidion supra-axillare* (Benth.) Domin, Biblioth. Bot. 89: 318 (1927). *Glochidion perakense* var. *supra-axillare* (Benth.) Airy Shaw, Kew Bull. 27: 72 (1972).
Glochidion brunneum Hook.f., Fl. Brit. India 5: 312 (1887).
Glochidion perakense Hook.f., Fl. Brit. India 5: 317 (1887).
Glochidion zeylanicum var. *malayanum* J.J.Sm. in S.H.Koorders & T.Valeton, Bijdr. Boomsoort. Java 12: 118 (1910).
Glochidion liukiuense Hayata, J. Coll. Sci. Imp. Univ. Tokyo 30: 265 (1911). *Glochidion lanceolatum* var. *liukiuense* (Hayata) Hurus., J. Fac. Sci. Univ. Tokyo, Sect. 3, Bot. 6: 332 (1954).
Glochidion pedunculatum Merr., Philipp. J. Sci., C 11: 67 (1916).
Glochidion sphaerostigmum Hayata, Icon. Pl. Formosan. 9: 96 (1920).
Glochidion glaberrimum Ridl., Bull. Misc. Inform. Kew 1923: 363 (1923).
Glochidion dasyphyllum var. *iriomatense* Hurus., J. Fac. Sci. Univ. Tokyo, Sect. 3, Bot. 6: 334 (1954).
Glochidion brunneum subsp. *andamanicum* N.P.Balakr. & Chakrab., Proc. Indian Acad. Sci., Pl. Sci. 92: 357 (1983).

Glochidion zollingeri Miq., Fl. Ned. Ind. 1(1): 686 (1856). *Phyllanthus zollingeri* (Miq.) Müll.Arg., Flora 48: 371 (1865).
Jawa. 42 JAW.

Synonyms:

Glochidion acutifolium Alston === **Glochidion nemorale** Thwaites

Glochidion adamii (Müll.Arg.) Gardner ex Beard === **Sauropus glaucus** (F.Muell.) Airy Shaw

Glochidion aeneum (Baill.) Müll.Arg. === **Phyllanthus aeneus** Baill. var. **aeneus**

Glochidion airyshawii N.P.Balakr. & Chakrab. === **Glochidion andamanicum** var. **desmogyne** (Hook.f.) Chakrab. & M.G.Gangop.

Glochidion anamiticum Kuntze === **Glochidion eriocarpum** Champ. ex Benth.

Glochidion annamense Beille === **Glochidion eriocarpum** Champ. ex Benth.

Glochidion arborescens Blume === **Glochidion zeylanicum** var. **arborescens** (Blume) Chakrab. & M.G.Gangop.

Glochidion arboreum Wight === **Glochidion candolleanum** (Wight & Arn.) Chakrab. & M.G.Gangop.

Glochidion arboreum var. *pauciflorum* Hook.f. === **Glochidion candolleanum** (Wight & Arn.) Chakrab. & M.G.Gangop.

Glochidion arnottianum Müll.Arg. === **Glochidion zeylanicum** var. **talbotii** (Hook.f.) Haines

Glochidion asperum (Müll.Arg.) Bedd. === **Glochidion heyneanum** (Wight & Arn.) Wight

Glochidion assamicum (Müll.Arg.) Hook.f. === **Glochidion ellipticum** Wight

Glochidion assamicum var. *brevipedicellatum* Hurus. & Yu.Tanaka === **Glochidion ellipticum** Wight

Glochidion baladense (Baill.) Müll.Arg. === **Phyllanthus baladensis** Baill.

Glochidion balsahanense Elmer === **Glochidion coronulatum** C.B.Rob.

Glochidion bancanum Miq. === **Glochidion zeylanicum** var. **arborescens** (Blume) Chakrab. & M.G.Gangop.

Glochidion bhutanicum D.G.Long === **Glochidion bourdillonii** var. **bhutanicum** (D.G.Long) Chakrab. & M.G.Gangop.

Glochidion bicolor (Müll.Arg.) Hayata === **Glochidion triandrum** (Blanco) C.B.Rob.

Glochidion bifarium Royle === ?

Glochidion bilobulatum Vasud.Nair & Chakrab. === **Glochidion khasicum** var. **bilobulatum** (Vasud.Nair & Chakrab.) Chakrab. & M.G.Gangop.

Glochidion blumeanum Müll.Arg. === **Glochidion obscurum** (Roxb. ex Willd.) Blume

Glochidion bodinieri H.Lév. === **Glochidion puberum** (L.) Hutch.

Glochidion botryanthum (Müll.Arg.) Pax & K.Hoffm. === **Phyllanthus botryanthus** Müll.Arg.

Glochidion bourgeoisii (Baill.) Müll.Arg. === **Phyllanthus bourgeoisii** Baill.

Glochidion brachylobum Müll.Arg. === **Glochidion coriaceum** Thwaites

Glochidion brachystylum Merr. === **Glochidion album** (Blanco) Boerl.

Glochidion breynioides C.B.Rob. === **Glochidion lutescens** Blume

Glochidion brunneum Hook.f. === **Glochidion zeylanicum** (Gaertn.) A.Juss. var. **zeylanicum**

Glochidion brunneum subsp. *andamanicum* N.P.Balakr. & Chakrab. === **Glochidion zeylanicum** (Gaertn.) A.Juss. var. **zeylanicum**

Glochidion bupleuroides (Baill.) Müll.Arg. === **Phyllanthus bupleuroides** Baill.

Glochidion canaranum (Müll.Arg.) Bedd. === **Glochidion zeylanicum** (Gaertn.) A.Juss. var. **zeylanicum**

Glochidion cantoniense Hance === **Glochidion lanceolarium** (Roxb.) Voigt

Glochidion capitatum J.J.Sm. === **Glochidion macrostigma** Hook.f.

Glochidion capitis-york Airy Shaw === **Glochidion benthamianum** Domin

Glochidion cataractarum Müll.Arg. === **Phyllanthus chamaecerasus** var. **vieillardii** (Baill.) M.Schmid

Glochidion cataractarum Airy Shaw === **Glochidion santisukii** Airy Shaw

Glochidion chamaecerasus (Baill.) Müll.Arg. === **Phyllanthus chamaecerasus** Baill.

Glochidion cinerascens Miq. === **Alphitonia cinerascens** (Miq.) Hoogl. (Rhamnaceae)

Glochidion compressicaule Kurz ex Teijsm. & Binn. === **Glochidion philippicum** (Cav.) C.B.Rob.

Glochidion concolor var. *obovatum* Müll.Arg. === **Glochidion concolor** Müll.Arg.

Glochidion cornutum (Baill.) Müll.Arg. === **Phyllanthus cornutus** Baill.

Glochidion coronatum Hook.f. === **Glochidion rubrum** Blume var. **rubrum**

Glochidion crassifolium (Müll.Arg.) Gardner ex Beard === **Sauropus crassifolius** (Müll.Arg.) Airy Shaw
Glochidion cumingii Müll.Arg. === **Glochidion album** (Blanco) Boerl.
Glochidion curtisii Hook.f. === **Glochidion glomerulatum** (Miq.) Boerl.
Glochidion dasyphyllum Miq. === **Glochidion superbum** Baill. ex Müll.Arg.
Glochidion dasyphyllum K.Koch === **Glochidion zeylanicum** var. **talbotii** (Hook.f.) Haines
Glochidion dasyphyllum var. *iriomatense* Hurus. === **Glochidion zeylanicum** (Gaertn.) A.Juss. var. **zeylanicum**
Glochidion desmocarpum Hook.f. === **Glochidion glomerulatum** (Miq.) Boerl.
Glochidion desmogyne Hook.f. === **Glochidion andamanicum** var. **desmogyne** (Hook.f.) Chakrab. & M.G.Gangop.
Glochidion diospyroides Schltr. === **Glochidion caledonicum** Müll.Arg.
Glochidion distichum Hance === **Glochidion puberum** (L.) Hutch.
Glochidion diversifolium (Miq.) Merr. === **Glochidion rubrum** Blume var. **rubrum**
Glochidion diversifolium Bedd. === **Glochidion ellipticum** Wight
Glochidion diversifolium var. *wightianum* (Müll.Arg.) Bedd. === **Glochidion ellipticum** Wight
Glochidion eleutherostylum Müll.Arg. === **Glochidion triandrum** (Blanco) C.B.Rob.
Glochidion ellipticum var. *wightiana* (Müll.Arg.) Hook. === **Glochidion ellipticum** Wight
Glochidion esquirolii H.Lév. === **Glochidion eriocarpum** Champ. ex Benth.
Glochidion fagifolium (Müll.Arg.) Miq. ex Bedd. === **Glochidion hohenackeri** (Müll.Arg.) Bedd. var. **hohenackeri**
Glochidion faguetii (Baill.) Müll.Arg. === **Phyllanthus faguetii** Baill.
Glochidion fenicis Merr. === **Glochidion urophylloides** Elmer
Glochidion ferdinandii var. *minor* Benth. === **Glochidion ferdinandii** (Müll.Arg.) F.M.Bailey
Glochidion ferdinandii var. *pubens* Maiden ex Airy Shaw === **Glochidion ferdinandii** (Müll.Arg.) F.M.Bailey
Glochidion ferdinandii var. *supra-axillare* (Benth.) F.M.Bailey === **Glochidion zeylanicum** (Gaertn.) A.Juss. var. **zeylanicum**
Glochidion flavum Ridl. === **Phyllanthus roseus** (Craib & Hutch.) Beille
Glochidion flexuosum (Siebold & Zucc.) F.Muell. ex Miq. === **Phyllanthus flexuosus** (Siebold & Zucc.) Müll.Arg.
Glochidion foliosum S.Moore === **Glochidion rubrum** Blume var. **rubrum**
Glochidion formosanum Hayata === **Glochidion philippicum** (Cav.) C.B.Rob.
Glochidion fortunei Hance === **Glochidion puberum** (L.) Hutch.
Glochidion fuscum (Müll.Arg.) Boerl. === **Glochidion varians** Miq.
Glochidion gamblei Hook.f. === **Glochidion daltonii** (Müll.Arg.) Kurz
Glochidion gardneri Thwaites === **Glochidion nemorale** Thwaites
Glochidion glaberrimum Ridl. === **Glochidion zeylanicum** (Gaertn.) A.Juss. var. **zeylanicum**
Glochidion glaucifolium Müll.Arg. === **Glochidion lutescens** Blume
Glochidion glaucogynum (Müll.Arg.) Bedd. === **Glochidion moonii** Thwaites var. **moonii**
Glochidion glaucum (Labill.) Müll.Arg. === **Glochidion billardieri** Baill.
Glochidion glaucum Blume === **Glochidion obscurum** (Roxb. ex Willd.) Blume
Glochidion globosum J.J.Sm. === **Glochidion novoguineense** K.Schum.
Glochidion glomerulatum var. *wallichianum* (Müll.Arg.) Chakrab. & M.G.Gangop. === **Glochidion glomerulatum** (Miq.) Boerl.
Glochidion granulare var. *heterogyne* Airy Shaw === **Glochidion granulare** Airy Shaw
Glochidion grave S.Moore === **Glochidion rubrum** Blume var. **rubrum**
Glochidion harveyanum var. *pubescens* Airy Shaw === **Glochidion harveyanum** Domin
Glochidion hayatae Croizat & Hara === **Glochidion triandrum** (Blanco) C.B.Rob.
Glochidion heterolobum Müll.Arg. === **Glochidion billardieri** Baill.
Glochidion hirsutum (Roxb.) Voigt === **Glochidion zeylanicum** var. **talbotii** (Hook.f.) Haines
Glochidion hirtellum (F.Muell.) H.Eichler === **Sauropus hirtellus** (F.Muell.) Airy Shaw
Glochidion hollandianum J.J.Sm. === **Glochidion lutescens** Blume
Glochidion hongkongense Müll.Arg. === **Glochidion zeylanicum** (Gaertn.) A.Juss. var. **zeylanicum**

Glochidion hongkongense var. *puberulum* Chakrab. & M.G.Gangop. === **Glochidion zeylanicum** var. **talbotii** (Hook.f.) Haines
Glochidion hypoleucum (Miq.) Boerl. === **Glochidion lutescens** Blume
Glochidion hypoleucum Hayata === **Glochidion triandrum** (Blanco) C.B.Rob.
Glochidion insulare Hook.f. === **Glochidion rubrum** Blume var. **rubrum**
Glochidion insulare var. *brevistylum* Chakrab. & M.G.Gangop. === **Glochidion rubrum** var. **brevistylum** (Chakrab. & Gang.) Chakrab. & M.G.Gangop.
Glochidion johnstonei Hook.f. === **Glochidion hohenackeri** var. **johnstonei** (Hook.f.) Chakrab. & M.G.Gangop.
Glochidion jussieuianum (Wight) Thwaites === **Glochidion stellatum** (Retz.) Bedd.
Glochidion kanalense Baill. === **Glochidion billardieri** Baill.
Glochidion kerrii Craib === **Glochidion dasystylum** var. **kerrii** (Craib) Chakrab. & M.G.Gangop.
Glochidion kollmannianum (Müll.Arg.) J.J.Sm. === **Glochidion lutescens** Blume
Glochidion kotoense Hayata === **Glochidion lanceolatum** Hayata
Glochidion laevigatum (Müll.Arg.) Hook.f. === **Glochidion lutescens** Blume
Glochidion laevigatum var. *cuspidatum* Ridl. === **Glochidion brooksii** Ridl.
Glochidion lambertianum Müll.Arg. === **Glochidion impuber** (Roxb.) Govaerts
Glochidion lanceifolium C.B.Rob. === **Glochidion urophylloides** Elmer
Glochidion lanceolatum var. *liukiuense* (Hayata) Hurus. === **Glochidion zeylanicum** (Gaertn.) A.Juss. var. **zeylanicum**
Glochidion laurifolium Heynh. === ?
Glochidion leiostylum Kurz === **Glochidion rubrum** Blume var. **rubrum**
Glochidion lenormandii Müll.Arg. === **Phyllanthus kanalensis** Baill.
Glochidion leptogynum (Müll.Arg.) Bedd. === **Glochidion nemorale** Thwaites
Glochidion leytense Elmer === **Glochidion album** (Blanco) Boerl.
Glochidion littorale Benth. === **Glochidion zeylanicum** (Gaertn.) A.Juss. var. **zeylanicum**
Glochidion littorale var. *caudatum* Airy Shaw === **Glochidion littorale** Blume
Glochidion littorale var. *culminicola* Airy Shaw === **Glochidion littorale** Blume
Glochidion liukiuense Hayata === **Glochidion zeylanicum** (Gaertn.) A.Juss. var. **zeylanicum**
Glochidion llanosii Müll.Arg. === **Sauropus villosus** (Blanco) Merr.
Glochidion longipedicellatum J.W.Moore === **Glochidion myrtifolium** J.W.Moore
Glochidion longipes P.T.Li === **Glochidion myrtifolium** J.W.Moore
Glochidion macrochorion (Baill.) Müll.Arg. === **Phyllanthus macrochorion** Baill.
Glochidion macrophyllum (Labill.) Müll.Arg. === **Glochidion macphersonii** Govaerts & Radcl.-Sm.
Glochidion macrophyllum Benth. === **Glochidion lanceolarium** (Roxb.) Voigt
Glochidion magnificum K.Schum. === **Glochidion submolle** (K.Schum. & Lauterb.) Airy Shaw
Glochidion majus Baill. === **Glochidion taitense** Baill. ex Müll.Arg.
Glochidion malabaricum (Müll.Arg.) Bedd. === **Glochidion ellipticum** Wight
Glochidion metanubigenum Hurus. & Yu.Tanaka === **Glochidion nubigenum** Hook.f.
Glochidion microbotrys Hook.f. === **Glochidion borneense** (Müll.Arg.) Boerl.
Glochidion microphyllum Ridl. === **Phyllanthus reticulatus** Poir. var. **reticulatus**
Glochidion microphyllum Müll.Arg. === **Phyllanthus pycnophyllus** Müll.Arg.
Glochidion mindanaense Elmer === **Glochidion album** (Blanco) Boerl.
Glochidion mindorense subsp. *glabrum* (J.J.Sm.) Airy Shaw === **Glochidion glabrum** J.J.Sm.
Glochidion mindorense subsp. *harveyanum* (Domin) Airy Shaw === **Glochidion harveyanum** Domin
Glochidion mindorense subsp. *paludicola* Airy Shaw === **Glochidion paludicola** (Airy Shaw) Airy Shaw
Glochidion mishmiense Hook.f. === **Glochidion zeylanicum** var. **talbotii** (Hook.f.) Haines
Glochidion molle Hook. & Arn. === **Glochidion zeylanicum** var. **talbotii** (Hook.f.) Haines
Glochidion montanum Thwaites === **Glochidion moonii** Thwaites var. **moonii**
Glochidion montanum f. *subglabrum* (Trimen) Alston === **Glochidion moonii** var. **subglabrum** Trimen

Glochidion moonii var. *glaucogynum* (Müll.Arg.) Trimen === **Glochidion moonii** Thwaites var. **moonii**
Glochidion moonii f. *glaucogynum* (Müll.Arg.) Alston === **Glochidion moonii** Thwaites var. **moonii**
Glochidion nanogynum (Müll.Arg.) Hook.f. === **Glochidion glomerulatum** (Miq.) Boerl.
Glochidion neilgherrense Wight === **Glochidion candolleanum** (Wight & Arn.) Chakrab. & M.G.Gangop.
Glochidion nepalense Kurz === **Glochidion heyneanum** (Wight & Arn.) Wight
Glochidion nitidum Merr. === **Glochidion stilpnophyllum** Govaerts & Radcl.-Sm.
Glochidion nitidum (Roxb.) Voigt === **Glochidion zeylanicum** (Gaertn.) A.Juss. var. **zeylanicum**
Glochidion obliquum (Willd.) Decne. === **Glochidion zeylanicum** (Gaertn.) A.Juss. var. **zeylanicum**
Glochidion obscurum var. *papuanum* J.J.Sm. === **Glochidion benthamianum** Domin
Glochidion ovalifolium Zipp. ex Span. === ?
Glochidion ovatum (Poir.) Müll.Arg. === **Phyllanthus ovatus** Poir.
Glochidion pachycarpum Alston === **Glochidion candolleanum** (Wight & Arn.) Chakrab. & M.G.Gangop.
Glochidion palustre Koord. === **Glochidion glomerulatum** (Miq.) Boerl.
Glochidion pancherianum (Baill.) Müll.Arg. === **Phyllanthus pancherianus** Baill.
Glochidion pauciflorum (Hook.f.) Gamble === **Glochidion candolleanum** (Wight & Arn.) Chakrab. & M.G.Gangop.
Glochidion pedunculatum Merr. === **Glochidion zeylanicum** (Gaertn.) A.Juss. var. **zeylanicum**
Glochidion penangense (Müll.Arg.) Airy Shaw === **Glochidion rubrum** Blume var. **rubrum**
Glochidion penangense var. *tenuistylum* (Stapf) Airy Shaw === **Glochidion tenuistylum** Stapf
Glochidion perakense Hook.f. === **Glochidion zeylanicum** (Gaertn.) A.Juss. var. **zeylanicum**
Glochidion perakense var. *supra-axillare* (Benth.) Airy Shaw === **Glochidion zeylanicum** (Gaertn.) A.Juss. var. **zeylanicum**
Glochidion perrottetianum Bedd. === **Glochidion candolleanum** (Wight & Arn.) Chakrab. & M.G.Gangop.
Glochidion philippinense Benth. === **Glochidion philippicum** (Cav.) C.B.Rob.
Glochidion physocarpum Airy Shaw === **Glochidion meijeri** Airy Shaw
Glochidion pinnatum Voigt === **Glochidion obscurum** (Roxb. ex Willd.) Blume
Glochidion platycalyx (Müll.Arg.) Pax & K.Hoffm. === **Phyllanthus platycalyx** Müll.Arg.
Glochidion platycladum Airy Shaw === **Glochidion novoguineense** K.Schum.
Glochidion poeppigianum Müll.Arg. === **Phyllanthus poeppigianus** (Müll.Arg.) Müll.Arg.
Glochidion pseudo-obscurum Pamp. === **Glochidion puberum** (L.) Hutch.
Glochidion puberulum Hosok. === ?
Glochidion pubicarpum Elmer === **Glochidion pubicapsa** Airy Shaw var. **pubicapsa**
Glochidion pulchellum Airy Shaw === **Phyllanthus stultitiae** Airy Shaw
Glochidion pusilliflorum S.Moore === **Glochidion lutescens** Blume
Glochidion pycnocarpum (Müll.Arg.) Bedd. === **Glochidion candolleanum** (Wight & Arn.) Chakrab. & M.G.Gangop.
Glochidion pycnocarpum var. *ellipticum* Hook.f. === **Glochidion candolleanum** (Wight & Arn.) Chakrab. & M.G.Gangop.
Glochidion quercinum (Müll.Arg.) Boerl. === **Glochidion philippicum** (Cav.) C.B.Rob.
Glochidion quinquestylum Elmer === **Glochidion triandrum** (Blanco) C.B.Rob.
Glochidion raiateense J.W.Moore === **Glochidion emarginatum** J.W.Moore
Glochidion ralphii Hook.f. === **Glochidion hohenackeri** (Müll.Arg.) Bedd. var. **hohenackeri**
Glochidion ramiflorum var. *lanceolatum* Müll.Arg. === **Glochidion concolor** Müll.Arg.
Glochidion ramiflorum var. *marquesanum* F.Br. === **Glochidion × marquesanum** (F.Br.) Croizat
Glochidion rhytidospermum (F.Muell. ex Müll.Arg.) H.Eichler === **Sauropus trachyspermus** (F.Muell.) Airy Shaw
Glochidion riedelianum (Müll.Arg.) Pax & K.Hoffm. === **Phyllanthus riedelianus** Müll.Arg.
Glochidion rigens (F.Muell.) H.Eichler === **Sauropus rigens** (F.Muell.) Airy Shaw

Glochidion rigidum (A.Juss.) Müll.Arg. === **Glochidion stellatum** (Retz.) Bedd.
Glochidion roxburghianum Müll.Arg. === **Glochidion obscurum** (Roxb. ex Willd.) Blume
Glochidion rubidulum T.L.Chin === **Glochidion thomsonii** (Müll.Arg.) Hook.f.
Glochidion rubrum f. *longistylis* J.J.Sm. === **Glochidion rubrum** Blume var. **rubrum**
Glochidion sablanense Elmer === **Glochidion benguetense** Elmer
Glochidion salicifolium J.W.Moore === **Glochidion moorei** P.T.Li
Glochidion salicifolium (Baill.) Müll.Arg. === **Phyllanthus salicifolius** Baill.
Glochidion salicifolium var. *dracunculoides* (Baill.) Müll.Arg. === **Phyllanthus dracunculoides** Baill.
Glochidion salomonis Briq. === **Glochidion fulvirameum** Miq.
Glochidion sclerophyllum Hook.f. === **Glochidion zeylanicum** var. **arborescens** (Blume) Chakrab. & M.G.Gangop.
Glochidion sessiliflorum var. *pedicellatum* Airy Shaw === **Glochidion sessiliflorum** Airy Shaw
Glochidion sessiliflorum var. *stylosum* Airy Shaw === **Glochidion sessiliflorum** Airy Shaw
Glochidion sieboldianum (Miq.) Koidz. === **Actinodaphne sieboldiana** Miq. (Lauraceae)
Glochidion silheticum (Müll.Arg.) Croizat === **Glochidion zeylanicum** var. **arborescens** (Blume) Chakrab. & M.G.Gangop.
Glochidion sinicum (Gaertn.) Hook. & Arn. === **Glochidion puberum** (L.) Hutch.
Glochidion sisparense Gamble === **Glochidion candolleanum** (Wight & Arn.) Chakrab. & M.G.Gangop.
Glochidion soegengii Airy Shaw === **Glochidion wisselense** Airy Shaw
Glochidion sorsogonense Elmer ex Merr. === **Glochidion robinsonii** Elmer
Glochidion sphaerostigmum Hayata === **Glochidion zeylanicum** (Gaertn.) A.Juss. var. **zeylanicum**
Glochidion subangulatum Elmer === **Glochidion urophylloides** Elmer
Glochidion subscandens Zoll. & Moritzi === **Glochidion zeylanicum** (Gaertn.) A.Juss. var. **zeylanicum**
Glochidion subscandens Kurz === **Glochidion helferi** (Müll.Arg.) Hook.f.
Glochidion subsessile var. *birmanicum* Chakrab. & M.G.Gangop. === **Glochidion lanceolarium** (Roxb.) Voigt
Glochidion sumatranum Miq. === **Glochidion zeylanicum** (Gaertn.) A.Juss. var. **zeylanicum**
Glochidion supra-axillare (Benth.) Domin === **Glochidion zeylanicum** (Gaertn.) A.Juss. var. **zeylanicum**
Glochidion symplocoides (Müll.Arg.) Bedd. === **Glochidion moonii** Thwaites var. **moonii**
Glochidion taitense var. *pitcairnense* F.Br. === **Glochidion pitcairnense** (F.Br.) H.St.John
Glochidion tannaense Guillaumin === **Glochidion ramiflorum** J.R.Forst. & G.Forst.
Glochidion thesioides H.Eichler === **Sauropus thesioides** (Benth.) Airy Shaw
Glochidion thorelii Beille === **Glochidion rubrum** Blume var. **rubrum**
Glochidion thwaitesii Müll.Arg. === **Glochidion stellatum** (Retz.) Bedd.
Glochidion tomentosum Dalzell === **Glochidion zeylanicum** var. **talbotii** (Hook.f.) Haines
Glochidion tomentosum var. *talbotii* Hook.f. === **Glochidion zeylanicum** var. **talbotii** (Hook.f.) Haines
Glochidion trachyspermum (F.Muell.) H.Eichler === **Sauropus trachyspermus** (F.Muell.) Airy Shaw
Glochidion triandrum var. *siamense* (Airy Shaw) P.T.Li === **Glochidion acuminatum** var. **siamense** Airy Shaw
Glochidion trilobum Ridl. === **Glochidion borneense** (Müll.Arg.) Boerl.
Glochidion tsusimense Nakai === ?
Glochidion umbratile Maiden & Betche === **Sauropus macranthus** Hassk.
Glochidion uniflorum J.F.Gmel. === **Glochidion ramiflorum** J.R.Forst. & G.Forst.
Glochidion vacciniifolium Müll.Arg. === **Phyllanthus vacciniifolius** (Müll.Arg.) Müll.Arg.
Glochidion velutinum Wight === **Glochidion heyneanum** (Wight & Arn.) Wight
Glochidion velutinum var. *nubigenum* (Hook.f.) Chakrab. & M.G.Gangop. === **Glochidion nubigenum** Hook.f.
Glochidion versicolor S.Moore === **Glochidion rubrum** Blume var. **rubrum**
Glochidion vespertilio (Baill.) Müll.Arg. === **Phyllanthus vespertilio** Baill.

Glochidion vieillardii Müll.Arg. === **Phyllanthus tritepalus** M.Schmid
Glochidion villicaule Hook.f. === **Glochidion eriocarpum** Champ. ex Benth.
Glochidion villosum Miq. === **Glochidion molle** Blume
Glochidion vinkiorum Airy Shaw === **Glochidion novoguineense** K.Schum.
Glochidion wagapense (Müll.Arg.) Briq. === **Glochidion billardieri** Baill.
Glochidion wallichianum Müll.Arg. === **Glochidion glomerulatum** (Miq.) Boerl.
Glochidion zeylanicum var. *lanceolatum* (Hayata) M.J.Deng & J.C.Wang === **Glochidion lanceolatum** Hayata
Glochidion zeylanicum var. *malayanum* J.J.Sm. === **Glochidion zeylanicum** (Gaertn.) A.Juss. var. **zeylanicum**
Glochidion zeylanicum var. *nitidum* (Roxb.) Haines === **Glochidion zeylanicum** (Gaertn.) A.Juss. var. **zeylanicum**
Glochidion zeylanicum var. *tomentosum* Trimen === **Glochidion zeylanicum** var. **talbotii** (Hook.f.) Haines
Glochidion zeylanicum var. *tomentosum* (Dalzell) Chakrab. & M.G.Gangop. === **Glochidion zeylanicum** var. **talbotii** (Hook.f.) Haines

Glochidinopsis

An orthographic variant of *Glochidionopsis*.

Glochidionopsis

Synonyms:
Glochidionopsis Steud. === **Glochidion** J.R.Forst. & G.Forst.
Glochidionopsis Blume === **Glochidion** J.R.Forst. & G.Forst.
Glochidionopsis sericea Blume === **Glochidion sericeum** (Blume) Zoll. & Moritzi

Glochisandra

Synonyms:
Glochisandra Wight === **Glochidion** J.R.Forst. & G.Forst.
Glochisandra acuminata Wight === **Glochidion lanceolarium** (Roxb.) Voigt

Glycydendron

2 species, Northern S America (Venezuela, Brazil and the Guianas); forest trees to 30 m or more with stout branches, hard wood and drupaceous fruit. Related to the African *Klaineanthus* and the Afro-American *Tetrorchidium* (Webster, Synopsis, 1994). (Crotonoideae)

Ducke, A. (1922). *Glycydendron.* Arq. Jard. Bot. Rio de Janeiro 3 (in Plantes nouvelles .. II): 199. Fr/La. — Protologue of genus and description of one species, *G. amazonicum.* [Additional note in *ibid.*, 4: 107 (1925), with botanical details in pl. 10.]

Glycydendron Ducke, Arch. Jard. Bot. Rio de Janeiro 3: 199 (1922).
N. South America, Brazil. 82 84.

Glycydendron amazonicum Ducke, Arch. Jard. Bot. Rio de Janeiro 3: 199 (1922).
Guianas, Venezuela (Amazonas), Brazil (Amazonas). 82 FRG GUY SUR VEN 84 BZN. Phan.

Glycydendron espiritosantense Kuhlm., Arq. Inst. Biol. Veg. 3: 46 (1936).
Brazil (Espírito Santo). 84 BZL. Phan.

Glyphostylus

Recently reduced to *Excoecaria*.

Synonyms:
Glyphostylus Gagnep. === **Excoecaria** L.
Glyphostylus laoticus Gagnep. === **Excoecaria laotica** (Gagnep.) Esser

Godefroya

Synonyms:
Godefroya Gagnep. === **Cleistanthus** Hook.f. ex Planch.
Godefroya rotundata (Jabl.) Gagnep. === **Cleistanthus rotundatus** Jabl.

Gonatogyne

1 species, Brazil; trees 10-15 m with nondescript distichously arranged foliage and flowers in small axillary fascicles. The disk in female flowers is glandular-lobed rather than entire as in *Savia*. Pax and Hoffmann initially (1922) included the single species in that genus before taking up Klotzsch's concept (1933). (Phyllanthoideae)

Pax, F. & K. Hoffmann (1922). *Savia*. In A. Engler (ed.), Das Pflanzenreich, IV 147 XV (Euphorbiaceae-Phyllanthoideae-Phyllantheae): 181-188. Berlin. (Heft 81.) La/Ge. — Includes (pp. 187-188) the single species of *Gonatogyne*, here at sectional rank.

Pax, F. & K. Hoffmann (1933). Über die Stellung der Gattung *Gonatogyne* innerhalb der Euphorbiaceae. Repert. Spec. Nov. Regni Veg. 31: 190-191. Ge. — Reinstatement of generic rank for *Savia* sect. *Gonatogyne*; discussion of placement of the genus within Phyllanthoideae and in relation to its more immediate allies.

Gonatogyne Klotzsch ex Müll.Arg. in C.F.P.von Martius, Fl. Bras. 11(2): 13 (1873).
Brazil. 84.

Gonatogyne brasiliensis (Baill.) Müll.Arg. in C.F.P.von Martius, Fl. Bras. 11(2): 14 (1873).
SE. Brazil. 84 BZL. Phan.
* *Amanoa brasiliensis* Baill., Étude Euphorb.: 581 (1858). *Savia brasiliensis* (Baill.) Pax & K.Hoffm. in H.G.A.Engler, Pflanzenr., IV, 147, XV: 187 (1922).
Gonatogyne lucens Klotzsch ex Baill., Adansonia 5: 345 (1865), nom. nud.

Synonyms:
Gonatogyne lucens Klotzsch ex Baill. === **Gonatogyne brasiliensis** (Baill.) Müll.Arg.

Grimmeodendron

2 species, Middle America (West Indies, in particular the Bahamas and the Greater Antilles); glabrous but poisonous shrubs or trees to 15 m with finely serrate leaves in loose rosettes, sometimes growing in coppices (*G. eglandulosum*, Bahamas). Related to *Bonania* but featuring terminal rather than axillary spikes; indeed, to one of the compilers *G. eglandulosum* is reminiscent of *Excoecaria agallocha*. (Euphorbioideae (except Euphorbieae))

Pax, F. (with K. Hoffmann) (1912). *Grimmeodendron*. In A. Engler (ed.), Das Pflanzenreich, IV 147 V (Euphorbiaceae-Hippomaneae): 258-259. Berlin. (Heft 52.) La/Ge. — 2 species, Bahamas and Greater Antilles (one first collected from the 'Cockpit Country' in Jamaica).

Grimmeodendron Urb., Symb. Antill. 5: 397 (1908).
Caribbean. 81.

Grimmeodendron eglandulosum (A.Rich.) Urb., Symb. Antill. 5: 398 (1908).
Bahamas, Cuba, Haiti. 81 BAH CUB HAI. Nanophan. or phan.
* *Stillingia eglandulosa* A.Rich. in R.de la Sagra, Hist. Fis. Cuba, Bot. 2: 202 (1850). *Excoecaria eglandulosa* (A.Rich.) Müll.Arg. in A.P.de Candolle, Prodr. 15(2): 1209 (1866).
Excoecaria sagraei Müll.Arg., Linnaea 32: 121 (1863).

Grimmeodendron jamaicense Urb., Symb. Antill. 5: 399 (1908).
Jamaica. 81 JAM. Phan.

Grossera

8 species, W. & WC. tropical Africa (7) and Madagascar (1); shrubs or small to medium trees in forest with terminal paniculate or racemose inflorescences. The Malagasy *G. perrieri*, the most distinctive species and moreover geographically disjunct there in dry deciduous forest, has a small inflorescence with long, lax pedicels; when described, Leandri suggested that it might deserve generic rank (for illustration, see his 'Arbres et grands arbustes malgaches .. Euphorbiacées' (1952; under **Malagassia**)). Some of the species in the treatment of Cavaco (1949) have been transferred to the similar-looking *Cavacoa*. (Crotonoideae)

Pax, F. (with K. Hoffmann) (1912). *Grossera*. In A. Engler (ed.), Das Pflanzenreich, IV 147 VI (Euphorbiaceae-Acalypheae-Chrozophorinae): 105-108. Berlin. (Heft 57.) La/Ge. — 3 species in 2 sections, Africa. [Sect. *Racemiformes* is now *Cavacoa*.]

Cavaco, A. (1949). Sur le genre *Grossera* (Euphorbiacées). Bull. Mus. Natl. Hist. Nat., II, 21: 272-278. Fr. — Historical introduction and general systematics; synopsis of genus (8 species in three subgenera) with key, descriptions of novelties, synonymy, references and citations, and indication of distribution; literature, p. 278. [Floral morphology and geographical distribution were the main criteria used here for subdivision of the genus, but evidence from vegetative anatomy was furnished for three species. The author's subgen. *Quadriloculastrum* and *C. quintasii* from sect. *Racemiformes* (subgen. *Grossera*) are now referred to *Cavacoa*.]

Léonard, J. (1955). À propos des genres africains *Grossera* Pax et *Cavacoa* J. Léonard. Bull. Jard. Bot. État 25: 315-324. Fr. —*Grossera* of Pax divided into 2 genera, one of them *Cavacoa*; *G. perrieri* perhaps also deserving of generic rank. A summary treatment of *Grossera* is provided (pp. 316-320), with descriptions of 2 new species. (For additional data, see Léonard (1958, 1959) under Africa.)

Grossera Pax, Bot. Jahrb. Syst. 33: 281 (1903).
W. & WC. Trop. Africa, Madagascar. 22 23 29.
Fourneaua Pierre ex Prain in D.Oliver, Fl. Trop. Afr. 6(1): 816 (1912), pro syn.

Grossera elongata Hutch. in A.W.Exell, Cat. Vasc. Pl. S. Tome: 294 (1944).
Principe. 23 GGI.

Grossera glomeratospicata J.Léonard, Bull. Jard. Bot. État 25: 319 (1955).
Zaire. 23 ZAI. Nanophan.

Grossera macrantha Pax in H.G.A.Engler, Pflanzenr., IV, 147, VII: 426 (1914).
Cameroon, Zaire. 23 CMN ZAI. Nanophan. or phan.

Grossera major Pax, Bot. Jahrb. Syst. 33: 282 (1903).
Cameroon. 23 CMN. Nanophan. – Close to *G. paniculata*.

Grossera multinervis J.Léonard, Bull. Jard. Bot. État 25: 317 (1955).
Zaire. 23 ZAI. Phan.

Grossera paniculata Pax, Bot. Jahrb. Syst. 33: 281 (1903).
Cameroon, Gabon. 23 CMN GAB. Nanophan.

Grossera perrieri Leandri, Bull. Soc. Bot. France 85: 524 (1938 publ. 1939).
Madagascar. 29 MDG.

Grossera vignei Hoyle, Bull. Misc. Inform. Kew 1935: 259 (1935).
Ivory Coast, Ghana. 22 GHA IVO. Nanophan.

Synonyms:
Grossera aurea Cavaco === **Cavacoa aurea** (Cavaco) J.Léonard
Grossera baldwinii Keay & Cavaco === **Cavacoa baldwinii** (Keay & Cavaco) J.Léonard
Grossera quintasii Pax & K.Hoffm. === **Cavacoa quintasii** (Pax & K.Hoffm.) J.Léonard

Guarania

Synonyms:
Guarania Wedd. ex Baill. === **Richeria** Vahl
Guarania gardneriana Baill. === **Richeria grandis** var. **gardneriana** (Baill.) Müll.Arg.
Guarania laurifolia Baill. === **Richeria grandis** Vahl var. **grandis**
Guarania longifolia Baill. === **Richeria grandis** var. **gardneriana** (Baill.) Müll.Arg.
Guarania purpurascens Wedd. ex Baill. === **Richeria grandis** Vahl var. **grandis**
Guarania ramiflora Wedd. ex Baill. === **Richeria grandis** Vahl var. **grandis**
Guarania spruceana Baill. === **Richeria grandis** Vahl var. **grandis**
Guarania suberosa Standl. === **Gurania suberosa** Standl. (Cucurbitaceae)

Gussonia

Synonyms:
Gussonia Spreng. === **Gymnanthes** Sw.
Gussonia concolor Spreng. === **Actinostemon concolor** (Spreng.) Müll.Arg.
Gussonia discolor Spreng. === **Gymnanthes discolor** (Spreng.) Müll.Arg.

*Gymnalypha**

Synonyms:
Gymnalypha Griseb. === **Acalypha** L.

Gymnanthes

24 species, mainly in Middle America, but extending to northern S America and Malesia; includes *Ateramnus*. Webster (in Webster & Huft 1988; see Americas) incorporated *Actinostemon* but this is here rejected on advice from Hajo Esser (personal communication). The two genera are in any case very closely related. *G. lucida* is widely distributed from Florida around the Caribbean; most of the other species are on present knowledge relatively local. No overall revision of this difficult genus of trees has yet succeeded that of Pax (1912). Esser (1994) and Esser et al. (1997; see **Euphorbioideae (except Euphorbieae)**) argue that the African *Duvigneaudia* should also be included and, furthermore, an extension be made to encompass the Malesian *Sebastiania borneensis* and *S. remota*. [Esser (1999) has made the necessary transfers for the Malesian species and they are accordingly here listed.] (Euphorbioideae (except Euphorbieae))

* Note: for *Guya* see Addendum, p. 1620.

Pax, F. (with K. Hoffmann) (1912). *Gymnanthes*. In A. Engler (ed.), Das Pflanzenreich, IV 147 V (Euphorbiaceae-Hippomaneae): 81-88. Berlin. (Heft 52.) La/Ge. — 11 species, mainly in Middle America. [Uncritical.]

Pax, F. (with K. Hoffmann) (1912). *Sebastiania*. In A. Engler (ed.), Das Pflanzenreich, IV 147 V (Euphorbiaceae-Hippomaneae): 88-153. Berlin. (Heft 52.) La/Ge. — No. 23 (pp. 122-123), *S. borneensis* (in sect. *Sarothorstachys*), is best placed in *Gymnanthes*.

Jablonski, E. (1967). *Gymnanthes*. Euphorbiaceae, Guayana Highland (Mem. New York Bot. Gard. 17(1)): 178-179. New York. En. — 1 species, *G. hypoleuca*.

Oe, E. (1988). Een revisie van *Ateramnus*. 47 pp. Utrecht. (Unpubl. thesis, University of Utrecht.) En. — [Not seen, but an unsatisfactory study according to Esser (1994: 45; see Euphorbioideae (except Euphorbieae)).]

• Esser, H..-J. (1999). A partial revision of the Hippomaneae (Euphorbiaceae) in Malesia. Blumea 44: 149-215, illus., maps. En. — *Gymnanthes*, pp. 165-172; treatment of 2 species with key, descriptions, synonymy, references, types, indication of distribution and habitat, illustration (*G. remota*), map and commentary; all general references, identification list and index to botanical names at end of paper. [The species are shrubs to small or (*G. borneensis*) medium trees, to 15 m. *G. remota* is limited to the Leuser massif in Sumatera.]

Gymnanthes Sw., Prodr.: 95 (1788).
SE. U.S.A., Mexico, S. America. 78 79 80 81 82 83 84 85.
Ateramnus P.Browne, Civ. Nat. Hist. Jamaica: 339 (1756).
Gussonia Spreng., Neue Entd. 2: 119 (1821).
Sarothrostachys Klotzsch, Arch. Naturgesch. 7: 185 (1841).
Gymnanthus Endl., Gen. Pl., Suppl. 5: 87 (1850).

Gymnanthes actinostemoides Müll.Arg., Linnaea 32: 103 (1863). *Sebastiania actinostemoides* (Müll.Arg.) Müll.Arg. in A.P.de Candolle, Prodr. 15(2): 1184 (1866).
Mexico (México State). 79 MXC. Phan.

Gymnanthes albicans (Griseb.) Urb., Symb. Antill. 3: 312 (1902).
Cuba. 81 CUB. Nanophan.
* *Excoecaria albicans* Griseb., Nachr. Königl. Ges. Wiss. Georg-Augusts-Univ. 1: 179 (1865). *Sebastiania albicans* (Griseb.) C.Wright, Anales Acad. Ci. Méd. Habana 7: 156 (1870).
Excoecaria venulosa C.Wright ex Griseb., Nachr. Königl. Ges. Wiss. Georg-Augusts-Univ. 1: 159 (1865).
Sebastiania grisebachiana Müll.Arg. in A.P.de Candolle, Prodr. 15(2): 1183 (1866).

Gymnanthes belizensis G.L.Webster, Ann. Missouri Bot. Gard. 54: 198 (1967).
Belize. 80 BLZ.
* *Phyllanthus longipes* Steyerm., Publ. Field Mus. Nat. Hist., Bot. Ser. 22: 153 (1940), nom. illeg.

Gymnanthes borneensis (Pax & K.Hoffm.) Esser, Blumea 44: 170 (1999).
Borneo. 42 BOR. Nanophan. or phan.
* *Sebastiania borneensis* Pax & K.Hoffm. in H.G.A.Engler, Pflanzenr., IV, 147, V: 122 (1912).

Gymnanthes discolor (Spreng.) Müll.Arg., Linnaea 32: 103 (1863).
Brazil (Rio de Janeiro), Paraguay. 84 BZL 85 PAR. Nanophan.
* *Gussonia discolor* Spreng., Neue Entd. 2: 119 (1821). *Stillingia discolor* (Spreng.) Baill., Adansonia 5: 326 (1865), nom. illeg. *Sebastiania discolor* var. *genuina* Müll.Arg. in A.P.de Candolle, Prodr. 15(2): 1185 (1866), nom. inval. *Sebastiania discolor* (Spreng.) Müll.Arg. in A.P.de Candolle, Prodr. 15(2): 1185 (1866).
Gymnanthes discolor var. *subconcolor* Müll.Arg., Linnaea 32: 103 (1863). *Sebastiania discolor* var. *subconcolor* (Müll.Arg.) Müll.Arg. in A.P.de Candolle, Prodr. 15(2): 1185 (1866).
Sebastiania discolor var. *fiebrigii* Pax & K.Hoffm. in H.G.A.Engler, Pflanzenr., IV, 147, V: 122 (1912).

Gymnanthes dressleri G.L.Webster, Ann. Missouri Bot. Gard. 75: 1130 (1988).
Panama. 80 PAN.

Gymnanthes elliptica Sw., Prodr.: 96 (1788). *Sebastiania elliptica* (Sw.) Müll.Arg. in A.P.de Candolle, Prodr. 15(2): 1181 (1866).
Jamaica. 81 JAM. Nanophan.
Excoecaria tinifolia Sw., Fl. Ind. Occid. 2: 1119 (1800).
Gymnanthes obtusa Baill., Étude Euphorb.: 530 (1858).

Gymnanthes farinosa (Griseb.) G.L.Webster, Ann. Missouri Bot. Gard. 75: 1131 (1988).
Guadeloupe, St. Lucia, Dominica. 81 LEE WIN. Phan.
* *Excoecaria farinosa* Griseb., Abh. Königl. Ges. Wiss. Göttingen 7: 169 (1857).

Gymnanthes gaudichaudii Müll.Arg., Linnaea 32: 96 (1863). *Stillingia gaudichaudii* (Müll.Arg.) Baill., Adansonia 5: 332 (1865). *Sebastiania gaudichaudii* (Müll.Arg.) Müll.Arg. in A.P.de Candolle, Prodr. 15(2): 1177 (1866).
Brazil (Bahia, Rio de Janeiro). 84 BZE BZL. Nanophan.
Excoecaria serrulata Miq., Linnaea 19: 446 (1847), provisional synonym.

Gymnanthes glabrata (Mart.) Govaerts in R.Govaerts, D.G.Frodin & A.Radcliffe-Smith, World Checklist Bibliogr. Euphorbiaceae: 979 (2000).
Brazil (Bahia, Rio de Janeiro). 84 BZE BZL. Nanophan.
* *Cnemidostachys glabrata* Mart., Nov. Gen. Sp. Pl. 1: 70 (1824). *Stillingia glabrata* (Mart.) Baill., Étude Euphorb., Atlas: 17 (1858). *Sebastiania corniculata* var. *glabrata* (Mart.) Müll.Arg. in A.P.de Candolle, Prodr. 15(2): 1172 (1866). *Sebastiania multiramea* var. *glabrata* (Mart.) Pax in H.G.A.Engler, Pflanzenr., IV, 147, V: 120 (1912).
Sarothrostachys multiramea Klotzsch, Arch. Naturgesch. 7: 185 (1841). *Sebastiania multiramea* (Klotzsch) Mart., Flora 24(2): 53 (1841). *Gymnanthes multiramea* (Klotzsch) Müll.Arg., Linnaea 32: 97 (1863). *Stillingia multiramea* (Klotzsch) Baill., Adansonia 5: 325 (1865). *Sebastiania multiramea* var. *genuina* Müll.Arg. in A.P.de Candolle, Prodr. 15(2): 1177 (1866), nom. inval.
Stillingia luschnathiana Baill., Étude Euphorb.: 525 (1858). *Sebastiania multiramea* var. *luschnathiana* (Baill.) Müll.Arg. in A.P.de Candolle, Prodr. 15(2): 1177 (1866).

Gymnanthes glandulosa (Sw.) Müll.Arg., Linnaea 32: 106 (1863).
Jamaica. 81 JAM. Phan.
* *Excoecaria glandulosa* Sw., Fl. Ind. Occid. 2: 1124 (1800). *Sebastiania glandulosa* (Sw.) Müll.Arg. in A.P.de Candolle, Prodr. 15(2): 1186 (1866).

Gymnanthes guyanensis Müll.Arg., Linnaea 32: 102 (1863). *Sebastiania guyanensis* (Müll.Arg.) Müll.Arg. in A.P.de Candolle, Prodr. 15(2): 1183 (1866).
Guyana (Rupununi). 82 GUY. Nanophan.

Gymnanthes hypoleuca Benth., Hooker's J. Bot. Kew Gard. Misc. 6: 325 (1854). *Stillingia hypoleuca* (Benth.) Baill., Adansonia 5: 330 (1865). *Sebastiania hypoleuca* (Benth.) Müll.Arg. in A.P.de Candolle, Prodr. 15(2): 1184 (1866).
S. Venezuela, Brazil (Pará, Amazonas). 82 VEN 84 BZN.

Gymnanthes insolita Ferris, Contr. Dudley Herb. 1: 75 (1927).
Mexico (Nayarit). 79 MXS. Phan.

Gymnanthes integra Fawc. & Rendle, Fl. Jamaica 4: 332 (1920).
Jamaica. 81 JAM.

Gymnanthes jamaicensis Urb., Symb. Antill. 7: 516 (1913).
Jamaica. 81 JAM.

Gymnanthes longipes Müll.Arg., Linnaea 34: 216 (1865). *Sebastiania longipes* (Müll.Arg.) Müll.Arg. in A.P.de Candolle, Prodr. 15(2): 1184 (1866).
Mexico (San Luis Potosí, Mexico). 79 MXC MXE. Phan.

Gymnanthes lucida Sw., Prodr.: 96 (1788). *Excoecaria lucida* (Sw.) Sw., Fl. Ind. Occid. 2: 1121 (1800). *Sebastiania lucida* (Sw.) Müll.Arg. in A.P.de Candolle, Prodr. 15(2): 1181 (1866).
S. Florida, Caribbean, Mexico, C. America. 78 FLA 79 MXE 80 BLZ COS GUA 81 BAH CUB DOM HAI JAM LEE PUE. Phan.

Gymnanthes nervosa Müll.Arg., Linnaea 23: 102 (1863). *Stillingia nervosa* (Müll.Arg.) Baill., Adansonia 5: 328 (1865). *Sebastiania nervosa* (Müll.Arg.) Müll.Arg. in A.P.de Candolle, Prodr. 15(2): 1183 (1866).
Bolivia, Paraguay, Brazil (Rio de Janeiro). 83 BOL 84 BZL 85 PAR. Phan. – Wood used locally.

Gymnanthes pallens (Griseb.) Müll.Arg., Linnaea 32: 106 (1863).
Cuba, Hispaniola. 81 CUB DOM HAI. Phan.
* *Excoecaria pallens* Griseb., Mem. Amer. Acad. Arts, n.s., 8: 161 (1861). *Sebastiania pallens* (Griseb.) Müll.Arg. in A.P.de Candolle, Prodr. 15(2): 1189 (1866). *Sapium pallens* (Griseb.) Borhidi, Acta Bot. Acad. Sci. Hung. 25: 17 (1979).
Excoecaria tenax Griseb., Nachr. Königl. Ges. Wiss. Georg-Augusts-Univ. 1: 179 (1865). *Sapium pallens* var. *tenax* (Griseb.) Borhidi, Acta Bot. Acad. Sci. Hung. 25: 17 (1979).
Sapium angustifolium Alain, Revista Soc. Cub. Bot. 10: 27 (1953).

Gymnanthes recurva Urb., Symb. Antill. 3: 312 (1902).
E. Cuba. 81 CUB. Nanophan.

Gymnanthes remota (Steenis) Esser, Blumea 44: 172 (1999).
Sumatera (Aceh). 42 SUM. Nanophan. or phan.
* *Sebastiania remota* Steenis, Bull. Jard. Bot. Buitenzorg, III, 17: 410 (1948).

Gymnanthes riparia (Schltdl.) Klotzsch, Arch. Naturgesch. 7: 182 (1841).
Mexico to S. Costa Rica. 79 MXG MXT 80 COS GUA. Phan.
* *Excoecaria riparia* Schltdl., Linnaea 7: 386 (1832).
Gymnanthes schlechtendaliana Müll.Arg., Linnaea 32: 100 (1863). *Sebastiania schlechtendaliana* (Müll.Arg.) Müll.Arg. in A.P.de Candolle, Prodr. 15(2): 1181 (1866).
Gymnanthes guatemalensis Standl. & Steyerm., Publ. Field Mus. Nat. Hist., Bot. Ser. 23: 122 (1944).

Gymnanthes widgrenii Müll.Arg., Linnaea 32: 97 (1863). *Sebastiania widgrenii* (Müll.Arg.) Müll.Arg. in A.P.de Candolle, Prodr. 15(2): 1178 (1866).
Brazil (Minas Gerais). 84 BZL. Nanophan.
Stillingia widgrenii (Müll.Arg.) Baill., Adansonia 5: 326 (1865).

Synonyms:
Gymnanthes angustifolia Müll.Arg. === **Sebastiania schottiana** (Müll.Arg.) Müll.Arg.
Gymnanthes bahiensis Müll.Arg. === **Sebastiania bahiensis** (Müll.Arg.) Müll.Arg.
Gymnanthes brachyclada Müll.Arg. === **Sebastiania klotzschiana** (Müll.Arg.) Müll.Arg.
Gymnanthes brachypoda (Griseb.) Pax & K.Hoffm. === **Actinostemon brachypodus** (Griseb.) Urb.
Gymnanthes brasiliensis (Spreng.) Müll.Arg. === **Sebastiania brasiliensis** Spreng.
Gymnanthes brasiliensis var. *divaricata* Müll.Arg. === **Sebastiania brasiliensis** Spreng.
Gymnanthes brasiliensis var. *erythroxyloides* Müll.Arg. === **Sebastiania brasiliensis** Spreng.
Gymnanthes brasiliensis f. *microphylla* Müll.Arg. === **Sebastiania brasiliensis** Spreng.
Gymnanthes brasiliensis var. *obovata* Müll.Arg. === **Sebastiania brasiliensis** Spreng.
Gymnanthes brasiliensis var. *robusta* Müll.Arg. === **Sebastiania brasiliensis** Spreng.
Gymnanthes brasiliensis f. *rufescens* Müll.Arg. === **Sebastiania brasiliensis** Spreng.
Gymnanthes brevifolia Müll.Arg. === **Sebastiania brevifolia** (Müll.Arg.) Müll.Arg.
Gymnanthes concolor (Spreng.) Müll.Arg. === **Actinostemon concolor** (Spreng.) Müll.Arg.

Gymnanthes discolor Baill. === **Sebastiania klotzschiana** (Müll.Arg.) Müll.Arg.
Gymnanthes discolor var. *subconcolor* Müll.Arg. === **Gymnanthes discolor** (Spreng.) Müll.Arg.
Gymnanthes granatensis Müll.Arg. === **Sebastiania granatensis** (Müll.Arg.) Müll.Arg.
Gymnanthes guatemalensis Standl. & Steyerm. === **Gymnanthes riparia** (Schltdl.) Klotzsch
Gymnanthes jacobinensis Müll.Arg. === **Sebastiania jacobinensis** (Müll.Arg.) Müll.Arg.
Gymnanthes klotzschiana Müll.Arg. === **Sebastiania klotzschiana** (Müll.Arg.) Müll.Arg.
Gymnanthes ligustrina (Michx.) Müll.Arg. === **Ditrysinia fruticosa** (Bartram) Govaerts & Frodin
Gymnanthes macrocarpa Müll.Arg. === **Sebastiania macrocarpa** Müll.Arg.
Gymnanthes marginata Baill. === **Sebastiania klotzschiana** (Müll.Arg.) Müll.Arg.
Gymnanthes multiramea (Klotzsch) Müll.Arg. === **Gymnanthes glabrata** (Mart.) Govaerts
Gymnanthes obtusa Baill. === **Gymnanthes elliptica** Sw.
Gymnanthes pachystachys (Klotzsch) Baill. === **Sebastiania pachystachys** (Klotzsch) Müll.Arg.
Gymnanthes pachystachys var. *pubescens* Müll.Arg. === **Sebastiania pachystachys** (Klotzsch) Müll.Arg.
Gymnanthes pavoniana Müll.Arg. === **Sebastiania pavoniana** (Müll.Arg.) Müll.Arg.
Gymnanthes polyandra (Griseb.) Benth. & Hook.f. ex B.D.Jacks. === **Forestiera rhamnifolia** (Oleaceae)
Gymnanthes pringlei S.Watson ex Pax === **Sebastiania pringlei** S.Watson
Gymnanthes pteroclada Müll.Arg. === **Sebastiania pteroclada** (Müll.Arg.) Müll.Arg.
Gymnanthes rigida Müll.Arg. === **Sebastiania rigida** (Müll.Arg.) Müll.Arg.
Gymnanthes schlechtendaliana Müll.Arg. === **Gymnanthes riparia** (Schltdl.) Klotzsch
Gymnanthes schomburgkii (Klotzsch) G.L.Webster === **Actinostemon schomburgkii** (Klotzsch) Hochr.
Gymnanthes schottiana Müll.Arg. === **Sebastiania schottiana** (Müll.Arg.) Müll.Arg.
Gymnanthes serrata Baill. ex Müll.Arg. === **Sebastiania serrata** (Baill. ex Müll.Arg.) Müll.Arg.
Gymnanthes serrata var. *pubescens* Müll.Arg. === **Sebastiania riedelii** Müll.Arg.
Gymnanthes stipulacea Müll.Arg. === **Sebastiania stipulacea** (Müll.Arg.) Müll.Arg.
Gymnanthes texana Standl. === **Forestiera reticulata** Torr. (Oleaceae)
Gymnanthes treculiana Müll.Arg. === **Stillingia treculiana** (Müll.Arg.) I.M.Johnst.
Gymnanthes trinervia Müll.Arg. === **Sebastiania trinervia** (Müll.Arg.) Müll.Arg.
Gymnanthes ypanemensis Müll.Arg. === **Sebastiania ypanemensis** (Müll.Arg.) Müll.Arg.

Gymnanthus

An apparent orthographic variant of *Gymnanthes*.

Synonyms:
Gymnanthus Endl. === **Gymnanthes** Sw.

Gymnobothrys

Synonyms:
Gymnobothrys Wall. ex Baill. === **Sapium** P.Browne

Gymnocarpus

Synonyms:
Gymnocarpus Thouars ex Baill. === **Uapaca** Baill.

Gymnostillingia

Synonyms:
Gymnostillingia Müll.Arg. === **Stillingia** L.
Gymnostillingia loranthacea Müll.Arg. === **Stillingia saxatilis** Müll.Arg.
Gymnostillingia macrantha Müll.Arg. === **Stillingia acutifolia** (Benth.) Benth. & Hook.f. ex Hemsl.

Gynamblosis

Synonyms:
Gynamblosis Torr. === **Croton** L.

Gynoon

Synonyms:
Gynoon A.Juss. === **Glochidion** J.R.Forst. & G.Forst.
Gynoon heyneanum Wight & Arn. === **Glochidion heyneanum** (Wight & Arn.) Wight
Gynoon jussieuianum Wight === **Glochidion stellatum** (Retz.) Bedd.
Gynoon rigidum A.Juss. === **Glochidion stellatum** (Retz.) Bedd.
Gynoon tetrandrum D.Dietr. === **Glochidion stellatum** (Retz.) Bedd.
Gynoon triandrum Wight & Arn. === **Glochidion stellatum** (Retz.) Bedd.

Haematospermum

Synonyms:
Haematospermum Wall. === **Homonoia** Lour.
Haematospermum salicinum (Hassk.) Baill. === **Homonoia riparia** Lour.

Haematostemon

2 species, N South America (Venezuela and the Guianas); small trees with long, spike-like inflorescences allied to *Angostyles* and *Astrococcus*, differing from the former genus in its staminate flowers and from both in the absence of a disk. The leaves of *H. guianensis* are, however, rather different from those of *H. coriaceus*, being more like those of *Angostyles longifolia* (Sandwith, 1950). (Acalyphoideae)

Pax, F. & K. Hoffmann (1919). *Haematostemon*. In A. Engler (ed.), Das Pflanzenreich, IV 147 IX (Euphorbiaceae-Acalypheae-Plukenetiinae): 31-32. Berlin. (Heft 68.) La/Ge. — 1 species.

Sandwith, N. Y. (1950). Contributions to the flora of tropical America, LI: On two Euphorbiaceae of British Guiana. Kew Bull. 5: 133-136. En. — Description of *H. guianensis* (pp. 133-134); discussion of relationships among this genus and its allies.

Jablonski, E. (1967). *Haematostemon*. Euphorbiaceae, Guayana Highland (Mem. New York Bot. Gard. 17(1)): 143, 145. New York. En. — 2 species.

Haematostemon Pax & K.Hoffm. in H.G.A.Engler, Pflanzenr., IV, 147, IX: 31 (1919).
Venezuela, Guyana. 82.

Haematostemon coriaceus (Baill.) Pax & K.Hoffm. in H.G.A.Engler, Pflanzenr., IV, 147, IX: 32 (1919).
Venezuela (Amazonas). 82 VEN. Hel.
* *Astrococcus coriaceus* Baill., Adansonia 5: 308 (1865).

Haematostemon guianensis Sandwith, Kew Bull. 5: 133 (1950).
Guyana (near Mahdia). 82 GUY.

Hamilcoa

1 species, W. & WC. Africa (Nigeria, Cameroon); climbing shrubs considered by Webster (Synopsis, 1994) to be related to the American *Nealchornea*. The leaves tend to be aggregated towards shoot apices. It was originally distinguished from *Plukenetia* by the presence of a 3-locular ovary; the two genera are, however, now regarded as being in different subfamilies, *Hamilcoa* being grouped by Webster with *Nealchornea* in the Stomatocalyceae (currently separate from Hippomaneae). (Euphorbioideae (except Euphorbieae))

Pax, F. (with K. Hoffmann) (1914). *Hamilcoa*. In A. Engler (ed.), Das Pflanzenreich, IV 147 VII [Euphorbiaceae-Additamentum V]: 419-420. Berlin. (Heft 63.) La/Ge. — 1 species, C. Africa; treated as the sole member of a new subtribe Hamilcoinae of tribe Hippomaneae. [In 1931 the subtribe (and genus) were transferred to the authors' Gelonieae.]

Stapf, O. (1915). *Hamilcoa zenkeri*. Ic. Pl. 31: pl. 3009. En. — Plant portrait with description and commentary.

Hamilcoa Prain, Bull. Misc. Inform. Kew 1912: 107 (1912).
W. & WC. Trop. Africa. 22 23.

Hamilcoa zenkeri (Pax) Prain, Bull. Misc. Inform. Kew 1912: 107 (1912). – FIGURE, p. 984.
Nigeria, Cameroon. 22 NGA 23 CMN. Cl. nanophan.
* *Plukenetia zenkeri* Pax, Bot. Jahrb. Syst. 43: 83 (1909).

Halecus

Synonyms:
Halecus Rumph. ex Raf. === **Croton** L.

Halliophytum

Synonyms:
Halliophytum I.M.Johnst. === **Tetracoccus** Engelm. ex Parry
Halliophytum capense (I.M.Johnst.) I.M.Johnst. === **Tetracoccus capensis** (I.M.Johnst.) Croizat
Halliophytum fasciculatum (S.Watson) I.M.Johnst. === **Tetracoccus fasciculatus** (S.Watson) Croizat
Halliophytum fasciculatum var. *hallii* (Brandegee) McMinn === **Tetracoccus fasciculatus** var. **hallii** (Brandegee) Dressler
Halliophytum hallii (Brandegee) I.M.Johnst. === **Tetracoccus fasciculatus** var. **hallii** (Brandegee) Dressler

Hancea

Synonyms:
Hancea Seem. === **Mallotus** Lour.
Hancea hookeriana Seem. === **Mallotus hookerianus** (Seem.) Müll.Arg.

Hasskarlia

Synonyms:
Hasskarlia Baill. === **Tetrorchidium** Poepp.
Hasskarlia didymostemon Baill. === **Tetrorchidium didymostemon** (Baill.) Pax & K.Hoffm.
Hasskarlia minor Prain === **Tetrorchidium didymostemon** (Baill.) Pax & K.Hoffm.
Hasskarlia oppositifolia Pax === **Tetrorchidium oppositifolium** (Pax) Pax
Hasskarlia tenuifolia Pax & K.Hoffm. === **Tetrorchidium tenuifolium** (Pax & K.Hoffm.) Pax & K.Hoffm.

Hebecocca

Synonyms:
Hebecocca Beurl. === **Omphalea** L.
Hebecocca panamensis (Klotzsch) Beurl. === **Omphalea diandra** L.

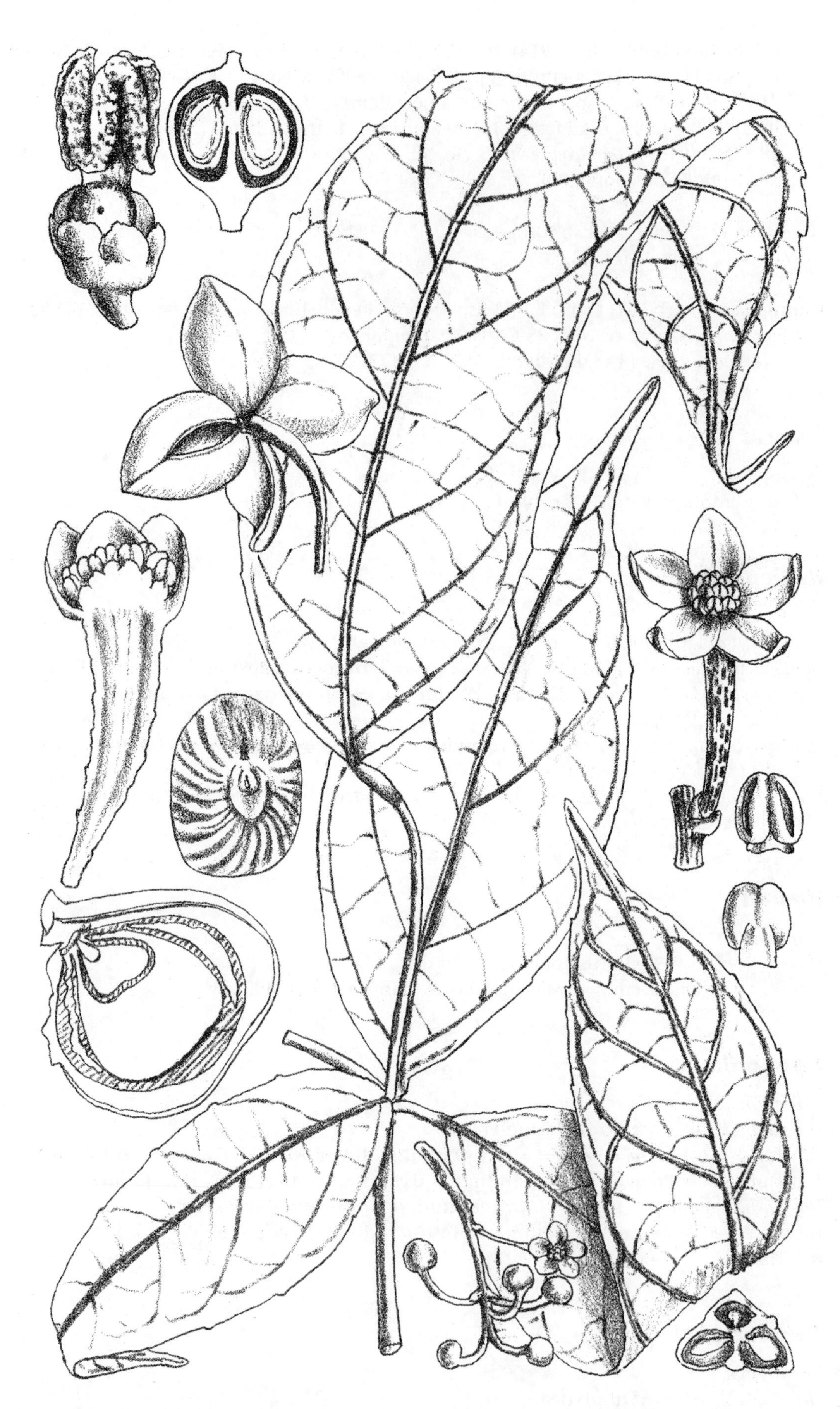

Hamilcoa zenkeri (Pax) Prain
Artist: ? Matilda Smith (some analyses by O. Stapf)
Ic. Pl. 31: pl. 3009 (1915)
KEW ILLUSTRATIONS COLLECTION

Hecatea

Synonyms:
Hecatea Thouars === **Omphalea** L.
Hecatea alternifolia Willd. === **Omphalea oppositifolia** (Willd.) L.J.Gillespie
Hecatea biglandulosa Pers. === **Omphalea oppositifolia** (Willd.) L.J.Gillespie
Hecatea oppositifolia Willd. === **Omphalea oppositifolia** (Willd.) L.J.Gillespie

Hecaterium

Synonyms:
Hecaterium Kunze ex Rchb. === **Omphalea** L.

Hedraiostylus

Synonyms:
Hedraiostylus Hassk. === **Pterococcus** Hassk.

Hedycarpus

Synonyms:
Hedycarpus Jack === **Baccaurea** Lour.
Hedycarpus javanicus Miq. === **Phyllanthus javanicus** (Miq.) Müll.Arg.
Hedycarpus lanceolatus Miq. === **Baccaurea lanceolata** (Miq.) Müll.Arg.
Hedycarpus malayanus Jack === **Baccaurea malayana** (Jack) King ex Hook.f.

Hemecyclia

An orthographic variant of *Hemicyclia.*

Synonyms:
Hemecyclia Wight & Arn. === **Drypetes** Vahl

Hemicicca

An occasionally adopted segregate of *Phyllanthus.*

Synonyms:
Hemicicca Baill. === **Phyllanthus** L.
Hemicicca flexuosa (Siebold & Zucc.) Hurus. === **Phyllanthus flexuosus** (Siebold & Zucc.) Müll.Arg.
Hemicicca glauca (Wall. ex Müll.Arg.) Hurus. & Yu.Tanaka === **Phyllanthus glaucus** Wall. ex Müll.Arg.
Hemicicca japonica Baill. === **Phyllanthus flexuosus** (Siebold & Zucc.) Müll.Arg.

Hemicyclia

A former segregate of *Drypetes.*

Synonyms:
Hemicyclia Wight & Arn. === **Drypetes** Vahl
Hemicyclia andamanica Kurz === **Drypetes andamanica** (Kurz) Pax & K.Hoffm.
Hemicyclia australasica Müll.Arg. === **Drypetes deplanchei** (Brongn. & Gris) Merr.

Hemicyclia deplanchei (Brongn. & Gris) Baill. ex Guillaumin === **Drypetes deplanchei** (Brongn. & Gris) Merr.
Hemicyclia elata Bedd. === **Drypetes elata** (Bedd.) Pax & K.Hoffm.
Hemicyclia gardneri Thwaites === **Drypetes gardneri** (Thwaites) Pax & K.Hoffm.
Hemicyclia hoaensis Pierre ex Gagnep. === **Drypetes hoaensis** Gagnep.
Hemicyclia lanceolata Thwaites === **Drypetes lanceolata** (Thwaites) Pax & K.Hoffm.
Hemicyclia lasiogyna F.Muell. === **Drypetes deplanchei** (Brongn. & Gris) Merr.
Hemicyclia ovalis J.J.Sm. ex Koord. & Valeton === **Drypetes ovalis** (J.J.Sm.) Pax & K.Hoffm.
Hemicyclia porteri Gamble === **Drypetes porteri** (Gamble) Pax & K.Hoffm.
Hemicyclia rhakodiskos Hassk. === **Drypetes rhakodiskos** (Hassk.) Bakh.f.
Hemicyclia sepiaria Wight & Arn. === **Drypetes sepiaria** (Wight & Arn.) Pax & K.Hoffm.
Hemicyclia serrata (Blume) J.J.Sm. === **Drypetes serrata** (Blume) Pax & K.Hoffm.
Hemicyclia subcrenata (Merr.) Merr. === **Drypetes subcrenata** (Merr.) Pax & K.Hoffm.
Hemicyclia sumatrana (Miq.) Müll.Arg. === **Drypetes sumatrana** (Miq.) Pax & K.Hoffm.
Hemicyclia travancorica Bourd. === **Drypetes travancorica** (Bourd.) Santapau & S.K.Jain
Hemicyclia venusta (Wight) Thwaites === **Drypetes venusta** (Wight) Pax & K.Hoffm.
Hemicyclia wightii Hook.f. === **Drypetes wightii** (Hook.f.) Pax & K.Hoffm.

Hemiglochidion

A New Guinean segregate of *Phyllanthus.*

Synonyms:
Hemiglochidion K.Schum. === **Phyllanthus** L.
Hemiglochidion cupuliforme K.Schum. === **Phyllanthus finschii** K.Schum.
Hemiglochidion finschii (K.Schum.) K.Schum. === **Phyllanthus finschii** K.Schum.
Hemiglochidion hylodendron K.Schum. === **Phyllanthus finschii** K.Schum.
Hemiglochidion warburgii (K.Schum.) K.Schum. === **Phyllanthus warburgii** K.Schum.

Hendecandra

Synonyms:
Hendecandra Eschsch. === **Croton** L.
Hendecandra divaricata Klotzsch === **Croton nitrariifolius** Baill.
Hendecandra texensis Klotzsch === **Croton texensis** (Klotzsch) Müll.Arg.
Hendecandra velleriflora Klotzsch === **Croton nitrariifolius** Baill.

Henribaillonia

Synonyms:
Henribaillonia Kuntze === **Thecacoris** A.Juss.

Heptallon

Synonyms:
Heptallon Raf. === **Croton** L.

Hermesia

Synonyms:
Hermesia Humb. & Bonpl. ex Willd. === **Alchornea** Sw.
Hermesia castaneifolia Humb. & Bonpl. ex Willd. === **Alchornea castaneifolia** (Humb. & Bonpl. ex Willd.) A.Juss.
Hermesia mexicana Hook. & Arn. === **Bernardia mexicana** (Hook. & Arn.) Müll.Arg.
Hermesia salicifolia Baill. === **Alchornea castaneifolia** (Humb. & Bonpl. ex Willd.) A.Juss.

Heterocalymnantha

Synonyms:
Heterocalymnantha Domin === **Sauropus** Blume
Heterocalymnantha minutifolia Domin === **Sauropus rigens** (F.Muell.) Airy Shaw

Heterocalyx

Synonyms:
Heterocalyx Gagnep. === **Agrostistachys** Dalzell

Heterochlamys

Synonyms:
Heterochlamys Turcz. === **Croton** L.

Heterocroton

Synonyms:
Heterocroton S.Moore === **Croton** L.
Heterocroton mentiens S.Moore === **Croton mentiens** (S.Moore) Pax

Hevea

9 species, S. America (centered on the Amazon Basin); small to medium forest trees with trifoliolate leaves. Two subgenera, *Microphyllae* (*H. microphylla*) and *Hevea* (all others) were proposed by Schultes (1977). This has superseded the former division into sects. *Hevea* and *Bisiphonia* first proposed by Baillon (1858; see General) and maintained by Mueller, Bentham, Pax and Huber (Schultes 1970). The literature on this genus is vast; a good recent introduction is Schultes (1990) who is responsible for most of the more recent taxonomic work. Schultes has argued that in fact hybridisation in natural conditions is rare; however, this still leaves somewhat unresolved the nature (and potential) of the many local variants within recognised species (a number of which were formerly themselves considered to be 'species'). Close relatives include *Cunuria*, *Micrandra* and, more controversially, *Vaupesia*. (Crotonoideae)

Huber, J. (1905). Ensaio d'uma synopse das especies do genero *Hevea* sob os pontos de vista systematico e geographico. Bol. Mus. Goeldi 4: 620-651. Pt. — Synopsis with partial keys.

Pax, F. (1910). *Hevea*. In A. Engler (ed.), Das Pflanzenreich, IV 147 [I] (Euphorbiaceae-Jatropheae): 117-128. Berlin. (Heft 42.) La/Ge. — 17 species in 2 sections. [Now obsolete. Ducke (1935) indicated that the key was 'not reliable'.]

Huber, J. (1913). Novas contribuções para o conhecimento do genero *Hevea*. Bol. Mus. Goeldi 7: 199-281. Pt/Ge. — Additions to 1905 paper. [In two parts, the first in German, the second in Portuguese; descriptions in Latin.]

Ducke, A. (1930). *Hevea*. Arq. Jard. Bot. Rio de Janeiro 5 (in Plantes nouvelles .. IV): 147-157, pls. 18-21. Fr/La. — Includes a synoptic key to known species of the genus with relatively detailed leads, followed by species commentaries (with descriptions and exsiccatae if new); illustrations at end of volume.

Ducke, A. (1933). *Hevea*. Arq. Jard. Bot. Rio de Janeiro 6 (in Plantes nouvelles .. V): 49-57. Fr/La. — Notes and novelties further to the work of Huber (1905) and the author. [Numerous infraspecific taxa down to forma level described.]

• Ducke, A. (1935). Revision of the genus *Hevea* Aubl., mainly the Brazilian species. Arch. Inst. Biol. Veg. Rio de Janeiro 2: 217-246, pl. 1-3. (Reissued 1939; 32 pp.) En. — Botanical history and geography (summary, p. 223-224); vernacular names and taxonomic concepts; key, descriptions, localities with exsiccatae, distribution and habitat, and commentary; doubtful species and nothospecies (pp. 244-245) and references (p. 246). [12 species accepted with many formal infraspecific taxa. The status of some named taxa within the author's species concepts was left unresolved.]

Ducke, A. (1939). Revision of the genus *Hevea*. 32 pp. Rio de Janeiro: Serviço de Publicidade Agrícola. En. — A reissue, with revisions, of Ducke (1935).

Schultes, R. E. (1943). The genus *Hevea* in Colombia. Bot. Mus. Leafl. 12: 1-19, 6 pls. En. — Notes on taxa and collecting localities, with key to species and varieties (pp. 16-17). [The author here argues for the possibility that the more primitive species and varieties were to be expected along the westernmost fringes of the range of the genus, i.e. along the eastern side of the Andes.]

Baldwin, J. T., Jr. (1947). *Hevea*: a first interpretation. J. Hered. 38: 54-64, map. En. — A discursive account, in the first instance on the occurrence and properties of hybrids in *Hevea*; speculations on the evolution of the genus and its relatives. [The complexity of the patterns observed in the genus was thought to be associated with the aged hills scattered through the Amazonian basin. Much scope existed for ecological selection. *H. pauciflora* was, of all the species, considered to be closest to *Cunuria*, a relationship also locally recognised. No formal taxonomy is presented.]

Schultes, R. E. (1947). Studies in the genus *Hevea*, I. The differentiation of *Hevea microphylla* and *H. minor*. Bot. Mus. Leafl. 13: 1-11. En. — Contribution towards a monograph, with extensive discussion; key, pp. 8-9.

Seibert, R. J. (1947). A study of *Hevea* (with its economic aspects) in the Republic of Peru. Ann. Missouri Bot. Gard. 34: 261-352, 13 plates, folding map. (Based on a Ph.D. dissertation, Washington University, St. Louis.) En. — Detailed study with extended general part covering a review of characters and key to species and infraspecific taxa and a descriptive treatment (5 species, one with an additional variety) with synonymy, references, vernacular names, indication of distribution, localities with exsiccatae, and often extended commentary covering critical features, hybrids, and habitat information; economic review (pp. 322-323); list of specimens seen, bibliography and complete index at end.

Schultes, R. E. (1948). Studies in the genus *Hevea*, II. The rediscovery of *H. rigidifolia*. Bot. Mus. Leafl. 13: 97-132, 1 pl., map. En. — Detailed discussion, with map; part of a contribution towards a monograph. [The author's review of distribution and habitat is particularly thorough.]

Schultes, R. E. (1950). Studies in the genus *Hevea*, III. On the use of the name *Hevea brasiliensis*. Bot. Mus. Leafl. 14: 79-86, 2 pls. En. — Extensive discussion; a contribution towards a monograph. A corrected authorship and full synonymy for *H. brasiliensis* (Willd. ex A. Juss.) Muell.-Arg. is included.

Schultes, R. E. (1952). Studies in the genus *Hevea*, IV. Notes on the range and variability of *Hevea microphylla*. Bot. Mus. Leafl. 15: 111-138, 5 pls. (incl. map). En/La. — Further particulars on *H. microphylla* (previously customarily united with *H. minor*); additions towards a monograph. An extended description in Latin is provided (pp. 118-120) along with a full list of localities with exsiccatae (see also the map, pl. 39). Variability, ecological relations and affinities are all considered in this detailed study.

Schultes, R. E. (1952). Studies in the genus *Hevea*, V. The status of the binomial *Hevea discolor*. Bot. Mus. Leafl. 15: 247-254. En. —*H. discolor* synonymised with *H. spruceana*, following Ducke (1930); critical discussion.

Schultes, R. E. (1952). Studies in the genus *Hevea*, VI. Notes, chiefly nomenclatural, on the *Hevea pauciflora* complex. Bot. Mus. Leafl. 15: 255-272, 3 pls. En. — Extensive discussion; reduction of *H. kunthiana* to *H. pauciflora* (Spruce ex Benth.) Muell.-Arg. and union of *H. confusa* and *H. minor* with *H. pauciflora* var. *coriacea* Ducke.

Schultes, R. E. (1952). Studies in the genus *Hevea*, VII. Bot. Mus. Leafl. 16: 21-44, 4 pls. En. — Miscellaneous notes; in 3 parts of which parts 1 (on *Hevea* collections in the De Candolle Herbarium) and 2 (*Hevea* in other European herbaria) are most relevant.

Jablonski, E. (1967). *Hevea*. Euphorbiaceae, Guayana Highland (Mem. New York Bot. Gard. 17(1)): 127-129. New York. En. — 5 species.

• Schultes, R. E. (1970). The history of taxonomic studies in *Hevea*. Bot. Rev. 36: 197-276. En. — As near a modern monograph as now available; 9 species and four additional varieties recognised, with key. Includes an important bibliography as well as a 'rogues' gallery' of those closely associated with collection and study of the genus. [See also Schultes, 1990.]
• Schultes, R. E. (1970). The history of taxonomic studies in *Hevea*. In P. Smit & R. J. Ch. V. ter Lange (eds), Essays in biohistory: 229-293, illus. Utrecht. (Regnum Vegetabile 71.) En. — A historical review of the genus and its 9 recognised species, followed by biographical sketches of persons particularly concerned with *Hevea* studies in the field or herbarium or both; references (pp. 291-293).

Schultes, R. E. (1977). A new infrageneric classification of *Hevea*. Bot. Mus. Leafl. 25: 243-257, illus., maps. En. — Historical notes; presentation of a new system (pp. 251-253). [*H. microphylla* (with map and figure) the only member of subgen. *Microphyllae*, with a distinct form of capsular dehiscence; all others in subgen. *Hevea*.]
• Schultes, R. E. (1990). A brief taxonomic view of the genus *Hevea*. vii, 57 pp., illus., map. Kuala Lumpur: Malaysian Rubber Research and Development Board. (MRRDB Monograph 14.) En. — Introduction with historical review and current opinion (including subdivision of the genus; the author – as in 1977 – accepts 2 subgenera, one of them limited to *H. microphylla*, the other encompassing the eight remaining accepted species); revisionary treatment (the species alphabetically arranged) with key, synonymy, detailed descriptions, distribution and habitat, derivation of epithets, observations on biology and other features, and illustrations including habit, botanical details, and seed morphology; concluding observations (pp. 48-50) and extensive bibliography (not limited to titles cited in the text). [The author notes that more detailed study of infraspecific forms was 'our present challenge'; current understanding was deficient (p. 3).]

Hevea Aubl., Hist. Pl. Guiane 2: 871 (1775).
S. Trop. America, introduced elsewere. (22) (23) (25) (40) (41) (42) (60) (80) (81) 82 83 84.
Siphonia Rich. in J.C.D.von Schreber, Gen. Pl.: 656 (1791).
Caoutchoua J.F.Gmel., Syst. Nat.: 1007 (1792).
Siphonanthus Schreb. ex Baill., Étude Euphorb.: 324 (1858).
Micrandra R.Br., Pterocymbium: 237 (4 June 1844), nom. rejic.

Hevea benthamiana Müll.Arg., Linnaea 34: 204 (1865). *Hevea benthamiana* var. *typica* Ducke, Arch. Jard. Bot. Rio de Janeiro 6: 54 (1933), nom. inval.
Brazil (Amazonas), Venezuela (Amazonas), Colombia. 82 VEN 83 CLM 84 BZN. Phan.
Hevea duckei Huber, Bol. Mus. Goeldi Paraense Hist. Nat. Ethnogr. 4: 631 (1906).
Hevea discolor Spruce ex Pax in H.G.A.Engler, Pflanzenr., IV, 147, I: 120 (1910).
Hevea benthamiana var. *huberiana* Ducke, Rev. Bot. Appl. Agric. Trop. 11: 29 (1931).
Hevea benthamiana f. *huberiana* (Ducke) Ducke, Arq. Inst. Biol. Veg. 2: 233 (1935).
Hevea benthamiana var. *obtusifolia* Ducke, Arch. Jard. Bot. Rio de Janeiro 6: 55 (1933).
Hevea benthamiana f. *obtusifolia* (Ducke) Ducke, Arq. Inst. Biol. Veg. 2: 234 (1935).
Hevea benthamiana var. *subglabrifolia* Ducke, Arch. Jard. Bot. Rio de Janeiro 6: 54 (1933).
Hevea benthamiana f. *subglabrifolia* (Ducke) Ducke, Arq. Inst. Biol. Veg. 2: 234 (1935).
Hevea benthamiana f. *caudata* (Ducke) Ducke, Arq. Inst. Biol. Veg. 2: 234 (1935).
Hevea benthamiana var. *caudata* Ducke, Arch. Jard. Bot. Rio de Janeiro 6: 55 (1935).
Hevea benthamiana f. *caudata* (Ducke) Ducke, Arq. Inst. Biol. Veg. 2: 234 (1935).

Hevea brasiliensis (Willd. ex A.Juss.) Müll.Arg., Linnaea 34: 204 (1865). – FIGURE, p. 990.
Peru, Colombia, Brazil (Amazonas, Pará, Mato Grosso, Paraná), Venezuela, Guiana, Bolivia. 82 FRG VEN 83 BOL CLM PER 84 BZC BZN BZS. Phan. – Source of the best natural rubber.
* *Siphonia brasiliensis* Willd. ex A.Juss., Euphorb. Gen.: t. 12 (1824).
Hevea janeirensis Müll.Arg. in C.F.P.von Martius, Fl. Bras. 11(2): 706 (1874). *Hevea brasiliensis* var. *janeirensis* (Müll.Arg.) Pax in H.G.A.Engler, Pflanzenr., IV, 147, I: 121 (1910).

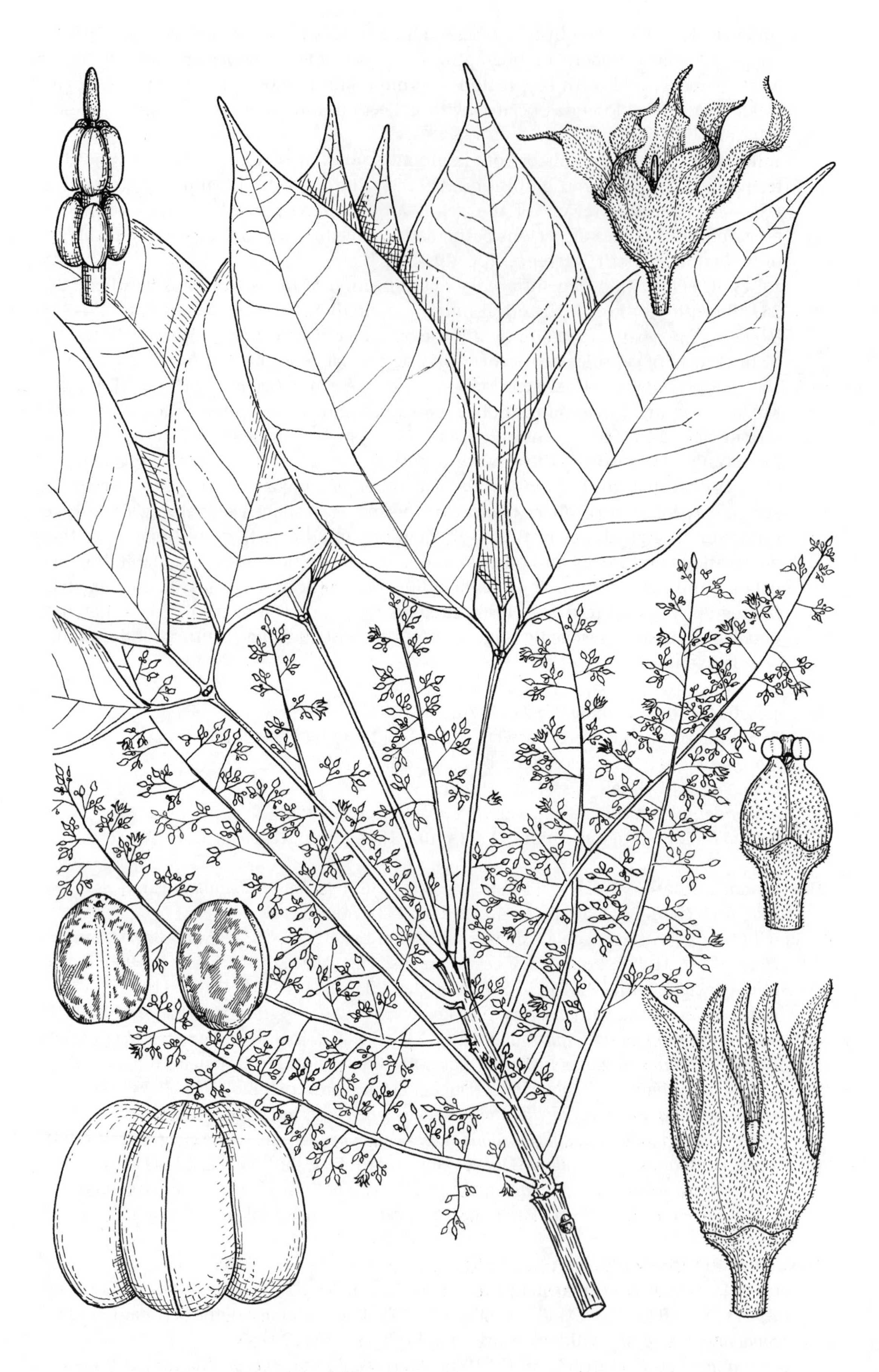

Hevea brasiliensis (Willd. ex A. Juss.) Müll. Arg.
Artist: P. Halliday
Fl. Trop. East Africa, Euphorbiaceae 1: 184, fig. 36 (1987)
KEW ILLUSTRATIONS COLLECTION

Hevea sieberi Warb., Kaoutschukpflanzen: 33 (1900).
Hevea brasiliensis var. *angustifolia* Ule ex Huber, Bol. Mus. Paraense Hist. Nat. Ethnogr. 3: 350 (1902).
Hevea brasiliensis var. *latifolia* Ule ex Huber, Bol. Mus. Paraense Hist. Nat. Ethnogr. 3: 350 (1902). *Hevea brasiliensis* f. *latifolia* (Ule ex Huber) Ule, Tropenpflanzer Beih. 6: 8 (1905).
Hevea brasiliensis f. *angustifolia* (Ule ex Huber) Ule, Tropenpflanzer Beih. 6: 8 (1905).
Hevea paludosa Ule, Bot. Jahrb. Syst. 35: 666 (1905).
Hevea brasiliensis var. *stylosa* Huber, Bol. Mus. Goeldi Paraense Hist. Nat. Ethnogr. 4: 640 (1906).
Hevea randiana Huber, Bol. Mus. Goeldi Paraense Hist. Nat. Ethnogr. 4: 636 (1906). *Hevea brasiliensis* var. *randiana* (Huber) Pax in H.G.A.Engler, Pflanzenr., IV, 147, I: 123 (1910). *Hevea brasiliensis* f. *randiana* (Huber) Ducke, Arq. Inst. Biol. Veg. 2: 224 (1935).
Hevea brasiliensis var. *acreana* Ule, Bot. Jahrb. Syst. 50(114): 14 (1914). *Hevea brasiliensis* f. *acreana* (Ule) Ducke, Bol. Técn. Inst. Agron. N. 10: 23 (1946).
Hevea brasiliensis mut. *granthamii* Barth, Bot. Gaz. 84: 200 (1927).
Hevea granthamii Bartlett, Bot. Gaz. 84: 200 (1927).

Hevea camporum Ducke, Arch. Jard. Bot. Rio de Janeiro 4: 111 (1925).
Brazil (SW. Amazonas). 84 BZN. Phan.

Hevea guianensis Aubl., Hist. Pl. Guiane 2: 871 (1775). *Hevea guianensis* subsp. *typica* Ducke, Arq. Inst. Biol. Veg. 2: 223 (1935), nom. inval. *Hevea guianensis* f. *typica* Ducke, Bol. Técn. Inst. Agron. N. 10: 8 (1946), nom. inval.
Colombia, Guiana, Venezuela, Peru, Brazil (Amazonas). 82 FRG VEN 83 CLM PER 84 BZN. Phan.

var. **guianensis**
Colombia, Guiana, Venezuela, Brazil (Amazonas). 82 FRG VEN 83 CLM 84 BZN. Phan.
Hevea peruviana Aubl., Hist. Pl. Guiane 4: 335 (1775).
Jatropha elastica L.f., Suppl. Pl.: 422 (1782). *Hevea elastica* (L.f.) H.Karst., Deut. Fl.: 589 (1882).
Hevea nigra Ule, Bot. Jahrb. Syst. 35: 667 (1905).
Hevea caucho Posada-Ar., Est. Cient.: 212 (1909).
Hevea collina Huber, Bol. Mus. Goeldi Paraense Hist. Nat. Ethnogr. 5: 249 (1909). *Hevea guianensis* var. *collina* (Huber) Ducke, Arch. Jard. Bot. Rio de Janeiro 4: 109 (1925).
Hevea guianensis subsp. *occidentalis* Ducke, Arq. Inst. Biol. Veg. 2: 223 (1935). *Hevea guianensis* var. *occidentalis* (Ducke) Ducke, Arq. Inst. Biol. Veg. 2: 229 (1935).

var. **lutea** (Spruce ex Benth.) Ducke & R.E.Schult., Caldesia 3: 249 (1945).
Brazil (Amazonas), Venezuela (Amazonas), Colombia (Vaupes), E. Peru. 82 VEN 83 CLM PER 84 BZN. Phan.
* *Siphonia lutea* Spruce ex Benth., Hooker's J. Bot. Kew Gard. Misc. 6: 370 (1854). *Hevea lutea* (Spruce ex Benth.) Müll.Arg., Linnaea 34: 204 (1865). *Hevea lutea* var. *typica* Ducke, Arch. Jard. Bot. Rio de Janeiro 6: 53 (1933), nom. inval.
Hevea apiculata Spruce ex Baill., Adansonia 4: 285 (1864). *Hevea lutea* var. *apiculata* (Spruce ex Baill.) Müll.Arg. in C.F.P.von Martius, Fl. Bras. 11(2): 302 (1874).
Hevea lutea var. *cuneata* Huber, Bol. Mus. Paraense Hist. Nat. Ethnogr. 3: 357 (1902). *Hevea cuneata* (Huber) Huber, Bol. Mus. Goeldi Paraense Hist. Nat. Ethnogr. 4: 578, 626 (1906). *Hevea brasiliensis* var. *cuneata* (Huber) Pax in H.G.A.Engler, Pflanzenr., IV, 147, I: 123 (1910). *Hevea guianensis* var. *cuneata* (Huber) Ducke, Arch. Jard. Bot. Rio de Janeiro 6: 51 (1933).
Hevea foxii Huber, Bol. Mus. Goeldi Paraense Hist. Nat. Ethnogr. 7: 228 (1913). *Hevea lutea* var. *foxii* (Huber) Ducke, Rev. Bot. Appl. Agric. Colon. 9: 627 (1929).
Hevea glabrescens Huber, Bol. Mus. Goeldi Paraense Hist. Nat. Ethnogr. 7: 230 (1913). *Hevea lutea* var. *glabrescens* (Huber) Ducke, Rev. Bot. Appl. Agric. Colon. 9: 627 (1929).
Hevea andinensis Preusse-Sperber, Tropenpflanzer 19: 192 (1916), nom. nud.

Hevea lutea var. *peruviana* Ducke, Rev. Bot. Appl. Agric. Colon. 9: 627 (1929). *Hevea guianensis* f. *peruviana* (Ducke) Ducke, Bol. Técn. Inst. Agron. N. 10: 24 (1946).
Hevea lutea var. *pilosula* Ducke, Arq. Inst. Biol. Veg. 6: 53 (1933). *Hevea lutea* f. *pilosula* (Ducke) Ducke, Arq. Inst. Biol. Veg. 2: 224 (1935). *Hevea guianensis* f. *pilosula* (Ducke) Ducke, Bol. Técn. Inst. Agron. N. 10: 9 (1946).

var. **marginata** (Ducke) Ducke, Arch. Jard. Bot. Rio de Janeiro 6: 51 (1933).
Brazil (Amazonas: lower Rio Negro). 84 BZN. Phan.
* *Hevea marginata* Ducke, Rev. Bot. Appl. Agric. Colon. 9: 624, 625 (1929). *Hevea guianensis* subsp. *marginata* (Ducke) Ducke, Arq. Inst. Biol. Veg. 2: 223 (1935).

Hevea microphylla Ule, Bot. Jahrb. Syst. 35: 669 (1905). *Hevea microphylla* var. *typica* Pax in H.G.A.Engler, Pflanzenr., IV, 147, I: 126 (1910), nom. inval.
Colombia, Venezuela, Brazil (Amazonas). 82 VEN 83 CLM 84 BZN. Phan.
Hevea microphylla var. *major* Pax in H.G.A.Engler, Pflanzenr., IV, 147, I: 126 (1910).

Hevea nitida Mart. ex Müll.Arg. in C.F.P.von Martius, Fl. Bras. 11(2): 301 (1874).
Colombia, Brazil (Amazonas). 83 CLM 84 BZN. Phan.

var. **nitida**
Colombia, Brazil (Amazonas). 83 CLM 84 BZN. Phan.
Hevea viridis Huber, Bull. Soc. Bot. France 49: 48 (1902).
Hevea brasiliensis var. *subconcolor* Ducke, Arch. Jard. Bot. Rio de Janeiro 6: 55 (1933). *Hevea brasiliensis* f. *subconcolor* (Ducke) Ducke, Arq. Inst. Biol. Veg. 2: 224 (1935).

var. **toxicodendroides** (R.E.Schult. & Vinton) R.E.Schult., Bot. Mus. Leafl. 13: 11 (1947).
Colombia (Apaporis-Vaupés Reg.). 83 CLM. Phan.
* *Hevea viridis* var. *toxicodendroides* R.E.Schult. & Vinton, Caldasia 3: 25 (1944).

Hevea pauciflora (Spruce ex Benth.) Müll.Arg., Linnaea 34: 203 (1865).
S. Trop. America. 82 FRG GUY SUR VEN 83 CLM PER 84 BZN. Phan.
* *Siphonia pauciflora* Spruce ex Benth., Hooker's J. Bot. Kew Gard. Misc. 6: 370 (1854). *Hevea pauciflora* var. *typica* Ducke, Bol. Técn. Inst. Agron. N. 10: 17 (1946), nom. inval.

var. **coriacea** Ducke, Arq. Inst. Biol. Veg. 2: 239 (1935).
Guyana, Venezuela (Amazonas), Colombia, Brazil (Amazonas), Peru. 82 GUY VEN 83 CLM PER 84 BZN. Phan.
Hevea confusa Hemsl., Hooker's Icon. Pl. 26: t. 2573-2574 (1898).
Hevea minor Hemsl., Hooker's Icon. Pl. 26: t. 2572 (1899).
Hevea humilior Ducke, Rev. Bot. Appl. Agric. Colon. 9: 624, 629 (1929).
Hevea pauciflora subsp. *coriacea* Ducke, Arq. Inst. Biol. Veg. 2: 225 (1935).

var. **pauciflora**
Guiana, Surinam, Venezuela, Colombia, Brazil (Amazonas). 82 FRG SUR VEN 83 CLM 84 BZN. Phan.
Siphonia kunthiana Baill., Étude Euphorb.: 326 (1858), nom. nud. *Hevea kunthiana* Huber, Bol. Mus. Paraense Hist. Nat. Ethnogr. 3: 349 (1902).
Hevea membranacea Müll.Arg. in C.F.P.von Martius, Fl. Bras. 11(2): 299 (1874).
Hevea membranacea var. *leiogyne* Ducke, Arch. Jard. Bot. Rio de Janeiro 6: 57 (1933). *Hevea membranacea* f. *leiogyne* (Ducke) Ducke, Arq. Inst. Biol. Veg. 2: 239 (1935). *Hevea pauciflora* f. *leiogyne* (Ducke) Ducke, Bol. Técn. Inst. Agron. N. 10: 17 (1947).

Hevea rigidifolia (Spruce ex Benth.) Müll.Arg., Linnaea 34: 203 (1865).
Brazil (Amazonas), Colombia (SE. Vaupés). 83 CLM 84 BZN. Phan.
* *Siphonia rigidifolia* Spruce ex Benth., Hooker's J. Bot. Kew Gard. Misc. 6: 371 (1854).

Hevea spruceana (Benth.) Müll.Arg., Linnaea 34: 204 (1865).
Guyana, Brazil (Amazonas). 82 GUY 84 BZN. Phan.

Siphonia discolor Spruce ex Benth., Hooker's J. Bot. Kew Gard. Misc. 6: 369 (1854). *Hevea discolor* (Spruce ex Benth.) Müll.Arg. in A.P.de Candolle, Prodr. 15(2): 717 (1866). *Hevea spruceana* f. *discolor* (Spruce ex Benth.) Ducke, Rev. Bot. Appl. Agric. Colon. 9: 630 (1929).

* *Siphonia spruceana* Benth., Hooker's J. Bot. Kew Gard. Misc. 6: 370 (1854).

Hevea paraensis Baill., Adansonia 4: 284 (1864).

Hevea similis Hemsl., Hooker's Icon. Pl. 26: t. 2576 (1899). *Hevea spruceana* var. *similis* (Hemsl.) Ducke, Arch. Jard. Bot. Rio de Janeiro 4: 109 (1925). *Hevea spruceana* f. *similis* (Hemsl.) Ducke, Rev. Bot. Appl. Agric. Colon. 9: 630 (1929).

Hevea spruceana var. *tridentata* Huber, Bol. Mus. Goeldi Paraense Hist. Nat. Ethnogr. 4: 644 (1906). *Hevea spruceana* f. *tridentata* (Huber) Ducke, Arq. Inst. Biol. Veg. 2: 242 (1935).

Micrandra ternata R.Br., Pterocymbium: 238 (4 June 1844).

Synonyms:

Hevea andinensis Preusse-Sperber === **Hevea guianensis** var. **lutea** (Spruce ex Benth.) Ducke & R.E.Schult.

Hevea apiculata Spruce ex Baill. === **Hevea guianensis** var. **lutea** (Spruce ex Benth.) Ducke & R.E.Schult.

Hevea benthamiana f. *caudata* (Ducke) Ducke === **Hevea benthamiana** Müll.Arg.

Hevea benthamiana var. *caudata* Ducke === **Hevea benthamiana** Müll.Arg.

Hevea benthamiana var. *huberiana* Ducke === **Hevea benthamiana** Müll.Arg.

Hevea benthamiana f. *huberiana* (Ducke) Ducke === **Hevea benthamiana** Müll.Arg.

Hevea benthamiana var. *obtusifolia* Ducke === **Hevea benthamiana** Müll.Arg.

Hevea benthamiana f. *obtusifolia* (Ducke) Ducke === **Hevea benthamiana** Müll.Arg.

Hevea benthamiana var. *subglabrifolia* Ducke === **Hevea benthamiana** Müll.Arg.

Hevea benthamiana f. *subglabrifolia* (Ducke) Ducke === **Hevea benthamiana** Müll.Arg.

Hevea benthamiana var. *typica* Ducke === **Hevea benthamiana** Müll.Arg.

Hevea brasiliensis f. *acreana* (Ule) Ducke === **Hevea brasiliensis** (Willd. ex A.Juss.) Müll.Arg.

Hevea brasiliensis var. *acreana* Ule === **Hevea brasiliensis** (Willd. ex A.Juss.) Müll.Arg.

Hevea brasiliensis var. *angustifolia* Ule ex Huber === **Hevea brasiliensis** (Willd. ex A.Juss.) Müll.Arg.

Hevea brasiliensis f. *angustifolia* (Ule ex Huber) Ule === **Hevea brasiliensis** (Willd. ex A.Juss.) Müll.Arg.

Hevea brasiliensis var. *cuneata* (Huber) Pax === **Hevea guianensis** var. **lutea** (Spruce ex Benth.) Ducke & R.E.Schult.

Hevea brasiliensis mut. *granthamii* Barth === **Hevea brasiliensis** (Willd. ex A.Juss.) Müll.Arg.

Hevea brasiliensis var. *janeirensis* (Müll.Arg.) Pax === **Hevea brasiliensis** (Willd. ex A.Juss.) Müll.Arg.

Hevea brasiliensis var. *latifolia* Ule ex Huber === **Hevea brasiliensis** (Willd. ex A.Juss.) Müll.Arg.

Hevea brasiliensis f. *latifolia* (Ule ex Huber) Ule === **Hevea brasiliensis** (Willd. ex A.Juss.) Müll.Arg.

Hevea brasiliensis var. *randiana* (Huber) Pax === **Hevea brasiliensis** (Willd. ex A.Juss.) Müll.Arg.

Hevea brasiliensis f. *randiana* (Huber) Ducke === **Hevea brasiliensis** (Willd. ex A.Juss.) Müll.Arg.

Hevea brasiliensis var. *stylosa* Huber === **Hevea brasiliensis** (Willd. ex A.Juss.) Müll.Arg.

Hevea brasiliensis var. *subconcolor* Ducke === **Hevea nitida** Mart. ex Müll.Arg. var. **nitida**

Hevea brasiliensis f. *subconcolor* (Ducke) Ducke === **Hevea nitida** Mart. ex Müll.Arg. var. **nitida**

Hevea camargoana Pires === ?

Hevea caucho Posada-Ar. === **Hevea guianensis** Aubl. var. **guianensis**

Hevea collina Huber === **Hevea guianensis** Aubl. var. **guianensis**

Hevea confusa Hemsl. === **Hevea pauciflora** var. **coriacea** Ducke

Hevea cuneata (Huber) Huber === **Hevea guianensis** var. **lutea** (Spruce ex Benth.) Ducke & R.E.Schult.

Hevea discolor Spruce ex Pax === **Hevea benthamiana** Müll.Arg.
Hevea discolor (Spruce ex Benth.) Müll.Arg. === **Hevea spruceana** (Benth.) Müll.Arg.
Hevea duckei Huber === **Hevea benthamiana** Müll.Arg.
Hevea elastica (L.f.) H.Karst. === **Hevea guianensis** Aubl. var. **guianensis**
Hevea foxii Huber === **Hevea guianensis** var. **lutea** (Spruce ex Benth.) Ducke & R.E.Schult.
Hevea glabrescens Huber === **Hevea guianensis** var. **lutea** (Spruce ex Benth.) Ducke & R.E.Schult.
Hevea gracilis Ducke === ?
Hevea granthamii Bartlett === **Hevea brasiliensis** (Willd. ex A.Juss.) Müll.Arg.
Hevea guianensis var. *collina* (Huber) Ducke === **Hevea guianensis** Aubl. var. **guianensis**
Hevea guianensis var. *cuneata* (Huber) Ducke === **Hevea guianensis** var. **lutea** (Spruce ex Benth.) Ducke & R.E.Schult.
Hevea guianensis subsp. *marginata* (Ducke) Ducke === **Hevea guianensis** var. **marginata** (Ducke) Ducke
Hevea guianensis subsp. *occidentalis* Ducke === **Hevea guianensis** Aubl. var. **guianensis**
Hevea guianensis var. *occidentalis* (Ducke) Ducke === **Hevea guianensis** Aubl. var. **guianensis**
Hevea guianensis f. *peruviana* (Ducke) Ducke === **Hevea guianensis** var. **lutea** (Spruce ex Benth.) Ducke & R.E.Schult.
Hevea guianensis f. *pilosula* (Ducke) Ducke === **Hevea guianensis** var. **lutea** (Spruce ex Benth.) Ducke & R.E.Schult.
Hevea guianensis subsp. *typica* Ducke === **Hevea guianensis** Aubl.
Hevea guianensis f. *typica* Ducke === **Hevea guianensis** Aubl.
Hevea huberiana Ducke === ?
Hevea humilior Ducke === **Hevea pauciflora** var. **coriacea** Ducke
Hevea janeirensis Müll.Arg. === **Hevea brasiliensis** (Willd. ex A.Juss.) Müll.Arg.
Hevea kunthiana Huber === **Hevea pauciflora** (Spruce ex Benth.) Müll.Arg. var. **pauciflora**
Hevea lutea (Spruce ex Benth.) Müll.Arg. === **Hevea guianensis** var. **lutea** (Spruce ex Benth.) Ducke & R.E.Schult.
Hevea lutea var. *apiculata* (Spruce ex Baill.) Müll.Arg. === **Hevea guianensis** var. **lutea** (Spruce ex Benth.) Ducke & R.E.Schult.
Hevea lutea var. *cuneata* Huber === **Hevea guianensis** var. **lutea** (Spruce ex Benth.) Ducke & R.E.Schult.
Hevea lutea var. *foxii* (Huber) Ducke === **Hevea guianensis** var. **lutea** (Spruce ex Benth.) Ducke & R.E.Schult.
Hevea lutea var. *glabrescens* (Huber) Ducke === **Hevea guianensis** var. **lutea** (Spruce ex Benth.) Ducke & R.E.Schult.
Hevea lutea var. *peruviana* Ducke === **Hevea guianensis** var. **lutea** (Spruce ex Benth.) Ducke & R.E.Schult.
Hevea lutea var. *pilosula* Ducke === **Hevea guianensis** var. **lutea** (Spruce ex Benth.) Ducke & R.E.Schult.
Hevea lutea f. *pilosula* (Ducke) Ducke === **Hevea guianensis** var. **lutea** (Spruce ex Benth.) Ducke & R.E.Schult.
Hevea lutea var. *typica* Ducke === **Hevea guianensis** var. **lutea** (Spruce ex Benth.) Ducke & R.E.Schult.
Hevea marginata Ducke === **Hevea guianensis** var. **marginata** (Ducke) Ducke
Hevea membranacea Müll.Arg. === **Hevea pauciflora** (Spruce ex Benth.) Müll.Arg. var. **pauciflora**
Hevea membranacea var. *leiogyne* Ducke === **Hevea pauciflora** (Spruce ex Benth.) Müll.Arg. var. **pauciflora**
Hevea membranacea f. *leiogyne* (Ducke) Ducke === **Hevea pauciflora** (Spruce ex Benth.) Müll.Arg. var. **pauciflora**
Hevea microphylla var. *major* Pax === **Hevea microphylla** Ule
Hevea microphylla var. *typica* Pax === **Hevea microphylla** Ule
Hevea minor Hemsl. === **Hevea pauciflora** var. **coriacea** Ducke
Hevea nigra Ule === **Hevea guianensis** Aubl. var. **guianensis**
Hevea paludosa Ule === **Hevea brasiliensis** (Willd. ex A.Juss.) Müll.Arg.

Hevea paraensis Baill. === **Hevea spruceana** (Benth.) Müll.Arg.
Hevea pauciflora subsp. *coriacea* Ducke === **Hevea pauciflora** var. **coriacea** Ducke
Hevea pauciflora f. *leiogyne* (Ducke) Ducke === **Hevea pauciflora** (Spruce ex Benth.) Müll.Arg. var. **pauciflora**
Hevea pauciflora var. *typica* Ducke === **Hevea pauciflora** (Spruce ex Benth.) Müll.Arg.
Hevea peruviana Aubl. === **Hevea guianensis** Aubl. var. **guianensis**
Hevea randiana Huber === **Hevea brasiliensis** (Willd. ex A.Juss.) Müll.Arg.
Hevea sieberi Warb. === **Hevea brasiliensis** (Willd. ex A.Juss.) Müll.Arg.
Hevea similis Hemsl. === **Hevea spruceana** (Benth.) Müll.Arg.
Hevea spruceana f. *discolor* (Spruce ex Benth.) Ducke === **Hevea spruceana** (Benth.) Müll.Arg.
Hevea spruceana var. *similis* (Hemsl.) Ducke === **Hevea spruceana** (Benth.) Müll.Arg.
Hevea spruceana f. *similis* (Hemsl.) Ducke === **Hevea spruceana** (Benth.) Müll.Arg.
Hevea spruceana f. *tridentata* (Huber) Ducke === **Hevea spruceana** (Benth.) Müll.Arg.
Hevea spruceana var. *tridentata* Huber === **Hevea spruceana** (Benth.) Müll.Arg.
Hevea viridis Huber === **Hevea nitida** Mart. ex Müll.Arg. var. **nitida**
Hevea viridis var. *toxicodendroides* R.E.Schult. & Vinton === **Hevea nitida** var. **toxicodendroides** (R.E.Schult. & Vinton) R.E.Schult.

Hexadena

Synonyms:
Hexadena Raf. === **Phyllanthus** L.

Hexadenia

Synonyms:
Hexadenia Klotzsch & Garcke === **Pedilanthus** Neck.

Hexakestra

An orthographic variant of *Hexakistra*.

Hexakistra

A name of J. D. Hooker; synonymous with *Andrachne*.

Hexaspermum

Synonyms:
Hexaspermum Domin === **Phyllanthus** L.
Hexaspermum paniculatum (Oliv.) Domin === **Phyllanthus clamboides** (F.Muell.) Diels

Heywoodia

1 species, East Tropical to South Africa, from Uganda and Kenya to KwaZulu-Natal and Eastern Cape. These trees exhibit foliar heteromorphy; in seedlings and adventitious shoots the leaves are usually peltate, while in adult shoots the blades are rhomboid-elliptic, the petiole being attached at the base. Webster (Synopsis, 1994: 36; see **General**) believed the genus to be 'perhaps nearest in morphological characters to the hypothetical ancestor of the family'; its nearest relative was possibly the American *Astrocasia* (Phyllantheae). (Phyllanthoideae)

Hutchinson, J. (1922). The genus *Heywoodia*. Bull. Misc. Inf. Kew 1922: 114-115, illus. En. — Good description and illustration of *H. lucens*; attention is drawn to the heteromorphy of the foliage.

Heywoodia lucens Sim
Artist: A. Kellett
Bull. Misc. Inform.: 116, fig. 1 (1922)
KEW ILLUSTRATIONS COLLECTION

Pax, F. & K. Hoffmann (1922). *Heywoodia*. In A. Engler (ed.), Das Pflanzenreich, IV 147 XV (Euphorbiaceae-Phyllanthoideae-Phyllantheae): 280. Berlin. (Heft 81.) La/Ge. — 1 species; Africa (S Africa).

Heywoodia Sim, Forest Fl. Cape: 326 (1907).
Kenya to S. Africa. 25 26 27. Phan.

Heywoodia lucens Sim, Forest Fl. Cape: 326 (1907). – FIGURE, p. 996.
Kenya to S. Africa. 25 KEN TAN UGA 26 MOZ 27 CPP NAT SWZ. Phan.

Hieronima

An orthographic variant of *Hieronyma*.

Hieronyma

21 species, Americas (Mexico to Bolivia and Brazil and in the West Indies); sometimes spelt 'Hieronima' or 'Hyeronima'. Small to large trees to 30 m or more with a lepidote indumentum, spirally arranged leaves and usually axillary inflorescences which may be important components of lowland and montane forest. *H. alchorneoides* is widely distributed in forests at lower elevations; it also appears rapidly in regrowth. The genus has usually been referred to the *Antidesma* alliance although particularly the indumentum is distinctive. The alliance (subtribe Antidesminae in the Webster system) has from time to time been regarded as part of a separate family (now known as Stilaginaceae) with a distinctive fruit and seed structure (cf. Meeuse 1990; see **General**). The hard wood of some species is valued timber for construction and furniture (Franco 1990). (Phyllanthoideae)

Tulasne, L. R. (1861). Antidesmeae. Fl. Brasiliensis 4(1): 329-336, pls. 87-90. Munich. La. — Flora treatment; coverage of 3 species of *Hieronyma*. [This treatment now superseded; see Franco 1990. It is included here for historical interest on account of the association of this genus with the Old World *Antidesma* in a distinct family.]

Pax, F. & K. Hoffmann (1922). *Hieronyma*. In A. Engler (ed.), Das Pflanzenreich, IV 147 XV (Euphorbiaceae-Phyllanthoideae-Phyllantheae): 31-40. Berlin. (Heft 81.) La/Ge. — 21 species, Americas. [Based on relatively little material.]

Jablonski, E. (1967). *Hieronyma*. Euphorbiaceae, Guayana Highland (Mem. New York Bot. Gard. 17(1)): 122-124. New York. En. — 3 species, 2 well-collected, 1 hardly entering area. *H. laxiflora* (=*H. alchorneoides*) considered to be widely distributed.

• Franco R., P. (1990). The genus *Hyeronima* (Euphorbiaceae) in South America. Bot. Jahrb. Syst. 111: 297-346, illus. En. — General survey with background, history and review of external and internal characters; well-illustrated revision of 10 species with key, descriptions, synonymy, references, types, vernacular names, indication of distribution and habitat, representative exsiccatae with localities, and commentary; two doubtful species; list of literature at end. [Much reduction of names, particularly in *H. alchorneoides*.]

Radcliffe-Smith, A. (1994). Proposal to conserve 4318 *Hieronyma* with a conserved spelling (Euphorbiaceae). Taxon 43: 485-486. En. — Nomenclatural.

Hieronyma Allemão, Pl. Novas Brasil: 1 (1848).
Mexico, Trop. America. 79 80 81 82 83 84. Nanophan. or phan.
Stilaginella Tul., Ann. Sci. Nat., Bot., III, 15: 240 (1851).

Hieronyma alchorneoides Allemão, Pl. Novas Brasil: 1 (1848). – FIGURE, p. 998.
SE. Mexico, Trop. America. 79 MXT 80 BLZ COS GUA PAN 81 TRT WIN 82 FRG GUY SUR VEN 83 BOL CLM ECU PER 84 BZC BZL BZN BZS. Phan.

Hieronyma alchorneoides Allemão var. *alchorneoides* (as *Hyeronimia alchorneoides*), ♂
Artist: 'duce Martio'
Martius, Fl. Bras. 4(1): pl. 88 (1861)
KEW ILLUSTRATIONS COLLECTION

var. **alchorneoides**

SE. Mexico, Trop. America. 79 MXT 80 BLZ COS GUA PAN 81 TRT WIN 82 FRG GUY SUR VEN 83 BOL CLM ECU PER 84 BZC BZL BZN BZS. Phan.

Stilaginella amazonica Tul., Ann. Sci. Nat., Bot., III, 15: 241 (1851).

Stilaginella ferruginea Tul., Ann. Sci. Nat., Bot., III, 15: 250 (1851). *Hieronyma ferruginea* (Tul.) Tul. in C.F.P.von Martius, Fl. Bras. 4(1): 334 (1861).

Stilaginella laxiflora Tul., Ann. Sci. Nat., Bot., III, 15: 244 (1851). *Hieronyma laxiflora* (Tul.) Müll.Arg., Linnaea 34: 67 (1865).

Hieronyma mollis Müll.Arg. in A.P.de Candolle, Prodr. 15(2): 269 (1866).
Hieronyma caribaea Urb., Repert. Spec. Nov. Regni Veg. 14: 139 (1919).
Hieronyma heterotricha Pax & K.Hoffm. in H.G.A.Engler, Pflanzenr., IV, 147, XV: 39 (1922).
Hieronyma mattogrossensis Pax & K.Hoffm. in H.G.A.Engler, Pflanzenr., IV, 147, XV: 39 (1922).
Hieronyma peruviana Pax & K.Hoffm. in H.G.A.Engler, Pflanzenr., IV, 147, XV: 37 (1922).
Hieronyma chocoensis Cuatrec., Revista Acad. Colomb. Ci. Exact. 7: 52 (1946).
Hieronyma ovatifolia Lundell, Wrightia 4: 134 (1970).

var. **stipulosa** P.Franco R., Bot. Jahrb. Syst. 111: 321 (1990).
Costa Rica to Peru. 80 COS PAN 81 WIN 82 FRG VEN 83 CLM ECU PER. Phan.

Hieronyma antioquensis Cuatrec., Revista Acad. Colomb. Ci. Exact. 8: 300 (1951).
Colombia (Antioquia). 83 CLM. Phan.

Hieronyma asperifolia Pax & K.Hoffm. in H.G.A.Engler, Pflanzenr., IV, 147, XV: 37 (1922).
N. Ecuador, Colombia. 83 CLM ECU. Phan.
Hieronyma sararita Cuatrec., Revista Acad. Colomb. Ci. Exact. 8: 298 (1951).

Hieronyma clusioides (Tul.) Griseb., Mem. Amer. Acad. Arts, n.s., 8: 157 (1861).
Puerto Rico, E. Cuba. 81 CUB PUE. Phan.
* *Stilaginella clusioides* Tul., Ann. Sci. Nat., Bot., III, 15: 245 (1851).
Hieronyma pallida Müll.Arg. in A.P.de Candolle, Prodr. 15(2): 270 (1866).
Antidesma rosaurianum M.Gómez, Dicc. Bot. Nom. Vulg. Cub. & Puerto-Riquenos: 29 (1889).

Hieronyma crassistipula Urb., Repert. Spec. Nov. Regni Veg. 28: 215 (1930).
Cuba (I. de la Juventud). 81 CUB.

Hieronyma cubana Müll.Arg. in A.P.de Candolle, Prodr. 15(2): 270 (1866).
W. & C. Cuba. 81 CUB.

Hieronyma domingensis Urb., Repert. Spec. Nov. Regni Veg. 16: 137 (1919).
Hispaniola. 81 DOM HAI. Phan.

Hieronyma duquei Cuatrec., Revista Acad. Colomb. Ci. Exact. 7: 52 (1946).
Venezuela, Ecuador, Colombia, Peru. 82 VEN 83 CLM ECU PER. Phan.

Hieronyma fendleri Briq., Annuaire Conserv. Jard. Bot. Genève 1900: 227 (1900).
Hieronyma moritziana var. *fendleri* (Briq.) Pax & K.Hoffm. in H.G.A.Engler, Pflanzenr., IV, 147, XV: 33 (1922).
S. Mexico to Bolivia. 79 MXT 80 COS GUA PAN 82 VEN 83 BOL CLM PER. Phan.
Hieronyma macrocarpa var. *moritziana* Müll.Arg., Linnaea 34: 66 (1865). *Hieronyma moritziana* (Müll.Arg.) Pax & K.Hoffm. in H.G.A.Engler, Pflanzenr., IV, 147, XV: 33 (1922).
Hieronyma buchtienii Pax & K.Hoffm. in H.G.A.Engler, Pflanzenr., IV, 147, XV: 33 (1922).
Hieronyma moritziana var. *yungasensis* Pax & K.Hoffm. in H.G.A.Engler, Pflanzenr., IV, 147, XV: 33 (1922).
Hieronyma nevadensis Cuetrec., Revista Acad. Colomb. Ci. Exact. 8: 300 (1951).

Hieronyma gentlei Lundell, Wrightia 5: 248 (1976).
Belize. 80 BLZ.

Hieronyma havanensis Urb., Repert. Spec. Nov. Regni Veg. 28: 215 (1930).
W. & C. Cuba. 81 CUB.

Hieronyma huilensis Cuatrec., Revista Acad. Colomb. Ci. Exact. 8: 299 (1951).
E. Colombia, Venezuela. 82 VEN 83 CLM. Phan.
Hieronyma pilifera Cuatrec., Revista Acad. Colomb. Ci. Exact. 8: 299 (1951).

Hieronyma jamaicensis Urb., Repert. Spec. Nov. Regni Veg. 16: 137 (1919).
Jamaica. 81 JAM. Phan.

Hieronyma macrocarpa Müll.Arg., Linnaea 34: 66 (1865).
Venezuela, Ecuador, Colombia, Peru. 82 VEN 83 CLM ECU PER. Phan.
Hieronyma macrocarpa var. *spruceana* Müll.Arg., Linnaea 34: 66 (1865).
Hieronyma colombiana Cuatrec., Revista Acad. Colomb. Ci. Exact. 8: 298 (1951).
Hieronyma croizatii Steyerm., Fieldiana, Bot. 21: 317 (1952).

Hieronyma montana Alain, Mem. New York Bot. Gard. 21: 123 (1971).
Dominican Rep. 81 DOM. Nanophan.

Hieronyma nipensis Urb., Repert. Spec. Nov. Regni Veg. 28: 216 (1930).
E. Cuba. 81 CUB.

Hieronyma oblonga (Tul.) Müll.Arg., Linnaea 34: 66 (1865).
SE. Mexico, Trop. America. 79 MXT 80 COS GUA PAN 82 FRG GUY VEN 83 CLM ECU PER 84 BZC BZE BZL. Phan.
Stilaginella benthamii Tul., Ann. Sci. Nat., Bot., III, 15: 248 (1851). *Hieronyma oblonga* var. *benthamii* (Tul.) Müll.Arg., Linnaea 34: 66 (1865).
Stilaginella blanchetiana Tul., Ann. Sci. Nat., Bot., III, 15: 249 (1851). *Hieronyma blanchetiana* (Tul.) Tul. in C.F.P.von Martius, Fl. Bras. 4(1): 333 (1861). *Hieronyma oblonga* var. *blanchetiana* (Tul.) Müll.Arg., Linnaea 34: 66 (1865).
* *Stilaginella oblonga* Tul., Ann. Sci. Nat., Bot., III, 15: 248 (1851). *Hieronyma oblonga* var. *genuina* Müll.Arg., Linnaea 34: 66 (1865), nom. inval.
Hieronyma oblonga var. *obtusata* Müll.Arg., Linnaea 34: 66 (1865).
Hieronyma guatemalensis Donn.Sm., Bot. Gaz. 54: 241 (1912).
Hieronyma andina Pax & K.Hoffm. in H.G.A.Engler, Pflanzenr., IV, 147, XV: 37 (1922).
Hieronyma dichrophylla J.J.Sm. ex Pax in H.G.A.Engler, Pflanzenr., IV, 147, XV: 34 (1922), pro syn.
Hieronyma poasana Standl., Publ. Field Mus. Nat. Hist., Bot. Ser. 18: 611 (1937).
Hieronyma oblonga var. *crassifolia* Cuatrec., Revista Acad. Colomb. Ci. Exact. 8: 300 (1951).
Hieronyma oblonga var. *nervata* Cuatrec., Revista Acad. Colomb. Ci. Exact. 8: 301 (1951).
Hieronyma oblonga f. *glabra* Steyerm., Fieldiana, Bot. 28: 951 (1957).

Hieronyma ovata Urb., Repert. Spec. Nov. Regni Veg. 28: 216 (1930).
E. Cuba. 81 CUB.

Hieronyma paucinervis Urb., Repert. Spec. Nov. Regni Veg. 28: 216 (1930).
E. Cuba. 81 CUB.

Hieronyma rufa P.Franco R., Bot. Jahrb. Syst. 111: 334 (1990).
Colombia. 83 CLM. Phan.

Hieronyma scabrida (Tul.) Müll.Arg., Linnaea 34: 66 (1865).
SW. Colombia, N. Ecuador. 83 CLM ECU. Phan.
* *Stilaginella scabrida* Tul., Ann. Sci. Nat., Bot., III, 15: 242 (1851).
Hieronyma hirtinervia Cuatrec., Revista Acad. Colomb. Ci. Exact. 8: 299 (1951).

Synonyms:
Hieronyma andina Pax & K.Hoffm. === **Hieronyma oblonga** (Tul.) Müll.Arg.
Hieronyma blanchetiana (Tul.) Tul. === **Hieronyma oblonga** (Tul.) Müll.Arg.
Hieronyma boliviana Pax === ?
Hieronyma buchtienii Pax & K.Hoffm. === **Hieronyma fendleri** Briq.
Hieronyma caribaea Urb. === **Hieronyma alchorneoides** Allemão var. **alchorneoides**
Hieronyma chocoensis Cuatrec. === **Hieronyma alchorneoides** Allemão var. **alchorneoides**
Hieronyma colombiana Cuatrec. === **Hieronyma macrocarpa** Müll.Arg.
Hieronyma croizatii Steyerm. === **Hieronyma macrocarpa** Müll.Arg.
Hieronyma dichrophylla J.J.Sm. ex Pax === **Hieronyma oblonga** (Tul.) Müll.Arg.
Hieronyma ferruginea (Tul.) Tul. === **Hieronyma alchorneoides** Allemão var. **alchorneoides**
Hieronyma guatemalensis Donn.Sm. === **Hieronyma oblonga** (Tul.) Müll.Arg.
Hieronyma heterotricha Pax & K.Hoffm. === **Hieronyma alchorneoides** Allemão var. **alchorneoides**
Hieronyma hirtinervia Cuatrec. === **Hieronyma scabrida** (Tul.) Müll.Arg.
Hieronyma laxiflora (Tul.) Müll.Arg. === **Hieronyma alchorneoides** Allemão var. **alchorneoides**
Hieronyma macrocarpa var. *moritziana* Müll.Arg. === **Hieronyma fendleri** Briq.
Hieronyma macrocarpa var. *spruceana* Müll.Arg. === **Hieronyma macrocarpa** Müll.Arg.
Hieronyma mattogrossensis Pax & K.Hoffm. === **Hieronyma alchorneoides** Allemão var. **alchorneoides**
Hieronyma mollis Müll.Arg. === **Hieronyma alchorneoides** Allemão var. **alchorneoides**
Hieronyma moritziana (Müll.Arg.) Pax & K.Hoffm. === **Hieronyma fendleri** Briq.
Hieronyma moritziana var. *fendleri* (Briq.) Pax & K.Hoffm. === **Hieronyma fendleri** Briq.
Hieronyma moritziana var. *yungasensis* Pax & K.Hoffm. === **Hieronyma fendleri** Briq.
Hieronyma nevadensis Cuetrec. === **Hieronyma fendleri** Briq.
Hieronyma oblonga var. *benthamii* (Tul.) Müll.Arg. === **Hieronyma oblonga** (Tul.) Müll.Arg.
Hieronyma oblonga var. *blanchetiana* (Tul.) Müll.Arg. === **Hieronyma oblonga** (Tul.) Müll.Arg.
Hieronyma oblonga var. *crassifolia* Cuatrec. === **Hieronyma oblonga** (Tul.) Müll.Arg.
Hieronyma oblonga var. *genuina* Müll.Arg. === **Hieronyma oblonga** (Tul.) Müll.Arg.
Hieronyma oblonga f. *glabra* Steyerm. === **Hieronyma oblonga** (Tul.) Müll.Arg.
Hieronyma oblonga var. *nervata* Cuatrec. === **Hieronyma oblonga** (Tul.) Müll.Arg.
Hieronyma oblonga var. *obtusata* Müll.Arg. === **Hieronyma oblonga** (Tul.) Müll.Arg.
Hieronyma ovatifolia Lundell === **Hieronyma alchorneoides** Allemão var. **alchorneoides**
Hieronyma pallida Müll.Arg. === **Hieronyma clusioides** (Tul.) Griseb.
Hieronyma peruviana Pax & K.Hoffm. === **Hieronyma alchorneoides** Allemão var. **alchorneoides**
Hieronyma pilifera Cuatrec. === **Hieronyma huilensis** Cuatrec.
Hieronyma poasana Standl. === **Hieronyma oblonga** (Tul.) Müll.Arg.
Hieronyma reticulata Britton ex Rusby === ?
Hieronyma sararita Cuatrec. === **Hieronyma asperifolia** Pax & K.Hoffm.

Hippocrepandra

Synonyms:
Hippocrepandra Müll.Arg. === **Monotaxis** Brongn.
Hippocrepandra gracilis Müll.Arg. === **Monotaxis gracilis** (Müll.Arg.) Baill.
Hippocrepandra lurida Müll.Arg. === **Monotaxis lurida** (Müll.Arg.) Benth.
Hippocrepandra neesiana Müll.Arg. === **Monotaxis grandiflora** Endl.

Hippomane

3 species, Florida, Middle America (Caribbean, Pacific Coast), S. America (Atlantic Colombia, Venezuela) and the Galápagos; shrubs or trees to 20 m with drupaceous fruits. The glabrous, laticiferous and very poisonous manzanillo, *H. mancinella,* is throughout the range

(generally on or near coasts), while the others, with spiny leaves, are in Hispaniola (for these latter, see Liogier in *Fl. Española* 4 (1989)). The large stone in the fruit of *H. mancinella* is distinctive; apart from this feature, surely related to marine dispersal, there are few differences from *Sapium*. (Euphorbioideae (except Euphorbieae))

Pax, F. (with K. Hoffmann) (1912). *Hippomane*. In A. Engler (ed.), Das Pflanzenreich, IV 147 V (Euphorbiaceae-Hippomaneae): 261-263. Berlin. (Heft 52.) La/Ge. — 1 species, circum-Caribbean from the Florida Keys and Mexico southwards; coastal or generally so. [1-2 additional species from Hispaniola are now usually included in the genus.]

Hippomane L., Sp. Pl.: 1191 (1753).
SE. U.S.A., Mexico, Trop. America. 78 79 80 81 82 83.
Mancanilla Plum. ex Adans., Fam. Pl. 2: 354 (1763).
Mancinella Tussac, Fl. Antill. 3: 21 (1824).

Hippomane horrida Urb. & Ekman, Ark. Bot. 22A(8): 64 (1929).
SW. Dominican Rep. (Pen. Barahona). 81 DOM. Nanophan.

Hippomane mancinella L., Sp. Pl.: 1191 (1753).
Florida (Key West), Caribbean, Mexico, C. America to N. Venezuela. 78 FLA 79 MXG 80 COS PAN 81 BAH CAY CUB DOM HAI JAM LEE PUE TRT WIN 82 guy VEN 83 CLM GAL. Nanophan. or phan.
Hippomane dioica Rottb., Acta Lit. Univ. Hafn. 1: 301 (1778).

Hippomane spinosa L., Sp. Pl.: 1191 (1753).
SW. Hispaniola. 81 DOM HAI. Nanophan. or phan.
Sapium ilicifolium Willd., Sp. Pl. 4: 573 (1805).

Synonyms:
Hippomane aucuparia (Jacq.) Crantz === **Sapium glandulosum** (L.) Morong
Hippomane biglandulosa L. === **Sapium glandulosum** (L.) Morong
Hippomane cerifera Sessé & Moç. === ?
Hippomane dioica Rottb. === **Hippomane mancinella** L.
Hippomane fruticosa Sessé & Moç. === ?
Hippomane glandulosa L. === **Sapium glandulosum** (L.) Morong
Hippomane zeocca L. ex B.D.Jacks. === **Sapium glandulosum** (L.) Morong

Holstia

Replaced by *Neoholstia* but now united with *Tannodia*.

Synonyms:
Holstia Pax === **Tannodia** Baill.
Holstia sessiliflora Pax === **Tannodia tenuifolia** var. **glabrata** Prain
Holstia tenuifolia Pax === **Tannodia tenuifolia** (Pax) Prain
Holstia tenuifolia var. *glabrata* (Prain) Pax === **Tannodia tenuifolia** var. **glabrata** Prain

Homalanthus

23 species, SE. Asia, Malesia and the Pacific, in the last-named east to Rapa Iti (SE. Polynesia) and south to the Kermadecs; naturalised in parts of Africa as well as regenerating freely from seed south of its natural limit in Australia. Laticiferous shrubs (*H. grandifolius*) or small to large trees (*H. arfakensis*, a mainly montane species, to as much as 40 m) with branching in pseudo-whorls, often prominent in secondary growth and conspicuous in foliage. *Omalanthus* is the original (and nomenclaturally correct) spelling,

but in view of the wide use of *Homalanthus* (also followed here) a proposal for conservation of that spelling has been made by Esser (1996). Three sections were recognised by Pax (1912); to these, a fourth (with bracts absent) was added by Airy-Shaw (1978). Esser (1997) in his revision has not accepted any subdivisions pending a phylogenetic analysis of the whole genus. This treatment also features much reduction of names and reveals frequent past misapplications. The latex is poisonous and particularly harmful to eyes; however, *H. populifolius* is in Australasia often cultivated. The genus is said to be without near relatives in the Hippomaneae (Webster, Synopsis, 1994); this aspect is, however, under further study. (Euphorbioideae (except Euphorbieae))

Pax, F. (with K. Hoffmann) (1912). *Homalanthus*. In A. Engler (ed.), Das Pflanzenreich, IV 147 V (Euphorbiaceae-Hippomaneae): 42-54. Berlin. (Heft 52.) La/Ge. — 19 species, mutually similar in appearance but relatively easily distinguishable. Additions in ibid., XIV (Additamentum VI): 56-57 (1919).

Airy-Shaw, H. K. (1968). Notes on Malesian and other Asiatic Euphorbiaceae, XCVI. New or noteworthy species of *Homalanthus* Juss. Kew Bull. 21: 409-418. En. — A substantial contribution on this genus, in five parts (with some novelties); in the second is a key distinguishing *H. populifolius, H. populneus* and *H. novoguineensis*.

Sykes, W. R. (1969). *Homalanthus* in New Zealand. New Zealand J. Bot. 7: 302-307, illus. En. — Review of *H. polyandrus* (Raoul I., Kermadec Islands) and the cultivated *H. populifolius* (now *H. nutans*) from Australia, with extensive discussion of distinguishing features.

Airy-Shaw, H. K. (1978). Notes on Malesian and other Asiatic Euphorbiaceae, CCVI. A new section in *Homalanthus* Juss. Kew Bull. 32: 418. En. — Delineation of sect. *Ebracteati* for *H. ebracteatus* of Vanuatu.

Airy-Shaw, H. K. (1979). Notes on Malesian and other Asiatic Euphorbiaceae, CCXXVII. *Homalanthus* Juss. Kew Bull. 33: 536-537. En. — Description of *H. xerocarpus* from Bougainville.

Airy-Shaw, H. K. (1980). Notes on Euphorbiaceae from Indomalesia, Australia and the Pacific, CCXLI. *Homalanthus* Juss. Kew Bull. 35: 398-399. En. — Description of *H. sulawesianus* from central Sulawesi.

Airy-Shaw, H. K. (1981). Notes on Asiatic, Malesian and Melanesian Euphorbiaceae, CCLII. *Homalanthus* Juss. Kew Bull. 36: 611-612. En. — Description of *H. milvus* from Luzon.

St. John, H. (1984). *Omalanthus* (Euphorbiaceae) in southeastern Polynesia. Nordic J. Bot. 4: 53-56, illus. (Pacific plant studies, 40.) En. — Synoptic treatment of 3 species (1 new) with key, types, synonymy, localities with exsiccatae, and notes.

Forster, P. I. (1994). *Omalanthus nutans* (Euphorbiaceae), the correct name for the 'native bleeding heart' or 'native poplar' of Australia. Telopea 6: 169-171. En. — Reduction of *Omalanthus populifolius* to *O. nutans*; the combined species ranges from the Southwestern Pacific to east-central Australia (Tonga, Fiji, Vanuatu, New Caledonia, Lord Howe Island, north, central and southern Queensland and eastern New South Wales). Records from further north in Australia formerly included here are *O. novoguineensis*. A key to all 3 species of the genus in Australia is also given. [These plants are frequently used in gardens and landscape rehabilitation in Australia. It may be noted here that *O. nutans* is also known from the Society and Caroline Islands.]

Esser, H.-J. (1996). Proposal to conserve the name *Homalanthus* (Euphorbiaceae) with a conserved spelling. Taxon 43: 555-556. En. — Proposal made for retention of *Homalanthus*, a widely used spelling, against the original (and nomenclaturally correct) *Omalanthus*.

- Esser, H.-J. (1997). A revision of *Omalanthus* (Euphorbiaceae) in Malesia. Blumea 42: 421-466, illus., maps. En. — General introduction and review of characters, phylogeny, biogeography and ecology; treatment of 13 species with keys, descriptions, indication of distribution, habitat, and ecology, vernacular names, and critical notes; literature cited and identification list at end.

Homalanthus A.Juss., Euphorb. Gen.: 50 (1824), orth. cons.
Trop. Asia to Pacific. 38 40 41 42 50 51 60 61 62.
Duania Noronha, Verh. Batav. Genootsch. Kunsten 5(4): 2 (1790).
Carumbium Reinw., Elench. Sem.: 319 (1823).
Omalanthus A.Juss., Euphorb. Gen.: 50 (1824), sphalm.
Garumbium Blume, Flora 8: t. 103 (1825).
Dibrachion Regel, Index Seminum (LE) 1865: 51 (1865).
Wartmannia Müll.Arg., Linnaea 34: 218 (1865).

Homalanthus acuminatus (Müll.Arg.) Pax in H.G.A.Engler, Nat. Pflanzenfam. ed. 2, 19c: 189 (1931).
Society Is. (Tahiti). 61 SCI.
* *Carumbium acuminatum* Müll.Arg. in A.P.de Candolle, Prodr. 15(2): 1144 (1866).
Macaranga reineckei Pax, Bot. Jahrb. Syst. 25: 646 (1898).

Homalanthus arfakiensis Hutch. in L.S.Gibbs, Fl. Arfak Mts.: 145 (1917).
Maluku, Irian Jaya. 42 MOL NWG. Nanophan. or phan.
Homalanthus agallochoides J.J.Sm., Nova Guinea 12: 547 (1917).
Homalanthus collinus Gage, Nova Guinea 12: 484 (1917).
Homalanthus megalanthus Gage, Nova Guinea 12: 482 (1917).
Homalanthus minutiflorus Airy Shaw, Kew Bull. 21: 414 (1968).

Homalanthus caloneurus Airy Shaw, Kew Bull. 21: 413 (1968).
Borneo (Brunei, Sabah). 42 BOR. Phan.
Homalanthus populneus var. *cordifolius* Heine, Repert. Spec. Nov. Regni Veg. 54: 235 (1951).

Homalanthus ebracteatua Guillaumin, J. Arnold Arbor. 13: 94 (1932).
Vanuatu. 60 VAN.

Homalanthus fastuosus (Linden) Villar in F.M.Blanco, Fl. Filip., ed. 3, 4(13A): 196 (1880).
Taiwan (Orchid I., Botel Tobago), Philippines. 38 TAI 42 PHI. Phan.
* *Mappa fastuosa* Linden, Ann. Hort. Belge Étrangère 15: 100 (1865).
Homalanthus alpinus Elmer, Leafl. Philipp. Bot. 1: 307 (1908).
Homalanthus bicolor Merr., Philipp. J. Sci., C 4: 282 (1909).
Homalanthus milvus Airy Shaw, Kew Bull. 36: 611 (1981).

Homalanthus giganteus Zoll. & Moritzi, Natuur.-Geneesk. Arch. Ned.-Indië 2: 584 (1845).
E. Jawa, Lesser Sunda Is. 42 JAW LSI. Phan.
Homalanthus niveus Pax & K.Hoffm. in H.G.A.Engler, Pflanzenr., IV, 147, V: 51 (1912).

Homalanthus grandifolius Ridl., J. Fed. Malay States Mus. 8:84 (1917).
Sumatera, Borneo. 42 BOR SUM. NAnophan. or phan.

Homalanthus longipes Pax & K.Hoffm. in H.G.A.Engler, Pflanzenr., IV, 147, V: 51 (1912).
Vanuatu. 60 VAN. Phan.

Homalanthus longistylus K.Schum. & Lauterb., Fl. Schutzgeb. Südsee: 407 (1900).
NE. Papua New Guinea, Bismarck Archip. 42 BIS NWG. Phan.
Homalanthus papuanus Pax & K.Hoffm. in H.G.A.Engler, Pflanzenr., IV, 147, V: 45 (1912).

Homalanthus macradenius Pax & K.Hoffm. in H.G.A.Engler, Pflanzenr., IV, 147, V: 51 (1912).
C. & S. Philippines. 42 PHI. Phan.
Homalanthus megaphyllus Merr., Philipp. J. Sci., C 9: 485 (1914 publ. 1915).
Homalanthus rotundifolius Merr., Philipp. J. Sci., C 9: 486 (1914 publ. 1915).
Homalanthus surigaoensis Elmer, Leafl. Philipp. Bot. 7: 2645 (1915).
Homalanthus concolor Merr., Philipp. J. Sci. 20: 398 (1922).

Homalanthus nervosus J.J.Sm., Nova Guinea 8: 792 (1912).
C. & NE. New Guinea. 42 NWG. Nanophan. or phan.
Homalanthus vernicosus Gage, Nova Guinea 12: 483 (1917).
Homalanthus deltoideus Airy Shaw, Kew Bull. 34: 594 (1980).

Homalanthus novoguineensis (Warb.) K.Schum. in K.M.Schumann & C.A.G.Lauterbach, Fl. Schutzgeb. Südsee: 407 (1900).
Lesser Sunda Is., Maluku, New Guinea, N. Australia, Solomon Is. 42 BIS LSI MOL NWG 50 NTA QLD WAU 60 SOL. Phan.
* *Carumbium novoguineense* Warb., Bot. Jahrb. Syst. 18: 199 (1894).
Homalanthus brachystachys Pax & K.Hoffm. in H.G.A.Engler, Pflanzenr., IV, 147, V: 47 (1912).
Homalanthus tetrandrus J.J.Sm., Nova Guinea 8: 791 (1912).
Homalanthus crinitus Gage, Nova Guinea 12: 484 (1917).
Homalanthus elegans Gage, Nova Guinea 12: 483 (1917).
Homalanthus beguinii J.J.Sm., Bull. Jard. Bot. Buitenzorg, III, 6: 98 (1924).
Homalanthus pachystylus Airy Shaw, Kew Bull. 34: 595 (1980).

Homalanthus nutans (G.Forst.) Guill., Zephyritis: 35 (1837).
Pacific. 60 FIJ NWC VAN 61 COO SCI TUB 62 CRL. Nanophan. or phan.
* *Croton nutans* G.Forst., Fl. Ins. Austr.: 67 (1786). *Stillingia nutans* (G.Forst.) Geiseler, Croton. Monogr.: 80 (1807). *Homalanthus nutans* var. *genuinus* Müll.Arg. in A.P.de Candolle, Prodr. 15(2): 1146 (1866), nom. inval. *Carumbium nutans* (G.Forst.) Müll.Arg. in A.P.de Candolle, Prodr. 15(2): 1146 (1866).
Homalanthus pedicellatus Benth., Hooker's J. Bot. Kew Gard. Misc. 2: 232 (1843).
Carumbium moerenhoutianum Müll.Arg. in A.P.de Candolle, Prodr. 15(2): 1146 (1866). *Homalanthus moerenhoutianus* (Müll.Arg.) Benth. & Hook.f. ex Drake, Ill. Fl. Ins. Pacif.: 293 (1892).
Homalanthus nutans var. *rhombifolius* Müll.Arg. in A.P.de Candolle, Prodr. 15(2): 1146 (1866).
Homalanthus nutans (Müll.Arg.) Benth. & Hook.f. ex Drake, Ill. Fl. Ins. Pacif.: 293 (1892).
Homalanthus nutans var. *major* Pax, Bot. Jahrb. Syst. 25: 648 (1898).
Homalanthus gracilis H.St.John, Nordic J. Bot. 4: 53 (1984).

Homalanthus polyadenius Pax & K.Hoffm. in H.G.A.Engler, Pflanzenr., IV, 147, XIV: 57 (1919).
NE. Papua New Guinea. 42 NWG. Nanophan. or phan. – Probably identical with *H. nervosus*.

Homalanthus polyandrus (Hook.f.) Cheeseman, Man. New Zealand Fl.: 630 (1906). – FIGURE, p. 1006.
Kermadec Is. 51 KER. Nanophan. or phan.
* *Carumbium polyandrum* Hook.f., Handb. N. Zeal. Fl.: 248 (1864).

Homalanthus populifolius Graham, Edinburgh New Philos. J. 3 : 175 (1827).
Papua New Guinea to Solomon Is., E. Australia. 42 BIS NWG 50 LHN NSW QLD VIC 60 SOL. Nanophan. or phan.
Homalanthus goodenoviensis Airy Shaw, Kew Bull. 21: 417 (1968).

Homalanthus populneus (Geiseler) Pax in H.G.A.Engler & K.A.E.Prantl, Nat. Pflanzenfam. 3(5): 96 (1892).
S. Thailand, W. & C. Malesia. 41 THA 42 BOR JAW LSI MLY MOL PHI SUL SUM. Nanophan. or phan.
* *Stillingia populnea* Geiseler, Croton. Monogr.: 80 (1807). *Carumbium populneum* (Geiseler) Müll.Arg. in A.P.de Candolle, Prodr. 15(2): 1144 (1866). *Homalanthus populneus* var. *genuinus* Pax in H.G.A.Engler, Pflanzenr., IV, 147, V: 46 (1912), nom. inval.

Homalanthus leschenaultianus A.Juss., Euphorb. Gen.: 50 (1824).
Excoecaria laevis Blanco, Fl. Filip.: 788 (1837). *Homalanthus populneus* var. *laevis* (Blanco) Merr., Sp. Blancoan.: 230 (1918).
Carumbium populneum var. *minus* Müll.Arg. in A.P.de Candolle, Prodr. 15(2): 1145 (1866).
Homalanthus populneus var. *minor* (Müll.Arg.) Merr., Philipp. J. Sci., C 7: 390 (1912 publ. 1913).
Homalanthus sulawesianus Airy Shaw, Kew Bull. 35: 398 (1980).

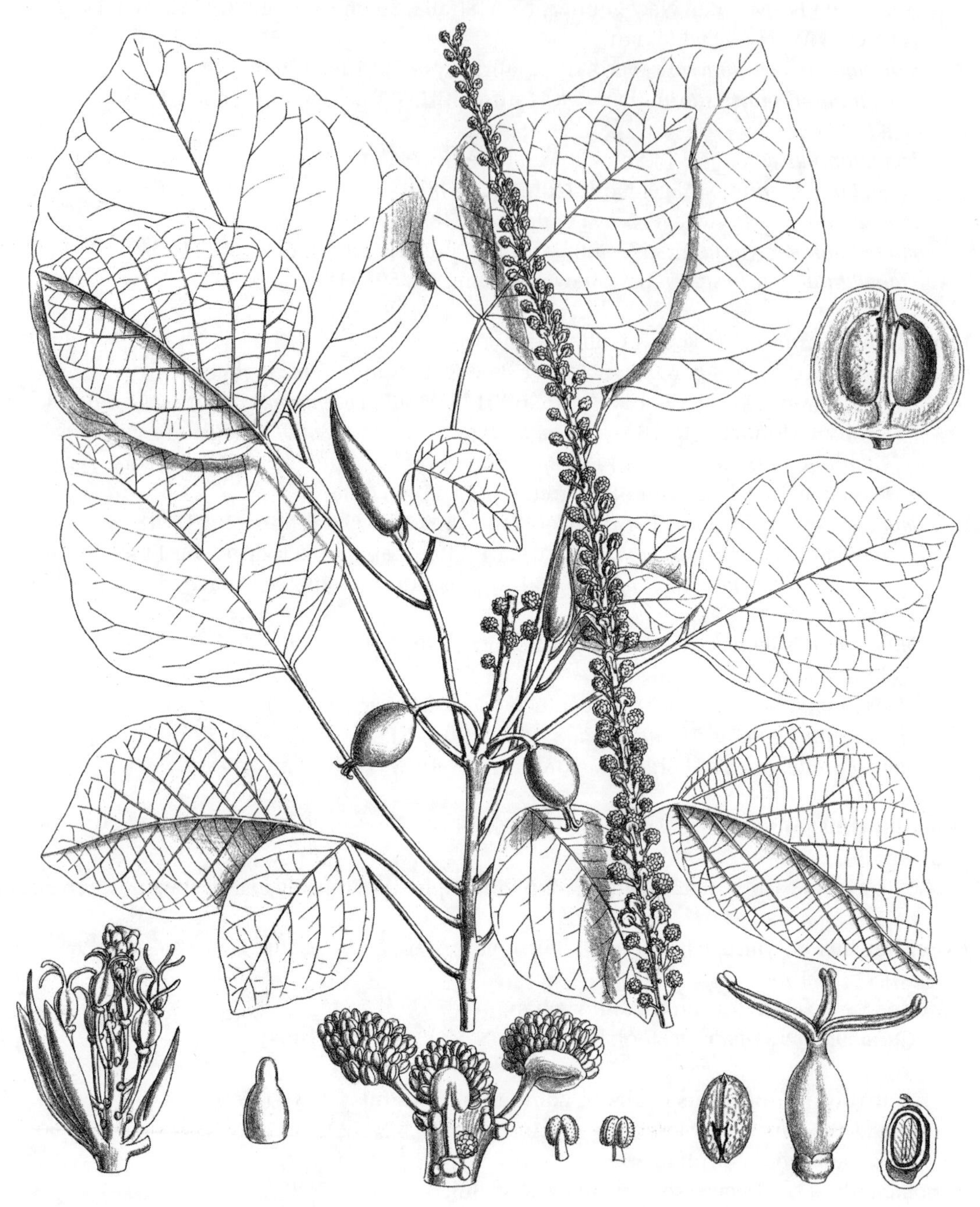

Homalanthus polyandrus (Hook.f.) Cheeseman
Artist: Matilda Smith
Cheeseman & Hemsley, Ill. New Zealand Fl. 2: pl. 179 (1914)
KEW ILLUSTRATIONS COLLECTION

Homalanthus remotus Esser, Blumea 42: 457 (1997).
Irian Jaya. 42 NWG. Nanophan. or phan.

Homalanthus repandus Schltr., Bot. Jahrb. Syst. 39: 154 (1906).
New Caledonia (incl. Loyalty Is.). 60 NWC. Nanophan. or phan.

Homalanthus schlechteri Pax & K.Hoffm. in H.G.A.Engler, Pflanzenr., IV, 147, V: 52 (1912).
New Caledonia (incl. Loyalty Is.). 60 NWC. Nanophan. or phan.

Homalanthus stillingifolius F.Muell., Fragm. 1: 32 (1858).
SE. Queensland, E. New South Wales. 50 NSW QLD. Nanophan.

Homalanthus stokesii F.Brown, Bernice P. Bishop Mus. Bull. 130: 151 (1935).
Tubuai Is. (Rapa). 61 TUB. Phan.

Homalanthus trivalvis Airy Shaw, Kew Bull. 21: 415 (1968).
Solomon Is. 60 SOL. Phan.
Homalanthus xerocarpus Airy Shaw, Kew Bull. 33: 536 (1979).

Synonyms:

Homalanthus agallochoides J.J.Sm. === **Homalanthus arfakiensis** Hutch.
Homalanthus alpinus Elmer === **Homalanthus fastuosus** (Linden) Villar
Homalanthus beguinii J.J.Sm. === **Homalanthus novoguineensis** (Warb.) K.Schum.
Homalanthus bicolor Merr. === **Homalanthus fastuosus** (Linden) Villar
Homalanthus brachystachys Pax & K.Hoffm. === **Homalanthus novoguineensis** (Warb.) K.Schum.
Homalanthus collinus Gage === **Homalanthus arfakiensis** Hutch.
Homalanthus concolor Merr. === **Homalanthus macradenius** Pax & K.Hoffm.
Homalanthus crinitus Gage === **Homalanthus novoguineensis** (Warb.) K.Schum.
Homalanthus deltoideus Airy Shaw === **Homalanthus nervosus** J.J.Sm.
Homalanthus elegans Gage === **Homalanthus novoguineensis** (Warb.) K.Schum.
Homalanthus goodenoviensis Airy Shaw === **Homalanthus populifolius** Graham
Homalanthus gracilis H.St.John === **Homalanthus nutans** (G.Forst.) Guill.
Homalanthus leschenaultianus A.Juss. === **Homalanthus populneus** (Geiseler) Pax
Homalanthus megalanthus Gage === **Homalanthus arfakiensis** Hutch.
Homalanthus megaphyllus Merr. === **Homalanthus macradenius** Pax & K.Hoffm.
Homalanthus milvus Airy Shaw === **Homalanthus fastuosus** (Linden) Villar
Homalanthus minutiflorus Airy Shaw === **Homalanthus arfakiensis** Hutch.
Homalanthus moerenhoutianus (Müll.Arg.) Benth. & Hook.f. ex Drake === **Homalanthus nutans** (G.Forst.) Guill.
Homalanthus niveus Pax & K.Hoffm. === **Homalanthus giganteus** Zoll. & Moritzi
Homalanthus nutans (Müll.Arg.) Benth. & Hook.f. ex Drake === **Homalanthus nutans** (G.Forst.) Guill.
Homalanthus nutans var. *genuinus* Müll.Arg. === **Homalanthus nutans** (G.Forst.) Guill.
Homalanthus nutans var. *major* Pax === **Homalanthus nutans** (G.Forst.) Guill.
Homalanthus nutans var. *rhombifolius* Müll.Arg. === **Homalanthus nutans** (G.Forst.) Guill.
Homalanthus pachystylus Airy Shaw === **Homalanthus novoguineensis** (Warb.) K.Schum.
Homalanthus papuanus Pax & K.Hoffm. === **Homalanthus longistylus** K.Schum. & Lauterb.
Homalanthus pedicellatus Benth. === **Homalanthus nutans** (G.Forst.) Guill.
Homalanthus populneus var. *cordifolius* Heine === **Homalanthus caloneurus** Airy Shaw
Homalanthus populneus var. *genuinus* Pax === **Homalanthus populneus** (Geiseler) Pax
Homalanthus populneus var. *laevis* (Blanco) Merr. === **Homalanthus populneus** (Geiseler) Pax
Homalanthus populneus var. *minor* (Müll.Arg.) Merr. === **Homalanthus populneus** (Geiseler) Pax
Homalanthus populneus var. *siccus* (Blanco) Pax === **Alchornea sicca** (Blanco) Merr.
Homalanthus rotundifolius Merr. === **Homalanthus macradenius** Pax & K.Hoffm.
Homalanthus sulawesianus Airy Shaw === **Homalanthus populneus** (Geiseler) Pax
Homalanthus surigaoensis Elmer === **Homalanthus macradenius** Pax & K.Hoffm.
Homalanthus tetrandrus J.J.Sm. === **Homalanthus novoguineensis** (Warb.) K.Schum.
Homalanthus vernicosus Gage === **Homalanthus nervosus** J.J.Sm.
Homalanthus xerocarpus Airy Shaw === **Homalanthus trivalvis** Airy Shaw

Homonoia

3 species, SE Asia, Malesia; riverine shrubs or small trees related to the formerly included *Spathiostemon* but differing in the presence of a lepidote indumentum. Both *H. riparia* and the less widely distributed and smaller *H. retusa* are distinctive 'rheophytes', found along river banks in many parts of SE. Asia and Malesia. This property makes them valuable in the control of riverbank and other erosion. A recent partial account is that of van Welzen (1998); further details may be had in the PROSEA-series (group 35). (Acalyphoideae)

Pax, F. & K. Hoffmann (1919). *Homonoia*. In A. Engler (ed.), Das Pflanzenreich, IV 147 XI (Euphorbiaceae-Acalypheae-Ricininae): 114-118. Berlin. (Heft 68.) La/Ge. — 3 species, of which one belongs to the now-distinct *Spathiostemon*. The remaining two are in S. and SE. Asia, Sunda, Philippines and Papuasia.

Airy-Shaw, H. K. (1978). Notes on Malesian and other Asiatic Euphorbiaceae, CCIV. *Homonoia* Lour. (s. str.) in New Guinea. Kew Bull. 32: 414-415. En. — Range extension (to Laloki River E. of Port Moresby, markedly disjunct from rest of range); first New Guinea record.

Welzen, P. C. van (1998). Revisions and phylogenies of Malesian Euphorbiaceae subtribe Lasiococcinae (*Homonoia, Lasiococca, Spathiostemon*) and *Clonostylis, Ricinus* and *Wetria*. Blumea 43: 131-164. (*Lasiococca* with Nguyen Nghia Thin & Vu Hoai Duc.) En. — General introduction, with history of studies; a note of caution on the use as a character of monoecy vs. dioecy given the imperfect state of much material; phylogeny with character table and cladogram; revision of Malesian *Homonoia* (pp. 136-141; 1 species) with description, synonymy, types, literature citations, vernacular names, indication of distribution, ecology and habitat, and notes on anatomy, uses and properties (where known), and systematics; identification list at end. [*H. riparia* is in Sumatera and Java used in the same way as willow in more temperate regions, against riverbank and slope erosion.]

Homonoia Lour., Fl. Cochinch.: 636 (1790).
S. China, Trop. Asia. 36 40 41 42.
Lumanaja Blanco, Fl. Filip.: 821 (1837).
Haematospermum Wall., Numer. List: 7953 (1847).

Homonoia intermedia Haines, Bot. Bihar Orissa 2: 111 (1921).
India. 40 IND.

Homonoia retusa (Graham ex Wight) Müll.Arg., Linnaea 34: 200 (1865). – FIGURE, p. 1009.
C. & SC. India. 40 IND 41 THA? VIE? Nanophan.
Adelia cuneata Wall., Numer. List: 7954 (1847), nom. inval.
* *Adelia retusa* Graham ex Wight, Icon. Pl. Ind. Orient. 5: t. 1869 (1852).

Homonoia riparia Lour., Fl. Cochinch.: 637 (1790).
Trop. & Subtrop. Asia. 36 CHC CHH CHS 38 TAI 40 ASS EHM IND 41 AND BMA CBD LAO NCB THA VIE 42 BOR JAW LSI MLY MOL NWG PHI SUL SUM. Nanophan. or phan.
Croton salicifolius Geiseler, Croton. Monogr.: 6 (1807).
Adelia neriifolia B.Heyne ex Roth, Nov. Pl. Sp.: 375 (1821).
Ricinus salicinus Hassk., Tijdschr. Ned.-Indië 10: 142 (1843). *Spathiostemon salicinus* (Hassk.) Hassk., Hort. Bogor. Desc.: 41 (1858). *Haematospermum salicinum* (Hassk.) Baill., Étude Euphorb.: 293 (1858).
Spathiostemon salicinus var. *angustifolius* Miq., Fl. Ned. Ind., Eerste Bijv.: 452 (1861).

Synonyms:
Homonoia comberi (Haines) Merr. === **Lasiococca comberi** Haines
Homonoia javensis (Blume) Müll.Arg. === **Spathiostemon javensis** Blume

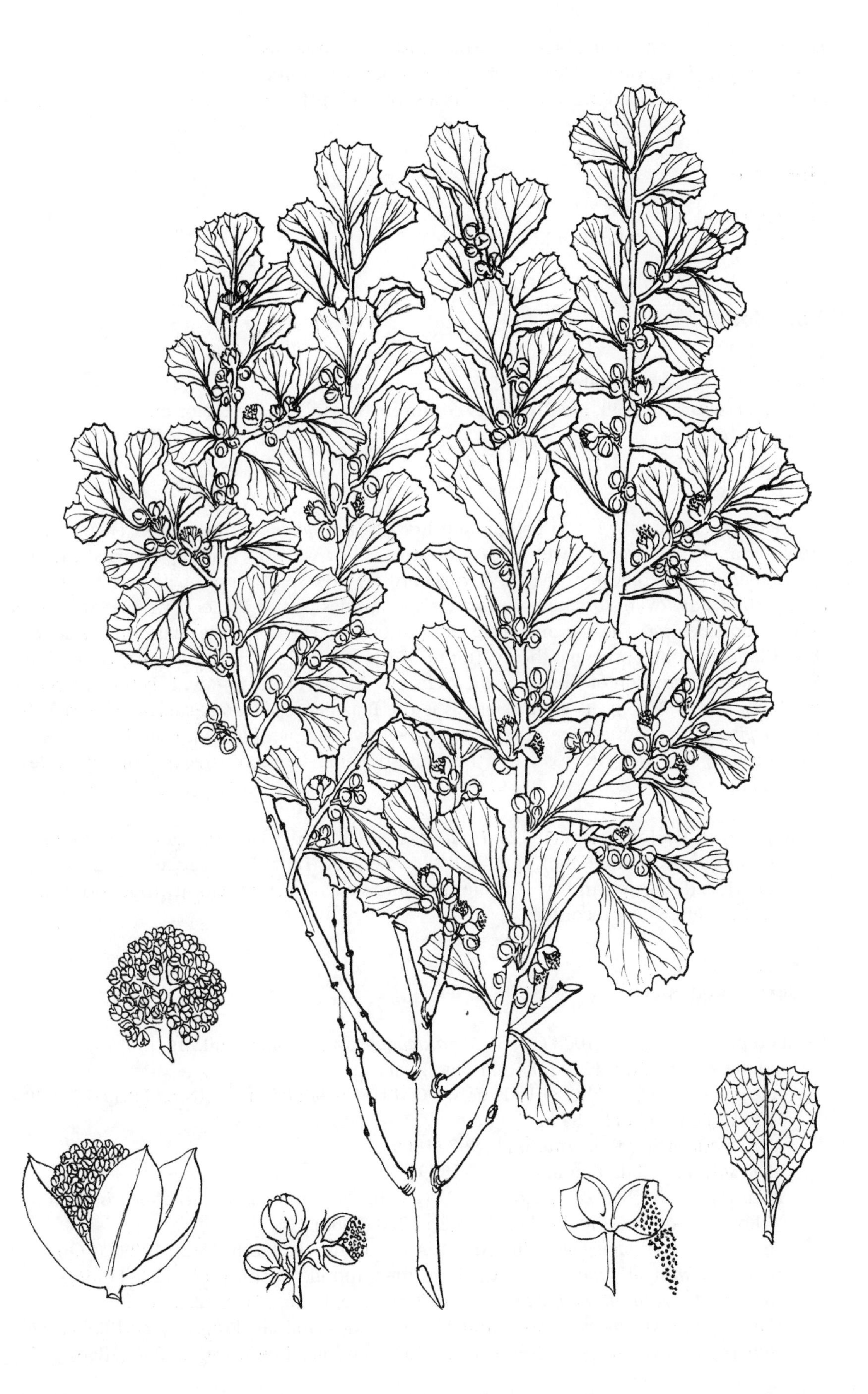

Homonoia retusa (Graham ex Wight) Müll. Arg. (as *Adelia ritusa* Graham ex Wight)
Artist: [Govindoo]
Wight, Ic. Pl. Ind. Irient. 5: pl. 1869 (1852)
KEW ILLUSTRATIONS COLLECTION

Homonoia javensis var. *ciliata* Merr. === **Spathiostemon javensis** Blume
Homonoia pseudoverticillata (Merr.) Merr. === **Lasiococca comberi** Haines
Homonoia symphyllifolia Kurz === **Lasiococca symphyllifolia** (Kurz) Hook.f.

Hotnima

Synonyms:
Hotnima A.Chev. === **Manihot** Mill.

Humblotia

Synonyms:
Humblotia Baill. === **Drypetes** Vahl
Humblotia comorensis Baill. === **Drypetes comorensis** (Baill.) Pax & K.Hoffm.

Hura

2 species, Americas (Mexico to Bolivia and Brazil); *H. crepitans* introduced elsewhere. Fast-growing laticiferous trees of wet or dry forest to 35 m with open rangy crowns, the two species being geographical vicariants. *H. crepitans* has large spiny trunks with serrate, spirally arranged, broadly ovate-cordate leaves and large capsules with c. 15 cocci which explosively dehisce on ripening. The sap and seeds are poisonous. The cone-like male inflorescences, their filaments united, presage the cyathium of the Euphorbieae. [These male 'cones' have determined the position of this genus in Webster's system of the family; it is the last before the Euphorbieae! Among its close relatives is the Brazilian *Ophthalmoblapton*. The capsules of *H. crepitans* were formerly emptied of seeds before ripening, filled with sand, and used as paperweights; this was the basis for the English name 'sandbox tree'.] (Euphorbioideae (except Euphorbieae))

Pax, F. (with K. Hoffmann) (1912). *Hura*. In A. Engler (ed.), Das Pflanzenreich, IV 147 V (Euphorbiaceae-Hippomaneae): 271-274. Berlin. (Heft 52.) La/Ge. — 2 species, Americas (S Mexico and West Indies to south-central S America); introductions of *H. crepitans* elsewhere recorded.

Hura L., Sp. Pl.: 1008 (1753).
Mexico, Trop. America. 79 80 81 82 83 84.

Hura crepitans L., Sp. Pl.: 1008 (1753). *Hura crepitans* var. *genuina* Müll.Arg. in A.P.de Candolle, Prodr. 15(2): 1229 (1866), nom. inval.
Trop. America. 80 COS NIC 81 BAH CUB DOM HAI JAM LEE TRT WIN 82 FRG GUY SUR VEN 83 BOL CLM PER 84 BZN. Phan.
Hura brasiliensis Willd., Enum. Pl.: 997 (1809).
Hura strepens Willd., Enum. Pl.: 997 (1809).
Hura senegalensis Baill., Adansonia 1: 77 (1860). *Hura crepitans* var. *senegalensis* (Baill.) Boiss. in A.P.de Candolle, Prodr. 15(2): 1230 (1866).
Hura crepitans f. *oblongifolia* Müll.Arg. in A.P.de Candolle, Prodr. 15(2): 1229 (1866).
Hura crepitans f. *orbicularis* Müll.Arg. in A.P.de Candolle, Prodr. 15(2): 1229 (1866).
Hura crepitans f. *ovata* Müll.Arg. in A.P.de Candolle, Prodr. 15(2): 1229 (1866).
Hura crepitans var. *membranacea* Müll.Arg. in A.P.de Candolle, Prodr. 15(2): 1229 (1866).
Hura crepitans var. *strepens* Müll.Arg. in A.P.de Candolle, Prodr. 15(2): 1230 (1866).

Hura polyandra Baill., Étude Euphorb.: 543 (1858).
W. Mexico, C. America. 79 MXN MXS 80 GUA HON NIC. Phan.

Synonyms:
Hura brasiliensis Willd. === **Hura crepitans** L.
Hura crepitans var. *genuina* Müll.Arg. === **Hura crepitans** L.
Hura crepitans var. *membranacea* Müll.Arg. === **Hura crepitans** L.
Hura crepitans f. *oblongifolia* Müll.Arg. === **Hura crepitans** L.
Hura crepitans f. *orbicularis* Müll.Arg. === **Hura crepitans** L.
Hura crepitans f. *ovata* Müll.Arg. === **Hura crepitans** L.
Hura crepitans var. *senegalensis* Müll.Arg. === **Hura crepitans** L.
Hura crepitans var. *senegalensis* (Baill.) Boiss. === **Hura crepitans** L.
Hura crepitans var. *strepens* Müll.Arg. === **Hura crepitans** L.
Hura senegalensis Baill. === **Hura crepitans** L.
Hura strepens Willd. === **Hura crepitans** L.

Hyaenanche

1 species, S. Africa; shrubs or small trees. The verticillate leaves recall *Mischodon* but taxonomically the genus is very isolated within the Caletieae (Webster, Synopsis, 1994). An unusual toxin, hyaenanchin, is produced in the pericarp of the fruit. (Oldfieldioideae)

Pax, F. & K. Hoffmann (1922). *Toxicodendron*. In A. Engler (ed.), Das Pflanzenreich, IV 147 XV (Euphorbiaceae-Phyllanthoideae-Phyllantheae): 284-287. Berlin. (Heft 81.) La/Ge. — 1 species; Africa (S Africa). [Now known under *Hyenanche*.]

Hyaenanche Lamb. & Vahl, Descr. Cinchona: 52 (1797).
S. Africa. 27. Nanophan.
Toxicodendrum Thunb., Kongl. Vetensk. Acad. Nya Handl. 17: 188 (1796).

Hyaenanche globosa (Gaertn.) Lamb. & Vahl, Descr. Cinchona: 52 (1797).
Cape Prov. 27 CPP. Nanophan.
* *Jatropha globosa* Gaertn., Fruct. Sem. Pl. 2: 122 (1790). *Toxicodendrum globosum* (Gaertn.) Pax & K.Hoffm. in H.G.A.Engler, Pflanzenr., IV, 147, XV: 285 (1922).
Toxicodendrum capense Thunb., Kongl. Vetensk. Acad. Nya Handl. 17: 188 (1796).
Hyaenanche capensis (Thunb.) Pers., Syn. Pl. 2: 627 (1807).

Synonyms:
Hyaenanche capensis (Thunb.) Pers. === **Hyaenanche globosa** (Gaertn.) Lamb. & Vahl

Hyeronima

An orthographic variant of *Hieronima*.

Hylandia

1 species, Australia (NE. Queensland); forest trees to 30 m. A part of the complex of Asiatic Crotonoideae (formerly all included in Jatropheae) including *Aleurites* and its relatives, *Ostodes, Dimorphocalyx, Oligoceras, Deutzianthus* and *Tapoides*. Webster (1994) relates this genus most closely to *Ostodes* and *Baloghia*, retaining them in Jatropheae but removing many of the other genera to different tribes. Airy-Shaw (1974) believed all these genera were the 'fragmented and largely isolated relics of a formerly much more extensive group'. (Crotonoideae)

Airy-Shaw, H. K. (1974). Notes on Malesian and other Asiatic Euphorbiaceae, CLXXXVI. A new Ostodoid genus from Queensland. Kew Bull. 29: 329-331. En. — Protologue of *Hylandia* and description of *H. dockrillii*, related to *Ostodes, Dimorphocalyx* and *Loerzingia* (now *Deutzianthus*), especially the last.

Hylandia dockrillii Airy Shaw
Artist: Ann Davies
Kew Bull. 35: 644, fig. 4 (1980)
KEW ILLUSTRATIONS COLLECTION

Hylandia Airy Shaw, Kew Bull. 29: 329 (1974).
Queensland. 50.

Hylandia dockrillii Airy Shaw, Kew Bull. 29: 329 (1974). – FIGURE, p. 1012.
Queensland (Cook). 50 QLD. Phan.

Hylococcus

Synonyms:
Hylococcus R.Br. ex Benth. === **Petalostigma** F.Muell.

Hymenocardia

6 species, Africa (5) and SE. Asia and W. Malesia (1); shrubs or small trees of woodland and forest with catkin-like male inflorescences and unmistakable 2-winged, samaroid fruits (illustrated in Pax & Hoffmann, 1922: 73). Some African species are quite widely distributed. Certain authorities have referred this genus to a distinct family, Hymenocardiaceae; among them have been Airy-Shaw and A.D.J. Meeuse. The pollen was long thought isolated in Euphorbiaceae, and arguments were made for a link with Urticales (and particularly Ulmaceae, on account of the fruits); however, Levin & Simpson (1994), from a palynological and trichological perspective, have suggested that its nearest relative is *Didymocistus* and, through it, provides a link with Antidesmeae (thus resembling the position of Pax & Hoffmann, who included it in their Antidesminae). A more extensive study, mainly from the palynological point of view, is that of Lobreau-Callen & Suárez Cervera (1994) who opt for a position amongst Oldfieldioideae. (Phyllanthoideae)

Pax, F. & K. Hoffmann (1922). *Hymenocardia.* In A. Engler (ed.), Das Pflanzenreich, IV 147 XV (Euphorbiaceae-Phyllanthoideae-Phyllantheae): 72-78. Berlin. (Heft 81.) La/Ge. — 8 species, all African except for 1 in SE Asia and W Malesia.

Dechamps, R., M. Mosango & E. Robbrecht (1985). Études systématiques sur les Hymenocardiaceae d'Afrique: la morphologie du pollen et l'anatomie du bois. Bull. Jard. Bot. Natl. Belg. 55: 473-485, illus. Fr. — Comprises two contributions, the first on pollen (all five African species in genus surveyed), the second on wood (three species examined). With respect to pollen, the five species were quite consistent.

Léonard, J. & M. Mosango (1985). *Hymenocardia.* Fl. Afrique Centrale: Hymenocardiaceae. Brussels. Fr. — Flora treatment. [*Hymenocardia* here treated as a separate family, following Airy-Shaw.]

Levin, G. A. & M. G. Simpson (1994). Phylogenetic relationships of *Didymocistus* and *Hymenocardia* (Euphorbiaceae). Ann. Missouri Bot. Gard. 81: 239-244, illus. En. — Palynological study; *Hymenocardia,* while sometimes made the type of a separate family associated with Urticales, shows in its pollen euphorbiaceous features. It is closest to the South American *Didymocistus*; both are best included in a tribe Hymenocardieae (cf. Webster 1994 under General). No taxonomic synopsis is presented.

Lobreau-Callen, D. & M. Suárez Cervera (1994). Pollen ultrastructure of *Hymenocardia* Wallich ex Lindley and comparison with other Euphorbiaceae. Rev. Palaeobot. Palynol. 81: 257-278, illus. En. — Pollen analysis, with direction towards a resolution of the much-discussed status of *Hymenocardia* in relation to Euphorbiaceae in general. [It was concluded that, while some exine features were similar to those in Urticales, the genus was 'more or less close' to Oldfieldioideae, perhaps near Hyenachneae and Picrodendreae (nos. 11 and 12 in Webster's system); the apparent urticalean similarities represented parallel developments.]

Hymenocardia Wall. ex Lindl., Intr. Nat. Syst. Bot., ed. 2: 441 (1836).
Trop. & S. Africa, Trop. Asia. 22 23 24 25 26 27 40 41 42.
Samaropyxis Miq., Fl. Ned. Ind., Eerste Bijv.: 464 (1861).

Hymenocardia acida Tul., Ann. Sci. Nat., Bot., III, 15: 256 (1851).
Trop. Africa. 22 BEN BKN GAM GHA GNB GUI IVO MLI NGA SEN SIE TOG 23 CAF CMN CON GAB RWA ZAI 24 CHA ETH SUD 25 KEN TAN UGA 26 ANG MLW MOZ ZAM ZIM. Nanophan. or phan.

var. **acida**
Trop. Africa. 22 BEN BKN GAM GHA GNB GUI IVO MLI NGA SEN SIE TOG 23 CAF CMN CON GAB ZAI 24 CHA ETH SUD 25 KEN TAN UGA 26 ANG MLW MOZ ZAM ZIM. Nanophan. or phan.
Hymenocardia mollis var. *glabra* Pax, Bot. Jahrb. Syst. 15: 528 (1893).
Hymenocardia granulata Beille, Bull. Soc. Bot. France 55(8): 62 (1908).
Hymenocardia lanceolata Beille, Bull. Soc. Bot. France 55(8): 63 (1908).
Hymenocardia obovata A.Chev. & Beille ex Beille, Bull. Soc. Bot. France 55(8): 62 (1908).

var. **mollis** (Pax) Radcl.-Sm., Kew Bull. 28: 324 (1973).
Zaire to Mozambique. 23 RWA ZAI 25 TAN 26 MLW MOZ ZAM ZIM. Nanophan. or phan.
* *Hymenocardia mollis* Pax, Bot. Jahrb. Syst. 15: 528 (1893). *Hymenocardia mollis* var. *genuina* Pax & K.Hoffm. in H.G.A.Engler, Pflanzenr., IV, 147, XV: 75 (1922), nom. inval.
Hymenocardia lasiophylla Pax, Bot. Jahrb. Syst. 19: 79 (1894). *Hymenocardia mollis* var. *lasiophylla* (Pax) Pax in H.G.A.Engler, Pflanzenr., IV, 147, XV: 76 (1922).

Hymenocardia heudelotii Planch. ex Müll.Arg., Flora 47: 518 (1864).
W. Trop. Africa. 22 GNB GNB GUI IVO LBR MLI NGA SEN SIE TOG. Nanophan. or phan.

var. **chevalieri** (Beille) J.Léonard, Bull. Jard. Bot. État 33: 405 (1963).
W. Trop. Africa. 22 GHA GNB GUI IVO MLI SIE. Nanophan. or phan.
* *Hymenocardia chevalieri* Beille, Bull. Soc. Bot. France 55(8): 61 (1908).

var. **heudelotii**
W. Trop. Africa. 22 GNB GUI IVO LBR MLI NGA SEN SIE TOG. Nanophan. or phan.
Hymenocardia guineensis Beille, Bull. Soc. Bot. France 55(8): 63 (1908).

Hymenocardia lyrata Tul., Ann. Sci. Nat., Bot., III, 15: 256 (1851).
W. Trop. Africa. 22 GHA GNB GUI IVO LBR SEN SIE. Nanophan.
Hymenocardia beillei A.Chev. ex Hutch. & Dalziel, Fl. W. Trop. Afr. 1: 286 (1928).

Hymenocardia punctata Wall. ex Lindl., Intr. Nat. Syst. Bot., ed. 2: 441 (1836).
Indo-China to Sumatera. 41 BMA CBD LAO THA VIE 42 MLY SUM. Nanophan. or phan.
Hymenocardia wallichii Tul., Ann. Sci. Nat., Bot., III, 15: 256 (1851).
Samaropyxis elliptica Miq., Fl. Ned. Ind., Eerste Bijv.: 465 (1861).
Hymenocardia laotica Gagnep., Bull. Soc. Bot. France 70: 436 (1923).
Hymenocardia wallichii var. *dasycarpa* Gagnep. in ?, : 544 (1927).

Hymenocardia ripicola J.Léonard, Bull. Jard. Bot. État 33: 406 (1963).
Cameroon, Congo, Zaire. 23 CMN CON ZAI. Nanophan. or phan.

Hymenocardia ulmoides Oliv., Hooker's Icon. Pl. 12: t. 1131 (1873). – FIGURE, p. 1015.
Trop. & S. Africa. 23 CAF CMN CON GAB ZAI 24 SUD 25 TAN 26 ANG MAL MOZ ZAM ZIM 27 CPP NAT TVL. Nanophan. or phan.
Hymenocardia poggei Pax, Bot. Jahrb. Syst. 15: 529 (1893).
Hymenocardia ulmoides var. *capensis* Pax, Bot. Jahrb. Syst. 28: 22 (1899). *Hymenocardia capensis* (Pax) Hutch., Bull. Misc. Inform. Kew 1920: 334 (1920).
Hymenocardia similis Pax & K.Hoffm. in H.G.A.Engler, Pflanzenr., IV, 147, XV: 74 (1922).
Hymenocardia ulmoides var. *longistyla* De Wild., Pl. Bequaert. 3: 447 (1926).

Synonyms:
Hymenocardia beillei A.Chev. ex Hutch. & Dalziel === **Hymenocardia lyrata** Tul.
Hymenocardia capensis (Pax) Hutch. === **Hymenocardia ulmoides** Oliv.

Hymenocardia ulmoides Oliv. (as *H. ulmoides*, Hymenocardiaceae)
Artist: P. Halliday
Fl. Trop. East Africa, Euphorbiaceae 2: 578, fig. 107 (1988)
KEW ILLUSTRATIONS COLLECTION

Hymenocardia chevalieri Beille === **Hymenocardia heudelotii** var. **chevalieri** (Beille) J.Léonard
Hymenocardia grandis Hutch. === **Holoptelea grandis** (Hutch.) Mildbr. (Ulmaceae)
Hymenocardia granulata Beille === **Hymenocardia acida** Tul. var. **acida**
Hymenocardia guineensis Beille === **Hymenocardia heudelotii** Planch. ex Müll.Arg. var. **heudelotii**
Hymenocardia intermedia Dinkl. ex Mildbr. === ?
Hymenocardia lanceolata Beille === **Hymenocardia acida** Tul. var. **acida**
Hymenocardia laotica Gagnep. === **Hymenocardia punctata** Wall. ex Lindl.
Hymenocardia lasiophylla Pax === **Hymenocardia acida** var. **mollis** (Pax) Radcl.-Sm.
Hymenocardia mollis Pax === **Hymenocardia acida** var. **mollis** (Pax) Radcl.-Sm.
Hymenocardia mollis var. *genuina* Pax & K.Hoffm. === **Hymenocardia acida** var. **mollis** (Pax) Radcl.-Sm.
Hymenocardia mollis var. *glabra* Pax === **Hymenocardia acida** Tul. var. **acida**
Hymenocardia mollis var. *lasiophylla* (Pax) Pax === **Hymenocardia acida** var. **mollis** (Pax) Radcl.-Sm.
Hymenocardia obovata A.Chev. & Beille ex Beille === **Hymenocardia acida** Tul. var. **acida**
Hymenocardia plicata (Müll.Arg.) Kurz === **Mallotus plicatus** (Müll.Arg.) Airy Shaw
Hymenocardia poggei Pax === **Hymenocardia ulmoides** Oliv.
Hymenocardia similis Pax & K.Hoffm. === **Hymenocardia ulmoides** Oliv.
Hymenocardia ulmoides var. *capensis* Pax === **Hymenocardia ulmoides** Oliv.
Hymenocardia ulmoides var. *longistyla* De Wild. === **Hymenocardia ulmoides** Oliv.
Hymenocardia wallichii Tul. === **Hymenocardia punctata** Wall. ex Lindl.
Hymenocardia wallichii var. *dasycarpa* Gagnep. === **Hymenocardia punctata** Wall. ex Lindl.

Hypocoton

Synonyms:
Hypocoton Urb. === **Bonania** A.Rich.
Hypocoton domingensis Urb. === **Bonania domingensis** (Urb.) Urb.

Ibina

A Noronha name; now in *Sauropus*.

Jablonskia

1 species, S. tropical America (Guianas and Amazon Basin SW. to Peru). A segregate of *Securinega* s.l., these virgate shrubs or small trees feature distichously arranged leaves on long slender branches. It would appear they are largely associated with watercourses. Webster (1994, Synopsis) suggested an affinity with the Old World *Aporusa* and *Ashtonia* but *Jablonskia congesta* differs in details of wood anatomy. (Phyllanthoideae)

Webster, G. (1984). *Jablonskia*, a new genus of Euphorbiaceae from South America. Syst. Bot. 9: 229-235, illus., map. En. — Protologue of genus (from Amazonian S America) and segregation of the single species from *Securinega* s.l. with full documentation and discussion of possible affinities. [See also following article by A. M. W. Mennega on wood anatomy: Syst. Bot. 9: 236-239. 1984.]

Jablonskia G.L.Webster, Syst. Bot. 9: 232 (1984).
S. Trop. America. 82 83 84. Nanophan. or phan.

Jablonskia congesta (Benth. ex Müll.Arg.) G.L.Webster, Syst. Bot. 9: 232 (1984).
Guyana, Surinam, Venezuela (Amazonas), Brazil (Amazonas), Peru. 82 FRG? GUY SUR VEN 83 PER 84 BZN. Nanophan. or phan.

* *Phyllanthus congestus* Benth. ex Müll.Arg., Linnaea 32: 25 (1863). *Securinega congesta* (Benth. ex Müll.Arg.) Müll.Arg. in C.F.P.von Martius, Fl. Bras. 11(2): 76 (1873). *Acidoton congestus* (Benth. ex Müll.Arg.) Kuntze, Revis. Gen. Pl. 2: 592 (1891).

Janipha

Synonyms:

Janipha Kunth === **Manihot** Mill.
Janipha foetida Kunth === **Manihot foetida** (Kunth) Pohl
Janipha loeflingii var. *multifida* Graham === **Manihot grahamii** Hook.
Janipha manihot var. *angustiloba* Torr. === **Manihot angustiloba** (Torr.) Müll.Arg.

Jatropa

An orthographic variant of *Jatropha*.

Jatropha

186 species, tropics and other warmer parts of the world, mainly in the Americas but also represented in Africa and Asia with one endemic Malagasy species. Shrubs (some caudiciform) or small trees, not unlike the related *Cnidoscolous* (although their apparent closeness has been questioned; cf. Miller & Webster 1962) but glabrous or, if hairy, the hairs not stinging. The palmately veined leaves are often more or less lobed. The seeds (notably *J. curcas*) yield a purgative oil, and in addition some species yield a rubber-forming latex known as 'chilte' (of interest during World War II; see McVaugh, 1943, 1945). Several taxa are moreover valued ornamentals, including some 'semi-succulents' (cf. Ingram 1957; Rodriguez 1993). Infrageneric revisions have been published by McVaugh (1945) and Dehgan & Webster (1976); the latter study was extended by Dehgan & Schutzman (1994) in preparation for a full treatment of tropical American species. In Africa there has been a concise revision of sect. *Robbrechiana*, a group mainly occurring from Kenya to Somalia (Gilbert & Thulin 1991); other treatments are related to floras. Studies of living plants in addition to herbarium material are essential towards a better understanding of this genus. A number of species are cultivated (cf. National Botanic Gardens, 1959; Rodriguez, 1992). (Crotonoideae)

Pax, F. (1910). *Jatropha*. In A. Engler (ed.), Das Pflanzenreich, IV 147 [I] (Euphorbiaceae-Jatropheae): 21-113. Berlin. (Heft 42.) La/Ge. — 156 species, in 3 subgenera and 14 sections (the last subgenus, with 6 sections (pp. 86-110) now referred to *Cnidoscolus*). See also Addenda (pp. 133-134) with 2 additional species and, in IV 147 II, Additamentum I (1910) for one further species.

Pax, F. (with K. Hoffmann) (1914). *Jatropha*. In A. Engler (ed.), Das Pflanzenreich, IV 147 VII [Euphorbiaceae-Additamentum V]: 397-401. Berlin. (Heft 63.) La/Ge. — Several additional species, some new.

Pax, F. & K. Hoffmann (1931). *Jatropha*. In A. Engler (ed.), Die natürlichen Pflanzenfamilien, 2. Aufl., 19c: 160-164. Leipzig. Ge. — Synopsis with description of genus; c. 150 species in 9 sections with representative species listed. [*Cnidoscolus* is here segregated.]

Croizat, L. (1943). New and critical Euphorbiaceae of Brazil. Tropical Woods 76: 11-14. En. — Includes discussion of mutual relationships among *Jatropha*, *Cnidoscolus* and *Manihot* (with a belief that they are interrelated and that *Cnidoscolus* merits only subgeneric rank within *Jatropha*).

McVaugh, R. (1943). The Mexican species of *Jatropha* (with special reference to possible sources of 'chilte' rubber). 23 pp., 14 text-fig. Washington: Rubber Development Corporation. (Mimeographed.) En. — Provisional revision and identification handbook. [Not seen; reference from Lundell, 1945 (see *Cnidoscolus*).]

McVaugh, R. (1945). The genus *Jatropha* in America: principal intrageneric groups. Bull. Torrey Bot. Club 72: 271-294. En. — Historical survey; presentation of a revised system with 4 sections with descriptions and keys to species in sects. *Macranthae*, *Adenorhopium* and *Curcas* (the latter having 2 subsections). Some species are described and critical notes given with reference to Mexican taxa, and an index to American species is also furnished (pp. 292-294). [Study undertaken as part of research into alternative sources of natural rubber.]

Wilbur, R. L. (1954). A synopsis of *Jatropha*, subsection *Eucurcas*, with the description of two new species from Mexico. J. Elisha Mitchell Soc. 70: 92-101. En. — Descriptive revision (5 species) with key, types, exsiccatae and commentary. [A relatively unspecialized taxon, now known to have 9 species; corresponds to sect. *Curcas* in the treatment of Dehgan and Webster (1979).]

National Botanic Gardens, Lucknow (1959). *Jatropha*. 5 pp., illus. (Bulletin 32). Lucknow. En. — An 'extension leaflet' covering six cultivated ornamental species, with brief descriptions of each along with soil and cultivation notes; no references.

Miller, K. I. & G. L. Webster (1962). Systematic position of *Cnidoscolus* and *Jatropha*. Brittonia 14: 174-180. En. — Study of additional evidence (anatomy, karology, palynology); support for separation of genera (which furthermore appear not to be that closely related). No key is furnished. [A range of species was chosen from each genus.]

Johnston, M. C. & B. H. Warnock (1963). The species of *Cnidoscolus* and *Jatropha* (Euphorbiaceae) in far western Texas. Southwestern Nat. 8: 121-126. En. — Includes treatment of two species of *Jatropha* with distribution map and key; no exsiccatae.

Jablonski, E. (1967). *Jatropha*. Euphorbiaceae, Guayana Highland (Mem. New York Bot. Gard. 17(1)): 155-156. New York. En. — 2 species, both of wide distribution.

Radcliffe-Smith, A. (1973). Notes on African Euphorbiaceae, II. *Jatropha*. Kew Bull. 28: 283-286. En. — Precursory for *Flora of Tropical East Africa*. Further notes in idem, IV (ibid): 521-523.

Dehgan, B. (1976). Experimental and evolutionary studies of relationships in the genus *Jatropha* L. (Euphorbiaceae). 436 pp. Davis, Calif. (Unpubl. Ph.D. dissertation, Univ. of California, Davis.) En. — Biosystematic treatment; partly publ. in Syst. Bot. 9: 467-478 (1984) as well as Dehgan & Webster (1979).

• Dehgan, B. & G. L. Webster (1979). Morphology and infrageneric relationships of the genus *Jatropha* (Euphorbiaceae). 73 pp., 33 pls. (Univ. Calif. Publ. Bot. 74). Berkeley. En. — Supraspecific revision, an outgrowth of hybridization and other studies; 2 subgenera and 10 sections recognized with keys, descriptions and indication of type species. Reviews of various character classes are included along with (pp. 49-66) a species checklist featuring references, synonymy, citations and type localities. A bibliography is given on pp. 67-73, but there is no index. A phylogram of putative relationships appears on p. 6 following a presentation of taxonomic history. [The authors consider the genus to be 'one of [the] least understood of all .. genera in the Euphorbiaceae'; an understanding of living plants was thought essential.]

Hemming, C. F. & A. Radcliffe-Smith (1987). A revision of the Somali species of *Jatropha* (Euphorbiaceae). Kew Bull. 42: 103-122. En. — Synoptic treatment, with key.

• Gilbert, M. G. & M. Thulin (1991). Synopsis of *Jatropha* sect. *Robecchiana* s.lat. (Euphorbiaceae). Nordic J. Bot. 11: 413-419, illus., En. — Descriptive treatment (7 species) with key, synonymy, references, types, descriptions of novelties with exsiccatae, notes on distribution, habitat and taxonomy, figures and map. [The group is largely restricted to northeastern tropical Africa from Kenya to Somalia.]

Radcliffe-Smith, A. (1991). Notes on African Euphorbiaceae, XXV. *Jatropha* (vi). Kew Bull. 46: 141-157. En. — Substantial contribution, with 6 new species; precursory to *Flora Zambesiaca* treatment.

Thulin, M. (1991). Four new species of *Jatropha* (Euphorbiaceae) from Somalia. Nordic J. Bot. 11: 527-533, illus. En. — Novelties with figures and notes; all in sect. *Spinosa*.

Rodriguez, A. (1992). An introduction to *Jatropha*. Plantsman 14: 48-53, illus. En. — General commentary; propagation and culture; notes on 6 species with illustration of *J. podagrica*; list of suppliers in Britain. [The listed *J. urens* should be in *Cnidoscolus*.]

- Dehgan, B. & B. Schutzman (1994). Contributions toward a monograph of neotropical *Jatropha*: phenetic and phylogenetic analyses. Ann. Missouri Bot. Gard. 81: 349-367. En. — A numerical taxonomic and phyogenetic analysis covering ultimately 77 taxa, with character tables and values assigned, matrix and results of principal components and cladistic analyses; presentation of a Nelsonian (strict consensus) tree (p. 361); comparison of the two forms of analysis and ramifications therefrom including speed of diversification, preferred habitats and probable origin.

Kamilya, P. & N. Paria (1994). Seedling morphology of some Indian species of *Jatropha* and its implications in taxonomy. Acta Bot. Indica 22: 251-256, illus. En. — Review and examination of seedling morphology in 5 species, with descriptions and voucher specimens; includes key. [Oriented towards practical identification; no thoughts on systemtatic and evolutionary questions.]

Sreenivasa Rao, E. & V. S. Raju (1995). The genus *Jatropha* L. in Andhra Pradesh, India. J. Econ. Taxon. Bot. 18: 585-589. En. — Enumeration of 7 species (3 native) with key (exclusively on vegetative features); includes synonymy, localities with exsiccatae and indication of distribution. [*JJ. curcas*, *integerrima*, *multifida* and *podagrica* are cultivated, the first commonly for its oil, the others as ornamentals.]

Heller, J. (1996). Physic nut: *Jatropha curcas* L. 66 pp., illus. Rome: International Plant Genetic Resources Institute [IPGRI]. (Promoting the conservation and use of underutilized and neglected crops, 1.) En. — Economic.

Jatropha L., Sp. Pl.: 1006 (1753).
Trop. & Subtrop. America, Trop. & S. Africa, Madagascar, Comoros, Arabian Pen., India, Sri Lanka. 22 23 24 25 26 27 29 35 40 76 77 79 80 81 82 83 84 85.
Ricinoides Mill., Gard. Dict. Abr. ed. 4(1754).
Curcas Adans., Fam. Pl. 2: 356 (1763).
Bromfeldia Neck., Elem. Bot. 2: 347 (1790).
Castiglionia Ruiz & Pav., Fl. Peruv. Prodr.: 139 (1794).
Mozinna Ortega, Nov. Pl. Descr. Dec.: 104 (1798).
Loureira Cav., Icon. 5: 17 (1799).
Mesandrinia Raf., Neogenyton: 3 (1825).
Adenoropium Pohl, Pl. Bras. Icon. Descr. 1: 12 (1826).
Adenorhopium Rchb., Consp. Regn. Veg.: 195 (1828).
Mazinna Spach, Hist. Nat. Vég. 2: 487 (1834).
Zimapania Engl. & Pax in H.G.A.Engler & K.A.E.Prantl, Nat. Pflanzenfam. 3(5): 119 (1892).
Collenucia Chiov., Fl. Somala 1: 177 (1929).

Jatropha aceroides (Pax & K.Hoffm.) Hutch. in D.Oliver, Fl. Trop. Afr. 6(1): 789 (1912).
Sudan. 24 SUD.
* *Jatropha lobata* subsp. *aceroides* Pax & K.Hoffm. in H.G.A.Engler, Pflanzenr., IV, 147, III: 34 (1910).

Jatropha aculeatissima Colla, Herb. Pedem. 5: 112 (1836).
Brazil (?). 84 +. – Perhaps a species of *Cnidoscolus*.

Jatropha aethiopica Müll.Arg., Flora 47: 485 (1864).
Ethiopia. 24 ETH. Cham.
Jatropha sabdariffa Schweinf., Beitr. Fl. Aethiop.: 37 (1867).

Jatropha afrotuberosa Radcl.-Sm. & Govaerts, Kew Bull. 52: 188 (1997).
Sudan, Uganda. 24 SUD 25 UGA. Tuber cham.
* *Jatropha tuberosa* Pax, Bot. Jahrb. Syst. 19: 111 (1894), nom. illeg.

Jatropha alamanii Müll.Arg., Linnaea 34: 207 (1865).
Mexico (Oaxaca). 79 MXS. Nanophan.

Jatropha andrieuxii Müll.Arg., Linnaea 34: 208 (1865).
Mexico (Puebla, Oaxaca). 79 MXC MXS. Nanophan.

Jatropha angustifolia Griseb., Nachr. Königl. Ges. Wiss. Georg-Augusts-Univ. 1: 171 (1865).
Jatropha angustifolia var. *genuina* Müll.Arg. in A.P.de Candolle, Prodr. 15(2): 1093 (1866), nom. inval.
W. Cuba (incl. I. de la Juventud). 81 CUB. Nanophan.
Jatropha glauca Griseb., Nachr. Königl. Ges. Wiss. Georg-Augusts-Univ. 1: 170 (1865), nom. illeg. *Jatropha angustifolia* var. *spathulata* Müll.Arg. in A.P.de Candolle, Prodr. 15(2): 1093 (1866).

Jatropha aspleniifolia Pax, Bot. Jahrb. Syst. 19: 108 (1894).
Somalia (Sanaag). 24 SOM. Nanophan.

Jatropha atacorensis A.Chev., Bull. Soc. Bot. France 58(8): 206 (1911 publ. 1912).
Benin (Atacora Mts.). 22 BEN. Nanophan.

Jatropha augustii Pax & K.Hoffm. in H.G.A.Engler, Pflanzenr., IV, 147, XVI: 191 (1924).
Peru. 83 PER. Nanophan.
Jatropha ciliata Müll.Arg., Linnaea 34: 209 (1865), nom. illeg.
Jatropha longipedunculata Pax & K.Hoffm. in H.G.A.Engler, Pflanzenr., IV, 147, XVI: 191 (1924), nom. illeg. *Jatropha hoffmanniae* Croizat, J. Arnold Arbor. 24: 168 (1943). *Jatropha ciliata* var. *longipedunculata* J.F.Macbr., Publ. Field Mus. Nat. Hist., Bot. Ser. 13(3A1): 159 (1951).

Jatropha bartlettii Wilbur, J. Elisha Mitchell Sci. Soc. 70: 99 (1954).
Mexico (Jalisco). 79 MXS. Nanophan.

Jatropha baumii Pax in O.Warburg (ed.), Kunene-Sambesi Exped.: 283 (1903).
Angola, SW. Zambia, S. Malawi ? 26 ANG MLW? ZAM. Tuber geophyte.

Jatropha bornmuelleri Pax in H.G.A.Engler, Pflanzenr., IV, 147, I: 133 (1910).
Brazil (Rio Grande do sul). 84 BZS. Cham.

Jatropha botswanica Radcl.-Sm., Kew Bull. 46: 142 (1991).
Botswana. 27 BOT. Tuber geophyte.

Jatropha breviloba (Morong) Pax & K.Hoffm. in H.G.A.Engler, Pflanzenr., IV, 147, XIV: 38 (1919).
Paraguay. 85 PAR. Nanophan.
* *Jatropha gossypiifolia* var. *breviloba* Morong, Ann. New York Acad. Sci. 7: 219 (1892).
Jatropha ribifolia var. *breviloba* (Morong) Pax in H.G.A.Engler, Pflanzenr., IV, 147, III: 28 (1910).

Jatropha brockmanii Hutch., Bull. Misc. Inform. Kew 1911: 360 (1911).
Somalia. 24 SOM.

Jatropha bullockii E.J.Lott, Madroño 31: 180 (1984).
Mexico (Jalisco). 79 MXS. Nanophan.

Jatropha campestris S.Moore, J. Linn. Soc., Bot. 40: 196 (1911).
Mozambique, Zimbabwe. 26 MOZ ZIM. Tuber geophyte.

Jatropha capensis (L.f.) Sond., Linnaea 23: 118 (1850).
Cape. 27 CPP. Cham. or nanophan.
* *Croton capensis* L.f., Suppl. Pl.: 422 (1782).

Jatropha cardiophylla (Torr.) Müll.Arg. in A.P.de Candolle, Prodr. 15(2): 1079 (1866).
S. Arizona, Mexico (Sonora). 76 ARI 79 MXN. Nanophan.
* *Mozinna cardiophylla* Torr. in W.H.Emory, Rep. U.S. Mex. Bound. 2(1): 198 (1858).

Jatropha cathartica Terán & Berland., Mem. Comision Limites: 9 (1832).
Texas, Mexico (Tamaulipas, Nuevo León). 77 TEX 79 MXE. Nanophan. – Caudiciform.
Jatropha berlandieri Torr. in W.H.Emory, Rep. U.S. Mex. Bound. 2(1): 198 (1858).

Jatropha catingae Ule, Bot. Jahrb. Syst. 42: 218 (1908).
Brazil (Bahia). 84 BZE. Nanophan. or phan.

Jatropha chacoana Fern.Casas, Candollea 39: 11 (1984).
Paraguay. 85 PAR.

Jatropha chamelensis Pérez-Jim., Bol. Soc. Bot. México 42: 35 (1982).
Mexico (Jalisco). 79 MXS. Phan.

Jatropha chevalieri Beille, Bull. Soc. Bot. France 55(8): 83 (1908).
Senegal. 22 SEN. Nanophan.
Jatropha lobata var. *senegalensis* Müll.Arg. in A.P.de Candolle, Prodr. 15(2): 1086 (1866).
Jatropha glauca var. *senegalensis* (Müll.Arg.) Hiern, Cat. Afr. Pl. 1: 969 (1900).
Jatropha lobata subsp. *senegalensis* (Müll.Arg.) Pax in H.G.A.Engler, Pflanzenr., IV, 147, III: 33 (1910).

Jatropha ciliata Sessé ex Cerv., Supl. Gaz. Lit. Mexico: 4 (2 July 1794).
Mexico (Mexico State, Puebla, Oaxaca). 79 MXC MXS. Nanophan.
Jatropha olivacea Müll.Arg., Linnaea 34: 207 (1865).
Jatropha grandifrons I.M.Johnst., Contr. Gray Herb. 68: 89 (1923).

Jatropha cinerea (Ortega) Müll.Arg. in A.P.de Candolle, Prodr. 15(2): 1079 (1866).
Mexico (Baja California, Sonora, Sinaloa, Magdalena I.). 79 MXN. Nanophan. – Caudiciform.
* *Mozinna cinerea* Ortega, Nov. Pl. Descr. Dec.: 108 (1798).
Mozinna canescens Benth., Bot. Voy. Sulphur: 52 (1844). *Jatropha canescens* (Benth.) Müll.Arg. in A.P.de Candolle, Prodr. 15(2): 1079 (1866).

Jatropha clavuligera Müll.Arg., Linnaea 34: 209 (1865).
Peru, Bolivia. 83 BOL PER. Nanophan.

Jatropha collina Thulin, Nordic J. Bot. 11: 531 (1991).
Somalia (NE. Sanaag). 24 SOM. Cham.

Jatropha confusa Hutch., Bull. Misc. Inform. Kew 1911: 361 (1911).
Angola. 26 ANG.

Jatropha contrerasii J.Jiménez Ram. & M.Martínez Gordillo, Anales Inst. Biol. Univ. Nac. Auton. Mexico, Bot. 65: 22 (1994).
Mexico (Guerrero). 79 MXS. Nanophan. or phan.

Jatropha conzattii J.Jiménez Ram., Anales Inst. Biol. Univ. Nac. Auton. Mexico, Bot. 62: 83 (1991).
Mexico (Oaxaca). 79 MXS. Nanophan. or phan.

Jatropha cordata (Ortega) Müll.Arg. in A.P.de Candolle, Prodr. 15(2): 1078 (1866).
Mexico (Sonora, Chihuahua, Durango, San Luis Potosí, Sinaloa, Jalisco). 79 MXE MXN MXS. Nanophan.
* *Mozinna cordata* Ortega, Nov. Pl. Descr. Dec.: 107 (1798).

Jatropha costaricensis G.L.Webster & Poveda, Brittonia 30: 265 (1978).
Costa Rica (Guanacaste). 80 COS. Nanophan. or phan.

Jatropha crinita Müll.Arg., Linnaea 34: 207 (1865).
Tanzania (Zanzibar) ? 25 TAN. Cham.

Jatropha cuneata Wiggins & Rollins, Contr. Dudley Herb. 3: 272 (1943).
Arizona, Mexico (Sonora, Baja California). 76 ARI 79 MXN. Nanophan.

Jatropha curcas L., Sp. Pl.: 1006 (1753). *Manihot curcas* (L.) Crantz, Inst. Rei Herb. 1: 167 (1766).
Mexico, C. America, Caribbean, Colombia, Venezuela, Guyana, Bolivia, Paraguay, S. Brazil, Chile. 79 MXC MXE MXG MXS MXT 80 COS GUA NIC 81 CUB DOM HAI JAM LEE PUE WIN 82 GUY VEN 83 BOL CLM 84 BZS 85 CLN PAR. Nanophan. or phan. – Yields a widely-used oil.
Ricinus americanus Mill., Gard. Dict. ed. 8: 2 (1768).
Jatropha acerifolia Salisb., Prodr. Stirp. Chap. Allerton: 389 (1796).
Ricinus jarak Thunb., Fl. Jav.: 23 (1825).
Jatropha condor Wall., Numer. List: 7799 (1847), nom. inval.
Jatropha tuberosa Elliot, Fl. Andhirica: 85 (1859).
Jatropha yucatanensis Briq., Annuaire Conserv. Jard. Bot. Genève 4: 230 (1900).
Jatropha afrocurcas Pax, Bot. Jahrb. Syst. 43: 83 (1909).
Jatropha edulis Cerv., Supl. Gaz. Lit. Mexico: 3 (2 July 1794).

Jatropha decipiens M.E.Jones, Contr. W. Bot. 18: 56 (1933).
Mexico (Jalisco). 79 MXS. Nanophn. or phan.

Jatropha decumbens Pax & K.Hoffm. in H.G.A.Engler, Pflanzenr., IV, 147, VII: 398 (1914).
Namibia. 27 NAM.

Jatropha dehganii J.Jiménez Ram., Acta Bot. Mexi. 30: 5 (1995).
Mexico (Jalisco). 79 MXS. Nanophan.

Jatropha deutziiflora Croizat, J. Wash. Acad. Sci. 33: 16 (1943).
Mexico (Oaxaca). 79 MXS.

Jatropha dhofarica Radcl.-Sm., Kew Bull. 35: 253 (1980). – FIGURE, p. 1023.
Oman. 35 OMA. Nanophan.

Jatropha dichtar J.F.Macbr., Candollea 5: 381 (1934).
E. Ethiopia, Somlia, Kenya. 24 ETH SOM 25 KEN. Nanophan.
Jatropha ferox Pax, Annuario Reale Ist. Bot. Roma 6: 185 (1896), nom. illeg.
Jatropha dichtar var. *gracilior* Radcl.-Sm., Kew Bull. 42: 116 (1987).

Jatropha dioica Cerv., Supl. Gaz. Lit. Mexico: 4 (2 July 1794). – FIGURE, p. 1026.
Texas, Mexico (Tamaulipas, Nuevo León, Coahuila, Chihuahua, Durango, Zacatecas to Oaxaca). 77 TEX 79 MXC MXE MXN MXS. Nanophan. or phan.
Mozinna spathulata Ortega, Nov. Pl. Descr. Dec.: 105 (1798). *Jatropha spathulata* var. *genuina* Müll.Arg. in A.P.de Candolle, Prodr. 15(2): 1081 (1866), nom. inval. *Jatropha spathulata* (Ortega) Müll.Arg. in A.P.de Candolle, Prodr. 15(2): 1081 (1866).
Jatropha cuneifolia Sessé & Moç., Tracts Bot.: 231 (1805).
Mozinna spathulata var. *sessiliflora* Hook., Hooker's Ic. Pl. 4: t. 357 (1841). *Jatropha spathulata* var. *sessiliflora* (Hook.) Müll.Arg. in A.P.de Candolle, Prodr. 15(2): 1082 (1866). *Jatropha dioica* var. *sessiliflora* (Hook.) McVaugh, Bull. Torrey Bot. Club 72: 37 (1945).
Jatropha dioica var. *graminea* McVaugh, Bull. Torrey Bot. Club 72: 39 (1945).

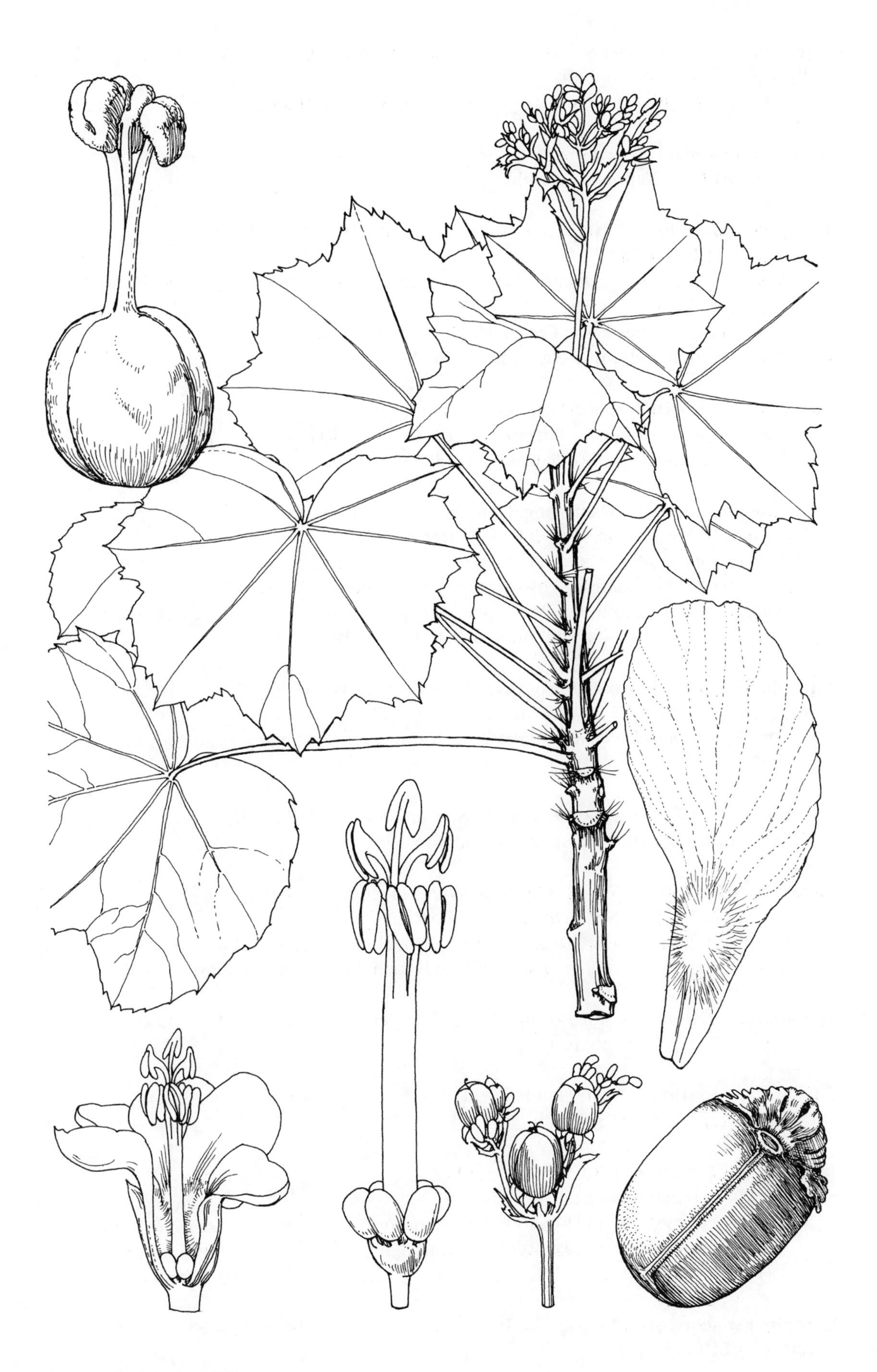

Jatropha dhofarica Radcl.-Sm.
Artist: Miss M. Millar Watt
Kew Bull. 35: 254, fig. 1 (1980)
KEW ILLUSTRATIONS COLLECTION

Jatropha dissecta (Chodat & Hassl.) Pax in H.G.A.Engler, Pflanzenr., IV, 147, I: 72 (1910).
Paraguay. 85 PAR. Nanophan.
* *Jatropha gossypiifolia* var. *dissecta* Chodat & Hassl., Bull. Herb. Boissier, II, 5: 611 (1905).

Jatropha divaricata Sw., Prodr.: 98 (1788).
Jamaica. 81 JAM. Nanophan. or phan.

Jatropha eglandulosa Pax in H.G.A.Engler, Pflanzenr., IV, 147, I: 63 (1910).
Paraguay. 85 PAR. Cham.
Jatropha elliptica var. *guarantica* Chodat & Hassl., Bull. Herb. Boissier, II, 5: 611 (1905).

Jatropha elbae J.Jiménez Ram., Cact. Suc. Mex. 31: 3 (1986).
Mexico (Guerrero). 79 MXS. Nanophan. or phan.

Jatropha ellenbeckii Pax, Bot. Jahrb. Syst. 33: 284 (1903).
Ethiopia, S. Somalia, Kenya, Tanzania. 24 ETH SOM 25 KEN TAN. Nanophan.
Jatropha fissispina Pax, Bot. Jahrb. Syst. 43: 83 (1909).

Jatropha elliptica (Pohl) Oken, Allg. Naturgesch. 3(3): 1595 (1841).
Brazil (Goiás, Minas Gerais). 84 BZC BZL. Hemicr.
* *Adenoropium ellipticum* Pohl, Pl. Bras. Icon. Descr. 1: 13 (1826).
Jatropha officinalis Mart. ex Pohl, Pl. Bras. Icon. Descr. 1: 13 (1826).
Jatropha opifera Mart. in C.F.P.von Martius & J.B.von Spix, Reise Bras. 2: 548 (1828).
Jatropha lacertii Silva Manso, Enum. Subst. Bras.: 8 (1836).

Jatropha erythropoda Pax & K.Hoffm. in H.G.A.Engler, Pflanzenr., IV, 147, III: 66 (1910).
SC. Trop. & S. Africa. 26 ZIM 27 BOT CPP NAM TVL.
Jatropha erythropoda var. *hirtula* Pax & K.Hoffm. in H.G.A.Engler, Pflanzenr., IV, 147, VII: 399 (1914).

Jatropha euarguta M.G.Gilbert & Thulin, Nordic J. Bot. 11: 418 (1991).
Kenya (Northen Frontier). 25 KEN. Nanophan.

Jatropha excisa Griseb., Abh. Königl. Ges. Wiss. Göttingen 19: 94 (1874).
Argentina (Catamarca, Salta). 85 AGW. Nanophan.
Jatropha excisa var. *pubescens* Lourteig & O'Donell, Lilloa 9: 123 (1943).
Jatropha excisa var. *viridiflora* Lourteig & O'Donell, Lilloa 9: 125 (1943).

Jatropha flavovirens Pax & K.Hoffm. in H.G.A.Engler, Pflanzenr., IV, 147, I: 30 (1910).
Paraguay. 85 PAR. Nanophan.

Jatropha fremontioides Standl., Publ. Field Mus. Nat. Hist., Bot. Ser. 22: 37 (1940).
Mexico (Oaxaca). 79 MXS. Nanophan.

Jatropha froesii Croizat, Trop. Woods 76: 13 (1943).
Brazil (Rio de Janeiro). 84 BZL.
* *Jatropha pubescens* Pax in H.G.A.Engler, Pflanzenr., IV, 147, I: 110 (1910), nom. illeg.
Cnidoscolus pubescens (Pax) Pax in H.G.A.Engler, Nat. Pflanzenfam. ed. 2, 19c: 166 (1931), nom. illeg.

Jatropha gallabatensis Schweinf., Verh. Zool.-Bot. Ges. Wien 18: 661 (1868).
Sudan. 24 SUD. Hemicr.

Jatropha galvanii J.Jiménez Ram. & L.M.Contr., Cact. Suc. Mex. 26: 3 (1981).
Mexico (Guerrero). 79 MXS.

Jatropha gaumeri Greenm., Publ. Field Columbian Mus., Bot. Ser. 2: 256 (1907).
Mexico (Yucatán), Belize, Guatemala. 79 MXT 80 BLZ GUA. Nanophan.

Jatropha giffordiana Dehgan & G.L.Webster, Madroño 25: 30 (1978).
Mexico (Baja California Sur). 79 MXN.

Jatropha glandulifera Roxb., Fl. Ind. ed. 1832, 3: 688 (1832).
E. India, Sri Lanka. 40 IND SRL. Succ. nanophan.

Jatropha glauca Vahl, Symb. Bot. 1: 78 (1790). *Jatropha lobata* subsp. *glauca* (Vahl) Pax in H.G.A.Engler, Pflanzenr., IV, 147, III: 32 (1910). *Jatropha lobata* var. *glauca* (Vahl) Pax ex Blatter, Rec. Bot. Surv. India 8: 435 (1923).
Yemen, Eritrea, Ethiopia, Somalia, Sudan. 24 ERI ETH SOM SUD 35 YEM. Nanophan.
Jatropha glauca Salisb., Prodr. Stirp. Chap. Allerton: 389 (1796), nom. illeg.
Jatropha ricinifolia Fenzl ex Baill., Adansonia 1: 63 (1860).
Jatropha lobata Müll.Arg. in A.P.de Candolle, Prodr. 15(2): 1085 (1866). *Jatropha lobata* var. *genuina* Müll.Arg. in A.P.de Candolle, Prodr. 15(2): 1085 (1866), nom. inval.
Jatropha lobata var. *richardiana* Müll.Arg. in A.P.de Candolle, Prodr. 15(2): 1086 (1866).

Jatropha gossypiifolia L., Sp. Pl.: 1006 (1753). *Manihot gossypiifolia* (L.) Crantz, Inst. Rei Herb. 1: 167 (1766).
S. Mexico, Trop. America. 79 MXT 80 COS GUA HON 81 ARU BAH CUB DOM HAI JAM NLA TRT VNA WIN 82 FRG GUY SUR VEN 83 CLM 84 BZE BZS 85 PAR. Nanophan. – Ornamental and also a green dye plant.

var. **elegans** (Pohl) Müll.Arg. in A.P.de Candolle, Prodr. 15(2): 1087 (1866).
S. Mexico, Trop. America. 79 MXT 80 GUA 81 BAH CUB 83 CLM 84 BZS 85 PAR.
* *Adenoropium elegans* Pohl, Pl. Bras. Icon. Descr. 1: 12 (1826). *Jatropha elegans* (Pohl) Klotzsch in B.Seemann, Bot. Voy. Herald: 102 (1853).

var. **gossypiifolia**
SE. Mexico, Trop. America. 79 MXT 80 COS GUA HON 81 ARU CUB JAM NLA TRT VNA WIN 82 GUY VEN 83 CLM 84 BZE BZS 85 PAR. Nanophan. – Ornamental and also a green dye plant.
Jatropha staphysagrifolia Mill., Gard. Dict. ed. 8: 9 (1768).
Jatropha jacquinii Baill., Adansonia 4: 268 (1864).
Jatropha gossypiifolia var. *staphysagriifolia* (Mill.) Müll.Arg. in A.P.de Candolle, Prodr. 15(2): 1087 (1866).

Jatropha grossidentata Pax & K.Hoffm. in H.G.A.Engler, Pflanzenr., IV, 147, VII: 398 (1914).
Bolivia, Paraguay, N. Argentina. 83 BOL 85 AGW PAR. Nanophan.

Jatropha guaranitica Speg., Anales Soc. Ci. Argent. 16: 93 (1883).
Paraguay. 85 PAR. Nanophan.

Jatropha hernandiifolia Vent., Jard. Malmaison: 52 (1803).
Caribbean. 81 DOM HAI JAM PUE. Nanophan. or phan.

var. **hernandiifolia**
Hispaniola, Jamaica. 81 DOM HAI JAM. Nanophan.
Loureira peltata Desv., Tabl. École Bot., ed. 3: 411 (1829). *Jatropha hernandiifolia* var. *peltata* (Desv.) Pax in H.G.A.Engler, Pflanzenr., IV, 147, III: 75 (1910).
Jatropha hernandiifolia var. *epeltata* Pax in H.G.A.Engler, Pflanzenr., IV, 147, III: 76 (1910). – Provisionally included here.
Jatropha heterophylla Sessé & Moç., Fl. Mexic., ed. 2: 224 (1894).

var. **portoricensis** (Millsp.) Urb., Symb. Antill. 4: 349 (1905).
Puerto Rico. 81 PUE.

Jatropha dioica Cerv. (as *Mozinna spathulata* Ortega var. *sessiliflora* Benth.)
Artist: [W.H. Fitch]
Ic. Pl. 4: pl. 357 (1841)
KEW ILLUSTRATIONS COLLECTION

Ricinus portoricensis Juss. ex Baill., Étude Euphorb.: 314 (1858), nom. illeg.
Jatropha portoricensis Millsp., Publ. Field Columbian Mus., Bot. Ser. 2: 59 (1900).

Jatropha heynei N.P.Balakr., Bull. Bot. Surv. India 3: 40 (1962).
S. India. 40 IND.
* *Jatropha heterophylla* B.Heyne ex Hook.f., Fl. Brit. India 5: 382 (1887).

Jatropha hieronymii Kuntze, Revis. Gen. Pl. 3(2): 287 (1898).
Bolivia, Argentina (Jujuy). 83 BOL 85 AGW. Nanophan. or phan.

Jatropha hildebrandtii Pax, Bot. Jahrb. Syst. 19: 108 (1894).
Kenya, Tanzania (incl. Zanzibar, Lamu I.), Comoros. 25 KEN TAN 29 COM. Cham. or nanophan.

var. **hildebrandtii**
Kenya, Tanzania (incl. Zanzibar, Lamu I.), Comoros. 25 KEN TAN 29 COM. Cham. or nanophan.
Jatropha pseudoglandulifera var. *zanguebarica* Hutch. in D.Oliver, Fl. Trop. Afr. 6(1): 1053 (1913).

var. **torrentis-lugardi** Radcl.-Sm., Kew Bull. 28: 283 (1973).
Kenya. 25 KEN.

Jatropha hintonii Wilbur, J. Elisha Mitchell Sci. Soc. 70: 95 (1954).
Mexico (Zacatecas). 79 MXE.

Jatropha hippocastanifolia Croizat, J. Arnold Arbor. 24: 168 (1943).
Bolivia, Paraguay. 83 BOL 85 PAR.

Jatropha hirsuta Hochst., Flora 28: 82 (1845).
Mozambique, S. Africa. 26 MOZ 27 NAT SWZ TVL. Hemicr. or cham.

var. **glabrescens** (Pax & K.Hoffm.) Prain in W.H.Harvey, Fl. Cap. 5(2): 424 (1920).
Natal. 27 NAT.
* *Jatropha glabrescens* Pax & K.Hoffm. in H.G.A.Engler, Pflanzenr., IV, 147, I: 62 (1910).

var. **hirsuta**
Mozambique. 26 MOZ. Hemicr. or cham.

var. **oblongifolia** Prain in W.H.Harvey, Fl. Cap. 5(2): 424 (1920).
Transvaal, Swaziland. 27 SWZ TVL.

Jatropha horizontalis M.G.Gilbert, Nordic J. Bot. 11: 416 (1991).
Ethiopia (Sidamo). 24 ETH. Cham.

Jatropha humboldtiana McVaugh, Bull. Torrey Bot. Club 72: 35 (1944).
Peru. 83 PER. Nanophan. or phan.

Jatropha humifusa Thulin, Nordic J. Bot. 11: 527 (1991).
Somalia (Mudug, Galguduud). 24 SOM. Cham.

Jatropha hypogyna Radcl.-Sm. & Thulin, Kew Bull. 46: 144 (1991).
Somalia (Galguduud). 24 SOM. Cham.

Jatropha inaequispina Thulin, Nordic J. Bot. 11: 530 (1991).
Somalia (Mudug, Mugaal). 24 SOM. Cham.

Jatropha × induta (Chodat & Hassl.) Pax in H.G.A.Engler, Pflanzenr., IV, 147, I: 72 (1910). J. dissecta × J. isabelliae.
Paraguay. 85 PAR.
* *Jatropha gossypiifolia* f. *induta* Chodat & Hassl., Bull. Herb. Boissier, II, 5: 611 (1905).
Jatropha × *brachypoda* Pax in H.G.A.Engler, Pflanzenr., IV, 147, I: 74 (1910).
Jatropha × *transiens* Pax in H.G.A.Engler, Pflanzenr., IV, 147, I: 73 (1910).

Jatropha integerrima Jacq., Enum. Syst. Pl.: 32 (1763).
Cuba, Hispaniola. (80) nic pan 81 CUB DOM HAI. Nanophan. or phan. – PHOTOGRAPH, COVER.
Jatropha hastata Jacq., Select. Stirp. Amer. Hist.: 256 (1763). *Jatropha integerrima* var. *hastata* (Jacq.) Fosberg, Rhodora 78(813): 102 (1976).
Jatropha acuminata Desr. in J.B.A.M.de Lamarck, Encycl. 4: 8 (1797).
Jatropha pandurifolia Andr., Bot. Repos. 4: t. 267 (1802). *Jatropha diversifolia* var. *pandurifolia* (Andr.) M.Gómez, Anales Hist. Nat. 23: 51 (1894).
Jatropha coccinea Link, Enum. Hort. Berol. Alt. 2: 406 (1822). *Jatropha pandurifolia* var. *coccinea* (Link) Pax in H.G.A.Engler, Pflanzenr., IV, 147, III: 50 (1910). *Jatropha integerrima* var. *coccinea* (Link) N.P.Balakr., Bull. Bot. Surv. India 22: 176 (1980 publ. 1982).
Jatropha diversifolia A.Rich. in R.de la Sagra, Hist. Fis. Cuba, Bot. 2: 207 (1850).
Jatropha pauciflora C.Wright ex Griseb., Nachr. Königl. Ges. Wiss. Georg-Augusts-Univ. 1: 170 (1865). *Jatropha diversifolia* var. *pauciflora* (C.Wright ex Griseb.) M.Gómez, Anales Hist. Nat. 23: 51 (1894).
Jatropha moluensis Sessé & Moç., Fl. Mexic., ed. 2: 224 (1894).
Jatropha glaucovirens Pax & K.Hoffm. in H.G.A.Engler, Pflanzenr., IV, 147, I: 51 (1910).
Jatropha pandurifolia var. *latifolia* Pax in H.G.A.Engler, Pflanzenr., IV, 147, III: 50 (1910). *Jatropha integerrima* var. *latifolia* (Pax) N.P.Balakr., Bull. Bot. Surv. India 22: 176 (1980 publ. 1982).

Jatropha intercedens Pax in H.G.A.Engler, Pflanzenr., IV, 147, I: 31 (1910).
Bolivia. 83 BOL. Nanophan.

Jatropha intermedia (Chodat & Hassl.) Pax in H.G.A.Engler, Pflanzenr., IV, 147, I: 63 (1910).
Paraguay. 85 PAR.
Jatropha gossypiifolia f. *glabrata* Chodat & Hassl., Bull. Herb. Boissier, II, 5: 612 (1905).
Jatropha gossypiifolia f. *latifolia* Chodat & Hassl., Bull. Herb. Boissier, II, 5: 612 (1905).
* *Jatropha gossypiifolia* var. *intermedia* Chodat & Hassl., Bull. Herb. Boissier, II, 5: 612 (1905).

Jatropha isabelliae Müll.Arg. in C.F.P.von Martius, Fl. Bras. 11(2): 489 (1874).
N. Argentina, Paraguay. 85 AGW PAR. Hemicr. or cham.
Jatropha antisyphilitica Speg., Anales Soc. Ci. Argent. 16: 31 (1883). *Jatropha isabelliae* var. *antisyphilitica* (Speg.) Pax in H.G.A.Engler, Pflanzenr., IV, 147, III: 72 (1910).
Jatropha gossypiifolia var. *grandifolia* Chodat & Hassl., Bull. Herb. Boissier, II, 5: 612 (1905). *Jatropha isabelliae* var. *grandifolia* (Chodat & Hassl.) Pax in H.G.A.Engler, Pflanzenr., IV, 147, III: 71 (1910).
Jatropha gossypiifolia var. *guaranitica* Chodat & Hassl., Bull. Herb. Boissier, II, 5: 612 (1905). *Jatropha isabelliae* var. *guaranitica* (Chodat & Hassl.) Pax in H.G.A.Engler, Pflanzenr., IV, 147, III: 612 (1905).
Jatropha gossypiifolia var. *palmata* Chodat & Hassl., Bull. Herb. Boissier, II, 5: 612 (1905). *Jatropha isabelliae* var. *palmata* (Chodat & Hassl.) Pax in H.G.A.Engler, Pflanzenr., IV, 147, III: 71 (1910).
Jatropha gossypiifolia var. *rhombifolia* Chodat & Hassl., Bull. Herb. Boissier, II, 5: 612 (1905). *Jatropha isabelliae* var. *rhombifolia* (Chodat & Hassl.) Pax in H.G.A.Engler, Pflanzenr., IV, 147, III: 72 (1910).
Jatropha isabelliae var. *cuneifolia* Pax in H.G.A.Engler, Pflanzenr., IV, 147, III: 71 (1910).

Jatropha rigidifolia Pax & K.Hoffm. in H.G.A.Engler, Pflanzenr., IV, 147, VII: 398 (1914).
Jatropha rigidifolia var. *glabrescens* Pax & K.Hoffm. in H.G.A.Engler, Pflanzenr., IV, 147, VII: 398 (1914).

Jatropha kamerunica Pax & K.Hoffm. in H.G.A.Engler, Pflanzenr., IV, 147, II: 102 (1910).
Senegal, Cameroon. 22 SEN 23 CMN.

var. **kamerunica**
Cameroon. 23 CMN.

var. **trochainii** Leandri, Bull. Soc. Bot. France 83: 525 (1936).
Senegal. 22 SEN.

Jatropha katharinae Pax in H.G.A.Engler, Pflanzenr., IV, 147, III: 28 (1910).
Paraguay. 85 PAR. Cham. or nanophan.

Jatropha lagarinthoides Sond., Linnaea 23: 118 (1850).
Cape, Transvaal. 27 CPP TVL. Cham.
Jatropha cluytioides Pax & K.Hoffm. in H.G.A.Engler, Pflanzenr., IV, 147, I: 65 (1910).
Jatropha lagarinthoides var. *cluytioides* (Pax & K.Hoffm.) Prain in W.H.Harvey, Fl. Cap. 5(2): 423 (1920), nom. illeg.
Jatropha latifolia var. *stenophylla* Pax in H.G.A.Engler, Pflanzenr., IV, 147, III: 133 (1910).

Jatropha latifolia Pax, Bot. Jahrb. Syst. 23: 531 (1897).
Mozambique, S. Africa. 26 MOZ 27 BOT CPP NAT SWZ TVL. Hemicr.

var. **angustata** Prain in W.H.Harvey, Fl. Cap. 5(2): 424 (1920).
Cape. 27 CPP.

var. **latifolia**
Botswana, Transvaal, KwaZulu-Natal. 27 BOT NAT TVL. Hemicr.

var. **subeglandulosa** Radcl.-Sm., Kew Bull. 46: 146 (1991).
Mozambique. 26 MOZ.

var. **swazica** Prain in W.H.Harvey, Fl. Cap. 5(2): 424 (1920).
Swaziland. 27 SWZ.

Jatropha loristipula Radcl.-Sm., Kew Bull. 46: 146 (1991).
S. Zimbabwe. 26 ZIM. Tuber geophyte.

Jatropha macrantha Müll.Arg., Linnaea 34: 209 (1865).
Peru. 83 PER. Nanophan.

Jatropha macrocarpa Griseb., Abh. Königl. Ges. Wiss. Göttingen 19: 94 (1874).
Argentina (Catamarca). 85 AGW. Nanophan.
Jatropha multiflora Pax & K.Hoffm. in H.G.A.Engler, Pflanzenr., IV, 147, XVI: 192 (1924).

Jatropha macrophylla Pax & K.Hoffm. in H.G.A.Engler, Pflanzenr., IV, 147, I: 80 (1910).
S. Tanzania, E. Zambia, Malawi. 25 TAN 26 MLW ZAM. Hemicr.

Jatropha macrorhiza Benth., Pl. Hartw.: 8 (1839).
Arizona, New Mexico, SW. Texas, Mexico (Chihuahua, Sonora). 76 ARI 77 NWM TEX 79 MXE MXN. Tuber geophyte.

var. **macrorhiza**
Arizona, New Mexico, SW. Texas, Mexico (Chihuahua, Sonora). 76 ARI 77 NWM TEX 79 MXE MXN. Tuber geophyte.

var. **septemfida** Engelm. in J.T.Rothrock, Rep. U.S. Geogr. Surv., Wheeler 6: 243 (1878).
Arizona, Mexico (Sonora, Chihuahua). 76 ARI 79 MXE MXN. Tuber geophyte.
Jatropha arizonica I.M.Johnst., Contr. Gray Herb. 68: 89 (1923).

Jatropha mahafalensis Jum. & H.Perrier, Bull. Écon. Madagascar: 179 (1910).
S. Madagascar. 29 MDG. Nanophan. or phan.

Jatropha maheshwarii Subram. & Nayar, Bull. Bot. Surv. India 6: 331 (1965).
S. India. 40 IND.

Jatropha malacophylla Standl., Proc. Biol. Soc. Wash. 37: 45 (1924).
W. & W. Mexico. 79 MXE MXN MXS. Nanophan. or phan.
Jatropha platanifolia Standl., Publ. Field Mus. Nat. Hist., Bot. Ser. 22: 38 (1940).

Jatropha malmeana Pax & K.Hoffm. in H.G.A.Engler, Pflanzenr., IV, 147, XVI: 191 (1924).
Brazil (Mato Grosso). 84 BZC.

Jatropha marginata Chiov., Fl. Somala 2: 391 (1932).
Somalia. 24 SOM. – Probably identical with *J. crinita*.

Jatropha martiusii (Pohl) Baill., Adansonia 4: 268 (1864).
Brazil (Bahia). 84 BZE. Nanophan.
* *Adenoropium martiusii* Pohl, Pl. Bras. Icon. Descr. 1: 16 (1826).

Jatropha matacensis Castell., Bol. Acad. Nac. Ci. 40: 255 (1958).
Argentina (Formosa). 85 AGE.

Jatropha mcvaughii Dehgan & G.L.Webster, Madroño 25: 36 (1978).
Mexico (Jalisco). 79 MXS.
Jatropha curcas var. *rufus* McVaugh, Bull. Torrey Bot. Club 72: 284 (1945).

Jatropha melanosperma Pax, Bot. Jahrb. Syst. 19: 110 (1894).
Sudan. 24 SUD.

Jatropha microdonta Radcl.-Sm., Kew Bull. 28: 284 (1973).
Kenya, Tanzania. 25 KEN TAN. Nanophan.

Jatropha minor Urb., Symb. Antill. 9: 213 (1924).
E. Cuba (Sierra de Nipe). 81 CUB. Nanophan.

Jatropha mollis Pax, Annuario Reale Ist. Bot. Roma 6: 184 (1896).
Ethiopia, Kenya. 24 ETH 25 KEN. Nanophan.

Jatropha mollissima (Pohl) Baill., Adansonia 4: 268 (1864).
Venezuela, Brazil (Pernambuco, Bahia, Minas Gerais). 82 VEN 84 BZE BZL. Nanophan. or phan.
* *Adenoropium mollissimum* Pohl, Pl. Bras. Icon. Descr. 1: 15 (1826). *Jatropha pohliana* var. *mollissima* (Pohl) Müll.Arg. in A.P.de Candolle, Prodr. 15(2): 1091 (1866).

var. **glabra** Müll.Arg. in A.P.de Candolle, Prodr. 15(2): 1092 (1866).
Venezuela. 82 VEN.
Jatropha pohliana var. *glabra* Müll.Arg. in A.P.de Candolle, Prodr. 15(2): 1092 (1866).

var. **mollissima**
Brazil (Pernambuco, Bahia, Minas Gerais). 84 BZE BZL. Nanophan. or phan.

Adenoropium divergens Pohl, Pl. Bras. Icon. Descr. 1: 12 (1826). *Jatropha divergens* (Pohl) Baill., Adansonia 4: 268 (1864), nom. illeg. *Jatropha pohliana* var. *divergens* (Pohl) Müll.Arg. in A.P.de Candolle, Prodr. 15(2): 1092 (1866). *Jatropha mollissima* var. *divergens* (Pohl) Müll.Arg. in A.P.de Candolle, Prodr. 15(2): 1092 (1866).
Adenoropium luxurians Pohl, Pl. Bras. Icon. Descr. 1: 16 (1826). *Jatropha luxurians* (Pohl) Baill., Adansonia 4: 268 (1864). *Jatropha pohliana* f. *luxurians* (Pohl) Müll.Arg. in A.P.de Candolle, Prodr. 15(2): 1091 (1866).
Adenoropium villosum Pohl, Pl. Bras. Icon. Descr. 1: 15 (1826). *Jatropha villosa* (Pohl) Baill., Adansonia 4: 268 (1864), nom. illeg. *Jatropha pohliana* var. *villosa* (Pohl) Müll.Arg. in A.P.de Candolle, Prodr. 15(2): 1091 (1866). *Jatropha mollissima* var. *villosa* (Pohl) Müll.Arg. in A.P.de Candolle, Prodr. 15(2): 1091 (1866).
Jatropha pohliana Müll.Arg., Mém. Soc. Phys. Genève 17(2): 449 (1864).
Jatropha mollissima var. *subglabra* Müll.Arg. in C.F.P.von Martius, Fl. Bras. 11(2): 494 (1874).
Jatropha pohliana var. *subglabra* Müll.Arg. in C.F.P.von Martius, Fl. Bras. 11(2): 494 (1874).
Jatropha mollissima var. *velutina* Pax & K.Hoffm. in H.G.A.Engler, Pflanzenr., IV, 147, XVI: 191 (1924).

Jatropha monroi S.Moore, J. Bot. 63: 147 (1925).
C. & S. Zimbabwe. 26 ZIM.
Jatropha cervicornis Suess., Trans. Rhodesia Sci. Assoc. 43: 10 (1951).

Jatropha moranii Dehgan & G.L.Webster, Madroño 25: 34 (1978).
Mexico (Baja California Sur). 79 MXN.

Jatropha multifida L., Sp. Pl.: 1006 (1753). *Manihot multifida* (L.) Crantz, Inst. Rei Herb. 1: 167 (1766).
SW. Texas, Mexico, Caribbean, Venezuela, S. Brazil, Paraguay. 77 TEX 79 MXE MXS 81 CUB DOM HAI LEE PUE TRT WIN 82 VEN 84 BZS 85 PAR. Nanophan. or phan. – Ornamental.
Jatropha janipha Blanco, Fl. Filip.: 758 (1837).

Jatropha mutabilis (Pohl) Baill., Adansonia 4: 267 (1864).
Brazil (Pernambuco, Bahia). 84 BZE. Nanophan.
* *Adenoropium mutabile* Pohl, Pl. Bras. Icon. Descr. 1: 14 (1826).

Jatropha nana Dalzell & Gibson, Bombay Fl.: 229 (1861).
W. India. 40 IND. Cham. or nanophan.

Jatropha natalensis Müll.Arg., Flora 47: 485 (1864).
Natal. 27 NAT. Tuber geophyte.

Jatropha neopauciflora Pax in H.G.A.Engler, Pflanzenr., IV, 147, I: 134 (1910).
Mexico (Puebla). 79 MXC. Nanophan. or phan.
* *Mozinna pauciflora* Rose, Contr. U. S. Natl. Herb. 12: 282 (1909). *Jatropha pauciflora* (Rose) Pax in H.G.A.Engler, Pflanzenr., IV, 147, I: 82 (1910), nom. illeg.
Jatropha harmsiana Mattf., Repert. Spec. Nov. Regni Veg. 19: 121 (1923).

Jatropha neriifolia Müll.Arg., Flora 47: 486 (1864).
Nigeria. 22 NGA. Cham.

Jatropha nogalensis Chiov., Fl. Somala 1: 306 (1929).
Somalia. 24 SOM.

Jatropha nudicaulis Benth., Bot. Voy. Sulphur: 165 (1846).
Colombia, Ecuador. 83 CLM ECU. Nanophan.

Jatropha oaxacana J.Jiménez Ram. & R.Torres, Anales Inst. Biol. Univ. Nac. Auton. Mexico, Bot. 65: 1 (1994).
Mexico (Oaxaca). 79 MXS.

Jatropha obbiadensis Chiov., Fl. Somala 1: 308 (1929).
Somalia. 24 SOM.

Jatropha oblanceolata Radcl.-Sm., Kew Bull. 28: 522 (1973 publ. 1974).
Kenya (Northern Frontier). 25 KEN. Nanophan.

Jatropha octandra Cerv., Supl. Gaz. Lit. Mexico: 4 (2 July 1794).
Mexico. 79 +.

Jatropha orangeana Dinter ex P.G.Mey., Mitt. Bot. Staatssamml. München 3: 612 (1960).
Namibia. 27 NAM.

Jatropha ortegae Standl., Publ. Field Mus. Nat. Hist., Bot. Ser. 22: 37 (1940).
Mexico (Sinaloa). 79 MXN.

Jatropha pachypoda Pax in H.G.A.Engler, Pflanzenr., IV, 147, I: 47 (1910).
Bolivia. 83 BOL. Nanophan.

Jatropha pachyrrhiza Radcl.-Sm., Kew Bull. 46: 150 (1991).
Zambia. 26 ZAM. Tuber geophyte.

Jatropha palmatifida Baker, Bull. Misc. Inform. Kew 1895: 227 (1895).
Somalia (Golis range). 24 SOM. Nanophan.

Jatropha palmatifolia Ule, Bot. Jahrb. Syst. 42: 219 (1908).
Brazil (Bahia). 84 BZE. Nanophan. or phan.

Jatropha papyrifera Pax & K.Hoffm., Notizbl. Bot. Gart. Berlin-Dahlem 10: 385 (1928).
Bolivia. 83 BOL.

Jatropha paradoxa (Chiov.) Chiov., Ann. Bot. (Rome) 18: 324 (1930).
Somalia. 24 SOM.
* *Collenucia paradoxa* Chiov., Fl. Somala 1: 177 (1929).

Jatropha paraguayensis Radcl.-Sm. & Govaerts, Kew Bull. 52: 189 (1997).
Paraguay. 85 PAR.
* *Jatropha ricinifolia* Pax in H.G.A.Engler, Pflanzenr., IV, 147, I: 34 (1910), nom. illeg.

Jatropha paxii Croizat, J. Arnold Arbor. 24: 168 (1943).
Cuba. 81 CUB.
* *Jatropha flabellifolia* Pax & K.Hoffm. in H.G.A.Engler, Pflanzenr., IV, 147, I: 52 (1910), nom. illeg.

Jatropha pedatipartita Kuntze, Revis. Gen. Pl. 3(2): 287 (1898).
Bolivia. 83 BOL. Nanophan.

Jatropha pedersenii Lourteig, Ark. Bot., n.s. 3: 83 (1955).
Argentina (Corrientes). 85 AGE. – Perhaps identical with *J. isabelliae*.

Jatropha peiranoi Lourteig & O'Donell, Lilloa 9: 135 (1943).
Argentina (Catamarca). 85 AGW.

Jatropha pelargoniifolia Courbai, Ann. Sci. Nat., Bot., IV, 18: 150 (1862). *Jatropha villosa* var. *pelargoniifolia* (Courbai) Chiov., Fl. Somala 1: 306 (1929).
Eritrea, Somalia, Oman, Yemen (incl. Dissée I.), Sudan, Kenya. 24 ERI ETH SOM SUD 25 KEN 35 OMA YEM. Nanophan.

var. **glabra** (Müll.Arg.) Radcl.-Sm., Kew Bull. 37: 683 (1983).
Kenya, Somalia, Yemen. 24 SOM 25 KEN 35 YEM.
* *Jatropha villosa* var. *glabra* Müll.Arg. in A.P.de Candolle, Prodr. 15(2): 1085 (1866). *Jatropha glandulosa* var. *glabra* (Müll.Arg.) Radcl.-Sm., Kew Bull. 35: 764 (1981).

var. **pelargoniifolia**
Eritrea, Somalia, Yemen (incl. Dissée I.), Sudan, Kenya. 24 ERI ETH SOM SUD 25 KEN 35 YEM. Nanophan.
Croton villosus Forssk., Fl. Aegypt.-Arab.: 163 (1775). *Jatropha glandulosa* Vahl, Symb. Bot. 1: 80 (1790), nom. illeg. *Jatropha villosa* var. *glandulosa* Paxin H.G.A.Engler, Pflanzenr., IV, 147, III: 45 (1910). *Jatropha villosa* var. *genuina* Müll.Arg. in A.P.de Candolle, Prodr. 15(2): 1085 (1866), nom. inval. *Jatropha villosa* (Forssk.) Müll.Arg. in A.P.de Candolle, Prodr. 15(2): 1085 (1866), nom. illeg.
Jatropha pelargoniifolia var. *glandulosa* (Pax) Radcl.-Sm., Kew Bull. 39: 788 (1984).

var. **sublobata** (O.Schwartz) Radcl.-Sm., Kew Bull. 37: 683 (1983).
Somalia, Yemen, Oman. 24 SOM 35 OMA YEM.
* *Jatropha villosa* var. *sublobata* O.Schwarz, Mitt. Inst. Allg. Bot. Hamburg 10: 140 (1939). *Jatropha glandulosa* var. *sublobata* (O.Schwarz) Radcl.-Sm., Kew Bull. 35: 764 (1981).

Jatropha peltata Cerv., Supl. Gaz. Lit. Mexico: 3 (2 July 1794).
Mexico. 79 +.

Jatropha pereziae J.Jiménez Ram., Acta Bot. Mex. 30: 2 (1995).
Mexico (Michoacán). 79 MXS. Phan.

Jatropha phillipseae Rendle, J. Bot. 36: 30 (1898).
Somalia (Wagga Mts.). 24 SOM. Nanophan.

Jatropha platyphylla Müll.Arg. in A.P.de Candolle, Prodr. 15(2): 1077 (1866).
Mexico (Sinaloa, Nayarit, Jalisco, Michoacán). 79 MXN MXS.

Jatropha podagrica Hook., Bot. Mag. 74: t. 4376 (1848).
Guatemala, NW. Honduras, Nicaragua. 80 cos GUA HON NIC pan. Succ. nanophan. – The most widely grown ornamental jatropha; illus. in Rodriguez (1993).

Jatropha prunifolia Pax ex Engl., Abh. Königl. Akad. Wiss. Berlin 1894: 14 (1894).
Kenya, Tanzania, Zambia. 25 KEN TAN 26 ZAM. Tuber cham.

Jatropha pseudocurcas Müll.Arg., Linnaea 24: 208 (1865).
Mexico (Oaxaca). 79 MXS. Nanophan.

Jatropha puncticulata Pax & K.Hoffm. in H.G.A.Engler, Pflanzenr., IV, 147, XVI: 192 (1924).
Paraguay. 85 PAR.

Jatropha purpurea Rose, Contr. U. S. Natl. Herb. 1: 357 (1895).
Mexico (Sonora, Sinaloa). 79 MXN. Nanophan.

Jatropha ribifolia (Pohl) Baill., Adansonia 4: 268 (1864).
Brazil (Pernambuco, Bahia), Paraguay. 84 BZE 85 PAR. Nanophan.
* *Adenoropium ribifolium* Pohl, Pl. Bras. Icon. Descr. 1: 15 (1826). *Jatropha gossypiifolia* var. *ribifolia* (Pohl) Müll.Arg. in C.F.P.von Martius, Fl. Bras. 11(2): 491 (1874).

var. **ambigua** Pax in H.G.A.Engler, Pflanzenr., IV, 147, III: 28 (1910).
Paraguay. 85 PAR.

var. **ribifolia**
Brazil (Pernambuco, Bahia). 84 BZE. Nanophan.

Jatropha riojae Miranda, Anales Inst. Biol. Univ. Nac. México 13: 456 (1942).
Mexico (Puebla). 79 MXC. Nanophan.

Jatropha rivae Pax, Annuario Reale Ist. Bot. Roma 6: 185 (1896).
Ethiopia to Tanzania. 24 ETH SOM 25 KEN TAN. Nanophan.

subsp. **parvifolia** (Chiov.) M.G.Gilbert & Thulin, Nordic J. Bot. 11: 414 (1991).
Ethiopia (Harerge Reg.), N. & C. Somalia. 24 ETH SOM. Nanophan.
Jatropha arguta Chiov., Fl. Somala 1: 307 (1929).
* *Jatropha parvifolia* Chiov., Atti Soc. Naturalisti Mat. Modena 66: 18 (1935 publ. 1934).

subsp. **quercifolia** M.G.Gilbert & Thulin, Nordic J. Bot. 11: 415 (1991).
Ethiopia (Gamo Gofa), Kenya, N. Tanzania. 24 ETH 25 KEN TAN. Nanophan.

subsp. **rivae**
Ethiopia (Bale, Sidamo), Somalia (Gedo), Kenya (Mandera). 24 ETH SOM 25 KEN.
Nanophan.

Jatropha robecchii Pax, Annuario Reale Ist. Bot. Roma 6: 184 (1896).
Somalia (Sanaag, Bari). 24 SOM. Nanophan.

Jatropha robertii S.Moore, J. Bot. 56: 212 (1918).
Brazil (Mato Grosso). 84 BZC.

Jatropha rosea Radcl.-Sm., Kew Bull. 42: 118 (1987).
Somalia. 24 SOM.

Jatropha rufescens Brandegee, Univ. Calif. Publ. Bot. 4: 88 (1910).
Mexico (Puebla). 79 MXC.

Jatropha rzedowskii J.Jiménez Ram., Cact. Suc. Mex. 30: 82 (1985).
Mexico (Puebla, Oaxaca). 79 MXC MXS.

Jatropha scaposa Radcl.-Sm., Kew Bull. 46: 151 (1991). – FIGURE, p. 1035.
Mozambique. 26 MOZ. Hemicr.

Jatropha schlechteri Pax, Bot. Jahrb. Syst. 28: 24 (1899).
Zimbabwe, Mozambique, Northern Prov. 26 MOZ ZIM 27 TVL. Tuber geophyte.

subsp. **schlechteri**
Mozambique, E. Northern Prov. 26 MOZ 27 TVL. Tuber geophyte.

subsp. **setifera** (Hutch.) Radcl.-Sm., Kew Bull. 46: 152 (1991).
Zimbabwe, N. Northern Prov. 26 ZIM 27 TVL. Tuber geophyte.
* *Jatropha setifera* Hutch., Botanist S. Afr.: 397 (1946).

Jatropha schweinfurthii Pax, Bot. Jahrb. Syst. 19: 110 (1894).
Sudan to Zimbabwe. 24 SUD 25 TAN UGA 26 ZAM ZIM. Hemicr.

subsp. **atrichocarpa** Radcl.-Sm., Kew Bull. 28: 285 (1973).
Tanzania, Zimbabwe. 25 TAN 26 ZIM. Hemicr.

subsp. **schweinfurthii**
Sudan, Uganda. 24 SUD 25 UGA. Hemicr.

Jatropha scaposa Radcl.-Sm.
Artist: Christine Grey-Wilson
Kew Bull. 46: 153, fig. 4 (1991)
KEW ILLUSTRATIONS COLLECTION

subsp. **zambica** Radcl.-Sm., Kew Bull. 46: 154 (1991).
Zambia. 26 ZAM. Hemicr.

Jatropha seineri Pax, Bot. Jahrb. Syst. 43: 84 (1909).
S. Angola, N. Namibia, Caprivi Strip, Botswana, Zimbabwe, Zambia. 26 ANG ZAM ZIM 27 BOT CPV NAM. Tuber geophyte.
Jatropha humilis N.E.Br., Bull. Misc. Inform. Kew 1909: 139 (1909).
Jatropha seineri var. *tomentella* Radcl.-Sm., Kew Bull. 46: 154 (1991).

Jatropha somalensis Pax in H.G.A.Engler, Pflanzenr., IV, 147, I: 68 (1910).
Somalia. 24 SOM. Nanophan.

Jatropha spicata Pax, Bot. Jahrb. Syst. 19: 109 (1894).
Kenya, Tanzania, Zimbabwe, Angola (Luanda), N. Transvaal. 25 KEN TAN 26 ANG ZIM 27 TVL. Succ. cham.
Jatropha kilimandscharica Pax & K.Hoffm. in H.G.A.Engler, Pflanzenr., IV, 147, I: 40 (1910).
Jatropha pseudoglandulifera Pax in H.G.A.Engler, Pflanzenr., IV, 147, I: 34 (1910).
Jatropha messinica E.A.Bruce, Bothalia 6: 226 (1951).

Jatropha spinosa Vahl, Symb. Bot. 1: 79 (1790).
Yemen, Somalia. 24 SOM 35 YEM. Nanophan. or phan.
* *Croton spinosus* Forssk., Fl. Aegypt.-Arab.: 163 (1775), nom. illeg.
Jatropha aculeata F.Dietr., Nachtr. Vollst. Lex. Gärtn. 4: 76 (1818).
Jatropha spinosa var. *crenulata* Pax in H.G.A.Engler, Pflanzenr., IV, 147, III: 56 (1910).
Jatropha spinosa var. *somaliensis* Pax in H.G.A.Engler, Pflanzenr., IV, 147, III: 56 (1910).

Jatropha spinosissima Thulin, Nordic J. Bot. 11: 532 (1991).
Somalia (Sanaag). 24 SOM. Cham.

Jatropha standleyi Steyerm., Publ. Field Mus. Nat. Hist., Bot. Ser. 22: 152 (1940).
Mexico (Oaxaca). 79 MXS.

Jatropha stephani J.Jiménez Ram. & M.Martínez Gordillo, Anales Inst. Biol. Univ. Nac. Auton. Mexico, Bot. 61: 1 (1991).
Mexico (Michoacán). 79 MXS. Phan.
Jatropha martinezii E.J.Lott & Dehgan, Syst. Bot. 17: 363 (1992).

Jatropha stevensii G.L.Webster, Ann. Missouri Bot. Gard. 74: 117 (1987).
Nicaragua. 80 NIC.

Jatropha stigmatosa Pax & K.Hoffm. in H.G.A.Engler, Pflanzenr., IV, 147, XVI: 192 (1924).
Brazil (Mato Grosso). 84 BZC.

Jatropha stuhlmannii Pax in H.G.A.Engler, Pflanzenw. Ost-Afrikas C: 240 (1895).
Somalia, Kenya, Tanzania. 24 SOM 25 KEN TAN. Tuber cham.

subsp. **somaliensis** Radcl.-Sm., Kew Bull. 42: 107 (1987).
Somalia. 24 SOM.

subsp. **stuhlmannii**
Kenya, Tanzania. 25 KEN TAN. Tuber cham.
Jatropha batawe Pax, Bot. Jahrb. Syst. 28: 420 (1900).

Jatropha subaequiloba Radcl.-Sm., Kew Bull. 46: 154 (1991).
Mozambique (Gaza, Inhambane, Bazaruto I.). 26 MOZ. Cham.

Jatropha sympetala S.F.Blake & Standl., Proc. Biol. Soc. Wash. 33: 118 (1920).
Mexico (Oaxaca). 79 MXS.

Jatropha tacumbensis Pax & K.Hoffm. in H.G.A.Engler, Pflanzenr., IV, 147, XVI: 191 (1924).
Paraguay. 85 PAR.

Jatropha tanjorensis J.L.Ellis & Saroja, J. Bombay Nat. Hist. Soc. 58: 834 (1962).
India (Tamil Nadu). 40 IND.

Jatropha tehuantepecana J.Jiménez Ram. & A.Campos Vilb., Anales Inst. Biol. Univ. Nac. Auton. Mexico, Bot. 65: 26 (1994).
Mexico (Oaxaca). 79 MXS.

Jatropha tenuicaulis Thulin, Nordic J. Bot. 11: 417 (1991).
Somalia (Bay). 24 SOM. Cham.

Jatropha tetracantha Chiov., Fl. Somala 2: 393 (1932).
Somalia. 24 SOM.

Jatropha thyrsantha Pax & K.Hoffm. in H.G.A.Engler, Pflanzenr., IV, 147, VII: 397 (1914).
Bolivia. 83 BOL.

Jatropha tlalcozotitlanensis J.Jiménez Ram., Cact. Suc. Mex. 30: 80 (1985).
Mexico (Guerrero). 79 MXS.

Jatropha trifida Chiov., Fl. Somala 2: 388 (1932).
Somalia. 24 SOM.
Jatropha brockmannii var. *leiosepala* Chiov., Pubbl. Ist. Stud. Sup. Firenze 1: 162 (1916).

Jatropha tropaeolifolia Pax in H.G.A.Engler, Pflanzenr., IV, 147, I: 56 (1910).
Somalia. 24 SOM.

Jatropha tupifolia Griseb., Nachr. Königl. Ges. Wiss. Georg-Augusts-Univ. 1: 170 (1865).
Jatropha diversifolia var. *tupifolia* (Griseb.) M.Gómez, Anales Hist. Nat. 23: 51 (1894).
C. & E. Cuba. 81 CUB. Nanophan.

Jatropha uncinulata Radcl.-Sm., Kew Bull. 46: 327 (1991).
S. Yemen. 35 YEM.

Jatropha unicostata Balf.f., Proc. Roy. Soc. Edinburgh 12: 94 (1884).
Socotra. 24 SOC. Nanophan.

Jatropha variabilis Radcl.-Sm., Kew Bull. 39: 789 (1984).
Somalia. 24 SOM.

Jatropha variegata Vahl, Symb. Bot. 1: 79 (1790).
Yemen. 35 YEM. Nanophan.

Jatropha variifolia Pax in H.G.A.Engler, Pflanzenr., IV, 147, I: 54 (1910).
Mozambique to KwaZulu-Natal. 26 MOZ 27 NAT TVL. Nanophan.
* *Jatropha heterophylla* Pax, Bot. Jahrb. Syst. 28: 25 (1899).

Jatropha velutina Pax & K.Hoffm. in H.G.A.Engler, Pflanzenr., IV, 147, I: 37 (1910).
Kenya (Kwale, Teita). 25 KEN. Succ. cham.
Jatropha acerifolia Pax, Bot. Jahrb. Syst. 19: 109 (1894), nom. illeg.

Jatropha vernicosa Brandegee, Zoe 5: 206 (1905).
Mexico (S. Baja California Sur). 79 MXN. Nanophan.

Jatropha villosa Wight, Icon. Pl. Ind. Orient. 3: t. 1159 (1846).
India. 40 IND.
Jatropha peltata Wight, Icon. Pl. Ind. Orient. 4: t. 1169 (1848), nom. illeg. *Jatropha wightiana* Müll.Arg. in A.P.de Candolle, Prodr. 15(2): 1080 (1866).

Jatropha weberbaueri Pax & K.Hoffm. in H.G.A.Engler, Pflanzenr., IV, 147, I: 45 (1910).
Peru, Ecuador. 83 ECU PER.

Jatropha websteri J.Jiménez Ram., Anales Inst. Biol. Univ. Nac. Auton. Mexico, Bot. 63: 25 (1992).
Mexico (Guerrero). 79 MXS.

Jatropha weddeliana Baill., Adansonia 4: 267 (1864).
Brazil (Mato Grosso), Paraguay. 84 BZC 85 PAR. Cham. or nanophan.

Jatropha woodii Kuntze, Revis. Gen. Pl. 3(2): 287 (1898).
KwaZulu-Natal. 27 NAT.
Jatropha woodii var. *vestita* Pax, Bot. Jahrb. Syst. 43: 84 (1909).
Jatropha woodii var. *kuntzei* Pax in H.G.A.Engler, Pflanzenr., IV, 147, III: 66 (1910).

Jatropha zeyheri Sond., Linnaea 23: 117 (1850).
S. Africa. 27 BOT CPP NAT SWZ TVL. Tuber geophyte.
Jatropha brachyadenia Pax & K.Hoffm. in H.G.A.Engler, Pflanzenr., IV, 147, I: 66 (1910).
Jatropha zeyheri var. *platyphylla* Pax in H.G.A.Engler, Pflanzenr., IV, 147, III: 66 (1910).
Jatropha zeyheri var. *subsimplex* Prain in W.H.Harvey, Fl. Cap. 5(2): 426 (1920).

Synonyms:
Jatropha acerifolia Pax === **Jatropha velutina** Pax & K.Hoffm.
Jatropha acerifolia Salisb. === **Jatropha curcas** L.
Jatropha aconitifolia Mill. === **Cnidoscolus aconitifolius** (Mill.) I.M.Johnst.
Jatropha aconitifolia var. *genuina* Müll.Arg. === **Cnidoscolus aconitifolius** (Mill.) I.M.Johnst.
Jatropha aconitifolia var. *multipartita* Müll.Arg. === **Cnidoscolus aconitifolius** (Mill.) I.M.Johnst. subsp. **aconitifolius**
Jatropha aconitifolia var. *palmata* (Willd.) Müll.Arg. === **Cnidoscolus aconitifolius** (Mill.) I.M.Johnst. subsp. **aconitifolius**
Jatropha aconitifolia var. *papaya* (Medik.) Pax === **Cnidoscolus aconitifolius** (Mill.) I.M.Johnst. subsp. **aconitifolius**
Jatropha acrandra Urb. === **Cnidoscolus acrandrus** (Urb.) Pax & K.Hoffm.
Jatropha aculeata F.Dietr. === **Jatropha spinosa** Vahl
Jatropha acuminata Desr. === **Jatropha integerrima** Jacq.
Jatropha adenophila Pax & K.Hoffm. === **Cnidoscolus urens** (L.) Arthur var. **urens**
Jatropha aesculifolia Kunth === **Manihot aesculifolia** (Kunth) Pohl
Jatropha afrocurcas Pax === **Jatropha curcas** L.
Jatropha aipi (Pohl) A.Moller === **Manihot esculenta** Crantz
Jatropha albomaculata Pax === **Cnidoscolus albomaculatus** (Pax) I.M.Johnst.
Jatropha albomaculata var. *nana* (Chodat & Hassl.) Pax === **Cnidoscolus albomaculatus** (Pax) I.M.Johnst.
Jatropha albomaculata var. *stimulosissima* (Chodat & Hassl.) Pax === **Cnidoscolus albomaculatus** (Pax) I.M.Johnst.
Jatropha albomaculata var. *subcuneata* Pax === **Cnidoscolus albomaculatus** (Pax) I.M.Johnst.
Jatropha angustidens (Torr.) Müll.Arg. === **Cnidoscolus angustidens** Torr.
Jatropha angustifolia Steud. === ?
Jatropha angustifolia var. *genuina* Müll.Arg. === **Jatropha angustifolia** Griseb.
Jatropha angustifolia var. *spathulata* Müll.Arg. === **Jatropha angustifolia** Griseb.
Jatropha anomala (Pohl) Steud. === **Manihot anomala** Pohl
Jatropha antisyphilitica Speg. === **Jatropha isabelliae** Müll.Arg.

Jatropha appendiculata Pax & K.Hoffm. === **Cnidoscolus appendiculatus** (Pax & K.Hoffm.) Pax & K.Hoffm.
Jatropha arborea Glaz. === ?
Jatropha arcuata (Pohl) Steud. === **Manihot divergens** Pohl
Jatropha arguta Chiov. === **Jatropha rivae** subsp. **parvifolia** (Chiov.) M.G.Gilbert & Thulin
Jatropha arizonica I.M.Johnst. === **Jatropha macrorhiza** var. **septemfida** Engelm.
Jatropha australis Lodd. ex G.Don === **Brachychiton populneus** (Schott & Endl.) R.Br. subsp. **populneus** (Sterculiaceae)
Jatropha bahiana Ule === **Cnidoscolus bahianus** (Ule) Pax & K.Hoffm.
Jatropha bahiana var. *genuina* Pax === **Cnidoscolus bahianus** (Ule) Pax & K.Hoffm.
Jatropha bahiana var. *rupestris* Ule === **Cnidoscolus bahianus** (Ule) Pax & K.Hoffm.
Jatropha basiacantha Pax & K.Hoffm. === **Cnidoscolus basiacanthus** (Pax & K.Hoffm.) J.F.Macbr.
Jatropha batawe Pax === **Jatropha stuhlmannii** Pax subsp. **stuhlmannii**
Jatropha bellatrix Ekman ex Urb. === **Cnidoscolus bellator** (Ekman ex Urb.) Léon
Jatropha berlandieri Torr. === **Jatropha cathartica** Terán & Berland.
Jatropha berteri Spreng. === ?
Jatropha brachyadenia Pax & K.Hoffm. === **Jatropha zeyheri** Sond.
Jatropha × *brachypoda* Pax === **Jatropha** × **induta** (Chodat & Hassl.) Pax
Jatropha brockmannii var. *leiosepala* Chiov. === **Jatropha trifida** Chiov.
Jatropha cajaniformis (Pohl) Steud. === **Manihot tripartita** (Spreng.) Müll.Arg. subsp. **tripartita**
Jatropha calyculata Vahl ex Steud. === ?
Jatropha calyculata Pax & K.Hoffm. === **Cnidoscolus calyculatus** (Pax & K.Hoffm.) I.M.Johnst.
Jatropha campanulata Pax === **Cnidoscolus campanulatus** (Pax) Pax
Jatropha canescens (Benth.) Müll.Arg. === **Jatropha cinerea** (Ortega) Müll.Arg.
Jatropha caricifolia Steud. === **Manihot anomala** Pohl subsp. **anomala**
Jatropha carpinifolia Pax === **Mildbraedia carpinifolia** (Pax) Hutch.
Jatropha carthaginensis Jacq. === **Manihot carthaginensis** (Jacq.) Müll.Arg.
Jatropha cecropiifolia (Pohl) Steud. === **Manihot cecropiifolia** Pohl
Jatropha cercidiphylla Standl. === **Astrocasia neurocarpa** (Müll.Arg.) I.M.Johnst. ex Standl.
Jatropha cervicornis Suess. === **Jatropha monroi** S.Moore
Jatropha ciliata Müll.Arg. === **Jatropha augustii** Pax & K.Hoffm.
Jatropha ciliata var. *longipedunculata* J.F.Macbr. === **Jatropha augustii** Pax & K.Hoffm.
Jatropha cleomifolia (Pohl) Steud. === **Manihot tripartita** (Spreng.) Müll.Arg. subsp. **tripartita**
Jatropha cluytioides Pax & K.Hoffm. === **Jatropha lagarinthoides** Sond.
Jatropha coccinea Link === **Jatropha integerrima** Jacq.
Jatropha coerulea Steud. === **Manihot caerulescens** Pohl subsp. **caerulescens**
Jatropha coerulescens (Pohl) Müll.Arg. === **Manihot caerulescens** Pohl
Jatropha condor Wall. === **Jatropha curcas** L.
Jatropha cordifolia Pax === **Cnidoscolus tubulosus** (Müll.Arg.) I.M.Johnst.
Jatropha crotalariiformis (Pohl) Steud. === **Manihot crotalariiformis** Pohl
Jatropha cuneifolia Sessé & Moç. === **Jatropha dioica** Cerv.
Jatropha curcas var. *rufus* McVaugh === **Jatropha mcvaughii** Dehgan & G.L.Webster
Jatropha dalechampiiformis (Pohl) Steud. === **Manihot tripartita** (Spreng.) Müll.Arg. subsp. **tripartita**
Jatropha diacantha Pax & K.Hoffm. === **Cnidoscolus diacanthus** (Pax & K.Hoffm.) J.F.Macbr.
Jatropha dichtar var. *gracilior* Radcl.-Sm. === **Jatropha dichtar** J.F.Macbr.
Jatropha diffusa (Pohl) Steud. === **Manihot esculenta** Crantz
Jatropha digitiformis (Pohl) Steud. === **Manihot esculenta** Crantz
Jatropha dioica var. *graminea* McVaugh === **Jatropha dioica** Cerv.
Jatropha dioica var. *sessiliflora* (Hook.) McVaugh === **Jatropha dioica** Cerv.
Jatropha divergens (Pohl) Steud. === **Manihot divergens** Pohl
Jatropha divergens (Pohl) Baill. === **Jatropha mollissima** (Pohl) Baill. var. **mollissima**
Jatropha diversifolia A.Rich. === **Jatropha integerrima** Jacq.
Jatropha diversifolia Steud. ===?
Jatropha diversifolia var. *pandurifolia* (Andr.) M.Gómez === **Jatropha integerrima** Jacq.

Jatropha diversifolia var. *pauciflora* (C.Wright ex Griseb.) M.Gómez === **Jatropha integerrima** Jacq.
Jatropha diversifolia var. *tupifolia* (Griseb.) M.Gómez === **Jatropha tupifolia** Griseb.
Jatropha dulcis J.F.Gmel. === **Manihot esculenta** Crantz
Jatropha edulis Cerv. === **Jatropha curcas** L.
Jatropha elastica L.f. === **Hevea guianensis** Aubl. var. **guianensis**
Jatropha elegans (Pohl) Klotzsch === **Jatropha gossypiifolia** var. **elegans** (Pohl) Müll.Arg.
Jatropha elliptica var. *guarantica* Chodat & Hassl. === **Jatropha eglandulosa** Pax
Jatropha erythropoda var. *hirtula* Pax & K.Hoffm. === **Jatropha erythropoda** Pax & K.Hoffm.
Jatropha excisa var. *pubescens* Lourteig & O'Donell === **Jatropha excisa** Griseb.
Jatropha excisa var. *viridiflora* Lourteig & O'Donell === **Jatropha excisa** Griseb.
Jatropha fallax Pax === **Mildbraedia carpinifolia** (Pax) Hutch. var. **carpinifolia**
Jatropha ferox Pax === **Jatropha dichtar** J.F.Macbr.
Jatropha ferox Müll.Arg. === **Cnidoscolus horridus** (Müll.Arg.) Pax & K.Hoffm.
Jatropha fischeri Steud. === ?
Jatropha fissispina Pax === **Jatropha ellenbeckii** Pax
Jatropha flabellifolia Pax & K.Hoffm. === **Jatropha paxii** Croizat
Jatropha flabellifolia (Pohl) Steud. === **Manihot esculenta** Crantz
Jatropha foetida (Kunth) Steud. === **Manihot foetida** (Kunth) Pohl
Jatropha fragrans Kunth === **Cnidoscolus fragrans** (Kunth) Pohl
Jatropha frutescens Loefl. === **Manihot carthaginensis** (Jacq.) Müll.Arg.
Jatropha glabrescens Pax & K.Hoffm. === **Jatropha hirsuta** var. **glabrescens** (Pax & K.Hoffm.) Prain
Jatropha glandulosa Vahl === **Jatropha pelargoniifolia** Courbai var. **pelargoniifolia**
Jatropha glandulosa var. *glabra* (Müll.Arg.) Radcl.-Sm. === **Jatropha pelargoniifolia** var. **glabra** (Müll.Arg.) Radcl.-Sm.
Jatropha glandulosa var. *sublobata* (O.Schwarz) Radcl.-Sm. === **Jatropha pelargoniifolia** var. **sublobata** (O.Schwartz) Radcl.-Sm.
Jatropha glauca Griseb. === **Jatropha angustifolia** Griseb.
Jatropha glauca A.Rich. === **Manihot esculenta** Crantz
Jatropha glauca Salisb. === **Jatropha glauca** Vahl
Jatropha glauca var. *senegalensis* (Müll.Arg.) Hiern === **Jatropha chevalieri** Beille
Jatropha glaucovirens Pax & K.Hoffm. === **Jatropha integerrima** Jacq.
Jatropha globosa Gaertn. === **Hyaenanche globosa** (Gaertn.) Lamb. & Vahl
Jatropha gossypiifolia var. *breviloba* Morong === **Jatropha breviloba** (Morong) Pax & K.Hoffm.
Jatropha gossypiifolia var. *dissecta* Chodat & Hassl. === **Jatropha dissecta** (Chodat & Hassl.) Pax
Jatropha gossypiifolia f. *glabrata* Chodat & Hassl. === **Jatropha intermedia** (Chodat & Hassl.) Pax
Jatropha gossypiifolia var. *grandifolia* Chodat & Hassl. === **Jatropha isabelliae** Müll.Arg.
Jatropha gossypiifolia var. *guaranitica* Chodat & Hassl. === **Jatropha isabelliae** Müll.Arg.
Jatropha gossypiifolia f. *induta* Chodat & Hassl. === **Jatropha** × **induta** (Chodat & Hassl.) Pax
Jatropha gossypiifolia var. *intermedia* Chodat & Hassl. === **Jatropha intermedia** (Chodat & Hassl.) Pax
Jatropha gossypiifolia f. *latifolia* Chodat & Hassl. === **Jatropha intermedia** (Chodat & Hassl.) Pax
Jatropha gossypiifolia var. *palmata* Chodat & Hassl. === **Jatropha isabelliae** Müll.Arg.
Jatropha gossypiifolia var. *rhombifolia* Chodat & Hassl. === **Jatropha isabelliae** Müll.Arg.
Jatropha gossypiifolia var. *ribifolia* (Pohl) Müll.Arg. === **Jatropha ribifolia** (Pohl) Baill.
Jatropha gossypiifolia var. *staphysagriifolia* (Mill.) Müll.Arg. === **Jatropha gossypiifolia** L. var. **gossypiifolia**
Jatropha gracilis (Pohl) Steud. === **Manihot gracilis** Pohl
Jatropha grandifrons I.M.Johnst. === **Jatropha ciliata** Sessé ex Cerv.
Jatropha hamosa (Pohl) Müll.Arg. === **Cnidoscolus hamosus** Pohl

Jatropha harmsiana Mattf. === **Jatropha neopauciflora** Pax
Jatropha hassleriana Pax === **Cnidoscolus hasslerianus** (Pax) Pax
Jatropha hastata Jacq. === **Jatropha integerrima** Jacq.
Jatropha hastata Griseb. === ?
Jatropha herbacea L. === **Cnidoscolus urens** (L.) Arthur var. **urens**
Jatropha hernandiifolia var. *epeltata* Pax === **Jatropha hernandiifolia** Vent.
Jatropha hernandiifolia var. *peltata* (Desv.) Pax === **Jatropha hernandiifolia** Vent. var. **hernandiifolia**
Jatropha heterophylla Sessé & Moç. === **Jatropha hernandiifolia** Vent. var. **hernandiifolia**
Jatropha heterophylla Pax === **Jatropha variifolia** Pax
Jatropha heterophylla B.Heyne ex Hook.f. === **Jatropha heynei** N.P.Balakr.
Jatropha heterophylla Steud. === **Manihot anomala** Pohl subsp. **anomala**
Jatropha heudelotii Baill. === **Ricinodendron heudelotii** (Baill.) Heckel
Jatropha hoffmanniae Croizat === **Jatropha augustii** Pax & K.Hoffm.
Jatropha horrida Müll.Arg. === **Cnidoscolus horridus** (Müll.Arg.) Pax & K.Hoffm.
Jatropha humilis N.E.Br. === **Jatropha seineri** Pax
Jatropha hypoleuca Pax === **Cnidoscolus hypoleucus** (Pax) Pax
Jatropha inermiflora Standl. === **Cnidoscolus rotundifolius** (Müll.Arg.) McVaugh
Jatropha integerrima var. *coccinea* (Link) N.P.Balakr. === **Jatropha integerrima** Jacq.
Jatropha integerrima var. *hastata* (Jacq.) Fosberg === **Jatropha integerrima** Jacq.
Jatropha integerrima var. *latifolia* (Pax) N.P.Balakr. === **Jatropha integerrima** Jacq.
Jatropha isabelliae var. *antisyphilitica* (Speg.) Pax === **Jatropha isabelliae** Müll.Arg.
Jatropha isabelliae var. *cuneifolia* Pax === **Jatropha isabelliae** Müll.Arg.
Jatropha isabelliae var. *grandifolia* (Chodat & Hassl.) Pax === **Jatropha isabelliae** Müll.Arg.
Jatropha isabelliae var. *guaranitica* (Chodat & Hassl.) Pax === **Jatropha isabelliae** Müll.Arg.
Jatropha isabelliae var. *palmata* (Chodat & Hassl.) Pax === **Jatropha isabelliae** Müll.Arg.
Jatropha isabelliae var. *rhombifolia* (Chodat & Hassl.) Pax === **Jatropha isabelliae** Müll.Arg.
Jatropha jacquinii Baill. === **Jatropha gossypiifolia** L. var. **gossypiifolia**
Jatropha jaenensis Pax & K.Hoffm. === **Cnidoscolus jaenensis** (Pax & K.Hoffm.) J.F.Macbr.
Jatropha janipha Lour. === **Manihot esculenta** Crantz
Jatropha janipha L. === **Manihot carthaginensis** (Jacq.) Müll.Arg.
Jatropha janipha Blanco === **Jatropha multifida** L.
Jatropha jurgensenii Briq. === **Cnidoscolus tubulosus** (Müll.Arg.) I.M.Johnst.
Jatropha kilimandscharica Pax & K.Hoffm. === **Jatropha spicata** Pax
Jatropha kunthiana Müll.Arg. === **Cnidoscolus kunthianus** (Müll.Arg.) Pax & K.Hoffm.
Jatropha lacertii Silva Manso === **Jatropha elliptica** (Pohl) Oken
Jatropha laciniosa (Pohl) Steud. === **Manihot tripartita** subsp. **laciniosa** (Pohl) D.J.Rogers & Appan
Jatropha lagarinthoides var. *cluytioides* (Pax & K.Hoffm.) Prain === **Jatropha lagarinthoides** Sond.
Jatropha lanciniosa (Pohl) Steud. === **Manihot tripartita** subsp. **laciniosa** (Pohl) D.J.Rogers & Appan
Jatropha latifolia var. *stenophylla* Pax === **Jatropha lagarinthoides** Sond.
Jatropha leuconeura Pax & K.Hoffm. === **Cnidoscolus leuconeurus** (Pax & K.Hoffm.) Pax & K.Hoffm.
Jatropha liebmannii Müll.Arg. === **Cnidoscolus liebmannii** (Müll.Arg.) Lundell
Jatropha loasoides Pax === **Cnidoscolus loasoides** (Pax) I.M.Johnst.
Jatropha lobata Müll.Arg. === **Jatropha glauca** Vahl
Jatropha lobata subsp. *aceroides* Pax & K.Hoffm. === **Jatropha aceroides** (Pax & K.Hoffm.) Hutch.
Jatropha lobata var. *genuina* Müll.Arg. === **Jatropha glauca** Vahl
Jatropha lobata subsp. *glauca* (Vahl) Pax === **Jatropha glauca** Vahl
Jatropha lobata var. *glauca* (Vahl) Pax ex Blatter === **Jatropha glauca** Vahl
Jatropha lobata var. *richardiana* Müll.Arg. === **Jatropha glauca** Vahl
Jatropha lobata var. *senegalensis* Müll.Arg. === **Jatropha chevalieri** Beille
Jatropha lobata subsp. *senegalensis* (Müll.Arg.) Pax === **Jatropha chevalieri** Beille

Jatropha loefgrenii Pax & K.Hoffm. === **Cnidoscolus loefgrenii** (Pax & K.Hoffm.) Pax & K.Hoffm.
Jatropha loeflingii Aresch. === ?
Jatropha longipedunculata Brandegee === **Cnidoscolus longipedunculatus** (Brandegee) Pax & K.Hoffm.
Jatropha longipedunculata Pax & K.Hoffm. === **Jatropha augustii** Pax & K.Hoffm.
Jatropha longipes Pax === **Cnidoscolus longipes** (Pax) I.M.Johnst.
Jatropha longipetiolata (Pohl) Steud. === **Manihot longipetiolata** Pohl
Jatropha loureirii (Pohl) Steud. === **Manihot esculenta** Crantz
Jatropha luxurians (Pohl) Baill. === **Jatropha mollissima** (Pohl) Baill. var. **mollissima**
Jatropha maculata Brandegee === **Cnidoscolus maculatus** (Brandegee) Pax & K.Hoffm.
Jatropha manihot L. === **Manihot esculenta** Crantz
Jatropha manihot Vell. === **Manihot palmata** Müll.Arg.
Jatropha maracayensis Chodat & Hassl. === **Cnidoscolus maracayensis** (Chodat & Hassl.) Pax & K.Hoffm.
Jatropha martinezii E.J.Lott & Dehgan === **Jatropha stephani** J.Jiménez Ram. & M.Martínez Gordillo
Jatropha messinica E.A.Bruce === **Jatropha spicata** Pax
Jatropha mitis Rottb. === **Manihot esculenta** Crantz
Jatropha mitis Sessé & Moç. === **Manihot esculenta** Crantz
Jatropha mollissima var. *divergens* (Pohl) Müll.Arg. === **Jatropha mollissima** (Pohl) Baill. var. **mollissima**
Jatropha mollissima var. *subglabra* Müll.Arg. === **Jatropha mollissima** (Pohl) Baill. var. **mollissima**
Jatropha mollissima var. *velutina* Pax & K.Hoffm. === **Jatropha mollissima** (Pohl) Baill. var. **mollissima**
Jatropha mollissima var. *villosa* (Pohl) Müll.Arg. === **Jatropha mollissima** (Pohl) Baill. var. **mollissima**
Jatropha moluccana L. === **Aleurites moluccana** (L.) Willd.
Jatropha moluensis Sessé & Moç. === **Jatropha integerrima** Jacq.
Jatropha montana Willd. === **Baliospermum montanum** (Willd.) Müll.Arg.
Jatropha multiflora Pax & K.Hoffm. === **Jatropha macrocarpa** Griseb.
Jatropha multiloba Pax === **Cnidoscolus multilobus** (Pax) I.M.Johnst.
Jatropha napeifolia Desr. === **Cnidoscolus aconitifolius** (Mill.) I.M.Johnst. subsp. **aconitifolius**
Jatropha neglecta (Pohl) Houst. ex Baill. === **Cnidoscolus urens** (L.) Arthur var. **urens**
Jatropha obtusifolia (Pohl ex Baill.) Müll.Arg. === **Cnidoscolus pubescens** Pohl
Jatropha obtusifolia var. *genuina* Müll.Arg. === **Cnidoscolus pubescens** Pohl
Jatropha obtusifolia var. *pubescens* (Pohl) Müll.Arg. === **Cnidoscolus pubescens** Pohl
Jatropha officinalis Mart. ex Pohl === **Jatropha elliptica** (Pohl) Oken
Jatropha oligandra Müll.Arg. === **Cnidoscolus oligandrus** (Müll.Arg.) Pax
Jatropha olivacea Müll.Arg. === **Jatropha ciliata** Sessé ex Cerv.
Jatropha opifera Mart. === **Jatropha elliptica** (Pohl) Oken
Jatropha orbicularis (Pohl) Steud. === **Manihot orbicularis** Pohl
Jatropha osteocarpa Schott ex Pax === **Cnidoscolus urens** (L.) Arthur var. **urens**
Jatropha palmata Sessé & Moç. ex Cerv. === **Manihot sp.**
Jatropha palmata Willd. === **Cnidoscolus aconitifolius** (Mill.) I.M.Johnst. subsp. **aconitifolius**
Jatropha palmata Vell. === **Manihot palmata** Müll.Arg.
Jatropha palmeri S.Watson === **Cnidoscolus palmeri** (S.Watson) Rose
Jatropha palustris Sessé & Moç. === ?
Jatropha pandurifolia Andr. === **Jatropha integerrima** Jacq.
Jatropha pandurifolia var. *coccinea* (Link) Pax === **Jatropha integerrima** Jacq.
Jatropha pandurifolia var. *latifolia* Pax === **Jatropha integerrima** Jacq.
Jatropha paniculata Ruiz & Pav. ex Pax === **Manihot esculenta** Crantz
Jatropha papaya Medik. === **Cnidoscolus aconitifolius** (Mill.) I.M.Johnst. subsp. **aconitifolius**
Jatropha parvifolia Chiov. === **Jatropha rivae** subsp. **parvifolia** (Chiov.) M.G.Gilbert & Thulin
Jatropha pauciflora C.Wright ex Griseb. === **Jatropha integerrima** Jacq.

Jatropha pauciflora (Rose) Pax === **Jatropha neopauciflora** Pax
Jatropha paucistaminea Pax === **Cnidoscolus paucistamineus** (Pax) Pax
Jatropha paviifolia (Pohl) Steud. === **Manihot paviifolia** Pohl
Jatropha pelargoniifolia var. *glandulosa* (Pax) Radcl.-Sm. === **Jatropha pelargoniifolia** Courbai var. **pelargoniifolia**
Jatropha peltata Kunth === ?
Jatropha peltata Wight === **Jatropha villosa** Wight
Jatropha peltata (Pohl) Steud. === **Manihot peltata** Pohl
Jatropha peltata C.Wright ex Sauvalle === ?
Jatropha pentaphylla (Pohl) Steud. === **Manihot pentaphylla** Pohl
Jatropha peruviana Müll.Arg. === **Cnidoscolus peruvianus** (Müll.Arg.) Pax & K.Hoffm.
Jatropha phyllacantha Müll.Arg. === **Cnidoscolus quercifolius** Pohl
Jatropha phyllacantha var. *lobata* (Pohl) Müll.Arg. === **Cnidoscolus quercifolius** Pohl
Jatropha phyllacantha var. *quercifolia* (Pohl) Müll.Arg. === **Cnidoscolus quercifolius** Pohl
Jatropha phyllacantha var. *repanda* (Pohl) Müll.Arg. === **Cnidoscolus quercifolius** Pohl
Jatropha pilosa (Pohl) Steud. === **Manihot pilosa** Pohl
Jatropha platanifolia Standl. === **Jatropha malacophylla** Standl.
Jatropha platyandra Pax === **Cnidoscolus rangel** (M.Gómez) McVaugh
Jatropha pohliana Müll.Arg. === **Jatropha mollissima** (Pohl) Baill. var. **mollissima**
Jatropha pohliana var. *divergens* (Pohl) Müll.Arg. === **Jatropha mollissima** (Pohl) Baill. var. **mollissima**
Jatropha pohliana var. *glabra* Müll.Arg. === **Jatropha mollissima** var. **glabra** Müll.Arg.
Jatropha pohliana f. *luxurians* (Pohl) Müll.Arg. === **Jatropha mollissima** (Pohl) Baill. var. **mollissima**
Jatropha pohliana var. *mollissima* (Pohl) Müll.Arg. === **Jatropha mollissima** (Pohl) Baill.
Jatropha pohliana var. *subglabra* Müll.Arg. === **Jatropha mollissima** (Pohl) Baill. var. **mollissima**
Jatropha pohliana var. *villosa* (Pohl) Müll.Arg. === **Jatropha mollissima** (Pohl) Baill. var. **mollissima**
Jatropha polyantha Pax & K.Hoffm. === **Cnidoscolus aconitifolius** subsp. **polyanthus** (Pax & K.Hoffm.) Breckon
Jatropha porrecta (Pohl) Steud. === **Manihot tripartita** (Spreng.) Müll.Arg. subsp. **tripartita**
Jatropha portoricensis Millsp. === **Jatropha hernandiifolia** var. **portoricensis** (Millsp.) Urb.
Jatropha pringlei (I.M.Johnst.) Standl. === **Cnidoscolus angustidens** Torr.
Jatropha pronifolia (Pohl) Steud. === **Manihot gracilis** Pohl subsp. **gracilis**
Jatropha pruinosa (Pohl) Steud. === **Manihot pruinosa** Pohl
Jatropha pseudoglandulifera Pax === **Jatropha spicata** Pax
Jatropha pseudoglandulifera var. *zanguebarica* Hutch. === **Jatropha hildebrandtii** Pax var. **hildebrandtii**
Jatropha pubescens (Pohl) Steud. === **Manihot anomala** subsp. **pubescens** (Pohl) D.J.Rogers & Appan
Jatropha pubescens Pax === **Jatropha froesii** Croizat
Jatropha pungens Forssk. === **Tragia pungens** (Forssk.) Müll.Arg.
Jatropha purpureocostata (Pohl) Steud. === **Manihot purpureocostata** Pohl
Jatropha pusilla (Pohl) Steud. === **Manihot pusilla** Pohl
Jatropha pyrophora Pax === **Cnidoscolus pyrophorus** (Pax) J.F.Macbr.
Jatropha quinquefolia (Pohl) Steud. === **Manihot quinquefolia** Pohl
Jatropha quinqueformis Steud. === **Manihot quinquefolia** Pohl
Jatropha quinqueloba (Pohl) Steud. === **Manihot quinqueloba** Pohl
Jatropha quinqueloba Cerv. === **Cnidoscolus quinquelobatus** (Mill.) Leon
Jatropha quinquelobata Mill. === **Cnidoscolus quinquelobatus** (Mill.) Leon
Jatropha rangel M.Gómez === **Cnidoscolus rangel** (M.Gómez) McVaugh
Jatropha regina Léon === **Cnidoscolus regina** (Léon) Radcl.-Sm. & Govaerts
Jatropha reniformis (Pohl) Steud. === **Manihot reniformis** Pohl
Jatropha ribifolia var. *breviloba* (Morong) Pax === **Jatropha breviloba** (Morong) Pax & K.Hoffm.
Jatropha ricinifolia Fenzl ex Baill. === **Jatropha glauca** Vahl

Jatropha ricinifolia Pax === **Jatropha paraguayensis** Radcl.-Sm. & Govaerts
Jatropha rigidifolia Pax & K.Hoffm. === **Jatropha isabelliae** Müll.Arg.
Jatropha rigidifolia var. *glabrescens* Pax & K.Hoffm. === **Jatropha isabelliae** Müll.Arg.
Jatropha rotundifolia Müll.Arg. === **Cnidoscolus rotundifolius** (Müll.Arg.) McVaugh
Jatropha sabdariffa Schweinf. === **Jatropha aethiopica** Müll.Arg.
Jatropha sagittatopartita (Pohl) Steud. === **Manihot sagittatopartita** Pohl
Jatropha salicifolia (Pohl) Steud. === **Manihot salicifolia** Pohl
Jatropha seineri var. *tomentella* Radcl.-Sm. === **Jatropha seineri** Pax
Jatropha sellowiana (Klotzsch ex Pax) Pax & K.Hoffm. === **Cnidoscolus sellowianus** Klotzsch ex Pax
Jatropha serrulata Pax & K.Hoffm. === **Cnidoscolus serrulatus** (Pax & K.Hoffm.) Pax & K.Hoffm.
Jatropha setifera Hutch. === **Jatropha schlechteri** subsp. **setifera** (Hutch.) Radcl.-Sm.
Jatropha silvestris Vell. === **Manihot esculenta** Crantz
Jatropha simayuca Ruiz & Pav. ex Pax === **Manihot anomala** subsp. **pavoniana** (Müll.Arg.) D.J.Rogers & Appan
Jatropha sinuata (Pohl) Steud. === **Manihot tripartita** (Spreng.) Müll.Arg. subsp. **tripartita**
Jatropha sparsifolia (Pohl) Steud. === **Manihot sparsifolia** Pohl
Jatropha spathulata (Ortega) Müll.Arg. === **Jatropha dioica** Cerv.
Jatropha spathulata var. *genuina* Müll.Arg. === **Jatropha dioica** Cerv.
Jatropha spathulata var. *sessiliflora* (Hook.) Müll.Arg. === **Jatropha dioica** Cerv.
Jatropha spinosa var. *crenulata* Pax === **Jatropha spinosa** Vahl
Jatropha spinosa var. *somaliensis* Pax === **Jatropha spinosa** Vahl
Jatropha staphysagrifolia Mill. === **Jatropha gossypiifolia** L. var. **gossypiifolia**
Jatropha stimulosa Michx. === **Cnidoscolus urens** var. **stimulosus** (Michx.) Govaerts
Jatropha stipulata Vell. === **Manihot esculenta** Crantz
Jatropha stipulosa Steud. === **Cnidoscolus urens** var. **stimulosus** (Michx.) Govaerts
Jatropha subintegra (Pax & K.Hoffm.) Pax & K.Hoffm. === **Cnidoscolus subinteger** (Pax & K.Hoffm.) Pax & K.Hoffm.
Jatropha tenerrima (Pohl) Steud. === **Manihot pentaphylla** subsp. **tenuifolia** (Pohl) D.J.Rogers & Appan
Jatropha tenuifolia (Pohl) Steud. === **Manihot pentaphylla** subsp. **tenuifolia** (Pohl) D.J.Rogers & Appan
Jatropha tenuifolia Pax & K.Hoffm. === **Cnidoscolus tenuifolius** (Pax & K.Hoffm.) I.M.Johnst.
Jatropha tepiquensis Costantin & Gallaud === **Cnidoscolus tubulosus** (Müll.Arg.) I.M.Johnst.
Jatropha texana Müll.Arg. === **Cnidoscolus texanus** (Müll.Arg.) Small
Jatropha tomentella (Pohl) Steud. === **Manihot tripartita** (Spreng.) Müll.Arg. subsp. **tripartita**
Jatropha tomentosa Spreng. === **Vitex cymosa** Bert. ex Spreng. (Lamiaceae)
Jatropha tomentosa (Pohl) Steud. === **Manihot tomentosa** Pohl
Jatropha × *transiens* Pax === **Jatropha** × **induta** (Chodat & Hassl.) Pax
Jatropha triloba Sessé ex Cerv. === **Manihot triloba** (Sessé ex Cerv.) McVaugh ex Miranda
Jatropha tripartita Spreng. === **Manihot tripartita** (Spreng.) Müll.Arg.
Jatropha triphylla (Pohl) Steud. === **Manihot triphylla** Pohl
Jatropha tuberosa Pax === **Jatropha afrotuberosa** Radcl.-Sm. & Govaerts
Jatropha tuberosa Elliot === **Jatropha curcas** L.
Jatropha tubulosa Müll.Arg. === **Cnidoscolus tubulosus** (Müll.Arg.) I.M.Johnst.
Jatropha tubulosa var. *quinqueloba* Müll.Arg. === **Cnidoscolus tubulosus** (Müll.Arg.) I.M.Johnst.
Jatropha tubulosa var. *septemloba* Müll.Arg. === **Cnidoscolus tubulosus** (Müll.Arg.) I.M.Johnst.
Jatropha tubulosa var. *triloba* Müll.Arg. === **Cnidoscolus tubulosus** (Müll.Arg.) I.M.Johnst.
Jatropha ulei Pax === **Cnidoscolus ulei** (Pax) Pax
Jatropha urens L. === **Cnidoscolus urens** (L.) Arthur
Jatropha urens var. *brachyloba* Müll.Arg. === **Cnidoscolus urens** (L.) Arthur var. **urens**
Jatropha urens var. *genuina* Müll.Arg. === **Cnidoscolus urens** (L.) Arthur

Jatropha urens var. *herbacea* (L.) Müll.Arg. === **Cnidoscolus urens** (L.) Arthur var. **urens**
Jatropha urens var. *inermis* Calvino === **Cnidoscolus chayamansa** McVaugh
Jatropha urens var. *longipedunculata* Brandegee === **Cnidoscolus longipedunculatus** (Brandegee) Pax & K.Hoffm.
Jatropha urens var. *marcgravii* (Pohl) Müll.Arg. === **Cnidoscolus urens** (L.) Arthur var. **urens**
Jatropha urens f. *osteocarpa* (Pohl) Müll.Arg. === **Cnidoscolus urens** (L.) Arthur var. **urens**
Jatropha urens var. *osteocarpa* (Pohl) Müll.Arg. === **Cnidoscolus urens** (L.) Arthur var. **urens**
Jatropha urens var. *stimulosa* (Michx.) Müll.Arg. === **Cnidoscolus urens** var. **stimulosus** (Michx.)
Jatropha urnigera Pax === **Cnidoscolus urnigerus** (Pax) Pax
Jatropha varians (Pohl) Steud. === **Manihot gracilis** subsp. **varians** (Pohl) D.J.Rogers & Appan
Jatropha villosa (Pohl) Baill. === **Jatropha mollissima** (Pohl) Baill. var. **mollissima**
Jatropha villosa (Forssk.) Müll.Arg. === **Jatropha pelargoniifolia** Courbai var. **pelargoniifolia**
Jatropha villosa var. *genuina* Müll.Arg. === **Jatropha pelargoniifolia** Courbai var. **pelargoniifolia**
Jatropha villosa var. *glabra* Müll.Arg. === **Jatropha pelargoniifolia** var. **glabra** (Müll.Arg.) Radcl.-Sm.
Jatropha villosa var. *glandulosa* Pax === **Jatropha pelargoniifolia** Courbai var. **pelargoniifolia**
Jatropha villosa var. *pelargoniifolia* (Courbai) Chiov. === **Jatropha pelargoniifolia** Courbai
Jatropha villosa var. *sublobata* O.Schwarz === **Jatropha pelargoniifolia** var. **sublobata** (O.Schwartz) Radcl.-Sm.
Jatropha viminea Retz. ex Steud. === ?
Jatropha violacea (Pohl) Steud. === **Manihot violacea** Pohl
Jatropha vitifolia Mill. === **Cnidoscolus vitifolius** (Mill.) Pohl
Jatropha vitifolia var. *cnicodendron* (Griseb.) Pax === **Cnidoscolus cnicodendron** Griseb.
Jatropha vitifolia var. *genuina* Müll.Arg. === **Cnidoscolus vitifolius** (Mill.) Pohl
Jatropha vitifolia var. *grisebachii* Pax === **Cnidoscolus cnicodendron** Griseb.
Jatropha vitifolia var. *maritima* Müll.Arg. === **Cnidoscolus cnicodendron** Griseb.
Jatropha vitifolia f. *nana* Chodat & Hassl. === **Cnidoscolus albomaculatus** (Pax) I.M.Johnst.
Jatropha vitifolia var. *obtusifolia* Müll.Arg. === **Cnidoscolus vitifolius** (Mill.) Pohl
Jatropha vitifolia var. *repanda* (Griseb.) Pax === **Cnidoscolus cnicodendron** Griseb.
Jatropha vitifolia f. *stimulosissima* Chodat & Hassl. === **Cnidoscolus albomaculatus** (Pax) I.M.Johnst.
Jatropha vitifolia f. *subintegra* Chodat & Hassl. === **Cnidoscolus subinteger** (Pax & K.Hoffm.) Pax & K.Hoffm.
Jatropha wightiana Müll.Arg. === **Jatropha villosa** Wight
Jatropha woodii var. *kuntzei* Pax === **Jatropha woodii** Kuntze
Jatropha woodii var. *vestita* Pax === **Jatropha woodii** Kuntze
Jatropha yucatanensis Briq. === **Jatropha curcas** L.
Jatropha zeyheri var. *platyphylla* Pax === **Jatropha zeyheri** Sond.
Jatropha zeyheri var. *subsimplex* Prain === **Jatropha zeyheri** Sond.

Joannesia

2 species, South America (Brazil); both introduced elsewhere (e.g. Venezuela, Madeira, Cairo, Calcutta, Dehra Dun, Bogor); trees to 20 m with distinctive palmately compound leaves. *J. insolita* Pittier, described from Venezuela, was in 1955 shown to have been based on cultivated examples of *J. princeps*. The genus appears to be most closely related to *Leeuwenbergia* (Africa) and *Annesijoa* (New Guinea); this is once more suggestive of now-disrupted past geographical links accompanied by extinction and character divergence. (Crotonoideae)

Pax, F. (1910). *Joannesia*. In A. Engler (ed.), Das Pflanzenreich, IV 147 [I] (Euphorbiaceae-Jatropheae): 116-117. Berlin. (Heft 42.) La/Ge. — 1 species, Brazil; extensively cultivated there and elsewhere.

Schultes, R. E. (1955). A note on the genus *Joannesia*. Bot. Mus. Leafl. 17: 25-26. En. —*J. princeps* identified as cultivated in Venezuela; the comment is additionally made that the tree has been grown elsewhere, even in the Old World (e.g. at the Indian Botanical Garden near Calcutta), from seeds or plants distributed from Rio de Janeiro in the nineteenth century.

Joannesia Vell., Alogr. Alkalis: 199 (1798).
Brazil. 84.
Anda A.Juss. in F.Cuvier, Dict. Sci. Nat. 2: 113 (1816).
Andicus Vell., Fl. Flumin.: 80 (1829).
Andiscus Vell., Fl. Flumin. 2: t. 86 (1829).

Joannesia heveoides Ducke, Arch. Jard. Bot. Rio de Janeiro 3: 198 (1922).
Brazil (Amazonas). 84 BZN. Phan.

Joannesia princeps Vell., Alogr. Alkalis: 199 (1798).
Brazil (Rio de Janeiro). 84 BZL. Phan.
Joannesia insolita Pittier, Bol. Soc. Venez. Ci. Nat. 6: 8 (1940).

Synonyms:
Joannesia insolita Pittier === **Joannesia princeps** Vell.

Julocroton

A long-recognised South American segregate of *Croton;* now reduced to that genus as sect. 29 in the system of Webster (1993; see there). The name is conserved against *Cieca* Adans.

Synonyms:
Julocroton Mart. === **Croton** L.
Julocroton abutiloides S.Moore === **Croton abutilopsis** G.L.Webster
Julocroton ackermannianus Müll.Arg. === **Croton ackermannianus** (Müll.Arg.) G.L.Webster
Julocroton acuminatissimus Pittier === **Croton acuminatissimus** (Pittier) G.L.Webster
Julocroton agrestis Pax & K.Hoffm. === **Croton agrestis** (Pax & K.Hoffm.) Radcl.-Sm. & Govaerts
Julocroton argenteus (L.) Didr. === **Croton argenteus** L.
Julocroton brittonianum Morong === **Croton subpannosus** Müll.Arg. ex Griseb.
Julocroton camporum Chodat & Hassl. === **Croton argenteus** L.
Julocroton chodatii Croizat === **Croton salzmannii** (Baill.) G.L.Webster
Julocroton conspurcatus (Schltdl.) Klotzsch === **Croton conspurcatus** Schltdl.
Julocroton cooperianus Croizat === **Croton cooperianus** (Croizat) Radcl.-Sm. & Govaerts
Julocroton crassirameus Croizat === **Croton spissirameus** Radcl.-Sm. & Govaerts
Julocroton decalobus (Müll.Arg.) Benth. & Hook.f. === **Croton decalobus** Müll.Arg.
Julocroton doratophylloides Croizat === **Croton doratophylloides** (Croizat) G.L.Webster
Julocroton doratophyllus (Baill.) Müll.Arg. === **Croton doratophyllus** Baill.
Julocroton elaeagnoides S.Moore === **Croton argenteus** L.
Julocroton fuscescens (Spreng.) Baill. === **Croton fuscescens** Spreng.
Julocroton gardneri Müll.Arg. === **Croton argentealbidus** Radcl.-Sm. & Govaerts
Julocroton geraesensis Baill. === **Croton geraesensis** (Baill.) G.L.Webster
Julocroton glazioui Croizat === **Croton tocantinsensis** Radcl.-Sm. & Govaerts
Julocroton herzogianus Pax & K.Hoffm. === **Croton herzogianus** (Pax & K.Hoffm.) Radcl.-Sm. & Govaerts
Julocroton holodiscus Croizat === **Croton holodiscus** (Croizat) Radcl.-Sm. & Govaerts
Julocroton hondensis (H.Karst.) Müll.Arg. === **Croton hondensis** (H.Karst.) G.L.Webster
Julocroton humilis Didr. === **Croton didrichsenii** G.L.Webster

Julocroton humilis var. *robustior* L.B.Sm. & Downs === **Croton robustior** (L.B.Sm. & Downs) Radcl.-Sm. & Govaerts
Julocroton humilis var. *solanaceus* Müll.Arg. === **Croton solanaceus** (Müll.Arg.) G.L.Webster
Julocroton integer Chodat === **Croton integer** (Chodat) Radcl.-Sm. & Govaerts
Julocroton lanatus Klotzsch ex Baill. === **Croton fuscescens** Spreng.
Julocroton lanceolatus Klotzsch ex Müll.Arg. === **Croton lanceolaris** G.L.Webster
Julocroton lepidus S.Moore === **Croton lepidus** (S.Moore) Radcl.-Sm. & Govaerts
Julocroton linearifolius (Chodat & Hassl.) Croizat === **Croton argenteus** L.
Julocroton lithrifolius Croizat === **Croton lithrifolius** (Croizat) Radcl.-Sm. & Govaerts
Julocroton malvoides Croizat === **Croton malvoides** (Croizat) Radcl.-Sm. & Govaerts
Julocroton microcalyx Müll.Arg. === **Croton microcalyx** (Müll.Arg.) G.L.Webster
Julocroton montevidensis Klotzsch ex Baill. === **Croton argenteus** L.
Julocroton montevidensis var. *linearifolius* Chodat & Hassl. === **Croton argenteus** L.
Julocroton montevidensis var. *pilosus* Müll.Arg. === **Croton argenteus** L.
Julocroton montevidensis var. *stipularis* Müll.Arg. === **Croton stipularis** (Müll.Arg.) G.L.Webster
Julocroton nervosus Baill. === **Croton calonervosus** G.L.Webster
Julocroton nigricans Mart. ex Schltdl. === **Croton nigricans** (Mart. ex Schltdl.) Radcl.-Sm. & Govaerts
Julocroton ostenii Herter === ?
Julocroton paniculatus Pax & K.Hoffm. === **Croton flavispicatus** Rusby
Julocroton paulensis Usteri === **Croton fuscescens** Spreng.
Julocroton peruvianus Müll.Arg. === **Croton flavispicatus** Rusby
Julocroton peruvianus var. *flavispicatus* (Rusby) Croizat === **Croton flavispicatus** Rusby
Julocroton phagedaenicus Mart. === **Croton triqueter** Lam.
Julocroton phyllanthus Chodat & Hassl. === **Croton phyllanthus** (Chodat & Hassl.) G.L.Webster
Julocroton pilosus (Müll.Arg.) Herter === **Croton argenteus** L.
Julocroton pulcher Croizat === **Croton rutilus** (Chodat & Hassl.) G.L.Webster
Julocroton pycnophyllus Schltdl. ex Müll.Arg. === **Croton salzmannii** (Baill.) G.L.Webster
Julocroton pycnophyllus f. *latifolius* Chodat & Hassl. === **Croton salzmannii** (Baill.) G.L.Webster
Julocroton pyrosoma Croizat === **Croton pyrosoma** (Croizat) Radcl.-Sm. & Govaerts
Julocroton quinquenervius Baill. === **Croton argenteus** L.
Julocroton ramboi L.B.Sm. & Downs === **Croton allemii** G.L.Webster
Julocroton riedelianus Müll.Arg. === **Croton cordeiroae** G.L.Webster
Julocroton rufescens Klotzsch ex Baill. ===?
Julocroton rupestris Chodat & Hassl. === **Croton rupestris** (Chodat & Hassl.) G.L.Webster
Julocroton rupestris var. *velutinus* Chodat & Hassl. === **Croton rupestris** (Chodat & Hassl.) G.L.Webster
Julocroton rutilus Chodat & Hassl. === **Croton rutilus** (Chodat & Hassl.) G.L.Webster
Julocroton salzmannii Baill. === **Croton salzmannii** (Baill.) G.L.Webster
Julocroton serratus Müll.Arg. === **Croton subpannosus** Müll.Arg. ex Griseb.
Julocroton solanaceus Klotzsch ex Baill. === **Croton solanaceus** (Müll.Arg.) G.L.Webster
Julocroton stipularis (Müll.Arg.) Müll.Arg. === **Croton stipularis** (Müll.Arg.) G.L.Webster
Julocroton subpannosus Müll.Arg. === **Croton subpannosus** Müll.Arg. ex Griseb.
Julocroton thellungianus Herter ex Arechav. === **Croton thellungianus** (Herter ex Arechav.) Radcl.-Sm. & Govaerts
Julocroton trichophilus Pax & K.Hoffm. === **Croton trichophilus** (Pax & K.Hoffm.) Radcl.-Sm. & Govaerts
Julocroton triqueter (Lam.) Didr. === **Croton triqueter** Lam.
Julocroton triqueter var. *conspurcatus* (Schltdl.) Müll.Arg. === **Croton conspurcatus** Schltdl.
Julocroton typhicephalus Croizat === **Croton solanaceus** (Müll.Arg.) G.L.Webster
Julocroton valenzuellae (Chodat & Hassl.) Croizat === ?
Julocroton velutinus (Chodat & Hassl.) Croizat === **Croton rupestris** (Chodat & Hassl.) G.L.Webster

Julocroton verbascifolius Müll.Arg. === **Croton verbascoides** G.L.Webster
Julocroton vergarenae Jabl. === **Croton vergarenae** (Jabl.) Gillespie
Julocroton villosissimus Chodat & Hassl. === **Croton villosissimus** (Chodat & Hassl.) Radcl.-Sm. & Govaerts
Julocroton villosissimus var. *valenzuellae* Chodat & Hassl. === **Croton** sp.
Julocroton viridulus Croizat === **Croton viridulus** (Croizat) Radcl.-Sm. & Govaerts
Julocroton vulpinus Croizat === **Croton apostolon** Radcl.-Sm. & Govaerts

Junghuhnia

Synonyms:
Junghuhnia Miq. === **Codiaeum** Rumph. ex A.Juss.

Jussieuia

Synonyms:
Jussieuia Houst. === **Cnidoscolus** Pohl

Kairothamnus

1 species, New Guinea (SE. part of Morobe Province in Papua New Guinea, where associated with the Papuan Ultramafic Belt). The habit of this small riverine tree is reminiscent of *Phyllanthus* but the floral characters, including the numerous free stamens, are similar to those of *Austrobuxus* within which it was originally described; in addition, the leaves show some similarity to those in *Petalostigma*. The genus, however, was tentatively placed by Webster (1994, Synopsis) in subtribe Pseuanthinae along with a miscellany of other Australasian and New Caledonian genera. (Oldfieldioideae)

Airy-Shaw, H. K. (1980). *Kairothamnus*. Euphorbiaceae of New Guinea: 121-122. London. (Kew Bull. Addit. Ser., 8.) En. — Regional revisionary treatment; for illustration see plate 1.
Airy-Shaw, H. K. (1980). *Kairothamnus*. Kew Bull. 34: 596-597 (in New Euphorbiaceae from New Guinea). En. — Monotypic, Papua New Guinea (Morobe Province, Paiawa-Buso coastal area SE of Lae).

Kairothamnus Airy Shaw, Kew Bull. 34: 596 (1980).
Papua New Guinea. 42. Nanophan. or phan.

Kairothamnus phyllanthoides (Airy Shaw) Airy Shaw, Kew Bull. 34: 596 (1980).
– FIGURE, p. 1049.
Papua New Guinea (Morobe). 42 NWG. Nanophan. or phan.
* *Austrobuxus phyllanthoides* Airy Shaw, Kew Bull. 29: 303 (1974).

Kaluhaburunghos

Synonyms:
Kaluhaburunghos L. ex Kuntze === **Cleistanthus** Hook.f. ex Planch.
Kaluhaburunghos stipularis Kuntze === **Cleistanthus stipularis** (Kuntze) Müll.Arg.

Kanopikon

Synonyms:
Kanopikon Raf. === **Euphorbia** L.

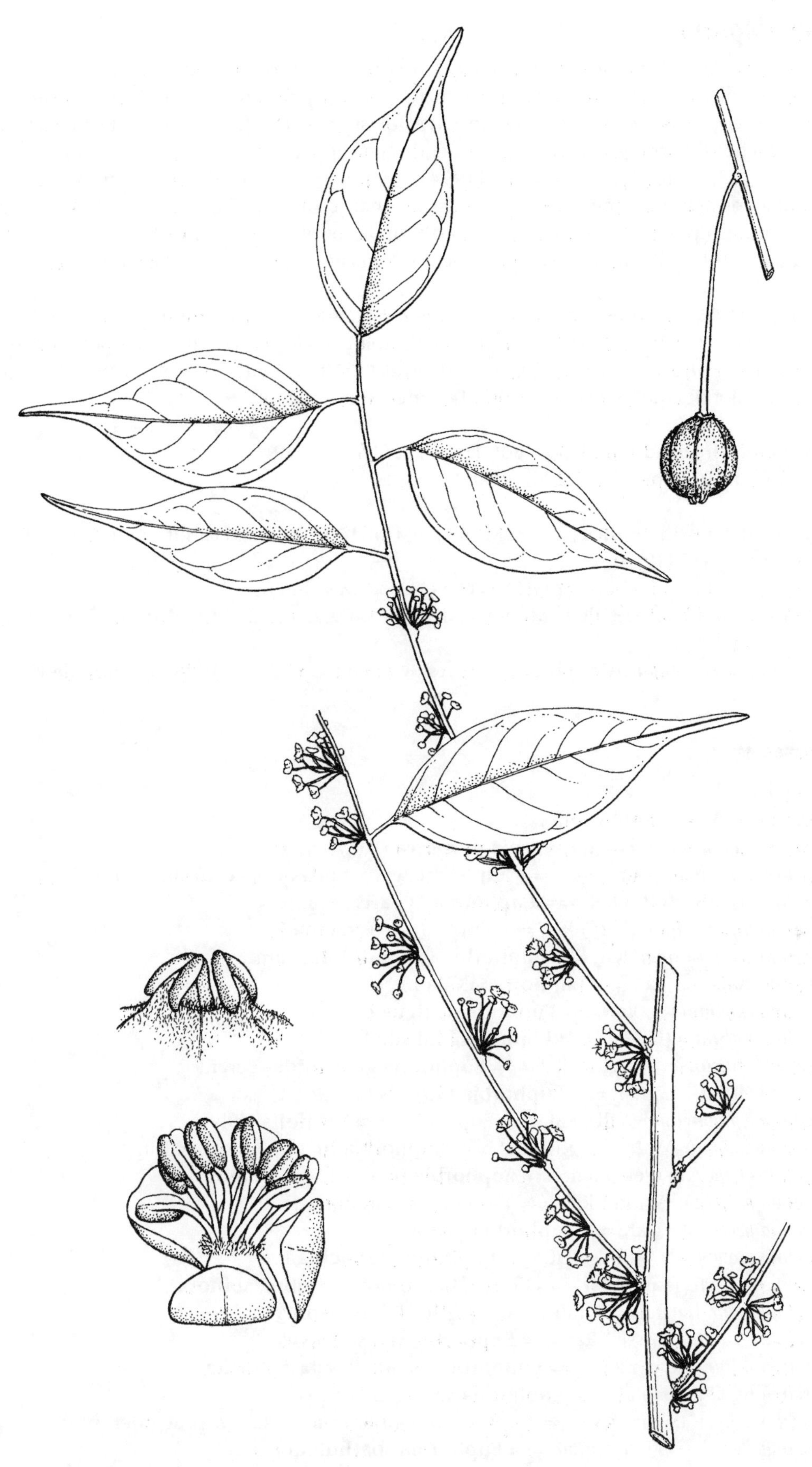

Kairothamnus phyllanthoides (Airy Shaw) Airy Shaw

Artist: Ann Davies
Airy Shaw, Euphorbiaceae of New Guinea: pl. 1, lower left (fig. 4 (1980), enlarged & rearranged
KEW ILLUSTRATIONS COLLECTION

Keayodendron

1 species, W. & WC. tropical Africa (Ivory Coast to Cameroon, but apparently very rare in Ghana); formerly in Flacourtiaceae (to which it is still referred (in error?) by Lebrun and Stork in their *Énumération*). Medium to large forest trees to 40 m without buttresses and with distichously arranged leaves (for a dendrological account, see Keay et al. in *Nigerian Trees* 1: 287 (1960)). Webster (1994, Synopsis) referred the genus to his Pseudolachnostylidinae but more by demission than design; earlier, Leandri (1959) argued for a relationship with *Drypetes*, at least partly on account of the drupaceous fruit (in other genera of Webster's Pseudolachnostylidinae the fruits are capsular). (Phyllanthoideae)

Leandri, J. (1959). Le problème du *Casearia bridelioides* Mildbr. ex Hutch. et Dalz. Bull. Soc. Bot. France 105: 512-517, illus. Fr. — Protologue of genus with incorporation of 1 species, *Keayodendron bridelioides* of W. and C. Africa (with its synonymy); extensive discussion including examination of a range of characters.

Keayodendron Leandri, Bull. Soc. Bot. France 105: 517 (1959).
Ivory Coast to Cameroon. 22 23. Phan.

Keayodendron bridelioides (Gilg & Mildbr. ex Hutch. & Dalziel) Leandri, Bull. Soc. Bot. France 105: 517 (1959).
Ivory Coast to Cameroon. 22 GHA IVO NGA 23 CMN. Phan.
* *Casearia bridelioides* Gilg & Mildbr. ex Hutch. & Dalziel, Bull. Misc. Inform. Kew 1928: 213 (1928).
Drypetes sassandraensis Aubrév., Fl. Forest. Côte d'Ivoire 2: 48 (1936), no latin descr.

Keraselma

Synonyms:
Keraselma Neck. === **Euphorbia** L.
Keraselma ciliatum Raf. === **Euphorbia purpurea** (Raf.) Fernald
Keraselma corifolium (Lam.) Raf. === **Euphorbia genistoides** var. **corifolia** (Lam.) N.E.Br.
Keraselma cyparissias (L.) Raf. === **Euphorbia cyparissias** L.
Keraselma diversifolium (Poir.) Raf. === **Euphorbia terracina** L.
Keraselma echinocarpum Raf. === **Euphorbia valerianifolia** Lam.
Keraselma esula (L.) Raf. === **Euphorbia esula** L.
Keraselma exiguum (L.) Raf. === **Euphorbia exigua** L.
Keraselma falcatum (L.) Raf. === **Euphorbia falcata** L.
Keraselma genistoides (Bergius) Raf. === **Euphorbia genistoides** Bergius
Keraselma lathyris (L.) Raf. === **Euphorbia lathyris** L.
Keraselma leptophyllum (Vill.) Raf. === **Euphorbia graminifolia** Vill.
Keraselma lucidum (Waldst. & Kit.) Raf. === **Euphorbia lucida** Waldst. & Kit.
Keraselma oleraceum (Pers.) Raf. === **Euphorbia peplus** L. var. **peplus**
Keraselma pallidum (Willd.) Raf. === **Euphorbia salicifolia** Host
Keraselma peplus (L.) Raf. === **Euphorbia peplus** L.
Keraselma provinciale (Willd.) Raf. === **Euphorbia terracina** L.
Keraselma pungens (Lam.) Raf. === **Euphorbia spinosa** L. subsp. **spinosa**
Keraselma reniforme Raf. === **Euphorbia peplus** L. var. **peplus**
Keraselma retusum (Forssk.) Raf. === **Euphorbia retusa** Forssk.
Keraselma rubrum (Cav.) Raf. === **Euphorbia falcata** L. subsp. **falcata**
Keraselma segetale (L.) Raf. === **Euphorbia segetalis** L.
Keraselma seguieri (Scop.) Raf. === **Euphorbia seguieriana** Neck. subsp. **seguieriana**
Keraselma spatulatum (Lam.) Raf. === **Euphorbia spathulata** Lam.
Keraselma squamosum (Willd.) Raf. === **Euphorbia squamosa** Willd.
Keraselma sylvaticum (L.) Raf. === **Euphorbia amygdaloides** L. subsp. **amygdaloides**
Keraselma virgatum (Haw.) Raf. === **Euphorbia esula** subsp. **tommasiniana** (Bertol.) Kuzmanov

Kirganelia

Sometimes, though mainly in the nineteenth century, segregated from *Phyllanthus*.

Synonyms:
Kirganelia Juss. === **Phyllanthus** L.
Kirganelia alba Blanco === **Glochidion album** (Blanco) Boerl.
Kirganelia boiviniana Baill. === **Phyllanthus decipiens** var. **boivinianus** (Baill.) Leandri
Kirganelia bojeriana Baill. === **Phyllanthus bojerianus** (Baill.) Müll.Arg.
Kirganelia decipiens Baill. === **Phyllanthus decipiens** (Baill.) Müll.Arg.
Kirganelia dubia (Blume) Baill. === **Phyllanthus reticulatus** Poir. var. **reticulatus**
Kirganelia dumetosa (Poir.) Spreng. === **Phyllanthus dumetosus** Poir.
Kirganelia eglandulosa Baill. === **Phyllanthus reticulatus** Poir. var. **reticulatus**
Kirganelia elegans Juss. ex Spreng. === **Phyllanthus casticum** P.Willemet. var. **casticum**
Kirganelia elliptica Spreng. === **Flueggea elliptica** (Spreng.) Baill.
Kirganelia floribunda Baill. === **Phyllanthus muellerianus** (Kuntze) Exell
Kirganelia floribunda (Kunth) Spreng. === **Phyllanthus salviifolius** Kunth
Kirganelia glaucescens Baill. === **Phyllanthus mocquerysianus** A.DC.
Kirganelia intermedia (Decne.) Baill. === **Phyllanthus reticulatus** Poir. var. **reticulatus**
Kirganelia lineata Alston === **Phyllanthus reticulatus** Poir. var. **reticulatus**
Kirganelia microcarpa (Benth.) Hurus. & Yu.Tanaka === **Phyllanthus reticulatus** Poir. var. **reticulatus**
Kirganelia multiflora (Willd.) Baill. === **Phyllanthus reticulatus** Poir. var. **reticulatus**
Kirganelia multiflora var. *glabra* Thwaites === **Phyllanthus reticulatus** var. **glaber** (Thwaites) Müll.Arg.
Kirganelia nigrescens Blanco === **Phyllanthus nigrescens** (Blanco) Müll.Arg.
Kirganelia pervilleana Baill. === **Phyllanthus pervilleanus** (Baill.) Müll.Arg.
Kirganelia phyllanthoides A.Juss. === **Phyllanthus casticum** P.Willemet. var. **casticum**
Kirganelia prieuriana Baill. === **Phyllanthus reticulatus** Poir. var. **reticulatus**
Kirganelia prieuriana var. *glabra* Baill. === **Phyllanthus reticulatus** var. **glaber** (Thwaites) Müll.Arg.
Kirganelia puberula (Miq. ex Baill.) Baill. === **Phyllanthus reticulatus** Poir. var. **reticulatus**
Kirganelia pumila Blanco === **Phyllanthus pumilus** (Blanco) Müll.Arg.
Kirganelia reticulata (Poir.) Baill. === **Phyllanthus reticulatus** Poir.
Kirganelia salviifolia (Kunth) Spreng. === **Phyllanthus salviifolius** Kunth
Kirganelia sinensis Baill. === **Phyllanthus reticulatus** Poir. var. **reticulatus**
Kirganelia triandra Blanco === **Glochidion triandrum** (Blanco) C.B.Rob.
Kirganelia trilocularis Baill. === **Phyllanthus decipiens** f. **trilocularis** (Baill.) Leandri
Kirganelia vieillardii Baill. === **Phyllanthus deplanchei** Müll.Arg.
Kirganelia villosa Blanco === **Sauropus villosus** (Blanco) Merr.
Kirganelia virginea J.F.Gmel. === **Phyllanthus casticum** P.Willemet. var. **casticum**
Kirganelia zanzibariensis Baill. === **Phyllanthus reticulatus** var. **glaber** (Thwaites) Müll.Arg.

Klaineanthus

1 species, W. & WC. tropical Africa (Nigeria to Gabon and Zaire); medium forest trees to 20-25 m with large axillary inflorescences (but sometimes appearing terminal above the leaves), most closely related to *Tetrorchidium* within Webster's subtribe Adenoclininae (Adenoclineae). Prain, in describing the genus, suggested an affinity with the Amazonian *Cunuria* in Micrandreae; however, here the laticifers are non-articulate. (Crotonoideae)

Prain, D. (1913). *Klaineanthus gaboniae*. Ic. Pl. 30: pl. 2985. La/En. — Plant portrait with description and commentary.

Pax, F. (with K. Hoffmann) (1914). *Klaineanthus*. In A. Engler (ed.), Das Pflanzenreich, IV 147 VII [Euphorbiaceae-Additamentum V]: 408-409. Berlin. (Heft 63.) La/Ge. — 1 species, Africa.

Klaineanthus Pierre ex Prain, Bull. Misc. Inform. Kew 1912: 105 (1912).
W. & WC. Trop. Africa. 22 23.

Klaineanthus gaboniae Pierre, Bull. Misc. Inform. Kew 1912: 106 (1912).
Nigeria, Cameroon, Gabon, Zaire. 22 NGA 23 CMN GAB ZAI. Phan.

Kleinodendron

Synonyms:
Kleinodendron L.B.Sm. & Downs === **Savia** Willd.
Kleinodendron riosulense L.B.Sm. & Downs === **Savia dictyocarpa** Müll.Arg.

Klotzschiphytum

Synonyms:
Klotzschiphytum Baill. === **Croton** L.

Kobiosis

Synonyms:
Kobiosis Raf. === **Euphorbia** L.

Koelera

Synonyms:
Koelera Willd. === **Drypetes** Vahl
Koelera serrata Maycock === **Drypetes glomerata** Griseb.

Koilodepas

11 species, India, Hainan and SE. Asia east to Borneo (10) and SE. New Guinea (1); shrubs or small trees with hard wood. Airy-Shaw recognised two sections, *Hyalodepas* and *Koilodepas*, based on the form of the fruiting calyx. The species within West Malesia are mutually rather closely related and may be difficult to distinguish unless fruits are available. Webster referred the genus to subtribe Epiprininae in tribe Epiprineae but according to Airy-Shaw the position of the genus in the family was 'isolated'. The most recent overall synopsis is by Airy-Shaw (1960) covering 8 species. A further revision, for *Flora Malesiana*, has been prepared by Muzzazinah. (Acalyphoideae)

Pax, F. (with K. Hoffmann) (1914). *Calpigyne*. In A. Engler (ed.), Das Pflanzenreich, IV 147 VII (Euphorbiaceae-Acalypheae-Mercurialinae): 254-255. Berlin. (Heft 63.) La/Ge. — Monotypic, Malesia. [Reduced to *Koilodepas* by Airy-Shaw.]

Pax, F. (with K. Hoffmann) (1914). *Coelodepas*. In A. Engler (ed.), Das Pflanzenreich, IV 147 VII (Euphorbiaceae-Acalypheae-Mercurialinae): 268-270. Berlin. (Heft 63.) La/Ge. — 6 species, S Asia, Malesia.

• Airy-Shaw, H. K. (1960). Notes on Malaysian Euphorbiaceae, XI. A synopsis of the genus *Koilodepas* (*Coelodepas* Hassk.), with a note on *Calpigyne frutescens* Bl. Kew Bull. 14: 382-391. En. — Account of 9 species (one with 2 varieties) in 2 sections; key, descriptions of new taxa, localities with exsiccatae and commentary. *Calpigyne frutescens* is here incorporated in *Koilodepas* and *'Coelodepas' hosei* Merr. is transferred to *Claoxylon*.

Airy-Shaw, H. K. (1963). Notes on Malaysian and other Asiatic Euphorbiaceae, XXXVI. Second thoughts on *Koilodepas* Hasskarl. Kew Bull. 16: 354-356. En. — Further notes on four species, including additional records.

Airy-Shaw, H. K. (1969). Notes on Malesian and other Asiatic Euphorbiaceae, CIX. New species of *Koilodepas* Hassk. Kew Bull. 23: 82-85. En. — Descriptions of two novelties from Borneo and notes on possible further ones, including one from southeastern New Guinea (see also Airy-Shaw 1981); sections indicated.

Airy-Shaw, H. K. (1981). Notes on Asiatic, Malesian and Melanesian Euphorbiaceae, CCXLIX. *Koilodepas* Hassk. Kew Bull. 36: 609-610. En. — Description of *K. homaliifolium* from the Central Province of Papua New Guinea (cf. Airy-Shaw 1969). [Recollected in 1984 by one of the authors of this *World Checklist and Bibliography of Euphorbiaceae*.]

Koilodepas Hassk., Verslagen Meded. Afd. Natuurk. Kon. Akad. Wetensch. 4: 139 (1855).
Hainan, Trop. Asia. 36 40 41 42.
Calpigyne Blume, Mus. Bot. 2: 193 (1856).
Coelodepas Hassk., Flora 40: 531 (1857).
Caelodepas Benth. & Hook.f., Gen. Pl. 3: 313 (1880).
Nephrostylus Gagnep., Bull. Soc. Bot. France 72: 467 (1925).

Koilodepas bantamense Hassk., Verslagen Meded. Afd. Natuurk. Kon. Akad. Wetensch. 4: 140 (1855).
Sumatera, Jawa. 42 JAW SUM. Nanophan. or phan.

Koilodepas brevipes Merr., Philipp. J. Sci. 30: 80 (1926).
Borneo (Sabah, Kalimantan). 42 BOR. Nanophan. or phan.

var. **brevipes**
Borneo (Sabah, N. Kalimantan). 42 BOR. Nanophan. or phan.

var. **stenosepalum** (Airy Shaw) Airy Shaw, Kew Bull., Addit. Ser. 4: 138 (1975).
Borneo (E. Kalimantan). 42 BOR. Nanophan.
* *Koilodepas stenosepalum* Airy Shaw, Kew Bull. 14: 390 (1960).

Koilodepas calycinum Bedd., Fl. Sylv. S. India: 207 (1872).
S. India. 40 IND. Nanophan. or phan.
Adenochlaena calycina Bedd., Fl. Sylv. S. India: 207 (1872).

Koilodepas ferrugineum Hook.f., Fl. Brit. India 5: 420 (1887).
Pen. Malaysia. 42 MLY.

Koilodepas frutescens (Blume) Airy Shaw, Kew Bull. 14: 385 (1960).
Borneo (SE. Kalimantan). 42 BOR. Nanophan.
* *Calpigyne frutescens* Blume, Mus. Bot. 2: 193 (1856). *Ptychopyxis frutescens* (Blume) Croizat, J. Arnold Arbor. 23: 49 (1942).

Koilodepas hainanense (Merr.) Croizat, J. Arnold Arbor. 23: 51 (1942).
N. Sumatera ?, Hainan. 36 CHH 42 SUM? Phan.
* *Calpigyne hainanensis* Merr., J. Arnold Arbor. 6: 135 (1925).

Koilodepas homaliifolium Airy Shaw, Kew Bull. 36: 609 (1981).
Papua New Guinea. 42 NWG. Phan.

Koilodepas laevigatum Airy Shaw, Kew Bull. 23: 83 (1969).
Borneo (Sarawak, Sabah). 42 BOR. Phan.
Koilodepas longifolium var. *integrifolium* Airy Shaw, Kew Bull. 14: 388 (1960).

Koilodepas longifolium Hook.f., Fl. Brit. India 5: 420 (1887).
S. Thailand, Pen. Malaysia, Vietnam, W. Sumatera (incl. Bangka), Borneo (Sarawak, Sabah). 41 THA VIE 42 BOR MLY SUM. Nanophan. or phan.

Koilodepas glanduligerum Pax & K.Hoffm. in H.G.A.Engler, Pflanzenr., IV, 147, VII: 270 (1914).
Koilodepas subcordatum Gage, Rec. Bot. Surv. India 9: 239 (1922).

Koilodepas pectinatum Airy Shaw, Kew Bull. 23: 82 (1969).
Borneo (Sabah). 42 BOR. Phan.

Koilodepas wallichianum Benth., Hooker's Icon. Pl. 13: t. 1288 (1879).
Pen. Malaysia. 42 MLY. Nanophan.

Synonyms:
Koilodepas glanduligerum Pax & K.Hoffm. === **Koilodepas longifolium** Hook.f.
Koilodepas hosei Merr. === **Claoxylon hosei** (Merr.) Airy Shaw
Koilodepas longifolium var. *integrifolium* Airy Shaw === **Koilodepas laevigatum** Airy Shaw
Koilodepas stenosepalum Airy Shaw === **Koilodepas brevipes** var. **stenosepalum** (Airy Shaw) Airy Shaw
Koilodepas subcordatum Gage === **Koilodepas longifolium** Hook.f.

Kunstlerodendron

Synonyms:
Kunstlerodendron Ridl. === **Chondrostylis** Boerl.

Kurkas

Synonyms:
Kurkas Raf. === **Croton** L.

Kurziodendron

Synonyms:
Kurziodendron N.P.Balakr. === **Trigonostemon** Blume

Lacanthis

Now formally accepted as a subgenus of *Euphorbia*.

Synonyms:
Lacanthis Raf. === **Euphorbia** L.

Lachnostylis

2 species, S Africa (Cape Province); a much-branched, densely foliaged shrub or small tree to 3 m, found mainly in the Knysna forest district. A second species, *L. bilocularis*, has sometimes been recognised for inland shrubby populations. In the past the genus sometimes was included with the South American *Discocarpus* (e.g. by Pax & Hoffmann 1922). Any connection between them is, however, surely ancient, possibly also involving now-extinct relatives. In addition, recent work suggests that the two genera are not as closely allied as previously believed. (Phyllanthoideae)

Bentham, G. (1879). *Lachnostylis capensis*. Ic. Pl. 13: 61-62, pl. 1279. En. — Plant portrait with text. [The author suggests a close affinity with *Discocarpus*.]

Pax, F. & K. Hoffmann (1922). *Discocarpus*. In A. Engler (ed.), Das Pflanzenreich, IV 147 XV (Euphorbiaceae-Phyllanthoideae-Phyllantheae): 203-205. Berlin. (Heft 81.) La/Ge. — 4 species; 3 in the Guianas and Brazil, 1 S Africa. [The S African species is that now usually referred to *Lachnostylis*.]

Dyer, R. A. (1943). The genus *Lachnostylis* Turcz. S. African J. Sci. 40: 123-126, illus. En. — General discussion; description of *Lachnostylis bilocularis* and amplification of generic description; localities of *L. bilocularis* with exsiccatae; no key. [Retained in southern Africa as distinct from *L. hirta*.]

Lachnostylis bilocularis R.A.Dyer, S. African J. Sci. 40: (1943).
Cape Prov. 27 CPP. Nanophan. or phan. – In inland habitats.

Lachnostylis Turcz., Bull. Soc. Imp. Naturalistes Moscou 19: 503 (1846).
S. Africa. 27. Nanophan.

Lachnostylis hirta (L.f.) Müll.Arg. in A.P.de Candolle, Prodr. 15(2): 224 (1866).
– FIGURE, p. 1056.
Cape Prov.. 27 CPP. Nanophan. – In coastal bush.
Clutia acuminata L.f., Suppl. Pl.: 432 (1782).
* *Clutia hirta* L.f., Suppl. Pl.: 432 (1782). *Lachnostylis hirta* var. *genuina* Müll.Arg. in A.P.de Candolle, Prodr. 15(2): 224 (1866), nom. inval. *Discocarpus hirtus* (L.f.) Pax & K.Hoffm. in H.G.A.Engler, Pflanzenr., IV, 147, XV: 204 (1922).
Clutia hirta Vahl, Symb. Bot. 2: 101 (1791), nom. illeg.
Clutia acuminata Thunb., Prodr. Pl. Cap.: 53 (1794), nom. illeg. *Lachnostylis hirta* var. *acuminata* Müll.Arg. in A.P.de Candolle, Prodr. 15(2): 224 (1866).
Lachnostylis capensis Turcz., Bull. Soc. Imp. Naturalistes Moscou 19: 503 (1846).
Lachnostylis minor Sond., Linnaea 23: 132 (1850).
Lachnostylis hirta var. *minor* Müll.Arg. in A.P.de Candolle, Prodr. 15(2): 224 (1866).

Synonyms:
Lachnostylis capensis Turcz. === **Lachnostylis hirta** (L.f.) Müll.Arg.
Lachnostylis hirta var. *acuminata* Müll.Arg. === **Lachnostylis hirta** (L.f.) Müll.Arg.
Lachnostylis hirta var. *genuina* Müll.Arg. === **Lachnostylis hirta** (L.f.) Müll.Arg.
Lachnostylis hirta var. *minor* Müll.Arg. === **Lachnostylis hirta** (L.f.) Müll.Arg.
Lachnostylis minor Sond. === **Lachnostylis hirta** (L.f.) Müll.Arg.

Lambertya

of F. Mueller = *Bertya*

Laneasagum

Synonyms:
Laneasagum Bedd. === **Drypetes** Vahl
Laneasagum oblongifolium Bedd. === **Drypetes oblongifolia** (Bedd.) Airy Shaw

Lasiococca

5 species, somewhat scattered in Asia (from the Himalaya to Vietnam and Hainan) and W. Malesia (Peninsular Malaysia); shrubs or small trees with leaves subopposite or pseudo-verticillate towards shoot apices. The three then-known species were keyed by Airy-Shaw (1968); two further were added by Nguyen Nghia Thin in 1986 of which one, *L. chanii*, is a calciphile. A close relationship with *Spathiostemon* and *Homonoia* is accepted by recent authorities. *Lasiococca* differs in being monoecious; the sepals in fruit are moreover persistent. A partial revision has been published by Peter van Welzen with Nguyen Nghia Thin and Vu Hoai Duc (1998). (Acalyphoideae)

Lachnostylis hirta (L.f.) Müll. Arg.
Artist: 'A.M.C.'
Ic. Pl. pl. 1279 (1879)
KEW ILLUSTRATIONS COLLECTION

Hooker, J. D. (1887). *Lasiococca symphilliaefolia.* Ic. Pl. 16: pl. 1587. La/En. — Plant portrait with description and commentary; protologue of genus.
Pax, F. & K. Hoffmann (1919). *Lasiococca.* In A. Engler (ed.), Das Pflanzenreich, IV 147 XI (Euphorbiaceae-Acalypheae-Ricininae): 118-119. Berlin. (Heft 68.) La/Ge. — 1 species, Himalaya.
Airy-Shaw, H. K. (1963). Notes on Malaysian and other Asiatic Euphorbiaceae, XXXIX. *Lasiococca* Hook.f. in Hainan. Kew Bull. 16: 358. En. — A substantial range extension.
Airy-Shaw, H. K. (1968). Notes on Malesian and other Asiatic Euphorbiaceae, XCIV. *Lasiococca* Hook.f. in the Malay Peninsula. Kew Bull. 21: 406-407. En. — Range extension for *L. malaccensis*; key to the three known species of the genus.
• Welzen, P. C. van (1998). Revisions and phylogenies of Malesian Euphorbiaceae subtribe Lasiococcinae (*Homonoia, Lasiococca, Spathiostemon*) and *Clonostylis, Ricinus* and *Wetria.* Blumea 43: 131-164. (*Lasiococca* with Nguyen Nghia Thin & Vu Hoai Duc.) En. — General introduction, with history of studies; a note of caution on the use as a character of monoecy vs. dioecy given the imperfect state of much material; phylogeny with character table and cladogram; revision of Malesian *Lasiococca* (pp. 141-144; 1 species) with description, synonymy, types, literature citations, vernacular names, indication of distribution, ecology and habitat, and notes on anatomy, uses and properties (where known), and systematics (including remarks on the species outside the Malesian region); identification list at end.

Lasiococca Hook.f., Hooker's Icon. Pl. 16: 1587 (1887).
India to Hainan. 37 40 42.

Lasiococca chanii Thin, J. Biol. (Vietnam) 8(3): 37 (1986).
Vietnam. 41 VIE.

Lasiococca comberi Haines, Bull. Misc. Inform. Kew 1920: 70 (1920). *Homonoia comberi* (Haines) Merr., Lingnan Sci. J. 19: 188 (1940).
E. India, Hainan, Vietnam. 36 CHH 40 IND 41 VIE. Phan.
Mallotus pseudoverticillatus Merr., Lingnan Sci. J. 14: 23 (1935). *Homonoia pseudoverticillata* (Merr.) Merr., Lingnan Sci. J. 19: 187 (1940). *Lasiococca comberi* var. *pseudoverticillata* (Merr.) H.S.Kiu, Acta Phytotax. Sin. 20: 108 (1982).

Lasiococca locii Thin, J. Biol. (Vietnam) 8(3): 37 (1986).
Vietnam. 41 VIE.

Lasiococca malaccensis Airy Shaw, Kew Bull. 21: 406 (1968).
Pen. Malaysia, Sulawesi, Lesser Sunda Is. (Flores). 42 LSI MLY SUL. Nanophan. or phan.

Lasiococca symphyllifolia (Kurz) Hook.f., Hooker's Icon. Pl. 16: 1587 (1887).
Sikkim. 40 EHM.
* *Homonoia symphyllifolia* Kurz, Flora 58: 32 (1875).

Synonyms:
Lasiococca comberi var. *pseudoverticillata* (Merr.) H.S.Kiu === **Lasiococca comberi** Haines

Lasiocroton

6 species, Caribbean (Bahamas, Cuba, Jamaica); shrubs or small trees to 10 m in dry habitats, mostly with thick, more or less tufted leaves. Closely related to *Leucocroton* in the Adelieae (Webster, Synopsis, 1994). No revision has appeared since 1914. (Acalyphoideae)

Pax, F. (with K. Hoffmann) (1914). *Lasiocroton.* In A. Engler (ed.), Das Pflanzenreich, IV 147 VII (Euphorbiaceae-Acalypheae-Mercurialinae): 60-61. Berlin. (Heft 63.) La/Ge. — 4 species, West Indies (Bahamas, Greater Antilles).

Lasiocroton Griseb., Fl. Brit. W. I.: 46 (1859).
N. Caribbean. 81.

Lasiocroton bahamensis Pax & K.Hoffm. in H.G.A.Engler, Pflanzenr., IV, 147, VII: 61 (1914).
Bahamas, Cuba, Haiti. 81 BAH CUB HAI. Nanophan. or phan.
Lasiocroton micranthus Pax & K.Hoffm. in H.G.A.Engler, Pflanzenr., IV, 147, VII: 61 (1914).

Lasiocroton fawcettii Urb., Symb. Antill. 6: 14 (1909).
Jamaica. 81 JAM. Phan.

Lasiocroton gracilis Britton & P.Wilson, Mem. Torrey Bot. Club 16: 76 (1920).
SE. Cuba. 81 CUB. Nanophan.

Lasiocroton harrisii Britton, Bull. Torrey Bot. Club 41: 16 (1914).
Jamaica. 81 JAM.

Lasiocroton macrophyllus (Sw.) Griseb., Fl. Brit. W. I.: 46 (1859).
Jamaica. 81 JAM. Nanophan.
* *Croton macrophyllus* Sw., Prodr.: 100 (1788).

Lasiocroton trelawniensis C.D.Adams, Phytologia 20: 312 (1970).
Jamaica. 81 JAM.

Synonyms:
Lasiocroton cordifolius Britton & P.Wilson === **Leucocroton cordifolius** (Britton & P.Wilson) Alain
Lasiocroton micranthus Pax & K.Hoffm. === **Lasiocroton bahamensis** Pax & K.Hoffm.
Lasiocroton prunifolius Griseb. === **Croton punctatus** Jacq.
Lasiocroton subpeltatus Urb. === **Leucocroton subpeltatus** (Urb.) Alain

Lasiogyne

Synonyms:
Lasiogyne Klotzsch === **Croton** L.
Lasiogyne pottsii Klotzsch === **Croton pottsii** (Klotzsch) Müll.Arg.

Lasiostyles

Synonyms:
Lasiostyles C.Presl === **Cleidion** Blume

Lasipana

Synonyms:
Lasipana Raf. === **Mallotus** Lour.

Lassia

Synonyms:
Lassia Baill. === **Tragia** Plum. ex L.
Lassia scandens Baill. === **Tragia lassia** Radcl.-Sm. & Govaerts

Lathyris

A Trew name = *Euphorbia*.

Lautembergia

Now united with *Orfilea* but in the past generally accepted as distinct.

Synonyms:
Lautembergia Baill. === **Orfilea** Baill.
Lautembergia multispicata Baill. === **Orfilea multispicata** (Baill.) G.L.Webster

Lebidiera

Synonyms:
Lebidiera Baill. === **Cleistanthus** Hook.f. ex Planch.
Lebidiera cunninghamii Müll.Arg. === **Cleistanthus cunninghamii** (Müll.Arg.) Müll.Arg.
Lebidiera malabarica Müll.Arg. === **Cleistanthus malabaricus** (Müll.Arg.) Müll.Arg.

Lebidieropsis

Synonyms:
Lebidieropsis Müll.Arg. === **Cleistanthus** Hook.f. ex Planch.
Lebidieropsis collina (Roxb.) Müll.Arg. === **Cleistanthus collinus** (Roxb.) Benth.
Lebidieropsis orbiculata (Roth) Müll.Arg. === **Cleistanthus collinus** (Roxb.) Benth.
Lebidieropsis orbiculata var. *collina* (Roxb.) Müll.Arg. === **Cleistanthus collinus** (Roxb.) Benth.
Lebidieropsis orbiculata var. *lambertii* Müll.Arg. === **Cleistanthus collinus** (Roxb.) Benth.

Leeuwenbergia

2 species, WC. Africa (Cameroon, Gabon, Zaire). This interesting Mayombe genus of laticiferous trees, one species of which reaches 40 m, is allied to *Joannesia* in the Americas and *Annesijoa* in New Guinea; on present evidence it is closer to the former than the latter (Letouzey & Hallé 1974). The wood of all is mutually similar and in addition shows affinities with that of *Hevea*. (Crotonoideae)

Letouzey, R. & N. Hallé (1974). *Leeuwenbergia*, genre nouveau d'Euphorbiacées (Crotonoidées-Joannesiées) d'Afrique centrale occidentale. Adansonia, II, 14: 379-388. Fr. — Protologue of genus with 2 species from west-central Africa, both also described for the first time. [See also A. Mariaux, 1974. Anatomie du bois de *Leeuwenbergia africana* R. Let. & N. Hallé. Adansonia, II, 14: 389-397, illus.]

Leeuwenbergia Letouzey & N.Hallé, Adansonia, n.s., 14: 380 (1974).
WC. Trop. Africa. 23.

Leeuwenbergia africana Letouzey & N.Hallé, Adansonia, n.s., 14: 386 (1974).
Cameroon, Zaire, Gabon. 23 CMN GAB ZAI.

Leeuwenbergia letestui Letouzey & N.Hallé, Adansonia, n.s., 14: 384 (1974).
Gabon. 23 GAB. – Provisionally accepted.

Leichhardtia

Synonyms:
Leichhardtia F.Muell. === **Phyllanthus** L.
Leichhardtia clamboides F.Muell. === **Phyllanthus clamboides** (F.Muell.) Diels

Leidesia

1 species, C. and S. Africa (Zaire to South Africa); delicate, anemophilous, diffusely branching annual herbs of woody ravines and forests. The genus is very closely related to *Seidelia* with which it partly overlaps in range; both are in turn related to the northern hemisphere *Mercurialis*. The three species recognised by Müller in the nineteenth century are now regarded as all parts of one variable taxon. (Acalyphoideae)

Prain, D. (1913). Mercurialineae and Adenoclineae of South Africa. Ann. Bot. 27: 371-410. En. — Includes (pp. 399-402) a synopsis of *Leidesia* (3 species), with key, synonymy, references and citations, localities with exsiccatae, and brief comments; biogeographical review at end of paper.
Pax, F. (with K. Hoffmann) (1914). *Leidesia*. In A. Engler (ed.), Das Pflanzenreich, IV 147 VII (Euphorbiaceae-Acalypheae-Mercurialinae): 284-286, illus. Berlin. (Heft 63.) La/Ge. — Monotypic, S Africa.

Leidesia Müll.Arg. in A.P.de Candolle, Prodr. 15(2): 792 (1866).
C. Trop. & S. Africa. 26 27.

Leidesia procumbens (L.) Prain, Ann. Bot. (Usteri) 27: 400 (1913 publ. 1914).
Zaire to S. Africa. 23 ZAI 26 MOZ ZIM 27 CPP NAT SWZ TVL. Ther.
* *Mercurialis procumbens* L., Sp. Pl.: 1036 (1753). *Adenocline procumbens* (L.) Druce, Bot. Soc. Exch. Club Brit. Isles 3: 413 (1913 publ. 1914). *Leidesia procumbens* var. *genuina* Pax & K.Hoffm. in H.G.A.Engler, Pflanzenr., IV, 147, VII: 285 (1914), nom. inval.
Croton ricinocarpus L., Sp. Pl. ed. 2: 1427 (1763).
Croton ricinokarpos Houtt., Nat. Hist. 2(6): 260 (1776).
Urtica capensis L.f., Suppl. Pl.: 417 (1782). *Mercurialis capensis* (L.f.) Spreng. ex Sond., Linnaea 23: 111 (1850). *Leidesia capensis* (L.f.) Müll.Arg. in A.P.de Candolle, Prodr. 15(2): 793 (1866).
Mercurialis androgyna Steud., Nomencl. Bot. 1: 524 (1821).
Acalypha obtusa Thunb., Fl. Cap., ed. 2: 546 (1823). *Leidesia obtusa* (Thunb.) Müll.Arg. in A.P.de Candolle, Prodr. 15(2): 793 (1866). *Leidesia procumbens* var. *obtusa* (Thunb.) Pax & K.Hoffm. in H.G.A.Engler, Pflanzenr., IV, 147, VII: 286 (1914).
Mercurialis tricocca Eckl. & Zeyh. ex Krauss, Flora 28: 85 (1845).
Leidesia sonderiana Müll.Arg. in A.P.de Candolle, Prodr. 15(2): 699 (1866).

Synonyms:
Leidesia capensis (L.f.) Müll.Arg. === **Leidesia procumbens** (L.) Prain
Leidesia firmula Prain === **Seidelia firmula** (Prain) Pax & K.Hoffm.
Leidesia obtusa (Thunb.) Müll.Arg. === **Leidesia procumbens** (L.) Prain
Leidesia procumbens var. *genuina* Pax & K.Hoffm. === **Leidesia procumbens** (L.) Prain
Leidesia procumbens var. *obtusa* (Thunb.) Pax & K.Hoffm. === **Leidesia procumbens** (L.) Prain
Leidesia sonderiana Müll.Arg. === **Leidesia procumbens** (L.) Prain

Leiocarpus

Synonyms:
Leiocarpus Blume === **Aporusa** Blume
Leiocarpus arborescens Hassk. === **Aporusa arborea** (Blume) Müll.Arg.

Leiocarpus arboreus Blume === **Aporusa arborea** (Blume) Müll.Arg.
Leiocarpus fruticosus Blume === **Aporusa frutescens** Blume
Leiocarpus quadrilocularis Miq. === **Aporusa quadrilocularis** (Miq.) Müll.Arg.
Leiocarpus serratus Hassk. === **Aporusa octandra** (Buch.-Ham. ex D.Don) Vickery var. **octandra**
Leiocarpus tinctorius Blume ex Pax & K.Hoffm. === **Aporusa octandra** (Buch.-Ham. ex D.Don) Vickery var. **octandra**

Leiopyxis

Synonyms:
Leiopyxis Miq. === **Cleistanthus** Hook.f. ex Planch.
Leiopyxis sumatrana Miq. === **Cleistanthus sumatranus** (Miq.) Müll.Arg.

Leontia

Synonyms:
Leontia Rchb. === **Croton** L.

Lepadena

Synonyms:
Lepadena Raf. === **Euphorbia** L.

Lepidanthus

Synonyms:
Lepidanthus Nutt. === **Andrachne** L.
Lepidanthus phyllanthoides Nutt. === **Andrachne phyllanthoides** (Nutt.) Müll.Arg.

Lepidococea

A Turczaninow name = *Caperonia*.

Lepidocroton

Synonyms:
Lepidocroton C.Presl === **Chrozophora** Neck. ex A.Juss.

Lepidostachys

Synonyms:
Lepidostachys Wall. ex Lindl. === **Aporusa** Blume
Lepidostachys grandifolia Planch. ex Müll.Arg. === **Aporusa fusiformis** Thwaites
Lepidostachys griffithii Planch. ex Pax & K.Hoffm. === **Aporusa microstachya** (Tul.) Müll.Arg.
Lepidostachys lanceolata Tul. === **Aporusa lanceolata** (Tul.) Thwaites
Lepidostachys macrophylla Tul. === **Aporusa macrophylla** (Tul.) Müll.Arg.
Lepidostachys oblonga Wall. === **Aporusa oblonga** Müll.Arg.
Lepidostachys parviflora Planch. ex Baill. === **Aporusa planchoniana** Baill.
Lepidostachys roxburghii Wall. ex Lindl. === **Aporusa octandra** (Buch.-Ham. ex D.Don) Vickery var. **octandra**
Lepidostachys roxburghii Hook. === **Aporusa wallichii** Hook.f. var. **wallichii**
Lepidostachys villosa Wall. === **Aporusa villosa** (Lindl.) Baill.

Lepidoturus

Synonyms:
Lepidoturus Bojer ex Baill. === **Alchornea** Sw.
Lepidoturus alnifolius Bojer ex Baill. === **Alchornea alnifolia** (Bojer ex Baill.) Pax & K.Hoffm.
Lepidoturus laxiflorus Benth. === **Alchornea laxiflora** (Benth.) Pax & K.Hoffm.
Lepidoturus occidentalis Müll.Arg. === **Alchornea occidentalis** (Müll.Arg.) Pax & K.Hoffm.

Leptemon

Synonyms:
Leptemon Raf. === **Croton** L.

Leptobotrys

Synonyms:
Leptobotrys Baill. === **Tragia** Plum. ex L.
Leptobotrys discolor Baill. === **Tragia urens** L.

Leptomeria

of Sieber (non R.Br.) = *Amperea*.

Synonyms:
Leptomeria xiphoclada Sieber ex Spreng. === **Amperea xiphoclada** (Sieber ex Spreng.) Druce

Leptonema

2 species, Madagascar; twiggy microphyllous shrubs to 3 m in rocky mountainous country. Never revised for *Pflanzenreich* but the species are covered in the precursor on Malagasy Phyllanthoideae (with illustration of *L. glabrum*) and flora treatment, both by Leandri (1957, 1958; see **Malagassia**). Referral of the genus to the Antidesminae by Webster (1994, Synopsis) was regarded as 'provisional'. (Phyllanthoideae)

Pax, F. & K. Hoffmann (1931). *Leptonema*. In A. Engler (ed.), Die natürlichen Pflanzenfamilien, 2. Aufl., 19c: 59. Leipzig. Ge. — Synopsis with brief description of genus; 1 species, Madagascar.

Leandri, J. (1958). *Leptonema*. Fl. Madag. Comores 111 (Euphorbiacées), I: 12-14. Paris. Fr. — Flora treatment (2 species); key.

Leptonema A.Juss., Euphorb. Gen.: 19 (1824).
Madagascar. 29. Nanophan.

Leptonema glabrum (Leandri) Leandri, Mém. Inst. Sci. Madagascar, Sér. B, Biol. Vég. 8: 212 (1957).
Madagascar. 29 MDG. Nanophan.
* *Leptonema venosum* var. *glabra* Leandri, Notul. Syst. (Paris) 6: 23 (1937).

Leptonema venosum (Poir.) A.Juss., Euphorb. Gen.: 19 (1824).
Madagascar. 29 MDG. Nanophan.
* *Acalypha venosa* Poir. in J.B.A.M.de Lamarck, Encycl. 6: 204 (1804). *Croton venosus* (Poir.) Geiseler, Croton. Monogr.: 42 (1807).

Synonyms:
Leptonema melanthesoides F.Muell. === **Flueggea virosa** subsp. **melanthesoides** (F.Muell.) G.L.Webster
Leptonema venosum var. *glabra* Leandri === **Leptonema glabrum** (Leandri) Leandri

Leptopus

Now united with *Andrachne*; formerly (and in some quarters still) accepted as distinct with a distribution in Europe, Asia and Malesia.

Synonyms:
Leptopus Decne. === **Andrachne** L.
Leptopus adiantoides (Lam.) Klotzsch & Garcke === **Euphorbia adiantoides** Lam.
Leptopus attenuatus (Hand.-Mazz.) Pojark. === **Andrachne esquirolii** H.Lév.
Leptopus australis (Zoll. & Moritzi) Pojark. === **Andrachne australis** Zoll. & Moritzi
Leptopus brasiliensis (Lam.) Klotzsch & Garcke === **Euphorbia hyssopifolia** L.
Leptopus calcareus (Ridl.) Pojark. === **Andrachne calcarea** Ridl.
Leptopus capillipes (Pax) Pojark. === **Andrachne chinensis** Bunge
Leptopus chinensis (Bunge) Pojark. === **Andrachne chinensis** Bunge
Leptopus chinensis var. *hirsutus* (Hutch.) P.T.Li === **Andrachne chinensis** Bunge
Leptopus chinensis var. *pubescens* (Hutch.) S.B.Ho === **Andrachne chinensis** Bunge
Leptopus clarkei (Hook.f.) Pojark. === **Andrachne clarkei** Hook.f.
Leptopus colchicus (Fisch. & C.A.Mey. ex Boiss.) Pojark. === **Andrachne colchica** Fisch. & C.A.Mey. ex Boiss.
Leptopus cordifolius Decne. === **Andrachne cordifolia** (Wall. ex Decne.) Müll.Arg.
Leptopus decaisnei (Benth.) Pojark. === **Andrachne decaisnei** Benth.
Leptopus decaisnei var. *orbicularis* (Benth.) Airy Shaw === **Andrachne decaisnei** var. **orbicularis** Benth.
Leptopus diplospermus (Airy Shaw) G.L.Webster === **Chorisandrachne diplosperma** Airy Shaw
Leptopus dominianus Pojark. === **Andrachne decaisnei** var. **orbicularis** Benth.
Leptopus emicans (Dunn) Pojark. === **Andrachne emicans** Dunn
Leptopus esquirolii (H.Lév.) P.T.Li === **Andrachne esquirolii** H.Lév.
Leptopus esquirolii var. *villosus* P.T.Li === **Andrachne esquirolii** H.Lév.
Leptopus hainanensis (Merr. & Chun) Pojark. === **Andrachne hainanensis** Merr. & Chun
Leptopus hartwegii (Boiss.) Klotzsch & Garcke === **Euphorbia adiantoides** Lam.
Leptopus hirsutus (Hutch.) Pojark. === **Andrachne chinensis** Bunge
Leptopus hirtus (Ridl.) Pojark. === **Andrachne hirta** Ridl.
Leptopus kwangsiensis Pojark. === **Andrachne esquirolii** H.Lév.
Leptopus lanceolatus (Pierre ex Beille) Pojark. === **Andrachne australis** Zoll. & Moritzi
Leptopus lolonus (Hand.-Mazz.) Pojark. === **Andrachne lolonum** Hand.-Mazz.
Leptopus montanus (Hutch.) Pojark. === **Andrachne chinensis** Bunge
Leptopus nanus P.T.Li === **Andrachne nana** (P.T.Li) Govaerts
Leptopus ocymoides Klotzsch & Garcke === **Euphorbia adenoptera** Bertol. subsp. **adenoptera**
Leptopus orbicularis (Benth.) Pojark. === **Andrachne decaisnei** var. **orbicularis** Benth.
Leptopus orinocensis Klotzsch & Garcke === **Euphorbia ?**
Leptopus pachyphyllus X.X.Chen === **Andrachne pachyphylla** (X.X.Chen) Govaerts
Leptopus philippinensis Pojark. === **Andrachne australis** Zoll. & Moritzi
Leptopus phyllanthoides (Nutt.) G.L.Webster === **Andrachne phyllanthoides** (Nutt.) Müll.Arg.
Leptopus poeppigii (Boiss.) Klotzsch & Garcke === **Euphorbia poeppigii** (Klotzsch & Garcke) Boiss.
Leptopus polypetalus (Kuntze) Pojark. === **Andrachne polypetala** Kuntze
Leptopus robinsonii Airy Shaw === **Andrachne robinsonii** (Airy Shaw) Govaerts
Leptopus segoviensis Klotzsch & Garcke === **Euphorbia segoviensis** (Klotzsch & Garcke) Boiss.
Leptopus yunnanensis P.T.Li === **Andrachne yunnanensis** (P.T.Li) Govaerts

Leptopus

A Klotzsch segregate of *Euphorbia*.

Synonyms:
Leptopus Klotzsch & Garcke === **Euphorbia** L.

Leptorhachis

Synonyms:
Leptorhachis Klotzsch === **Tragia** Plum. ex L.
Leptorhachis hastata Klotzsch === **Tragia leptorhachis** Radcl.-Sm. & Govaerts

Leucadenia

Synonyms:
Leucadenia Klotzsch ex Baill. === **Croton** L.

Leucandra

Synonyms:
Leucandra Klotzsch === **Tragia** Plum. ex L.
Leucandra betonicifolia Klotzsch === **Tragia leucandra** Pax & K.Hoffm.

Leucocroton

28 species, West Indies (27 in Cuba, 1 in Hispaniola); shrubs or small trees to 10 m, the leaves more or less tufted and in some species quite narrow. All are indicator species (2 on limestone, the rest on serpentines), with several of the latter being nickel hyperaccumulators (Borhidi, 1991). Three sections are recognised by Borhidi, based on leaf venation and floral ch characters. Along with *Lasiocroton* the genus has been referred by Webster (1994, Synopsis) to his Adelieae. (Acalyphoideae)

Pax, F. (with K. Hoffmann) (1914). *Leucocroton*. In A. Engler (ed.), Das Pflanzenreich, IV 147 VII (Euphorbiaceae-Acalypheae-Mercurialinae): 62-64. Berlin. (Heft 63.) La/Ge. — 5 species, Cuba and Hispaniola.

- Borhidi, A. (1991). Taxonomic revision of genus *Leucocroton* (Euphorbiaceae). Acta Bot. Hung. 36: 13-40, illus. En. — Synoptic treatment with key, descriptions of novelties (with types, exsiccatae and commentary); list of all species (with types), references, and illustrations at end.

Leucocroton Griseb., Abh. Königl. Ges. Wiss. Göttingen 9: 20 (1861).
Cuba, Hispaniola. 81.

Leucocroton acunae Borhidi, Acta Bot. Hung. 36: 20 (1991).
Cuba. 81 CUB.

Leucocroton anomalus Borhidi, Acta Bot. Hung. 36: 17 (1991).
Cuba. 81 CUB.

Leucocroton bracteosus Urb., Symb. Antill. 9: 204 (1924).
E. Cuba. 81 CUB. Nanophan.

Leucocroton brittonii Alain, Contr. Ocas. Mus. Hist. Nat. Colegio "De La Salle" 11: 5 (1952).
E. Cuba. 81 CUB. Nanophan.

Leucocroton comosus Urb., Symb. Antill. 9: 201 (1924).
E. Cuba (Sierra de Nipe). 81 CUB. Nanophan.
Leucocroton dictyophyllus Urb., Symb. Antill. 9: 202 (1924).

Leucocroton cordifolius (Britton & P.Wilson) Alain, Contr. Ocas. Mus. Hist. Nat. Colegio "De La Salle" 11: 5 (1952).
NE. Cuba. 81 CUB. Nanophan.
* *Lasiocroton cordifolius* Britton & P.Wilson, Mem. Torrey Bot. Club 16: 76 (1920).

Leucocroton discolor Urb., Symb. Antill. 9: 203 (1924).
E. Cuba (Sierra de Nipe). 81 CUB. Nanophan.

Leucocroton ekmanii Urb., Symb. Antill. 9: 199 (1924).
E. Cuba (Sierra Sagua Baracoa). 81 CUB. Nanophan.

Leucocroton flavicans Müll.Arg. in A.P.de Candolle, Prodr. 15(2): 757 (1866).
C. Cuba. 81 CUB. Nanophan. or phan.
Leucocroton flavicans var. *angustifolius* Müll.Arg. in A.P.de Candolle, Prodr. 15(2): 757 (1866). *Leucocroton angustifolius* (Müll.Arg.) Pax & K.Hoffm. in H.G.A.Engler, Pflanzenr., IV, 147, VII: 63 (1914).
Leucocroton revolutus C.Wright, Anales Acad. Ci. Méd. Habana 7: 153 (1870), nom. illeg.
Leucocroton flavescens Benth. & Hook.f., Gen. Pl. 3: 312 (1880), nom. illeg.

Leucocroton havanensis Borhidi, Acta Bot. Hung. 36: 21 (1991).
Cuba. 81 CUB.

Leucocroton incrustatus Borhidi, Acta Bot. Hung. 36: 18 (1991).
Cuba. 81 CUB.

Leucocroton leprosus (Willd.) Pax & K.Hoffm. in H.G.A.Engler, Pflanzenr., IV, 147, VII: 64 (1914).
Hispaniola. 81 DOM HAI. Nanophan.
* *Croton leprosus* Willd., Sp. Pl. 4: 553 (1805). *Bernardia leprosa* (Willd.) Müll.Arg., Linnaea 34: 172 (1865). *Adelia leprosa* (Willd.) Moscoso, Cat. Fl. Doming. Pt. 1: 302 (1943).
Adelia ferruginea Poit. ex Baill., Étude Euphorb.: 418 (1858).

Leucocroton linearifolius Britton, Bull. Torrey Bot. Club 44: 14 (1917).
E. Cuba (Sierra de Moa). 81 CUB. Nanophan.

Leucocroton longibracteatus Borhidi, Acta Bot. Hung. 36: 19 (1991).
Cuba. 81 CUB.

Leucocroton microphyllus (A.Rich.) Pax & K.Hoffm. in H.G.A.Engler, Pflanzenr., IV, 147, VII: 64 (1914).
C. & E. Cuba. 81 CUB. Nanophan.
* *Adelia microphylla* A.Rich. in R.de la Sagra, Hist. Fis. Cuba, Bot. 11: 209 (1850). *Bernardia microphylla* (A.Rich.) Müll.Arg., Linnaea 34: 172 (1865).
Bernardia lycioides Müll.Arg. ex Pax & K.Hoffm. in H.G.A.Engler, Pflanzenr., IV, 147, VII: 64 (1914), pro syn.

Leucocroton moaensis Borhidi & O.Muñiz, Acta Bot. Acad. Sci. Hung. 18: 30 (1973).
Cuba. 81 CUB.

Leucocroton moncadae Borhidi, Acta Bot. Acad. Sci. Hung. 21: 222 (1975 publ. 1976).
Cuba. 81 CUB. Nanophan.

Leucocroton obovatus Urb., Symb. Antill. 9: 200 (1924).
E. Cuba (Sierra del Cristal). 81 CUB. Nanophan.

Leucocroton pachyphylloides Borhidi, Acta Bot. Hung. 36: 18 (1991).
Cuba. 81 CUB.

Leucocroton pachyphyllus Urb., Symb. Antill. 9: 202 (1924).
E. Cuba (Sierra de Moa). 81 CUB. Nanophan.

Leucocroton pallidus Britton & P.Wilson, Bull. Torrey Bot. Club 53: 461 (1926).
E. Cuba. 81 CUB. Nanophan.

Leucocroton revolutus C.Wright ex Sauvalle, Anales Acad. Ci. Méd. Habana 7: 154 (1870).
W. & C. Cuba. 81 CUB. Nanophan.

Leucocroton sameki Borhidi, Acta Bot. Hung. 36: 20 (1991).
Cuba. 81 CUB.

Leucocroton saxicola Britton, Bull. Torrey Bot. Club 44: 13 (1917).
E. Cuba (Sierra de Nipe). 81 CUB. Nanophan.

Leucocroton stenophyllus Urb., Deutsche Bot. Ges. 36: 505 (1918 publ. 1919).
E. Cuba (Sierra de Nipe). 81 CUB. Nanophan.
* *Leucocroton angustifolius* Britton, Bull. Torrey Bot. Club 44: 14 (1917), nom. illeg.

Leucocroton subpeltatus (Urb.) Alain, Contr. Ocas. Mus. Hist. Nat. Colegio "De La Salle" 11: 5 (1952).
NE. Cuba. 81 CUB. Nanophan.
Lasiocroton subpeltatus Urb., Symb. Antill. 9: 205 (1924).

Leucocroton virens Griseb., Nachr. Königl. Ges. Wiss. Georg-Augusts-Univ. 1: 175 (1865).
EC. & NE. Cuba. 81 CUB. Nanophan.

var. **glaber** Borhidi, Acta Bot. Hung. 36: 21 (1991).
Cuba. 81 CUB. Nanophan.

var. **virens**
Cuba. 81 CUB. Nanophan.

Leucocroton wrightii Griseb., Abh. Königl. Ges. Wiss. Göttingen 9: 21 (1861).
Cuba. 81 CUB. Nanophan.
Croton wrightii Griseb., Mem. Amer. Acad. Arts, n.s. 8: 160 (1861).
Leucocroton flavicans var. *latifolius* Müll.Arg. in A.P.de Candolle, Prodr. 15(2): 757 (1866).

Synonyms:
Leucocroton angustifolius Britton === **Leucocroton stenophyllus** Urb.
Leucocroton angustifolius (Müll.Arg.) Pax & K.Hoffm. === **Leucocroton flavicans** Müll.Arg.
Leucocroton dictyophyllus Urb. === **Leucocroton comosus** Urb.
Leucocroton flavescens Benth. & Hook.f. === **Leucocroton flavicans** Müll.Arg.
Leucocroton flavicans var. *angustifolius* Müll.Arg. === **Leucocroton flavicans** Müll.Arg.
Leucocroton flavicans var. *latifolius* Müll.Arg. === **Leucocroton wrightii** Griseb.
Leucocroton revolutus C.Wright === **Leucocroton flavicans** Müll.Arg.

Lingelsheimia

7 species, Africa (3, Gabon, Zaïre, Tanzania) and Madagascar (4); a genus of forest shrubs or small trees with distichously arranged foliage now inclusive of *Aerisilvaea* and

Danguyodrypetes. Some further species, formerly included here, are presently referred to *Drypetes*. From the latter genus, with which it remains associated in the tribe Drypeteae, *Lingelsheimia* differs in having bifid styles and capsular, 3-locular fruits. [Separate family status has been proposed for Drypeteae; for further discussion, see *Drypetes*. On the other hand, some doubt has been expressed as to whether or not *Lingelsheimia* is properly in the tribe (Webster, Synopsis, 1994).] (Phyllanthoideae)

Pax, F. & K. Hoffmann (1922). *Lingelsheimia*. In A. Engler (ed.), Das Pflanzenreich, IV 147 XV (Euphorbiaceae-Phyllanthoideae-Phyllantheae): 279-280. Berlin. (Heft 81.) La/Ge. — 1 species; Africa (W Africa).

Leandri, J. (1958). *Danguyodrypetes*. Fl. Madag. Comores 111 (Euphorbiacées), I: 158-163. Paris. Fr. — Flora treatment (4 species) with key. [All species are now considered to be in *Lingelsheimia* (Radcliffe-Smith, 1997).]

Léonard, J. (1962). Notulae systematicae, XXXIII. Sur les limites entre les genres *Drypetes* et *Lingelsheimia* Pax. Bull. Jard. Bot. État 32: 513-516. Fr. — Includes a detailed key for distinguishing the two genera and a diagnosis of a new species, *L. longipedicellata*; historical survey. [With this paper 2 species are credited to *Lingelsheimia*.]

Radcliffe-Smith, A. & M. M. Harley (1990). Notes on African Euphorbiaceae, XXI. *Aerisilvaea* and *Zimmermanniopsis*, two new phyllanthoid genera for the flora of Tanzania. Kew Bull. 45: 147-156. En. — Includes protologue of *Aerisilvaea*, with 1 species, *A. sylvestris* (also new). [Genus now reduced to *Lingelsheimia*, with a consequent range extension for the latter.]

Radcliffe-Smith, A. (1997). Notes on African and Madagascan Euphorbiaceae. Kew Bull. 52: 171-176. En. — Pp. 171-173, 'An evaluation of the genus *Aerisilvaea*', provides evidence for the reduction of *Danguyodrypetes* and *Aerisilvaea* to *Lingelsheimia*. In addition, *A. serrata* (the second species in that genus) was transferred to Celastraceae and included with *Maytenus undata* (Thunb.) Blakelock.

Lingelsheimia Pax, Bot. Jahrb. Syst. 43: 317 (1909).
WC. & E. Trop. Africa, Madagascar. 23 25 29.
Danguyodrypetes Leandri, Bull. Soc. Bot. France 85: 524 (1938 publ. 1939).
Aerisilvaea Radcl.-Sm., Kew Bull. 45: 149 (1990).

Lingelsheimia abbayesii (Leandri) Radcl.-Sm., Kew Bull. 52: 172 (1997).
SE. Madagascar. 29 MDG. Nanophan.
* *Danguyodrypetes abbayesii* Leandri, Bull. Mus. Natl. Hist. Nat., II, 29: 509 (1958).

Lingelsheimia ambigua (Leandri) Radcl.-Sm., Kew Bull. 52: 172 (1997).
NW. Madagascar. 29 MDG. Nanophan.
* *Danguyodrypetes ambigua* Leandri, Mém. Inst. Sci. Madagascar, Sér. B, Biol. Vég. 8: 252 (1957).

Lingelsheimia fiherenensis (Leandri) Radcl.-Sm., Kew Bull. 52: 172 (1997).
WC. Madagascar. 29 MDG.
* *Danguyodrypetes fiherenensis* Leandri, Mém. Inst. Sci. Madagascar, Sér. B, Biol. Vég. 8: 253 (1957).

Lingelsheimia frutescens Pax, Bot. Jahrb. Syst. 43: 317 (1909).
Zaire. 23 ZAI. Nanophan.

Lingelsheimia longipedicellata J.Léonard, Bull. Jard. Bot. État 32: 515 (1962).
Gabon. 23 GAB.

Lingelsheimia manongarivensis (Leandri) G.L.Webster, Ann. Missouri Bot. Gard. 81: 47 (1994).
NW. Madagascar. 29 MDG. Nanophan.
* *Danguyodrypetes manongarivensis* Leandri, Bull. Soc. Bot. France 85: 524 (1938 publ. 1939).

Lingelsheimia sylvestris (Radcl.-Sm.) Radcl.-Sm., Kew Bull. 52: 172 (1997).
E. Tanzania. 25 TAN. Phan.
* *Aerisilvaea sylvestris* Radcl.-Sm., Kew Bull. 45: 151 (1990).

Synonyms:
Lingelsheimia capillipes Pax === **Drypetes capillipes** (Pax) Pax & K.Hoffm.
Lingelsheimia gilgiana (Pax) Hutch. === **Drypetes gilgiana** (Pax) Pax & K.Hoffm.
Lingelsheimia parvifolia (Müll.Arg.) Hutch. === **Drypetes parvifolia** (Müll.Arg.) Pax & K.Hoffm.
Lingelsheimia tessmanniana (Pax) Hutch. === **Drypetes tessmanniana** (Pax) Pax & K.Hoffm.

Linostachys

Synonyms:
Linostachys Klotzsch ex Schltdl. === **Acalypha** L.

Liodendron

Recently re-combined with *Putranjiva* (Radcliffe Smith 1995; see that genus).

Synonyms:
Liodendron H.Keng === **Putranjiva** Wall.
Liodendron formosanum (Kaneh. & Sasaki ex Shimada) H.Keng === **Putranjiva formosana** Kaneh. & Sasaki ex Shimada
Liodendron integerrimum (Koidz.) H.Keng === **Drypetes integerrima** (Koidz.) Hosok.
Liodendron matsumurae (Koidz.) H.Keng === **Putranjiva matsumurae** Koidz.

Liparena

Synonyms:
Liparena Poit. ex Leman === **Drypetes** Vahl

Liparene

An orthographic variant of *Liparena*.

Lithoxylon

Synonyms:
Lithoxylon Endl. === **Actephila** Blume
Lithoxylon grandifolium Müll.Arg. === **Actephila lindleyi** (Steud.) Airy Shaw
Lithoxylon lindleyi Steud. === **Actephila lindleyi** (Steud.) Airy Shaw
Lithoxylon nitidum Müll.Arg. === **Actephila lindleyi** (Steud.) Airy Shaw

Lobanilia

7 species, Madagascar; shrubs or small trees with penniveined leaves, some at relatively high altitudes (*L. claoxyloides* from 2000-2875 m in the Tsaratanana complex). The genus is distinguished from *Claoxylon* and its relatives by virtue of a stellate indumentum; it was assigned by Radcliffe-Smith (1989) to a distinct subtribe Lobaniliinae within the large tribe Acalypheae. (Acalyphoideae)

Pax, F. (with K. Hoffmann) (1914). *Claoxylon.* In A. Engler (ed.), Das Pflanzenreich, IV 147 VII (Euphorbiaceae-Acalypheae-Mercurialinae): 100-131, illus. Berlin. (Heft 63.) La/Ge. — Sect. 12, *Luteobrunnea* (pp. 126-127, with 2 species) provides the basis for *Lobanilia.* [Two of the doubtful species on pp. 127-128 are also in this genus.]

Radcliffe-Smith, A. (1989). Notes on Madagascan Euphorbiaceae, II. *Claoxylon* sect. *Luteobrunnea.* Kew Bull. 44: 333-340, illus. En. — Excision from *Claoxylon* of this most strongly distinct section (no. 12 of Pax and Hoffmann, 1914) and erection of a new genus, here described as *Lobanilia*; 7 species accepted (three described as new) with key (p. 339).

Lobanilia Radcl.-Sm., Kew Bull. 44: 334 (1989).
Madagascar. 29.

Lobanilia asterothrix Radcl.-Sm., Kew Bull. 44: 335 (1989).
C. Madagascar. 29 MDG. Nanophan.

Lobanilia bakeriana (Baill.) Radcl.-Sm., Kew Bull. 44: 335 (1989).
Madagascar. 29 MDG. Nanophan.
* *Claoxylon bakerianum* Baill., Bull. Mens. Soc. Linn. Paris 2: 995 (1892).

Lobanilia claoxyloides Radcl.-Sm., Kew Bull. 44: 337 (1989).
N. Madagascar. 29 MDG. Nanophan.

Lobanilia crotonoides Radcl.-Sm., Kew Bull. 44: 337 (1989).
SE. Madagascar. 29 MDG. Nanophan. or phan.

Lobanilia hirtella (Baill.) Radcl.-Sm., Kew Bull. 44: 338 (1989).
C. Madagascar. 29 MDG. Nanophan.
* *Claoxylon hirtellum* Baill., Adansonia 1: 284 (1861).

Lobanilia luteobrunnea (Baker) Radcl.-Sm., Kew Bull. 44: 338 (1989).
C. Madagascar. 29 MDG. Nanophan.
* *Croton luteobrunneus* Baker, J. Linn. Soc., Bot. 20: 254 (1883). *Claoxylon luteobrunneum* (Baker) Baill., Bull. Mens. Soc. Linn. Paris 2: 995 (1892).

Lobanilia ovalis (Baill.) Radcl.-Sm., Kew Bull. 44: 338 (1989).
C. Madagascar. 29 MDG. Nanophan.
Croton baronii Baill., Bull. Mens. Soc. Linn. Paris 2: 863 (1890), nom. inval.
* *Claoxylon ovale* Baill., Bull. Mens. Soc. Linn. Paris 2: 996 (1892).

Lobocarpus

Synonyms:
Lobocarpus Wight & Arn. === **Glochidion** J.R.Forst. & G.Forst.
Lobocarpus candolleanus Wight & Arn. === **Glochidion candolleanum** (Wight & Arn.) Chakrab. & M.G.Gangop.

Loerzingia

Synonyms:
Loerzingia Airy Shaw === **Deutzianthus** Gagnep.
Loerzingia thyrsiflora Airy Shaw === **Deutzianthus thyrsiflorus** (Airy Shaw) G.L.Webster

Lomanthes

Synonyms:
Lomanthes Raf. === **Phyllanthus** L.

Longetia

1 species, New Caledonia; shrub to 4 m of maquis on serpentine soils with opposite leaves resembling *Austrobuxus*. In a strict sense, maintained here, the genus is limited to *L. buxoides*; *L. depauperata* has been transferred to *Scagea* and Malesian and other New Caledonian species to *Austrobuxus*. *Longetia* is distinguished from *Austrobuxus* partly by virtue of its being monoecious rather than dioecious but also due to shorter spines in the pollen exine. [In Brummitt (1992) entirely synonymised with *Austrobuxus* following Airy-Shaw (for references, see that genus).] (Oldfieldioideae)

Pax, F. & K. Hoffmann (1922). *Longetia*. In A. Engler (ed.), Das Pflanzenreich, IV 147 XV (Euphorbiaceae-Phyllanthoideae-Phyllantheae): 289-291. Berlin. (Heft 81.) La/Ge. — 6 species; New Caledonia and W. Malesia. [Most now in *Austrobuxus*; 1 in *Scagea*.]

McPherson, G. & C. Tirel (1987). *Longetia*. Fl. Nouvelle-Calédonie, 14 (Euphorbiacées, I): 188-193. Paris. Fr. — Flora treatment. [A monotypic, endemic genus as now interpreted.]

Longetia Baill., Adansonia 6: 352 (1866).
New Caledonia. 60. Nanophan.

Longetia buxoides Baill., Adansonia 2: 228 (1862). *Austrobuxus buxoides* (Baill.) Airy Shaw, Kew Bull. 25: 507 (1971).
New Caledonia. 60 NWC. Nanophan.
Austrobuxus gracilis Airy Shaw, Kew Bull. 37: 378 (1982).
Austrobuxus pisocarpus Airy Shaw, Kew Bull. 37: 378 (1982).

Synonyms:
Longetia carunculata (Baill.) Pax & K.Hoffm. === **Austrobuxus carunculatus** (Baill.) Airy Shaw
Longetia clusiacea (Baill.) Pax & K.Hoffm. === **Austrobuxus clusiaceus** (Baill.) Airy Shaw
Longetia depauperata Baill. === **Scagea depauperata** (Baill.) McPherson
Longetia eugeniifolia Guillaumin === **Austrobuxus eugeniifolius** (Guillaumin) Airy Shaw
Longetia gynotricha Guillaumin === **Scagea oligostemon** (Guillaumin) McPherson
Longetia malayana (Benth.) Pax & K.Hoffm. === **Austrobuxus nitidus** Miq. var. **nitidus**
Longetia montana (Ridl.) Pax & K.Hoffm. === **Austrobuxus nitidus** var. **montanus** (Ridl.) Whitmore
Longetia nitida (Miq.) Steenis === **Austrobuxus nitidus** Miq.
Longetia swainii Beuzev. & C.T.White === **Austrobuxus swainii** (Beuzev. & C.T.White) Airy Shaw

Lophobios

Synonyms:
Lophobios Raf. === **Euphorbia** L.

Lortia

Synonyms:
Lortia Rendle === **Monadenium** Pax
Lortia erubescens Rendle === **Monadenium erubescens** (Rendle) N.E.Br.
Lortia major Pax === **Monadenium erubescens** (Rendle) N.E.Br.

Loureira

Synonyms:
Loureira Cav. === **Jatropha** L.
Loureira peltata Desv. === **Jatropha hernandiifolia** Vent. var. **hernandiifolia**

Luntia

Synonyms:
Luntia Neck. === **Croton** L.

Lyciopsis

Presently viewed as an infrageneric taxon in *Euphorbia*.

Synonyms:
Lyciopsis (Boiss.) Schweinf. === **Euphorbia** L.

Mabea

38 species, Americas; laticiferous shrubs or small to medium trees, sometimes as tall as 30 m (*MM. excelsa* and *fistulifera*), or more or less climbing (a feature unusual in Hippomaneae). *M. frutescens*, a species of white sand in the upper Orinoco, is recorded as having a caudiciform stock. 2 sections have been accepted by Esser (1994: 112) in place of the 4 of Pax (1912) although the monophyly of sect. *Mabea* is as yet not effectively established. A diverse range of habitats is recorded, although humid rain and gallery forest is favoured (Esser 1994: 103). The least advanced species, however, are inhabitants of open places such as savanna. (Euphorbioideae (except Euphorbieae))

Pax, F. (with K. Hoffmann) (1912). *Mabea*. In A. Engler (ed.), Das Pflanzenreich, IV 147 V (Euphorbiaceae-Hippomaneae): 26-42. Berlin. (Heft 52.) La/Ge. — 29 species, Americas; mutually very similar and not easily distinguishable. Additions in ibid., XIV (Additamentum VI): 55-56 (1919). [Now superseded.]

Jablonski, E. (1967). *Mabea*. Euphorbiaceae, Guayana Highland (Mem. New York Bot. Gard. 17(1)): 164-177. New York. En. — 18 species, 7 new; 7 interpreted only from type photographs or descriptions.

Emmerich, M. (1987). Nova conceituação de *Mabea fistulifera* Mart. (Euphorbiaceae). Bradea 4(47): 370-376, illus., map. Pt. — Proposal of a new subspecies with analysis, key, description, commentary, illustrations and map. [The species proper is in eastern Brazil; the subspecies from western São Paulo, Goiás and Tocatins to Pará, Mato Grosso and Bolivia.]

Huft, M. J. (1987). Notes on *Mabea* (Euphorbiaceae) in Central America, together with comments on sect. *Apodae* in Brazil. Phytologia 62: 339-343. En. — Re-evaluation of *M. excelsa* Standley & Steyerm. with notes on sect. *Apodae* (but no key); description of *M. jefensis*.

Esser, H.-J. (1993). New species and a new combination in *Mabea* (Euphorbiaceae) from South America. Novon 3(4): 341-351, illus. En. — Novelties and notes; 5 species newly described and 1 reduced to subspecific rank. All were judged to be of comparatively limited distribution. An estimate of 40 species for the genus was given.

• Esser, H.-J. (1994). Systematische Studien an den Hippomaneae Adr. Jussieu ex Bartling (Euphorbiaceae) insbesondere den Mabeinae Pax et K. Hoffm. 305 pp., pp., illus., maps. Hamburg. (Unpubl. Ph.D. dissertation, Univ. of Hamburg.) Ge. —*Mabea*, pp. 76-221 (39 species); includes introduction, key, descriptions, synonymy, types, distribution and habitat, commentary, and localities with exsiccatae. Appendices A-C include for all genera an index to specimens seen, collections in herb. HBG, and illustrations, cladograms and distribution maps. [As of compilation of the *World Checklist* in 1997-98 not yet formally published save for novelties and notes.]

Mabea Aubl., Hist. Pl. Guiane 2: 867 (1775).
Mexico, S. America. 79 80 81 82 83 84 85.

Mabea anadena Pax & K.Hoffm. in H.G.A.Engler, Pflanzenr., IV, 147, XIV: 55 (1919).
Peru, Bolivia, Brazil (Amazonas). 83 BOL PER 84 BZN.

Mabea angularis Hollander, Proc. Kon. Ned. Akad. Wetensch., C 89: 147 (1986).
Peru, N. Brazil, Guyana. 82 GUY 83 PER 84 BZN.

Mabea angustifolia Spruce ex Benth., Hooker's J. Bot. Kew Gard. Misc. 6: 365 (1854). *Mabea angustifolia* var. *genuina* Müll.Arg. in A.P.de Candolle, Prodr. 15(2): 1149 (1866), nom. inval.
Peru, Bolivia, Brazil (Amazonas, Pará, Rio de Janeiro, Goiás, Mato Grosso). 83 BOL PER 84 BZC BZL BZN. Phan.
Mabea angustifolia var. *oblonga* Benth., Hooker's J. Bot. Kew Gard. Misc. 6: 365 (1854).
Mabea fistulifera Benth., Hooker's J. Bot. Kew Gard. Misc. 6: 365 (1854), nom. nud.
Mabea angustifolia var. *major* Müll.Arg. in C.F.P.von Martius, Fl. Bras. 11(2): 520 (1874).
Mabea angustifolia var. *myrtifolia* Müll.Arg. in C.F.P.von Martius, Fl. Bras. 11(2): 519 (1874).
Mabea riedelii Müll.Arg. in C.F.P.von Martius, Fl. Bras. 11(2): 518 (1874).
Mabea angustifolia var. *longifolia* Britton, Mem. Torrey Bot. Club 4: 258 (1895). *Mabea longifolia* (Britton) Pax & K.Hoffm. in H.G.A.Engler, Pflanzenr., IV, 147, V: 30 (1912).
Mabea juinensis Emmerich, Bradea 5(26): 285 (1989).

Mabea anomala Müll.Arg. in C.F.P.von Martius, Fl. Bras. 11(2): 526 (1874).
Brazil (Amazonas), Venezuela (Amazonas), Colombia (Guainia). 82 VEN 83 CLM 84 BZN. Phan.

Mabea arenicola Esser, Novon 3: 341 (1993).
Amazon Basin. 82 VEN 83 CLM 84 BZN. Phan.

Mabea biglandulosa Baill. ex Müll.Arg. in A.P.de Candolle, Prodr. 15(2): 1151 (1866).
Mabea montana subsp. *biglandulosa* (Baill. ex Müll.Arg.) Hollander, Proc. Kon. Ned. Akad. Wetensch., C 89: 149 (1986).
Guyana (S. Mt. Roraima), Brazil (Roraima). 82 GUY 84 BZN. Cl. nanophan.
Mabea volubilis Klotzsch in M.R.Schomburgk, Fauna Fl. Brit. Gui.: 1098 (1848), nom. nud.
Mabea piriri var. *laevigata* Müll.Arg. in A.P.de Candolle, Prodr. 15(2): 1150 (1866). *Mabea occidentalis* var. *laevigata* (Müll.Arg.) Müll.Arg. in C.F.P.von Martius, Fl. Bras. 11(2): 522 (1874).

Mabea caudata Pax & K.Hoffm. in H.G.A.Engler, Pflanzenr., IV, 147, V: 282 (1912).
Guyana, Surinam. 82 GUY SUR. Nanophan. or phan.

Mabea chocoensis Croizat, Caldasia 2: 361 (1944).
Colombia. 83 CLM.

Mabea elata Steyerm., Publ. Field Mus. Nat. Hist., Bot. Ser. 17: 418 (1938).
Ecuador, Peru. 83 ECU PER. Phan.
Mabea rhynchophylla Diels, Notizbl. Bot. Gart. Berlin-Dahlem 14: 335 (1939), nom. inval.

Mabea elegans Rusby, Mem. New York Bot. Gard. 7: 288 (1927).
Bolivia, Brazil (Amazonas). 83 BOL 84 BZN. Phan.
Mabea prancei Emmerich, Bol. Mus. Nac. Rio de Janeiro, Bot. 62: 4 (1981).

Mabea excelsa Standl. & Steyerm., Publ. Field Mus. Nat. Hist., Bot. Ser. 23: 123 (1944).
Mexico (Chiapas), Guatemala, Costa Rica, Nicaragua. 79 MXT 80 COS GUA NIC. Phan.

Mabea fistulifera Mart. in C.F.P.von Martius & J.B.von Spix, Reise Bras. 2: 479 (1828).
Brazil, Bolivia, Paraguay. 83 BOL 84 BZC BZE BZL BZN 85 PAR. Phan.

subsp. **bahiensis** (Emmerich) Esser, Novon 3: 349 (1993).
Brazil (Bahia, Minas Gerais). 84 BZE BZL. Phan.
* *Mabea bahiensis* Emmerich, Bradea 5(26): 289 (1989).

subsp. **fistulifera**
Brazil, Bolivia, Paraguay. 83 BOL 84 BZC BZE BZL BZN 85 PAR. Phan.
Mabea ferruginea Benth., Hooker's J. Bot. Kew Gard. Misc. 6: 366 (1854), nom. nud.

subsp. **robusta** Emmerich, Bradea 4: 370 (1987).
Brazil, Bolivia, Paraguay. 83 BOL 84 BZC BZL 85 PAR. Phan.

Mabea frutescens Jabl., Mem. New York Bot. Gard. 17: 175 (1967).
Colombia (Guainia), Venezuela (Amazonas). 82 VEN 83 CLM. Nanophan. or phan.
Mabea orbiculata Jabl., Mem. New York Bot. Gard. 17: 176 (1967).

Mabea gaudichaudiana Baill., Adansonia 4: 372 (1864).
Brazil (São Paulo). 84 BZL. Nanophan. – Provisionally accepted.

Mabea glaziovii Pax & K.Hoffm. in H.G.A.Engler, Pflanzenr., IV, 147, V: 37 (1912).
Brazil (Bahia, Minas Gerais). 84 BZE BZL. Phan.

Mabea jefensis Huft, Phytologia 62: 341 (1987).
Panama. 80 PAN. Nanophan. or phan.

Mabea klugii Steyerm., Publ. Field Mus. Nat. Hist., Bot. Ser. 17: 416 (1938).
Nicaragua to Ecuador. 80 COS NIC PAN 83 CLM ECU. Phan.

Mabea linearifolia Jabl., Mem. New York Bot. Gard. 17: 176 (1967).
Venezuela (Amazonas: near Yapacana). 82 VEN. Cham.

Mabea longibracteata Esser, Novon 3: 343 (1993).
Venezuela (Amazonas), Brazil (Amazonas). 82 VEN 84 BZN. Phan.

Mabea macbridei I.M.Johnst., Contr. Gray Herb. 75: 27 (1925).
Colombia, Ecuador, Peru. 83 CLM ECU PER. Phan.

Mabea macrocalyx Esser, Novon 3: 345 (1993).
Venezuela. 82 VEN. Phan.

Mabea montana Müll.Arg. in A.P.de Candolle, Prodr. 15(2): 1151 (1866).
Costa Rica, Panama, Trinidad, Colombia, Venezuela, Guyana. 80 COS PAN 81 TRT 82 GUY VEN 83 CLM.
Mabea lucida Pax & K.Hoffm. in H.G.A.Engler, Pflanzenr., IV, 147, V: 36 (1912). *Mabea montana* subsp. *lucida* (Pax & K.Hoffm.) Hollander, Proc. Kon. Ned. Akad. Wetensch., C 89: 150 (1986).
Mabea verrucosa Pax & K.Hoffm. in H.G.A.Engler, Pflanzenr., IV, 147, V: 37 (1912).
Mabea costata Pax & K.Hoffm. in H.G.A.Engler, Pflanzenr., IV, 147, VII: 419 (1914).
Mabea atroviridis Pax & K.Hoffm. in H.G.A.Engler, Pflanzenr., IV, 147, XIV: 55 (1919).
Mabea ciliata Gleason, Bull. Torrey Bot. Club 54: 610 (1927).
Mabea longepedicellata Pittier, J. Wash. Acad. Sci. 19: 353 (1929).
Mabea floribunda Jabl., Mem. New York Bot. Gard. 17: 174 (1967).
Mabea parguazae Jabl., Mem. New York Bot. Gard. 17: 168 (1967).

Mabea nitida Spruce ex Benth., Hooker's J. Bot. Kew Gard. Misc. 6: 367 (1854).
C. & S. Trop. America. 80 BLZ 82 VEN 83 BOL CLM ECU PER 84 BZN. Phan.
Mabea pallida Baill., Adansonia 4: 371 (1864), nom. nud.
Mabea nitida var. *albiflora* Müll.Arg. in A.P.de Candolle, Prodr. 15(2): 1152 (1866).
Mabea nitida var. *purpurascens* Müll.Arg. in A.P.de Candolle, Prodr. 15(2): 1152 (1866).
Mabea depauperata Pax & K.Hoffm. in H.G.A.Engler, Pflanzenr., IV, 147, XIV: 56 (1919).
Mabea muricata Jabl., Mem. New York Bot. Gard. 17: 169 (1967).

Mabea occidentalis Benth., Hooker's J. Bot. Kew Gard. Misc. 6: 364 (1854). *Mabea occidentalis* var. *genuina* Müll.Arg. in C.F.P.von Martius, Fl. Bras. 11(2): 521 (1874), nom. inval.
Mexico to Venezuela and Ecuador. 79 MXT 80 COS GUA PAN 82 VEN 83 CLM ECU. Nanophan. or phan.
Mabea pallida Müll.Arg. in A.P.de Candolle, Prodr. 15(2): 1150 (1866).
Mabea glauca Klotzsch ex Pax in H.G.A.Engler, Pflanzenr., IV, 147, V: 37 (1912), pro syn.
Mabea microcarpa Pittier, J. Wash. Acad. Sci. 19: 353 (1929).
Mabea acutissima Killip, J. Wash. Acad. Sci. 24: 49 (1934).
Mabea belizensis Lundell, Field & Lab. 13: 4 (1945).

Mabea ovata Esser, Novon 3: 347 (1993).
Brazil (Pará). 84 BZN. Phan.

Mabea paniculata Spruce ex Benth., Hooker's J. Bot. Kew Gard. Misc. 6: 367 (1854).
Brazil (Pará, Mato Grosso), Bolivia (La Paz), Paraguay. 83 BOL 84 BZC BZN 85 PAR. Phan.
Mabea rubrinervis Baill., Adansonia 4: 371 (1864), nom. nud.
Mabea weddelliana Baill., Adansonia 4: 371 (1864), nom. nud.
Mabea paniculata var. *oblongifolia* Müll.Arg. in C.F.P.von Martius, Fl. Bras. 11(2): 526 (1874).
Mabea paniculata var. *ovata* Müll.Arg. in C.F.P.von Martius, Fl. Bras. 11(2): 526 (1874).
Mabea paraguensis Müll.Arg. in C.F.P.von Martius, Fl. Bras. 11(2): 527 (1874).
Mabea crenulata S.Moore, Trans. Linn. Soc. London, Bot. 4: 470 (1895).
Mabea indorum S.Moore, Trans. Linn. Soc. London, Bot. 4: 469 (1895).

Mabea piriri Aubl., Hist. Pl. Guiane 2: 867 (1775). *Mabea piriri* var. *genuina* Müll.Arg. in A.P.de Candolle, Prodr. 15(2): 1150 (1866), nom. inval.
C. & S. Trop. America. 80 COS PAN 82 FRG GUY SUR VEN 83 CLM PER 84 BZE BZL BZN. Phan.
Omphalea lactescens Vell., Fl. Flumin. 10: 11 (1831). *Mabea lactescens* (Vell.) Müll.Arg. in C.F.P.von Martius, Fl. Bras. 11(2): 523 (1874).
Mabea surinamensis Klotzsch in M.R.Schomburgk, Fauna Fl. Brit. Gui.: 1185 (1848), nom. nud.
Mabea maynensis Spruce, J. Linn. Soc., Bot. 5: 7 (1861).
Mabea brasiliensis Müll.Arg. in A.P.de Candolle, Prodr. 15(2): 1151 (1866). *Mabea brasiliensis* var. *genuina* Pax & K.Hoffm. in H.G.A.Engler, Pflanzenr., IV, 147, V: 36 (1912), nom. inval.
Mabea piriri var. *concolor* Müll.Arg. in A.P.de Candolle, Prodr. 15(2): 1150 (1866). *Mabea occidentalis* var. *concolor* (Müll.Arg.) Müll.Arg. in C.F.P.von Martius, Fl. Bras. 11(2): 522 (1874). *Mabea speciosa* subsp. *concolor* (Müll.Arg.) Hollander, Proc. Kon. Ned. Akad. Wetensch., C 89: 154 (1986).
Mabea brasiliensis var. *intermedia* Pax & K.Hoffm. in H.G.A.Engler, Pflanzenr., IV, 147, V: 36 (1912).
Mabea piririoides Steyerm., Publ. Field Mus. Nat. Hist., Bot. Ser. 17: 419 (1938).

Mabea pohliana (Benth.) Müll.Arg. in A.P.de Candolle, Prodr. 15(2): 1152 (1866).
Brazil (Acre, Bahia, Goiás, Maranhão), Bolivia (La Paz, Santa Cruz). 83 BOL 84 BZC BZE BZN. Nanophan.
* *Mabea paniculata* var. *pohliana* Benth., Hooker's J. Bot. Kew Gard. Misc. 6: 368 (1854).

Mabea pulcherrima Müll.Arg., Flora 55: 44 (1872).
N. Brazil, Guianas, Venezuela (Bolívar), Peru. 82 FRG GUY SUR VEN 83 PER 84 BZN. Cl. nanophan.
Mabea eximia Ducke, Arch. Jard. Bot. Rio de Janeiro 4: 107 (1925).

Mabea rubicunda Jabl., Mem. New York Bot. Gard. 17: 170 (1967).
Guyana (S. Pakaraima Mts.). 82 GUY. Phan.

Mabea salicoides Esser, Novon 3: 349 (1993).
Guiana, Brazil (Amazonas). 82 FRG 84 BZN. Phan.

Mabea speciosa Müll.Arg. in C.F.P.von Martius, Fl. Bras. 11(2): 520 (1874).
Brazil, Colombia, Ecuador, Guianas, Peru. 82 FRG GUY SUR 83 CLM ECU PER 84 BZC BZE BZN. Phan.
Mabea piriri var. *obovata* Müll.Arg. in A.P.de Candolle, Prodr. 15(2): 1150 (1866). *Mabea occidentalis* var. *obovata* (Müll.Arg.) Müll.Arg. in C.F.P.von Martius, Fl. Bras. 11(2): 521 (1874).
Mabea piriri var. *purpurascens* Müll.Arg. in A.P.de Candolle, Prodr. 15(2): 1150 (1866). *Mabea occidentalis* var. *purpurascens* (Müll.Arg.) Müll.Arg. in C.F.P.von Martius, Fl. Bras. 11(2): 521 (1874).
Mabea occidentalis var. *setulosa* Müll.Arg. in C.F.P.von Martius, Fl. Bras. 11(2): 522 (1874). *Mabea setulosa* (Müll.Arg.) Hollander, Proc. Kon. Ned. Akad. Wetensch., C 89: 153 (1986).
Mabea caudata var. *concolor* Lanj., Euphorb. Surinam: 39 (1931).
Mabea saramaccensis Croizat, Bull. Torrey Bot. Club 75: 405 (1948).

Mabea standleyi Steyerm., Publ. Field Mus. Nat. Hist., Bot. Ser. 17: 417 (1938).
W. South America, Brazil (Acre). 83 CLM ECU PER 84 BZN. Phan.

Mabea subserrulata Spruce ex Benth., Hooker's J. Bot. Kew Gard. Misc. 6: 366 (1854).
Brazil (Amazonas). 84 BZN. Phan.

Mabea subsessilis Pax & K.Hoffm. in H.G.A.Engler, Pflanzenr., IV, 147, V: 282 (1912).
Colombia, S. Venezuela, Guianas, Peru (Loreto, San Martin), Brazil (Amazonas, Pará, Roraima). 82 FRG GUY SUR VEN 83 CLM PER 84 BZN. Phan.
Mabea argutissima Croizat, Bull. Torrey Bot. Club 67: 288 (1940).
Mabea subsessilis var. *peruviana* J.F.Macbr., Field Mus. Nat. Hist., Bot. Ser. 13: 188 (1951).

Mabea taquari Aubl., Hist. Pl. Guiane 2: 870 (1775). *Mabea taquari* var. *genuina* Müll.Arg. in A.P.de Candolle, Prodr. 15(2): 1149 (1866), nom. inval.
Trinidad, Venezuela, Guianas, Brazil (Amapá, Pará, Roraima). 81 TRT 82 FRG GUY SUR VEN 84 BZN. (Cl.) nanophan. or phan.
Maprounea glauca Desv. in W.Hamilton, Prodr. Pl. Ind. Occid.: 54 (1825).
Mabea schomburgkii Benth., Hooker's J. Bot. Kew Gard. Misc. 6: 365 (1854).
Mabea taquari var. *angustifolia* Müll.Arg. in A.P.de Candolle, Prodr. 15(2): 1149 (1866).

Mabea trianae Pax, Bot. Jahrb. Syst. 26: 506 (1899).
Colombia, Venezuela. 82 VEN 83 CLM. Phan.
Mabea parvifolia Pax & K.Hoffm. in H.G.A.Engler, Pflanzenr., IV, 147, V: 282 (1912).

Mabea uleana Pax & K.Hoffm. in H.G.A.Engler, Pflanzenr., IV, 147, XIV: 56 (1919). Brazil (Amazonas). 84 BZN.

Synonyms:

Mabea acutissima Killip === **Mabea occidentalis** Benth.
Mabea angustifolia var. *genuina* Müll.Arg. === **Mabea angustifolia** Spruce ex Benth.
Mabea angustifolia var. *longifolia* Britton === **Mabea angustifolia** Spruce ex Benth.
Mabea angustifolia var. *major* Müll.Arg. === **Mabea angustifolia** Spruce ex Benth.
Mabea angustifolia var. *myrtifolia* Müll.Arg. === **Mabea angustifolia** Spruce ex Benth.
Mabea angustifolia var. *oblonga* Benth. === **Mabea angustifolia** Spruce ex Benth.
Mabea argutissima Croizat === **Mabea subsessilis** Pax & K.Hoffm.
Mabea atroviridis Pax & K.Hoffm. === **Mabea montana** Müll.Arg.
Mabea bahiensis Emmerich === **Mabea fistulifera** subsp. **bahiensis** (Emmerich) Esser
Mabea belizensis Lundell === **Mabea occidentalis** Benth.
Mabea brasiliensis Müll.Arg. === **Mabea piriri** Aubl.
Mabea brasiliensis var. *genuina* Pax & K.Hoffm. === **Mabea piriri** Aubl.
Mabea brasiliensis var. *intermedia* Pax & K.Hoffm. === **Mabea piriri** Aubl.
Mabea caudata var. *concolor* Lanj. === **Mabea speciosa** Müll.Arg.
Mabea ciliata Gleason === **Mabea montana** Müll.Arg.
Mabea costata Pax & K.Hoffm. === **Mabea montana** Müll.Arg.
Mabea crenulata S.Moore === **Mabea paniculata** Spruce ex Benth.
Mabea depauperata Pax & K.Hoffm. === **Mabea nitida** Spruce ex Benth.
Mabea eximia Ducke === **Mabea pulcherrima** Müll.Arg.
Mabea ferruginea Benth. === **Mabea fistulifera** Mart. subsp. **fistulifera**
Mabea fistulifera Benth. === **Mabea angustifolia** Spruce ex Benth.
Mabea floribunda Jabl. === **Mabea montana** Müll.Arg.
Mabea glauca Klotzsch ex Pax === **Mabea occidentalis** Benth.
Mabea indorum S.Moore === **Mabea paniculata** Spruce ex Benth.
Mabea juinensis Emmerich === **Mabea angustifolia** Spruce ex Benth.
Mabea lactescens (Vell.) Müll.Arg. === **Mabea piriri** Aubl.
Mabea longepedicellata Pittier === **Mabea montana** Müll.Arg.
Mabea longifolia (Britton) Pax & K.Hoffm. === **Mabea angustifolia** Spruce ex Benth.
Mabea lucida Pax & K.Hoffm. === **Mabea montana** Müll.Arg.
Mabea maynensis Spruce === **Mabea piriri** Aubl.
Mabea microcarpa Pittier === **Mabea occidentalis** Benth.
Mabea montana subsp. *biglandulosa* (Baill. ex Müll.Arg.) Hollander === **Mabea biglandulosa** Baill. ex Müll.Arg.
Mabea montana subsp. *lucida* (Pax & K.Hoffm.) Hollander === **Mabea montana** Müll.Arg.
Mabea muricata Jabl. === **Mabea nitida** Spruce ex Benth.
Mabea nitida var. *albiflora* Müll.Arg. === **Mabea nitida** Spruce ex Benth.
Mabea nitida var. *purpurascens* Müll.Arg. === **Mabea nitida** Spruce ex Benth.
Mabea occidentalis var. *concolor* (Müll.Arg.) Müll.Arg. === **Mabea piriri** Aubl.
Mabea occidentalis var. *genuina* Müll.Arg. === **Mabea occidentalis** Benth.
Mabea occidentalis var. *laevigata* (Müll.Arg.) Müll.Arg. === **Mabea biglandulosa** Baill. ex Müll.Arg.
Mabea occidentalis var. *obovata* (Müll.Arg.) Müll.Arg. === **Mabea speciosa** Müll.Arg.
Mabea occidentalis var. *purpurascens* (Müll.Arg.) Müll.Arg. === **Mabea speciosa** Müll.Arg.
Mabea occidentalis var. *setulosa* Müll.Arg. === **Mabea speciosa** Müll.Arg.
Mabea orbiculata Jabl. === **Mabea frutescens** Jabl.
Mabea pallida Müll.Arg. === **Mabea occidentalis** Benth.
Mabea pallida Baill. === **Mabea nitida** Spruce ex Benth.
Mabea paniculata var. *oblongifolia* Müll.Arg. === **Mabea paniculata** Spruce ex Benth.
Mabea paniculata var. *ovata* Müll.Arg. === **Mabea paniculata** Spruce ex Benth.
Mabea paniculata var. *pohliana* Benth. === **Mabea pohliana** (Benth.) Müll.Arg.
Mabea paraguensis Müll.Arg. === **Mabea paniculata** Spruce ex Benth.
Mabea parguazae Jabl. === **Mabea montana** Müll.Arg.

Mabea parvifolia Pax & K.Hoffm. === **Mabea trianae** Pax
Mabea piriri var. *concolor* Müll.Arg. === **Mabea piriri** Aubl.
Mabea piriri var. *genuina* Müll.Arg. === **Mabea piriri** Aubl.
Mabea piriri var. *laevigata* Müll.Arg. === **Mabea biglandulosa** Baill. ex Müll.Arg.
Mabea piriri var. *obovata* Müll.Arg. === **Mabea speciosa** Müll.Arg.
Mabea piriri var. *purpurascens* Müll.Arg. === **Mabea speciosa** Müll.Arg.
Mabea piririoides Steyerm. === **Mabea piriri** Aubl.
Mabea prancei Emmerich === **Mabea elegans** Rusby
Mabea rhynchophylla Diels === **Mabea elata** Steyerm.
Mabea riedelii Müll.Arg. === **Mabea angustifolia** Spruce ex Benth.
Mabea rubrinervis Baill. === **Mabea paniculata** Spruce ex Benth.
Mabea saramaccensis Croizat === **Mabea speciosa** Müll.Arg.
Mabea schomburgkii Benth. === **Mabea taquari** Aubl.
Mabea setulosa (Müll.Arg.) Hollander === **Mabea speciosa** Müll.Arg.
Mabea speciosa subsp. *concolor* (Müll.Arg.) Hollander === **Mabea piriri** Aubl.
Mabea subsessilis var. *peruviana* J.F.Macbr. === **Mabea subsessilis** Pax & K.Hoffm.
Mabea surinamensis Klotzsch === **Mabea piriri** Aubl.
Mabea taquari var. *angustifolia* Müll.Arg. === **Mabea taquari** Aubl.
Mabea taquari var. *genuina* Müll.Arg. === **Mabea taquari** Aubl.
Mabea verrucosa Pax & K.Hoffm. === **Mabea montana** Müll.Arg.
Mabea volubilis Klotzsch === **Mabea biglandulosa** Baill. ex Müll.Arg.
Mabea weddelliana Baill. === **Mabea paniculata** Spruce ex Benth.

Maborea

An orthographic variant of *Meborea*.

Macaranga

282 species, Africa (40), Madagascar (c. 13), and S. Asia to Japan (Nansei-shoto), Australia and the Pacific; key centres are West Malesia, the Philippines and Papuasia. Shrubs or trees, generally small to medium-sized (though sometimes reaching 30 m) and usually with more or less coarse, palmately veined leaves, often in secondary vegetation and all looking more or less alike to the uninitiated. Many species, however, have leaves at least sometimes peltate, and one soon learns at least to recognise *M. tanarius* which ranges widely from Asia to eastern Australia (in sect. *Tanarius*, no. 19 of Pax & Hoffmann (1931)). Some are myrmecophytes (cf. Ridley, 1910; Pax & Hoffmann, 1931: 128). Current studies by Ferry Slik (Leiden) and Stuart Davies (Kota Samarahan, Sarawak) suggest that, at least in Borneo, different stages of forest succession are characterised by differing ranges of species, with Davies focussing on those in the largely myrmecophilous sect. *Pachystemon* (nos. 30 (with 31) of Pax & Hoffmann, a characteristically West Malesian group of about 38 species with 19 endemic to Borneo). Other sections also have primary or secondary geographical foci, even within single land masses such as Borneo. In addition, the species in West Africa are not closely related to those in East Africa. Much evolution within the genus has involved *in situ* modifications and adaptations (Davies, unpubl.). Slik (unpubl.) has suggested that the quantity of *M. gigantea* (sect. *Pachystemon*) in study sites in E. Kalimantan (Borneo) is a potential inverse indicator of forest integrity. Early successional species tend to be more widely-ranging than those of later successional stages. – No overall revision has yet succeeded that of Pax (1914) who recognised 32 sections (these grouped into three 'supersections': *Laeves*, *Echinatae* and *Tuberculatae*). In their 1931 survey Pax & Hoffmann increased the number of sections to 36; these numbers are used here and in our bibliographic references. Some regional accounts have been published, notably for Papuasia (Perry, 1953; Whitmore, 1980), Peninsular Malaysia (Whitmore, 1967, and in *Tree Flora of Malaya*), and South Asia (Whitmore, 1978). A first approach to a modern treatment of Malagasy species is given in the key in McPherson (1996). The genus has also been treated in other floras as well

as in Airy-Shaw's family accounts for different parts of Malesia. Sect. *Pseudo-rottlera* (no. 21 of Pax & Hoffmann; preliminary revision by Airy-Shaw, 1965) is apparently transitional to *Mallotus* (although in the Webster system the two genera are in different subtribes of Acalypheae on account of differences in indumentum). Active students of the Malesian species at present include Davies, Slik and T. C. Whitmore. (Acalyphoideae)

Ridley, H. N. (1910). Symbiosis of ants and plants. Ann. Bot. 24: 457-483, 2 pls. En. — Includes (pp. 470-482) observations on *Macaranga*, with detailed studies of three myrmecophilous species. [Different stages towards development of a true symbiosis were detected among the different plants examined in the paper, with the highest level being achieved in *Macaranga triloba* (mainly of west Malesia), a level comparable to that in the swollen-thorn acacias of Central America.]

- Pax, F. (with K. Hoffmann) (1914). *Macaranga*. In A. Engler (ed.), Das Pflanzenreich, IV 147 VII (Euphorbiaceae-Acalypheae-Mercurialinae): 298-395, map. Berlin. (Heft 63.) La/Ge. — 160-170 species in 32 sections (these in turn grouped into three 'supersections': *Laeves*, *Echinatae* and *Tuberculatae*), Africa, Madagascar and from S Asia to Australia and the Pacific. For additions, see Pax and Hoffmann (1919). A table of characters discriminating the genus from *Mallotus* appears on p. 299, and there also sections . Within the genus a primary character for subdivision was ovary morphology. A map (pl. 1 at end of volume) depicts the distribution of the genus and some of its sections and species.

Pax, F. & K. Hoffmann (1919). *Macaranga*. In A. Engler (ed.), Das Pflanzenreich, IV 147 XIV (Euphorbiaceae-Additamentum VI): 24-32. Berlin. (Heft 68.) La/Ge. — Includes 20 additional species from New Guinea.

Pax, F. & K. Hoffmann (1924). *Macaranga*. In A. Engler (ed.), Das Pflanzenreich, IV 147 XVII (Euphorbiaceae-Additamentum VII): 184-186. Berlin. (Heft 85.) La/Ge. — Miscellaneous notes and novelties (including two from New Guinea).

Perry, L. M. (1953). The Papuasian species of *Macaranga*. J. Arnold Arbor. 34: 191-257. (Plantae papuanae archboldianae, XXI.) En. — Regional revision with keys, synonymy, localities with exsiccatae and commentary, covering 54 species; one further species doubtful. [The sectional system of Pax was not here introduced, as noted by the author (p. 193). The key was intended purely for identification. Most collections cited date from after 1920.]

Airy-Shaw, H. K. (1965). Notes on Malaysian and other Asiatic Euphorbiaceae, LV. A preliminary survey of *Macaranga* sect. *Pseudo-rottlera* (Reichb. f. et Zoll.) Pax et K. Hoffm. Kew Bull. 19: 315-328. En. — Synopsis of 13 species with key, descriptions of novelties, synonymy, localities with exsiccatae, and commentary. [Sect. 21 in Pax & Hoffmann's system.]

Whitmore, T. C. (1967). Studies in *Macaranga*, an easy genus of Malayan wayside trees. Malayan Nat. J. 20: 89-99, illus. En. — General introduction; keys to species and alphabetically arranged notes (generally on habitat and distinctive features).

Airy-Shaw, H. K. (1968). Notes on Malesian and other Asiatic Euphorbiaceae, XCIII. Notes on *Macaranga* Thou. Kew Bull. 21: 403-406. En. — Three parts, one an addition to the author's treatment of sect. *Pseudo-rottlera* (1965).

Airy-Shaw, H. K. (1969). Notes on Malesian and other Asiatic Euphorbiaceae, CXI. New or noteworthy species of *Macaranga* Thou. Kew Bull. 23: 88-114. En. — Treatment of many species (some new), the sections indicated. Includes key to 9 species of sect. *Tanarius* (no. 19 in Pax & Hoffmann's system) in E Malesia and the Solomons. [Sect. *Tanarius* is represented in W. Malesia largely by *M. tanarius*, but there is significant speciation in E. Malesia.]

Whitmore, T. C. (1969). Studies in *Macaranga*, III. First thoughts on species evolution in Malayan *Macaranga*. Biol. J. Linn. Soc. 1: 223-231, illus., maps. En. — Geographical and systematic observations (with some maps). [The suggestion is made, among others, that *M. quadricornis* Ridl. (in sect. 30 of Pax & Hoffmann, *Pachystemon*) has evolved twice, a distinct possibility as all the species in its group are mutually very closely related.]

Airy-Shaw, H. K. (1971). Notes on Malesian and other Asiatic Euphorbiaceae, CXLI. New or noteworthy species of *Macaranga* Thou. Kew Bull. 25: 529-543. En. — Novelties and notes covering several species; also includes discussion of sect. *Pachystemon* (no. 30 of Pax & Hoffmann along with their no. 31, *Caladiifoliae,* here reduced) with acceptance of 20 species in the first instance.

Whitmore, T. C. & H. K. Airy Shaw (1971). Studies in *Macaranga,* IV. New and notable records for Malaya. Kew Bull. 25: 237-242. En. — Novelties and notes, precursory to an account of the genus in *Tree Flora of Malaya* 1 (1972).

Airy-Shaw, H. K. (1972). Notes on Malesian and other Asiatic Euphorbiaceae, CLXV: Further note on *Macaranga winkleri,* and on sect. *Winklerianae.* Kew Bull. 27: 90-91. En. — Includes citation of all exsiccatae at Kew of the Bornean *M. winkleri* (Pax & Hoffmann's sect. 32, *Winklerianae*).

Whitmore, T. C. (1972). Studies in *Macaranga,* V. *M. lowii.* Gardens Bull. Singapore 26: 62. En. — Reduction of *M. auriculata. M. lowii* is the commonest member in Peninsular Malaysia of sect. *Pseudo-rottlera.*

Airy-Shaw, H. K. (1974). Notes on Malesian and other Asiatic Euphorbiaceae, CLXXXII. New species of *Macaranga* Thou. Kew Bull. 29: 322-325. En. — Three novelties, one from Borneo, the others from New Guinea; sections given. The author indicates that in New Guinea there is a 'constellation' of species in sect. *Tanarius* (no. 19 of Pax & Hoffmann) of which possibly others remain to be made known.

Whitmore, T. C. (1974). Studies in *Macaranga,* VI. Novelties from Borneo and a reduction in Malaya. Kew Bull. 29: 445-450. En. — Additions to sections *Pachystemon, Adenoceras* (no. 12 of Pax & Hoffmann, 1931), *Winklerianae* and *Pseudo-rottlera;* reduction of *M. griffithiana* (sect. *Pachystemon*) to *M. motleyana.* [A precursor to Airy-Shaw, *Euphorbiaceae of Borneo* (1975; see Malesia).]

Airy-Shaw, H. K. (1978). Notes on Malesian and other Asiatic Euphorbiaceae, CCIII. New species of *Macaranga* from the Bismarcks, Solomons and New Hebrides. Kew Bull. 32: 410-414. En. — Descriptions of three new species, with indication of sections (one new, but without indication of affinities).

Airy-Shaw, H. K. (1978). Notes on Malesian and other Asiatic Euphorbiaceae, CCXVIII. *Macaranga* Thou. Kew Bull. 33: 67-71. En. — Novelties and notes; 3 species, New Guinea.

Whitmore, T. C. (1978). Studies in *Macaranga,* VII. The genus in "Greater India". Gardens Bull. Singapore 31: 51-56. En. — Includes key to and alphabetically arranged enumeration of 12 species with synonymy, citations, indication of distribution, and commentary. Covers Pakistan, India, Nepal, Bhutan, Myanmar, and Andaman and Nicobar Islands.

Whitmore, T. C. (1980). Studies in *Macaranga,* VIII. *Macaranga* in New Guinea and the Bismarck Archipelago. Kew Bull. 34: 599-606. En. — Precursory to treatment in Airy-Shaw, *Euphorbiaceae of New Guinea* (1980; see Malesia); includes a new review of the genus in Papuasia (75 species, with 19 additional infraspecific taxa) featuring taxonomic history, statistics, natural groups, variation, and ecology followed by descriptions of nine new species. Nine natural groups are provisionally distinguished, of which 'Dioica' (formerly no. 21, sect. *Pseudo-rottlera*), 'Longistipulata' (based on no. 23 of Pax & Hoffmann, sect. *Longistipulata*) and 'Tanarius' (no. 19 of Pax & Hoffmann) are the largest.

McPherson, G. (1996). A new species of *Macaranga* (Euphorbiaceae) from Madagascar. Bull. Mus. Natl. Hist. Nat., IV, B (Adansonia, III), 18: 275-278, illus. En. — An addition to the Malagasy species in this genus: *M. grallata.* A provisional key to known species is also included (p. 278; some appear twice and the author indicates that reduction of some names is likely).

Slik, J. W. F. (1998). Keys to the taxa of *Macaranga* and *Mallotus* (Euphorbiaceae) of East Kalimantan (Indonesia). *Flora Males. Bull.* 12(4): 157-178. En.

Macaranga Thouars, Gen. Nov. Madag.: 26 (1806).

Trop. & Subtrop. Old World. 22 23 24 25 26 27 29 36 38 40 41 42 50 60 61 62 (63).

Panopia Noronha ex Thouars, Gen. Nov. Madag.: 26 (1806).

Mappa A.Juss., Euphorb. Gen.: 44 (1824).

Pachystemon Blume, Bijdr.: 626 (1826).

Adenoceras Rchb. & Zoll. ex Baill., Étude Euphorb.: 430 (1858).
Mecostylis Kurz ex Teijsm. & Binn., Tijdschr. Ned.-Indië 27: 44 (1864).
Phocea Seem., J. Bot. 8: 68 (1870).
Tanarius Rumph. ex Kuntze, Revis. Gen. Pl. 2: 619 (1891).

Macaranga acerifolia Airy Shaw, Kew Bull. 33: 67 (1978). – FIGURE, p. 1081.
Irian Jaya. 42 NWG. Nanophan.

Macaranga adenophila Pax & K.Hoffm. in H.G.A.Engler, Pflanzenr., IV, 147, VII: 310 (1914).
Thailand (Koh Shan I.). 41 THA. – Close to *M. griffithiana.*

Macaranga advena Pax & K.Hoffm. in H.G.A.Engler, Pflanzenr., IV, 147, XIV: 31 (1919).
NW. Papua New Guinea (Mt. Felsspitze). 42 NWG. Phan. – Close to *M. misimae.*

Macaranga aetheadenia Airy Shaw, Kew Bull. 29: 322 (1974).
Borneo (C. Sarawak). 42 BOR. Phan.

Macaranga albescens L.M.Perry, J. Arnold Arbor. 34: 244 (1953).
New Guinea. 42 NWG. Phan.

Macaranga alchorneifolia Baker, J. Linn. Soc., Bot. 25: 344 (1890).
Madagascar. 29 MDG. Nanophan.

Macaranga alchorneoides Pax & Lingelsh., Repert. Spec. Nov. Regni Veg. 3:25 (1906).
New Caledonia. 60 NWC. Nanophan. or phan.
Macaranga longispica S.Moore, J. Linn. Soc., Bot. 45: 407 (1921).

Macaranga aleuritoides F.Muell., Papuan Pl.: 21 (1876).
Maluku (Kai I.), New Guinea, Bismarck Archip., Solomon Is. 42 BIS MOL NWG 60 SOL. Phan.
Macaranga riparia Engl., Bot. Jahrb. Syst. 7: 463 (1886).

Macaranga alnifolia Baker, J. Linn. Soc., Bot. 20: 256 (1883).
C. Madagascar. 29 MDG. Phan.

Macaranga amissa Airy Shaw, Kew Bull. 21: 403 (1968).
Pen. Malaysia, S. Sumatera (incl. Bangka), Borneo (Sabah). 42 BOR MLY SUM. Phan.
* *Endospermum perakense* King ex Hook.f., Fl. Brit. India 5: 458 (1887).

Macaranga amplifolia Merr., Philipp. J. Sci., C 7: 392 (1912 publ. 1913).
Philippines. 42 PHI.

Macaranga anceps Airy Shaw, Kew Bull. 19: 324 (1965).
Sumatera, Borneo. 42 BOR SUM. Phan.

subsp. **anceps**
N. Sumatera. 42 SUM. Phan.

subsp. **puncticulata** Whitmore, Kew Bull. 29: 450 (1974).
E. Sumatera, Borneo (Sarawak). 42 BOR SUM. Phan.

Macaranga andamanica Kurz, Forest Fl. Burma 2: 389 (1877).
Andaman Is., S. Burma, S. Thailand, Pen. Malaysia, Vietnam, Hainan, SW. China (Yunnan). 36 CHC CHH 41 AND BMA MLY THA VIE. Nanophan. or phan.
Macaranga brandisii King ex Hook.f., Fl. Brit. India 5: 453 (1887).
Macaranga bracteata Merr., Lingnan Sci. J. 6: 281 (1928 publ. 1930).
Macaranga rosuliflora Croizat, J. Arnold Arbor. 23: 51 (1942).

Macaranga acerifolia Airy Shaw (1), *M. clavata* Warb. (2), *M. clemensiae* L.M. Perry (3), *M. domatiosa* Airy-Shaw (4)
Artist: Ann Davies
Airy-Shaw, Euphorbiaceae of New Guinea: pl. 6 (1980)
KEW ILLUSTRATIONS COLLECTION

Macaranga angolensis (Müll.Arg.) Müll.Arg. in A.P.de Candolle, Prodr. 15(2): 994 (1866).
E. Cameroon, Congo, Zaire, Uganda, Angola. 23 CMN CON ZAI 25 UGA 26 ANG. Nanophan. or phan.
* *Mappa angolensis* Müll.Arg., J. Bot. 2: 337 (1864).
Macaranga mollis Pax, Bot. Jahrb. Syst. 19: 93 (1894). *Macaranga angolensis* var. *mollis* (Pax) Prain in D.Oliver, Fl. Trop. Afr. 6(1): 937 (1912).
Macaranga guignardii Beille, Bull. Soc. Bot. France 55(8): 77 (1908).
Macaranga mildbraediana Pax, Bot. Jahrb. Syst. 43: 322 (1909).

Macaranga angustifolia K.Schum. & Lauterb., Fl. Schutzgeb. Südsee: 398 (1900).
Papua New Guinea. 42 NWG. Phan.

Macaranga assas Amougou, Bull. Jard. Bot. Belg. 60: 279 (1990).
Cameroon, Central African Rep., Zaire. 23 CAF CMN ZAI.

Macaranga astrolabica Pax & K.Hoffm. in H.G.A.Engler, Pflanzenr., IV, 147, VII: 343 (1914).
N. Papua New Guinea (Astrolabe Bay). 42 NWG. Phan. – Type destroyed.

Macaranga attenuata J.W.Moore, Bernice P. Bishop Mus. Bull. 102: 31 (1933).
Society Is. (Moorea, Raiatea). 61 SCI. Phan.

Macaranga baccaureifolia Airy Shaw, Kew Bull. 19: 316 (1965).
Pen. Malaysia, Borneo (Sabah). 42 BOR MLY. Phan.

Macaranga bailloniana Müll.Arg. in A.P.de Candolle, Prodr. 15(2): 1013 (1866).
Comores (incl. Mayotte). 29 COM.
Macaranga cordifolia Boivin ex Baill., Étude Euphorb.: 432 (1858), nom. nud.
Macaranga eglandulosa Baill., Étude Euphorb.: 432 (1858), nom. nud.

Macaranga balabacensis Pax & K.Hoffm. in H.G.A.Engler, Pflanzenr., IV, 147, VII: 368 (1914).
Philippines. 42 PHI.

Macaranga balakrishnanii B.Mitra & Chakrab., J. Econ. Taxon. Bot. 15: 465 (1991).
Sikkim. 40 EHM.

Macaranga balansae Gagnep., Bull. Soc. Bot. France 69: 701 (1922 publ. 1923).
Vietnam. 41 VIE.

Macaranga bancana (Miq.) Müll.Arg. in A.P.de Candolle, Prodr. 15(2): 990 (1866).
Thailand, W. Malesia. 41 THA 42 BOR MLY SUM.
* *Pachystemon bancanus* Miq., Fl. Ned. Ind., Eerste Bijv.: 462 (1861).
Macaranga calcicola var. *calcifuga* Whitmore, Kew Bull. 29: 445 (1974). *Macaranga calcifuga* (Whitmore) R.I.Milne, Kew Bull. 49: 453 (1994).

Macaranga barteri Müll.Arg., Flora 47: 535 (1864).
W. Trop. Africa to Uganda. 22 GHA GUI IVO LBR NGA SIE TOG 23 CMN EQG ZAI 25 UGA. Phan.
Macaranga heudelotii var. *nitida* Beille, Mém. Soc. Bot. France 55(8b): 79 (1908).
Macaranga lancifolia Pax, Bot. Jahrb. Syst. 43: 322 (1909).
Macaranga rowlandii Prain, Bull. Misc. Inform. Kew 1910: 127 (1910).

Macaranga beccariana Merr., Webbia 7: 315 (1950).
Borneo (Sarawak, Brunei, Sabah). 42 BOR. Phan.
Macaranga hypoleuca var. *borneensis* Hutch. ex Gibbs, J. Linn. Soc., Bot. 42: 136 (1914).

Macaranga beillei Prain, Bull. Misc. Inform. Kew 1910: 239 (1910).
Ivory Coast. 22 IVO. Nanophan. or phan.

Macaranga belensis L.M.Perry, J. Arnold Arbor. 34: 200 (1953).
Irian Jaya (NE. of Lake Habbema). 42 NWG. Phan.

Macaranga bicolor Müll.Arg., Linnaea 34: 199 (1865).
Philippines. 42 PHI.
Macaranga utilis Elmer ex Merr., Enum. Philipp. Fl. Pl. 2: 440 (1923), pro syn.

Macaranga bifoveata J.J.Sm., Nova Guinea 8: 790 (1912).
New Guinea. 42 NWG. Phan.
Macaranga latifolia L.M.Perry, J. Arnold Arbor. 34: 217 (1953).

Macaranga boutonioides Baill., Adansonia 1: 265 (1861).
Madagascar (Nosi Bé I.), Comoros (Mayotte). 29 COM MDG. Nanophan.
Macaranga ovata Boivin ex Baill., Adansonia 1: 266 (1861).
Macaranga rottleroides Baill., Adansonia 1: 262 (1861).
Macaranga hildebrandtii Baill., Bull. Mens. Soc. Linn. Paris 2: 996 (1892).
Macaranga humblotiana Baill., Bull. Mens. Soc. Linn. Paris 2: 996 (1892).

Macaranga brachythyrsa Pax & K.Hoffm. in H.G.A.Engler, Pflanzenr., IV, 147, VII: 320 (1914).
Borneo (SE. Kalimantan). 42 BOR. Phan. – Perhaps identical with *M. petanostyla*.

Macaranga brachytricha Airy Shaw, Kew Bull. 23: 103 (1969).
NE. Papua New Guinea. 42 NWG. Phan.

Macaranga brevipetiolata Airy Shaw, Kew Bull. 19: 326 (1965).
Borneo (Sarawak, Sabah, E. Kalimantan). 42 BOR. Phan.

Macaranga brooksii Ridl., Bull. Misc. Inform. Kew 1925: 90 (1925).
W. & S. Sumatera. 42 SUM.

Macaranga brunneofloccosa Pax & K.Hoffm. in H.G.A.Engler, Pflanzenr., IV, 147, XIV: 28 (1919).
New Guinea. 42 NWG. Phan.
Macaranga brunneofloccosa var. *calvescens* Pax & K.Hoffm. in H.G.A.Engler, Pflanzenr., IV, 147, XVI: 28 (1924).

Macaranga bullata Pax & K.Hoffm. in H.G.A.Engler, Pflanzenr., IV, 147, XIV: 31 (1919).
N. Papua New Guinea (Mt. Schrader). 42 NWG. – Perhaps identical with *M. intonsa*.

Macaranga caesariata A.C.Sm., Contr. U. S. Natl. Herb. 37: 73 (1967).
Fiji (Viti Levu). 60 FIJ. Phan.

Macaranga caladiifolia Becc., Malesia 2: 46 (1884).
Pen. Malaysia, E. Sumatera, Borneo. 42 BOR MLY SUM. Phan.
Macaranga caladiifolia var. *pilosula* Pax & K.Hoffm. in H.G.A.Engler, Pflanzenr., IV, 147, VII: 384 (1914).
Macaranga caladiifolia var. *truncata* Pax & K.Hoffm. in H.G.A.Engler, Pflanzenr., IV, 147, VII: 384 (1914).
Macaranga tenuiramea Pax & K.Hoffm. in H.G.A.Engler, Pflanzenr., IV, 147, VII: 384 (1914).
Macaranga myrmecophila Diels, Bot. Jahrb. Syst. 60: 309 (1926).
Macaranga eloba Pax & K.Hoffm., Mitt. Bot. Mus. Hamburg 7: 228 (1931).

Macaranga calcicola Airy Shaw, Kew Bull. 25: 536 (1971).
Sumatera, Borneo (SW. Sarawak, E. Kalimantan). 42 BOR SUM. Phan.

Macaranga capensis (Baill.) Sim, Forest Fl. Cape: 314 (1907).
Ethiopia to S. Africa. 23 ZAI 24 ETH 25 KEN TAN UGA 36 MLW MOZ ZAM ZIM 27 CPP NAT TVL. Phan.
* *Mappa capensis* Baill., Adansonia: 155 (1862). *Mallotus capensis* (Baill.) Müll.Arg., Linnaea 34: 189 (1865).
Macaranga bachmannii Pax, Bot. Jahrb. Syst. 23: 525 (1897).
Macaranga ruwenzorica Pax, Bot. Jahrb. Syst. 43: 322 (1909).
Macaranga inopinata Prain in D.Oliver, Fl. Trop. Afr. 6(1): 914 (1912).
Macaranga multiglandulosa Pax & K.Hoffm. in H.G.A.Engler, Pflanzenr., IV, 147, VII: 343 (1914).
Macaranga usambarica Pax & K.Hoffm. in H.G.A.Engler, Pflanzenr., IV, 147, VII: 344 (1914).

Macaranga carolinensis Volkens, Bot. Jahrb. Syst. 31: 466 (1901).
Caroline Is. (Yap, Palau). 62 CRL. Phan.

Macaranga carrii L.M.Perry, J. Arnold Arbor. 34: 235 (1953).
New Guinea. 42 NWG. Phan.
Macaranga leonardii L.M.Perry, J. Arnold Arbor. 34: 234 (1953). *Macaranga carrii* var. *leonardii* (L.M.Perry) Whitmore, Kew Bull., Addit. Ser. 8: 141 (1980).
Macaranga womersleyi L.M.Perry, J. Arnold Arbor. 34: 232 (1953). *Macaranga carrii* var. *womersleyi* (L.M.Perry) Whitmore, Kew Bull., Addit. Ser. 8: 142 (1980).
Macaranga carrii var. *laevis* Whitmore, Kew Bull., Addit. Ser. 8: 141 (1980).
Macaranga carrii var. *myolensis* Whitmore, Kew Bull., Addit. Ser. 8: 142 (1980).

Macaranga caudata Pax & K.Hoffm. in H.G.A.Engler, Pflanzenr., IV, 147, XIV: 30 (1919).
New Guinea. 42 NWG. Phan.

Macaranga caudatifolia Elmer, Leafl. Philipp. Bot. 2: 427 (1908).
Philippines. 42 PHI.
Macaranga cuneata Elmer, Leafl. Philipp. Bot. 2: 428 (1908).
Macaranga sibuyanensis Elmer, Leafl. Philipp. Bot. 3: 922 (1910).
Macaranga apoensis Elmer, Leafl. Philipp. Bot. 7: 2616 (1915).

Macaranga celebica Koord., Meded. Lands Plantentuin 19: 626 (1898).
Sulawesi. 42 SUL.

Macaranga chatiniana (Baill.) Müll.Arg. in A.P.de Candolle, Prodr. 15(2): 996 (1866).
? 4 +. Phan.
* *Mappa chatiniana* Baill., Adansonia 1: 349 (1861).

Macaranga chlorolepis Airy Shaw, Kew Bull. 25: 530 (1971).
Papua New Guinea, Bismarck Archip. (New Britain). 42 BIS NWG. Phan.

Macaranga choiseuliana Airy Shaw, Kew Bull. 32: 412 (1978).
Solomon Is. (Bougainville, Choiseul). 60 SOL.

Macaranga chrysotricha K.Schum. & Lauterb., Fl. Schutzgeb. Südsee: 399 (1900).
Papua New Guinea. 42 NWG. Phan.
Macaranga chrysotricha var. *glaucescens* Mansf., J. Arnold Arbor. 10: 78, 233 (1929).

Macaranga cissifolia (Zoll. & Rchb.f.) Müll.Arg. in A.P.de Candolle, Prodr. 15(2): 993 (1866).
Sumatera. 42 SUM.
* *Mappa cissifolia* Zoll. & Rchb.f., Linnaea 29: 466 (1858).

Macaranga clavata Warb., Bot. Jahrb. Syst. 13: 349 (1891). – FIGURE, p. 1081.
New Guinea, Maluku (Kai Is.), Bismarck Archip. (New Ireland). 42 BIS MOL NWG. Phan.

Macaranga clemensiae L.M.Perry, J. Arnold Arbor. 34: 232 (1953). – FIGURE, p. 1081.
Papua New Guinea. 42 NWG. Phan.

Macaranga coggygria Airy Shaw, Kew Bull. 29: 324 (1974).
Papua New Guinea. 42 NWG. Phan.

Macaranga congestiflora Merr., Philipp. J. Sci., C 4: 282 (1909).
Philippines. 42 PHI.

Macaranga conglomerata Brenan, Kew Bull. 4: 94 (1949).
SE. Kenya, NE Tanzania. 25 KEN TAN. Phan.

Macaranga conifera (Zoll.) Müll.Arg. in A.P.de Candolle, Prodr. 15(2): 1005 (1866).
Andaman Is., Pen. Malaysia, Sumatera (incl. Bangka), Borneo (SW. Sarawak, Brunei, Sabah, E. Kalimantan). 41 AND 42 BOR MLY SUM. Nanophan. or phan.
* *Mappa conifera* Zoll., Linnaea 29: 466 (1858).
Pachystemon populifolius Miq., Fl. Ned. Ind., Eerste Bijv.: 462 (1861). *Macaranga populifolia* (Miq.) Müll.Arg. in A.P.de Candolle, Prodr. 15(2): 1006 (1866).

Macaranga constricta Whitmore & Airy Shaw, Kew Bull. 25: 238 (1971).
Pen. Malaysia. 42 MLY.

Macaranga cordifolia (Roxb.) Müll.Arg. in A.P.de Candolle, Prodr. 15(2): 1002 (1866).
Maluku. 42 MOL. – Provisionally accepted.
* *Adelia cordifolia* Roxb., Fl. Ind. ed. 1832, 3: 849 (1832). *Macaranga moluccana* Radcl.-Sm. & Govaerts, Kew Bull. 52: 480 (1997), nom. illeg.

Macaranga coriacea (Baill.) Müll.Arg. in A.P.de Candolle, Prodr. 15(2): 1006 (1866).
New Caledonia (incl. Loyalty Is.). 60 NWC. Nanophan. or phan.
* *Cleidion coriaceum* Baill., Adansonia 2: 218 (1862).

Macaranga corymbosa (Müll.Arg.) Müll.Arg. in A.P.de Candolle, Prodr. 15(2): 1006 (1866).
New Caledonia. 60 NWC. Phan.
* *Mappa corymbosa* Müll.Arg., Linnaea 34: 199 (1865).
Macaranga fulvescens Schltr., Bot. Jahrb. Syst. 39: 151 (1906).
Macaranga insularis Schltr., Bot. Jahrb. Syst. 39: 151 (1906).
Macaranga oreophila Pax & K.Hoffm. in H.G.A.Engler, Pflanzenr., IV, 147, VII: 363 (1914).

Macaranga costulata Pax & K.Hoffm. in H.G.A.Engler, Pflanzenr., IV, 147, VII: 338 (1914).
Borneo (Sarawak, Sabah, E. Kalimantan). 42 BOR. Phan.

Macaranga crassistipulosa Pax & K.Hoffm. in H.G.A.Engler, Pflanzenr., IV, 147, XIV: 26 (1919).
Sulawesi. 42 SUL.

Macaranga cucullata J.J.Sm., Nova Guinea 8: 237 (1910).
New Guinea. 42 NWG. Phan.

Macaranga cuernosensis Elmer, Leafl. Philipp. Bot. 2: 429 (1908).
Philippines. 42 PHI.

Macaranga cumingii (Baill.) Müll.Arg. in A.P.de Candolle, Prodr. 15(2): 1005 (1866).
Philippines. 42 PHI.
* *Mappa cumingii* Baill. in A.P.de Candolle, Prodr. 15(2): 1005 (1866).

Macaranga cuneifolia (Zoll.) Müll.Arg. in A.P.de Candolle, Prodr. 15(2): 1004 (1866).
Sumatera (Bangka). 42 SUM. – Provisionally accepted.
* *Mappa cuneifolia* Zoll., Linnaea 29: 467 (1858).

Macaranga cupularis Müll.Arg., Flora 47: 466 (1864).
Madagascar. 29 MDG. Phan.
Macaranga ciliata Bojer ex Baker, J. Linn. Soc., Bot. 30: 259 (1883).

Macaranga curtisii Hook.f., Fl. Brit. India 5: 448 (1887).
S. Thailand, Pen. Malaysia, Sumatera, N. Borneo. 41 THA 42 BOR MLY SUM. Phan.

var. **curtisii**
S. Thailand, Pen. Malaysia, Sumatera, N. Borneo. 41 THA 42 BOR MLY SUM. Phan.

var. **glabra** Whitmore, Kew Bull. 25: 240 (1971).
Pen. Malaysia. 42 MLY. Phan.

Macaranga cuspidata Boivin ex Baill., Adansonia 1: 260 (1861).
Madagascar. 29 MDG. Phan.
Macaranga peltata Boivin ex Baill., Étude Euphorb.: 432 (1858), nom. nud.

Macaranga dallachyana (Baill.) Airy Shaw, Kew Bull. 23: 90 (1969).
Queensland (Cape York Pen.). 50 QLD. Nanophan. or phan.
* *Echinus dallachyanus* Baill., Adansonia 6: 314 (1866).
Mallotus dallachyi F.Muell., Fragm. 6: 184 (1868). *Macaranga dallachyi* (F.Muell.) F.Muell. ex Benth., Fl. Austral. 6: 144 (1873).

Macaranga darbyshirei Airy Shaw, Kew Bull. 29: 323 (1974).
New Guinea. 42 NWG. Phan.

Macaranga decaryana Leandri, Notul. Syst. (Paris) 10: 149 (1942).
Madagascar. 29 MDG.

Macaranga decipiens L.M.Perry, J. Arnold Arbor. 34: 212 (1953).
New Guinea. 42 NWG. Nanophan. or phan.

Macaranga densiflora Warb., Bot. Jahrb. Syst. 13: 350 (1891).
New Guinea, Bismarck Archip., Solomon Is. 42 BIS NWG 60 SOL. Nanophan. or phan.
Macaranga acuminata Warb. ex J.J.Sm., Nova Guinea 8: 238 (1910).
Macaranga acuminata Ridl., Trans. Linn. Soc. London, Bot. 9: 147 (1916), nom. illeg.

Macaranga denticulata (Blume) Müll.Arg. in A.P.de Candolle, Prodr. 15(2): 1000 (1866).
Assam, Burma, Andaman Is., S. China, Thailand, Pen. Malaysia, Sumatera, Jawa. 36 CHS 40 ASS 41 AND BMA MLY THA 42 JAW SUM. Nanophan. or phan.
* *Mappa denticulata* Blume, Bijdr.: 625 (1826).
Mappa gummiflua Miq., Fl. Ned. Ind., Eerste Bijv.: 458 (1861). *Macaranga gummiflua* (Miq.) Müll.Arg. in A.P.de Candolle, Prodr. 15(2): 1000 (1866).
Macaranga perakensis Hook.f., Fl. Brit. India 5: 447 (1887).
Macaranga henricorum Hemsl., J. Linn. Soc., Bot. 26: 442 (1894).

Macaranga depressa (Müll.Arg.) Müll.Arg. in A.P.de Candolle, Prodr. 15(2): 989 (1866).
Borneo (Sarawak, Sabah, Kalimantan). 42 BOR. Phan.
* *Pachystemon depressus* Müll.Arg., Flora 47: 465 (1864). *Macaranga depressa* var. *genuina* Müll.Arg. in A.P.de Candolle, Prodr. 15(2): 989 (1866), nom. inval.

f. **depressa**
Borneo (Sarawak, Sabah, E. & SE. Kalimantan). 42 BOR. Phan.

f. **glabra** Whitmore, Kew Bull. 29: 446 (1974).
N. Borneo. 42 BOR. Phan.

f. **strigosa** Whitmore, Kew Bull. 29: 446 (1974).
Borneo (Sarawak, N. Kalimantan). 42 BOR. Phan.
Macaranga divergens Müll.Arg. in A.P.de Candolle, Prodr. 15(2): 989 (1866).

Macaranga dibeleensis De Wild., Ann. Mus. Congo Belge, Bot., V, 2: 281 (1908).
Zaire. 23 ZAI. Phan.

Macaranga didymocarpa Whitmore, Kew Bull. 39: 607 (1984).
Borneo (Sabah). 42 BOR.

Macaranga diepenhorstii (Miq.) Müll.Arg. in A.P.de Candolle, Prodr. 15(2): 998 (1866).
Pen. Malaysia, Sumatera. 42 MLY SUM.
* *Mappa diepenhorstii* Miq., Fl. Ned. Ind., Eerste Bijv.: 457 (1861).
Macaranga pachyphylla Müll.Arg. in A.P.de Candolle, Prodr. 15(2): 999 (1866).

Macaranga digyna (Wight) Müll.Arg. in A.P.de Candolle, Prodr. 15(2): 1007 (1866).
Sri Lanka. 40 SRL.
* *Claoxylon digynum* Wight, Icon. Pl. Ind. Orient. 5: t. 1884 (1852).

Macaranga dioica (G.Forst.) Müll.Arg., Linnaea 34: 199 (1865).
NE. New Guinea, Bismarck Archip., Solomon Is., Santa Cruz Is., Vanuatu. 42 BIS NWG 60 SCZ SOL VAN. Nanophan. or phan.
* *Ricinus dioicus* G.Forst., Fl. Ins. Austr.: 67 (1786).
Macaranga urophylla Pax & K.Hoffm. in H.G.A.Engler, Pflanzenr., IV, 147, VII: 377 (1914).

Macaranga dipterocarpifolia Merr., Philipp. J. Sci. 1(Suppl.): 205 (1906).
Philippines. 42 PHI.

Macaranga domatiosa Airy Shaw, Kew Bull. 25: 540 (1971). – FIGURE, p. 1081.
New Guinea. 42 NWG. Nanophan. or phan.
Macaranga caryocarpa Airy Shaw, Kew Bull. 25: 541 (1971).

Macaranga ducis Whitmore, Kew Bull. 34: 602 (1980).
Papua New Guinea. 42 NWG. Phan.

Macaranga ebolowana Pax & K.Hoffm., Bot. Jahrb. Syst. 58(130): 40 (1923).
Cameroon. 23 CMN.

Macaranga echinocarpa Baker, J. Linn. Soc., Bot. 20: 255 (1883).
C. Madagascar. 29 MDG. Nanophan.

Macaranga endertii Whitmore, Kew Bull. 29: 449 (1974).
Borneo (Sarawak, E. Kalimantan). 42 BOR. Phan.

Macaranga esquirolii (H.Lév.) Rehder, J. Arnold Arbor. 18: 214 (1937).
China (Guizhou). 36 CHC.
* *Morinda esquirolii* H.Lév., Fl. Kouy-Tchéou: 368 (1915).

Macaranga eymae L.M.Perry, J. Arnold Arbor. 34: 200 (1953).
Irian Jaya (Wissel lakes). 42 NWG. Phan.

Macaranga faiketo Whitmore, Kew Bull. 34: 607 (1980).
Solomon Is., Santa Cruz Is. 60 SCZ SOL. Phan.

Macaranga fallacina Pax & K.Hoffm. in H.G.A.Engler, Pflanzenr., IV, 147, XIV: 31 (1919).
New Guinea. 42 NWG. Nanophan. or phan.

Macaranga ferruginea Baker, J. Linn. Soc., Bot. 22: 521 (1887).
C. Madagascar. 29 MDG. Phan.
Macaranga platyphylla Baker ex Baill., Bull. Mens. Soc. Linn. Paris 2: 991 (1892).

Macaranga fragrans L.M.Perry, J. Arnold Arbor. 34: 235 (1953).
New Guinea. 42 NWG. Phan.

Macaranga fulva Airy Shaw, Kew Bull. 19: 322 (1965).
Borneo (E. Kalimantan). 42 BOR. Nanophan. or phan.

Macaranga gabunica Prain, Bull. Misc. Inform. Kew 1912: 104 (1912).
Gabon, Congo. 23 CON GAB. Cl. nanophan.

Macaranga galorei Whitmore, Kew Bull. 34: 602 (1980).
Papua New Guinea. 42 NWG. Nanophan. or phan.

Macaranga gamblei Hook.f., Fl. Brit. India 5: 445 (1887).
Sikkim. 40 EHM.

Macaranga gigantea (Rchb.f. & Zoll.) Müll.Arg. in A.P.de Candolle, Prodr. 15(2): 995 (1866).
Thailand, Pen. Malaysia, Sumatera, Borneo, Sulawesi. 41 THA 42 BOR MLY SUL SUM. Phan.
* *Mappa gigantea* Zoll., Linnaea 29: 465 (1858).
Mappa megalophylla Müll.Arg., Flora 47: 467 (1864). *Macaranga megalophylla* (Müll.Arg.) Müll.Arg. in A.P.de Candolle, Prodr. 15(2): 995 (1866).
Macaranga rugosa Müll.Arg. in A.P.de Candolle, Prodr. 15(2): 995 (1866).
Macaranga incisa Gage, Rec. Bot. Surv. India 9: 245 (1922).

Macaranga gigantifolia Merr., Philipp. J. Sci., C 7: 391 (1912 publ. 1913).
Philippines, Sulawesi ?, Maluku ? 42 MOL? PHI SUL? Nanophan. or phan.

Macaranga glaberrima (Hassk.) Airy Shaw, Kew Bull. 19: 322 (1965).
Jawa, Lesser Sunda Is. (Flores), New Guinea. 42 JAW LSI NWG. Nanophan. or phan.
* *Rottlera glaberrima* Hassk., Flora 25(2): 41 (1842). *Mallotus glaberrimus* (Hassk.) Müll.Arg., Linnaea 34: 192 (1865).

var. **glaberrima**
Jawa, Lesser Sunda Is. (Flores), Irian Jaya. 42 JAW LSI NWG. Nanophan. or phan.
Rottlera subfalcata Rchb.f. & Zoll., Acta Soc. Regiae Sci. Indo-Neerl. 1: 11 (1856). *Macaranga subfalcata* (Rchb.f. & Zoll.) Müll.Arg. in A.P.de Candolle, Prodr. 15(2): 1007 (1866).
Macaranga haplostachya Pax & K.Hoffm. in H.G.A.Engler, Pflanzenr., IV, 147, XIV: 25 (1919).

var. **schoddei** Airy Shaw, Kew Bull. 23: 107 (1969).
Papua New Guinea. 42 NWG. Nanophan. or phan.

Macaranga glabra (Juss.) Pax & K.Hoffm. in H.G.A.Engler, Pflanzenr., IV, 147, VII: 355 (1914).
Lesser Sunda Is. (Timor). 42 LSI.
* *Mappa glabra* Juss., Nouv. Arch. Mus. Paris 3: 487 (1834).

Macaranga glandulifera L.M.Perry, J. Arnold Arbor. 34: 218 (1953).
Irian Jaya (NE. of Habbema lake). 42 NWG. Nanophan. or phan.

Macaranga gracilis Pax & K.Hoffm. in H.G.A.Engler, Pflanzenr., IV, 147, XIV: 31 (1919).
New Guinea. 42 NWG. Phan.

Macaranga graeffeana Pax & K.Hoffm., Notizbl. Bot. Gart. Berlin-Dahlem 10: 384 (1928).
Fiji. 60 FIJ. Nanophan.

var. **crenata** (A.C.Sm.) A.C.Sm., J. Arnold Arbor. 33: 384 (1952).
Fiji. 60 FIJ. Phan.
* *Macaranga crenata* A.C.Sm., Bernice P. Bishop Mus. Bull. 141: 86 (1936).

var. **graeffeana**
Fiji. 60 FIJ. Nanophan.

var. **major** A.C.Sm., J. Arnold Arbor. 33: 384 (1952).
Fiji. 60 FIJ. Phan.

Macaranga grallata McPherson, Bull. Mus. Natl. Hist. Nat., B, Adansonia 18: 275 (1996).
E. Madagascar. 29 MDG. Phan.

Macaranga grandifolia (Blanco) Merr., Philipp. J. Sci., C 7: 394 (1912 publ. 1913).
Philippines. 42 PHI.
* *Croton grandifolius* Blanco, Fl. Filip.: 753 (1873).
Macaranga porteana André, Rev. Hort. 60: 175 (1888).

Macaranga grayana Müll.Arg. in A.P.de Candolle, Prodr. 15(2): 1001 (1866).
Samoa. 60 SAM. Phan.

Macaranga hageniana Gilli, Ann. Naturhist. Mus. Wien 83: 437 (1979 publ. 1980).
Papua New Guinea. 42 NWG.

Macaranga harveyana (Müll.Arg.) Müll.Arg. in A.P.de Candolle, Prodr. 15(2): 998 (1866).
Fiji, Tonga, Samoa, Cook Is. 60 FIJ SAM TON 61 COO. Phan.
* *Mappa harveyana* Müll.Arg., Flora 47: 467 (1864).
Macaranga harveyana var. *glabrata* Pax & K.Hoffm. in H.G.A.Engler, Pflanzenr., IV, 147, VII: 357 (1914), nom. illeg.
Macaranga harveyana var. *puberula* Pax & K.Hoffm. in H.G.A.Engler, Pflanzenr., IV, 147, VII: 357 (1914).

Macaranga havilandii Airy Shaw, Kew Bull. 23: 112 (1969).
Borneo (Sarawak). 42 BOR. Nanophan. or phan.
Macaranga havilandii var. *resecta* Airy Shaw, Kew Bull. 25: 534 (1971).

Macaranga hemsleyana Pax & K.Hoffm. in H.G.A.Engler, Pflanzenr., IV, 147, VII: 322 (1914).
S. China, Hainan. 36 CHC CHH CHS. Phan.
* *Mallotus populifolius* Hemsl., J. Linn. Soc., Bot. 26: 441 (1891).

Macaranga henryi (Pax & K.Hoffm.) Rehder, Sunyatsenia 3: 240 (1936).
China (Yunnan). 36 CHC.
* *Mallotus henryi* Pax & K.Hoffm. in H.G.A.Engler, Pflanzenr., IV, 147, VII: 177 (1914).

Macaranga herculis Whitmore, Kew Bull. 39: 608 (1984).
Papua New Guinea. 42 NWG.

Macaranga heterophylla (Müll.Arg.) Müll.Arg. in A.P.de Candolle, Prodr. 15(2): 993 (1866).
W. Trop. Africa. 22 GHA GNB GUI IVO LBR SEN SIE TOG. Nanophan. or phan.
* *Mappa heterophylla* Müll.Arg., J. Bot. 2: 336 (1864).
Macaranga quinquelobata Beille, Bull. Soc. Bot. France 55(8): 78 (1908).

Macaranga heudelotii Baill., Adansonia 1: 69 (1860).
W. Trop. Africa. 22 GAM GHA GNB GUI IVO LBR MLI NGA SEN SIE TOG. Nanophan. or phan.
Macaranga apicifera Beille, Bull. Soc. Bot. France 55(8): 79 (1908).

Macaranga hexandra (Roxb.) Müll.Arg. in A.P.de Candolle, Prodr. 15(2): 1003 (1866).
Pen. Malaysia. 42 MLY.
* *Rottlera hexandra* Roxb., Fl. Ind. ed. 1832, 3: 829 (1832).

Macaranga heynei I.M.Johnst., Contr. Gray Herb. 68: 90 (Aug. 1923).
Pen. Malaysia, Sumatera. 42 MLY SUM.
Rottlera montana B.Heyne ex Baill., Étude Euphorb.: 430 (1858), pro syn.
* *Macaranga javanica* var. *montana* Müll.Arg. in A.P.de Candolle, Prodr. 15(2): 1005 (1866). *Macaranga montana* (Müll.Arg.) Pax & K.Hoffm. in H.G.A.Engler, Pflanzenr., IV, 147, VII: 321 (1914), nom. illeg.
Macaranga robiginosa Ridl., Bull. Misc. Inform. Kew 1923: 367 (Dec. 1923).

Macaranga hispida (Blume) Müll.Arg. in A.P.de Candolle, Prodr. 15(2): 990 (1866).
Sulawesi, Maluku, Philippines. 42 MOL PHI SUL.
* *Mappa hispida* Blume, Bijdr.: 624 (1826).
Macaranga rotundifolia Elmer ex Pax & K.Hoffm. in H.G.A.Engler, Pflanzenr., IV, 147, VII: 368 (1914), pro syn.

Macaranga hoffmannii L.M.Perry, J. Arnold Arbor. 34: 242 (1953).
NW. Papua New Guinea (Felsspitze). 42 NWG. – Provisionally accepted.
* *Macaranga acuminata* Pax & K.Hoffm. in H.G.A.Engler, Pflanzenr., IV, 147, XIV: 28 (1919), nom. illeg.

Macaranga hosei King ex Hook.f., Fl. Brit. India 5: 449 (1887).
S. Thailand, Pen. Malaysia, Sumatera (incl. Bangka), Borneo. 41 THA 42 BOR MLY SUM. Phan.
Macaranga gossypiifolia Pax & K.Hoffm. in H.G.A.Engler, Pflanzenr., IV, 147, VII: 311 (1914).
Macaranga pseudopruinosa Pax & K.Hoffm. in H.G.A.Engler, Pflanzenr., IV, 147, VII: 308 (1914).

Macaranga huahineensis Florence, Bull. Mus. Natl. Hist. Nat., B, Adansonia 18: 263 (1996).
Society Is. (Huahine). 61 SCI. Phan.

Macaranga hullettii King ex Hook.f., Fl. Brit. India 5: 452 (1887).
Pen. Malaysia, Sumatera, Borneo. 42 BOR MLY SUM. Nanophan. or phan.

subsp. **borneensis** Whitmore, Kew Bull. 29: 446 (1974).
Borneo. 42 BOR. Nanophan. or phan.

subsp. **hullettii**
Pen. Malaysia, Sumatera. 42 MLY SUM. Nanophan. or phan.
Macaranga bartlettii Merr., Pap. Michigan Acad. Sci. 19: 161 (1933 publ. 1934).

Macaranga humbertii Leandri, Notul. Syst. (Paris) 10: 160 (1942).
Madagascar. 29 MDG.

Macaranga hurifolia Beille, Bull. Soc. Bot. France 55(8): 80 (1908).
W. Trop. Africa to Cameroon. 22 GHA IVO LBR NGA SIE TOG 23 CMN. Nanophan. or phan.
Macaranga togoensis Pax, Bot. Jahrb. Syst. 43: 221 (1909).

Macaranga hypoleuca (Rchb.f. & Zoll.) Müll.Arg. in A.P.de Candolle, Prodr. 15(2): 992 (1866).
S. Thailand, Pen. Malaysia, Sumatera, Borneo. 41 THA 42 BOR MLY SUM. Phan.
* *Mappa hypoleuca* Rchb.f. & Zoll., Acta Soc. Regiae Sci. Indo-Neerl. 1: 30 (1856).

Macaranga hystrichogyne Airy Shaw, Kew Bull. 33: 68 (1978).
New Guinea. 42 NWG. Phan.

Macaranga inamoena F.Muell. ex Benth., Fl. Austral. 6: 145 (1873).
Queensland (Cape York Pen.). 50 QLD. Nanophan. or phan.
Mallotus inamoenus F.Muell. ex Benth., Fl. Austral. 6: 146 (1873).

Macaranga indica Wight, Icon. Pl. Ind. Orient. 5: t. 1883 (1852).
SW. India, Sri Lanka, Andaman Is., Bhutan, Assam, SW. China, Thailand, Pen. Malaysia. 36 CHC 40 ASS EHM IND SRL 41 AND THA 42 MLY. Phan.
Trewia hernandifolia Roth, Nov. Pl. Sp.: 374 (1821). Provisional synonym.
Macaranga flexuosa Wight, Icon. Pl. Ind. Orient. 5(2): 23 (1852).
Macaranga adenantha Gagnep., Bull. Soc. Bot. France 69: 701 (1922 publ. 1923).

Macaranga indistincta Whitmore, Kew Bull. 29: 447 (1974).
Sumatera, Borneo (Sarawak, Sabah). 42 BOR SUM. Phan.
Pachystemon depressus var. *mollis* Müll.Arg., Flora 47: 465 (1864). *Macaranga depressa* var. *mollis* (Müll.Arg.) Müll.Arg. in A.P.de Candolle, Prodr. 15(2): 989 (1866).

Macaranga induta L.M.Perry, J. Arnold Arbor. 34: 244 (1953).
New Guinea, Bismarck Archip. (New Britain). 42 BIS NWG. Phan.

subsp. **induta**
New Guinea, Bismarck Archip. (New Britain). 42 BIS NWG. Phan.

subsp. **paucinervis** Airy Shaw, Kew Bull. 33: 70 (1978).
Papua New Guinea. 42 NWG. Phan.

Macaranga inermis Pax & K.Hoffm. in H.G.A.Engler, Pflanzenr., IV, 147, VII: 333 (1914).
Sulawesi ?, Maluku ?, New Guinea, Bismarck Archip. (New Britain), Queensland. 42 BIS MOL? NWG SUL? 50 QLD. Phan.
Macaranga effusa Pax & K.Hoffm. in H.G.A.Engler, Pflanzenr., IV, 147, XIV: 26 (1919).
Macaranga mallotiformis Pax & K.Hoffm. in H.G.A.Engler, Pflanzenr., IV, 147, XIV: 27 (1919). *Macaranga inermis* var. *mallotiformis* (Pax & K.Hoffm.) L.M.Perry, J. Arnold Arbor. 34: 252 (1953).
Macaranga penninervia Pax & K.Hoffm. in H.G.A.Engler, Pflanzenr., IV, 147, XIV: 26 (1919). *Macaranga inermis* var. *penninervia* (Pax & K.Hoffm.) L.M.Perry, J. Arnold Arbor. 34: 253 (1953).
Macaranga inermis var. *plurifoveata* L.M.Perry, J. Arnold Arbor. 34: 254 (1953).

Macaranga intonsa Whitmore, Kew Bull. 34: 603 (1980).
Papua New Guinea. 42 NWG. Phan.

Macaranga involucrata (Roxb.) Baill., Étude Euphorb.: 432 (1858).
Sulawesi to N. Australia. 42 BIS MOL NWG SUL 50 NTA QLD. Nanophan. or phan.
* *Urtica involucrata* Roxb., Fl. Ind. ed. 1832, 3: 592 (1832).

var. **involucrata**
Sulawesi to Bismarck Archip. 42 BIS MOL NWG SUL. Nanophan. or phan.
Mappa amboinensis Müll.Arg., Linnaea 34: 197 (1865). *Macaranga amboinensis* (Müll.Arg.) Müll.Arg. in A.P.de Candolle, Prodr. 15(2): 1002 (1866).
Macaranga schleinitziana K.Schum., Bot. Jahrb. Syst. 9: 207 (1888).

Macaranga involucrata var. *keyensis* Warb., Bot. Jahrb. Syst. 13: 352 (1891). *Macaranga keyensis* (Warb.) Pax & K.Hoffm. in H.G.A.Engler, Pflanzenr., IV, 147, VII: 376 (1914).
Macaranga schleinitziana var. *lobulata* Pax & K.Hoffm. in H.G.A.Engler, Pflanzenr., IV, 147, VII: 373 (1914).
Macaranga dalechampioides S.Moore, J. Bot. 61(Suppl.): 48 (1923).
Macaranga seta-felis Whitmore, Kew Bull. 36: 423 (1981).

var. **mallotoides** (F.Muell.) L.M.Perry, J. Arnold Arbor. 34: 223 (1953).
Irian Jaya, Northern Territory, Queensland. 42 NWG 50 NTA QLD. Nanophan. or phan.
Macaranga asterolasia F.Muell., Fragm. 4: 140 (1864).
* *Macaranga mallotoides* F.Muell., Fragm. 4: 139 (1864).

Macaranga javanica (Blume) Müll.Arg. in A.P.de Candolle, Prodr. 15(2): 1004 (1866).
Sumatera (incl. Bangka), Jawa. 42 JAW SUM.
* *Mappa javanica* Blume, Bijdr.: 625 (1826).

Macaranga johannium Whitmore, Kew Bull. 34: 603 (1980).
Papua New Guinea. 42 NWG. Phan.

Macaranga kampotensis Gagnep., Bull. Soc. Bot. France 69: 702 (1922 publ. 1923).
Cambodia. 41 CBD.

Macaranga kanehirae Hosok., Trans. Nat. Hist. Soc. Taiwan 25: 27 (1935).
Caroline Is. (Ponape). 62 CRL.

Macaranga kilimandscharica Pax in H.G.A.Engler, Pflanzenw. Ost-Afrikas C: 238 (1895).
Macaranga capensis var. *kilimandscharica* (Pax) Friis & M.G.Gilbert, Kew Bull. 41: 68 (1986).
Ethiopia to Zambia. 23 BUR RWA ZAI 24 ETH SUD 25 KEN TAN UGA 26 MLW ZAM. Phan.
Macaranga mildbraediana Pax & K.Hoffm. in H.G.A.Engler, Pflanzenr., IV, 147, VII: 343 (1914), nom. inval.
Macaranga nyassae Pax & K.Hoffm. in H.G.A.Engler, Pflanzenr., IV, 147, VII: 343 (1914).
Macaranga neomildbraediana Lebrun, Ann. Soc. Sci. Bruxelles, Sér. B 54: 160 (1934).

Macaranga kinabaluensis Airy Shaw, Kew Bull. 23: 91 (1969).
Borneo (Sabah). 42 BOR. Phan.

Macaranga kingii Hook.f., Fl. Brit. India 5: 451 (1887).
Pen. Malaysia, Sumatera (Riau I.), Borneo (Sarawak). 42 BOR MLY SUM. Phan.
Macaranga insignis Merr., Philipp. J. Sci., C 11: 69 (1916).
Macaranga kingii var. *platyphylla* Airy Shaw, Kew Bull. 25: 533 (1971).

Macaranga klaineana Pierre ex Prain, Bull. Misc. Inform. Kew 1912: 104 (1912).
Gabon. 23 GAB. Cl. nanophan.

Macaranga kostermansii L.M.Perry, J. Arnold Arbor. 34: 251 (1953).
Irian Jaya (Vogelkop Pen.). 42 NWG. Nanophan. or phan.

Macaranga kurzii (Kuntze) Pax & K.Hoffm. in H.G.A.Engler, Pflanzenr., IV, 147, VII: 360 (1914).
S. Burma, Thailand, S. China, Vietnam. 36 CHC 41 BMA THA VIE. Phan.
Macaranga membranacea Kurz, J. Asiat. Soc. Bengal, Pt. 2, Nat. Hist. 42(2): 246 (1873), nom. illeg.
* *Tanarius kurzii* Kuntze, Revis. Gen. Pl. 2: 620 (1891).
Macaranga andersonii Craib, Bull. Misc. Inform. Kew 1911: 466 (1911).

Macaranga laciniata Whitmore & Airy Shaw, Kew Bull. 25: 241 (1971).
S. Thailand, NE. Pen. Malaysia. 41 THA 42 MLY. Phan.

Macaranga lamellata Whitmore, Kew Bull. 29: 448 (1974).
Borneo (Sabah). 42 BOR.

Macaranga lanceolata Pax & K.Hoffm. in H.G.A.Engler, Pflanzenr., IV, 147, XIV: 25 (1919).
NW. Papua New Guinea. 42 NWG. Phan.

Macaranga letestui Pellegr., Bull. Mus. Natl. Hist. Nat. 34: 229 (1928).
Congo. 23 CON.

Macaranga leytensis Merr., Philipp. J. Sci., C 7: 393 (1912 publ. 1913).
Philippines. 42 PHI.

Macaranga lineata Airy Shaw, Kew Bull. 33: 104 (1969).
NE. Papua New Guinea. 42 NWG. Phan.

Macaranga loheri Elmer, Leafl. Philipp. Bot. 2: 432 (1908).
Philippines, Sulawesi. 42 PHI SUL.

Macaranga longicaudata L.M.Perry, J. Arnold Arbor. 34: 233 (1953).
Papua New Guinea. 42 NWG. Phan.

Macaranga longipetiolata De Wild., Pl. Bequaert. 3: 478 (1926).
Zaire. 23 ZAI.

Macaranga longistipulata (Kurz ex Tejism. & Binn.) Müll.Arg. in A.P.de Candolle, Prodr. 15(2): 991 (1866).
Maluku (Ceram). 42 MOL.
* *Rottlera longistipulata* Kurz ex Teijsm. & Binn., Tijdschr. Ned.-Indië 27: 43 (1864).

Macaranga lophostigma Chiov., Atti Reale Accad. Ital., Mem. Cl. Sci. Fis. 11: 55 (1940).
Ethiopia. 24 ETH.

Macaranga louisiadum Airy Shaw, Kew Bull. 23: 106 (1969).
SE. Papua New Guinea (incl. Louisiade Archip.). 42 NWG. Phan.

Macaranga lowii King ex Hook.f., Fl. Brit. India 5: 453 (1887).
S. Thailand, Pen. Malaysia, Vietnam, Hainan, Borneo (Sarawak, Sabah, Kalimantan), Philippines. 36 CHH 41 THA VIE 42 BOR MLY PHI. Nanophan. or phan.

var. **kostermansii** Airy Shaw, Kew Bull. 23: 107 (1969).
Borneo (Kalimantan). 42 BOR. Nanophan. or phan.

var. **lowii**
S. Thailand, Pen. Malaysia, Vietnam, Hainan, Borneo (Sarawak, Sabah, E. Kalimantan), Philippines. 36 CHH 41 THA VIE 42 BOR MLY PHI. Nanophan. or phan.
Mallotus auriculatus Merr., Philipp. J. Sci., C 7: 369 (1912 publ. 1913). *Macaranga auriculata* (Merr.) Airy Shaw, Kew Bull. 19: 325 (1965).
Mallotus affinis Merr., Philipp. J. Sci., C 13: 82 (1918).
Macaranga poilanei Gagnep., Bull. Soc. Bot. France 69: 703 (1922 publ. 1923).
Mallotus tsiangii Merr. & Chun, Sunyatsenia 1: 63 (1930).

Macaranga lugubris Whitmore, Kew Bull. 39: 609 (1984).
Papua New Guinea. 42 NWG.

Macaranga lutescens (Pax & Lingelsh.) Pax in H.G.A.Engler, Pflanzenr., IV, 147, VII: 361 (1914).
New Caledonia. 60 NWC.
* *Cleidion lutescens* Pax & Lingelsh., Repert. Spec. Nov. Regni Veg. 3: 25 (1906).

Macaranga macropoda Baker, J. Linn. Soc., Bot. 20: 257 (1883).
C. Madagascar. 29 MDG. Phan.

Macaranga magna Turrill, Bull. Misc. Inform. Kew 1924: 393 (1924).
Fiji (Viti Levu). 60 FIJ. Phan.
* *Macaranga grandifolia* Turrill, J. Linn. Soc., Bot. 43: 38 (1915), nom. illeg.

Macaranga magnifolia L.M.Perry, J. Arnold Arbor. 34: 236 (1953).
Sulawesi, New Guinea. 42 NWG SUL. Phan.

Macaranga magnistipulosa Pax, Bot. Jahrb. Syst. 43: 222 (1909).
Equatorial Guinea. 23 EQG. Nanophan.

Macaranga mappa (L.) Müll.Arg. in A.P.de Candolle, Prodr. 15(2): 1000 (1866).
N. Sulawesi, Maluku. 42 MOL SUL.
* *Ricinus mappa* L., Sp. Pl. ed. 2: 1430 (1763). *Acalypha mappa* (L.) Willd., Sp. Pl. 4: 526 (1805). *Alchornea mappa* (L.) Oken, Allg. Naturgesch. 3(3): 159 (1841).

Macaranga marikoensis A.C.Sm., J. Arnold Arbor. 33: 385 (1952).
Fiji. 60 FIJ. Phan.

Macaranga mauritiana Bojer ex Baill., Étude Euphorb.: 432 (1859).
Mauritius. 29 MAU. Phan.

Macaranga megacarpa Airy Shaw, Kew Bull. 32: 413 (1978).
Vanuatu. 60 VAN.

Macaranga meiophylla S.Moore, J. Linn. Soc., Bot. 45: 408 (1921).
New Caledonia. 60 NWC.

Macaranga melanosticta Airy Shaw, Kew Bull. 25: 539 (1971).
Papua New Guinea. 42 NWG. Phan.

Macaranga mellifera Prain, J. Linn. Soc., Bot. 11: 201 (1911).
Malawi, Zimbabwe, Mozambique. 26 MLW MOZ ZIM. Phan.

Macaranga membranacea Müll.Arg. in A.P.de Candolle, Prodr. 15(2): 996 (1866).
Fiji (N. & W. Vanua Levu). 60 FIJ. Phan.

Macaranga misimae Airy Shaw, Kew Bull. 23: 109 (1969).
New Guinea. 42 NWG. Nanophan. or phan.

Macaranga mista S.Moore, J. Linn. Soc., Bot. 45: 407 (1921).
New Caledonia. 60 NWC.

Macaranga modesta Pax & K.Hoffm. in H.G.A.Engler, Pflanzenr., IV, 147, VII: 375 (1914).
Jawa. 42 JAW.

Macaranga monandra Müll.Arg., J. Bot. 2: 337 (1864).
Trop. Africa. 22 NGA 23 CMN CON EQG GAB ZAI 25 TAN UGA 26 ANG. Phan.
Macaranga zenkeri Pax, Bot. Jahrb. Syst. 23: 526 (1897).

Macaranga montana Merr., Philipp. J. Sci., C 7: 394 (1912 publ. 1913).
Philippines. 42 PHI.
Macaranga merrilliana Pax & K.Hoffm. in H.G.A.Engler, Pflanzenr., IV, 147, VII: 334 (1914).

Macaranga motleyana (Müll.Arg.) Müll.Arg. in A.P.de Candolle, Prodr. 15(2): 994 (1866).
S. Thailand to Sumatera, Borneo (Sabah, E. & SE. Kalimantan). 41 THA 42 BOR MLY SUM.
* *Mappa motleyana* Müll.Arg., Flora 47: 467 (1864).

subsp. **griffithiana** (Müll.Arg.) Whitmore, Kew Bull. 29: 448 (1974).
S. Thailand, Pen. Malaysia, Sumatera. 41 THA 42 MLY SUM. Phan.
* *Macaranga griffithiana* Müll.Arg. in A.P.de Candolle, Prodr. 15(2): 998 (1866).

subsp. **motleyana**
E. Sumatera, Borneo (Sabah, E. & SE. Kalimantan). 42 BOR SUM.

Macaranga myriantha Müll.Arg. in A.P.de Candolle, Prodr. 15(2): 1008 (1866).
Maluku (Ambon). 42 MOL.

Macaranga myriolepida Baker, J. Linn. Soc., Bot. 21: 442 (1885).
C. Madagascar. 29 MDG. Nanophan.

Macaranga necopina Whitmore, Kew Bull. 36: 423 (1981).
Sumatera. 42 SUM.

Macaranga neobritannica Airy Shaw, Kew Bull. 25: 531 (1971).
Papua New Guinea, Bismarck Archip. (New Britain). 42 BIS NWG. Phan.

Macaranga nicobarica N.P.Balakr. & Chakr., Gard. Bull. Singapore 31: 57 (1978).
Nicobar Is., Andaman Is., Burma (Amherst) ? 41 AND BMA? NCB. Phan.

Macaranga noblei Elmer, Leafl. Philipp. Bot. 2: 678 (1910).
Philippines. 42 PHI.

Macaranga novoguineensis J.J.Sm., Nova Guinea 8: 789 (1912).
New Guinea, Bismarck Archip. (New Britain). 42 BIS NWG. Nanophan. or phan.

Macaranga oblongifolia Baill., Étude Euphorb.: 432 (1858).
Madagascar. 29 MDG. Phan.
Macaranga hildebrandtii Pax & K.Hoffm. in H.G.A.Engler, Pflanzenr., IV, 147, XVII: 185 (1924), nom. provis.

Macaranga obovata Boivin ex Baill., Adansonia 1: 263 (1861).
Madagascar. 29 MDG. Phan.
Macaranga reticulata Baill., Étude Euphorb.: 432 (1858), nom. nud.

Macaranga occidentalis (Müll.Arg.) Müll.Arg. in A.P.de Candolle, Prodr. 15(2): 994 (1866).
Nigeria, Cameroon, Bioko. 22 NGA 23 CMN GGI. Phan.
* *Mappa occidentalis* Müll.Arg., Flora 47: 467 (1864).
Macaranga preussii Pax, Bot. Jahrb. Syst. 19: 92 (1894).

Macaranga ovatifolia Merr., Philipp. J. Sci. 16: 562 (1920).
Philippines. 42 PHI.

Macaranga palustris Whitmore, Kew Bull. 34: 604 (1980).
Papua New Guinea. 42 NWG. Phan.

Macaranga papuana (J.J.Sm.) Pax & K.Hoffm. in H.G.A.Engler, Pflanzenr., IV, 147, VII: 368 (1914).
New Guinea. 42 NWG. Nanophan. or phan.
* *Macaranga hispida* var. *papuana* J.J.Sm., Nova Guinea 8: 234 (1910).

var. **glabristipulata** Whitmore, Kew Bull., Addit. Ser. 8: 152 (1980).
Irian Jaya. 42 NWG. Nanophan. or phan.
* *Macaranga ovalifolia* Ridl., Trans. Linn. Soc. London, Bot. 9: 148 (1916).

var. **papuana**
New Guinea. 42 NWG. Nanophan. or phan.

Macaranga parvibracteata Pax & K.Hoffm. in H.G.A.Engler, Pflanzenr., IV, 147, XIV: 24 (1919).
Maluku (Ambon). 42 MOL.

Macaranga paxii Prain, Bull. Misc. Inform. Kew 1910: 127 (1910).
Nigeria, Cameroon. 22 NGA 23 CMN. Nanophan. or phan.

Macaranga pearsonii Merr., Philipp. J. Sci. 29: 383 (1926).
Borneo (Sabah, E. & S. Kalimantan). 42 BOR. Phan.

Macaranga peltata (Roxb.) Müll.Arg. in A.P.de Candolle, Prodr. 15(2): 1010 (1866).
Sikkim, India, Sri Lanka, Andaman Is., S. Burma. 40 EHM IND SRL 41 AND BMA.
* *Osyris peltata* Roxb., Fl. Ind. ed. 1832, 3: 755 (1832). *Macaranga wightiana* Baill., Étude Euphorb.: 432 (1858), nom. illeg.
Macaranga roxburghii Wight, Icon. Pl. Ind. Orient. 5(2): 23 (1852).
Macaranga tomentosa Wight, Icon. Pl. Ind. Orient. 5(2): 23 (1852).

Macaranga petanostyla Airy Shaw, Kew Bull. 25: 537 (1971).
Borneo (Sabah ?, E. Kalimantan). 42 BOR. Phan.

Macaranga pierreana Prain, Bull. Misc. Inform. Kew 1912: 105 (1912).
Gabon (Sierra de Cristal). 23 GAB. Nanophan. or phan.

Macaranga pilosula Airy Shaw, Kew Bull. 23: 105 (1969).
Papua New Guinea. 42 NWG. Phan.

Macaranga platyclada Pax & K.Hoffm. in H.G.A.Engler, Pflanzenr., IV, 147, XIV: 30 (1919).
W. & C. New Guinea. 42 NWG. Phan.
Macaranga roemeri Pax & K.Hoffm. in H.G.A.Engler, Pflanzenr., IV, 147, XVII: 185 (1924).

Macaranga pleioneura Airy Shaw, Kew Bull. 23: 110 (1969).
New Guinea. 42 NWG. Nanophan. or phan.

var. **pleioneura**
New Guinea. 42 NWG. Nanophan. or phan.

var. **velutina** Whitmore, Kew Bull., Addit. Ser. 8: 153 (1980).
Papua New Guinea. 42 NWG. Nanophan. or phan.

Macaranga pleiostemon Pax & K.Hoffm. in H.G.A.Engler, Pflanzenr., IV, 147, XIV: 24 (1919).
New Guinea, Solomon Is. (Bougainville). 42 NWG 60 SOL. Nanophan. or phan.
Macaranga pleiostemon f. *pubescens* L.M.Perry, J. Arnold Arbor. 34: 256 (1953).

Macaranga pleytei L.M.Perry, J. Arnold Arbor. 24: 203 (1953).
Irian Jaya. 42 NWG. Phan.

Macaranga poggei Pax, Bot. Jahrb. Syst. 19: 94 (1894).
Congo, Zaire. 23 CON ZAI. Nanophan. or phan.
Macaranga gilletii De Wild., Ann. Mus. Congo Belge, Bot., V, 1: 276 (1906).
Macaranga laurentii De Wild., Ann. Mus. Congo Belge, Bot., V, 2: 282 (1908).
Macaranga chevalieri Beille, Bull. Soc. Bot. France 57(8): 125 (1910). *Macaranga poggei* var. *chevalieri* (Beille) Prain in D.Oliver, Fl. Trop. Afr. 6(1): 941 (1912).

Macaranga polyadenia Pax & K.Hoffm. in H.G.A.Engler, Pflanzenr., IV, 147, XIV: 25 (1919).
New Guinea, Bismarck Archip., Solomon Is., Santa Cruz Is., Queensland (Cape York pen.). 42 BIS NWG 50 QLD 60 SCZ SOL. Phan.
Macaranga fimbriata S.Moore, J. Bot. 61(Suppl.): 48 (1923).
Macaranga multiflora C.T.White, Proc. Roy. Soc. Queensland 55: 83 (1944).
Macaranga fimbriata var. *doctersii* L.M.Perry, J. Arnold Arbor. 34: 249 (1953).

Macaranga polyneura Gilli, Ann. Naturhist. Mus. Wien 83: 438 (1979 publ. 1980).
Papua New Guinea. 42 NWG.

Macaranga praestans Airy Shaw, Kew Bull. 19: 318 (1965).
Borneo (Sarawak, Brunei). 42 BOR. Nanophan. or phan.

Macaranga pruinosa (Miq.) Müll.Arg. in A.P.de Candolle, Prodr. 15(2): 992 (1866).
Pen. Malaysia, E. Sumatera (incl. Siberut), Borneo (Sarawak, W. Kalimantan). 42 BOR MLY SUM. Phan.
* *Mappa pruinosa* Miq., Fl. Ned. Ind., Eerste Bijv.: 457 (1861).
Macaranga maingayi Hook.f., Fl. Brit. India 5: 449 (1887).
Mallotus maingayi Durand & Jacks., Index Kew., Suppl. 1: 262 (1906), nom. illeg.
Macaranga formicarum Pax & K.Hoffm. in H.G.A.Engler, Pflanzenr., IV, 147, VII: 308 (1914).

Macaranga puberula Heine, Repert. Spec. Nov. Regni Veg. 54: 232 (1951).
Borneo (Sabah). 42 BOR. Phan.

Macaranga punctata K.Schum. in K.M.Schumann & U.M.Hollrung, Fl. Kais. Wilh. Land: 80 (1889).
New Guinea. 42 NWG. (Cl.) nanophan. or phan.
Macaranga isadenia Pax & K.Hoffm. in H.G.A.Engler, Pflanzenr., IV, 147, VII: 377 (1914).
Macaranga ovalifolia Pax & K.Hoffm. in H.G.A.Engler, Pflanzenr., IV, 147, XIV: 29 (1919), nom. illeg.
Macaranga pseudopeltata Pax & K.Hoffm. in H.G.A.Engler, Pflanzenr., IV, 147, XIV: 29 (1919).
Macaranga maluensis Pax & K.Hoffm. in H.G.A.Engler, Pflanzenr., IV, 147, XVII: 185 (1924).
Macaranga punctata var. *whitei* L.M.Perry, J. Arnold Arbor. 34: 229 (1953).

Macaranga puncticulata Gage, Rec. Bot. Surv. India 9: 246 (1922).
Pen. Malaysia. 42 MLY.

Macaranga pustulata King ex Hook.f., Fl. Brit. India 5: 445 (1887).
Uttrakhand, Nepal, Sikkim. 40 EHM NEP WHM.
Macaranga gmelinifolia King ex Hook.f., Fl. Brit. India 5: 445 (1887).

Macaranga quadricornis Ridl., Bull. Misc. Inform. Kew 1923: 367 (1923).
S. Burma, S. Thailand, Pen. Malaysia, Sumatera (incl. Bangka), Borneo (Sarawak, Sabah). 41 BMA THA 42 BOR MLY SUM. Phan.

Macaranga quadriglandulosa Warb., Bot. Jahrb. Syst. 13: 350 (1891). *Macaranga quadriglandulosa* var. *genuina* Pax & K.Hoffm. in H.G.A.Engler, Pflanzenr., IV, 147, VII: 356 (1914), nom. inval.
New Guinea, Bismarck Archip., Solomon Is. 42 BIS NWG 60 SOL. Phan.
Macaranga tanarius var. *abbreviata* J.J.Sm., Nova Guinea 8: 238 (1910). *Macaranga quadriglandulosa* var. *abbreviata* (J.J.Sm.) Pax & K.Hoffm. in H.G.A.Engler, Pflanzenr., IV, 147, VII: 356 (1914).
Macaranga quadriglandulosa var. *digyna* Pax & K.Hoffm. in H.G.A.Engler, Pflanzenr., IV, 147, VII: 356 (1914).
Macaranga quadriglandulosa var. *variabilis* L.M.Perry, J. Arnold Arbor. 34: 239 (1953).

Macaranga racemosa Baker, J. Linn. Soc., Bot. 22: 520 (1887).
C. Madagascar. 29 MDG.

Macaranga raivavaeensis H.St.John, Nordic J. Bot. 3: 452 (1983).
Tubuai Is. (Raivavae, Rimatara). 61 TUB. Phan.

Macaranga ramiflora Elmer, Leafl. Philipp. Bot. 2: 433 (1908).
Philippines. 42 PHI.

Macaranga rarispina Whitmore, Kew Bull. 29: 450 (1974).
Borneo (Sarawak). 42 BOR. Nanophan. or phan.

Macaranga recurvata Gage, Rec. Bot. Surv. India 9: 246 (1922).
Pen. Malaysia, Borneo (Brunei, Sabah, E. Kalimantan), Sulawesi ? 42 BOR MLY SUL? Phan.

Macaranga reiteriana Pax & K.Hoffm. in H.G.A.Engler, Pflanzenr., IV, 147, XVII: 185 (1924).
New Guinea. 42 NWG. Phan.

Macaranga repandodentata Airy Shaw, Kew Bull. 19: 321 (1965).
Borneo (Sarawak, Brunei, Sabah, E. Kalimantan). 42 BOR. Nanophan. or phan.

Macaranga rhizinoides (Blume) Müll.Arg. in A.P.de Candolle, Prodr. 15(2): 1011 (1866).
Jawa. 42 JAW. Phan.
* *Zanthoxylum rhizinoides* Blume, Bijdr.: 248 (1825).
Macaranga blumeana Müll.Arg. in H.G.A.Engler, Pflanzenr., IV, 147, VII: 349 (1914), pro syn.

Macaranga rhodonema Airy Shaw, Kew Bull. 25: 542 (1971).
NE. Papua New Guinea. 42 NWG. Nanophan.

Macaranga ribesioides Baker, J. Linn. Soc., Bot. 21: 442 (1885).
C. Madagascar. 29 MDG. Phan.

Macaranga robinsonii Merr., Philipp. J. Sci., C 11: 284 (1916).
Maluku (Morotai, Ambon). 42 MOL.

Macaranga rostrata Heine, Repert. Spec. Nov. Regni Veg. 54: 233 (1951).
Borneo (Sarawak, Sabah). 42 BOR. Phan.

Macaranga rufibarbis Warb., Bot. Jahrb. Syst. 16: 21 (1893).
New Guinea. 42 NWG. Nanophan. or phan.
Macaranga tenella Pax & K.Hoffm. in H.G.A.Engler, Pflanzenr., IV, 147, VII: 379 (1914). *Macaranga rufibarbis* var. *tenella* (Pax & K.Hoffm.) L.M.Perry, J. Arnold Arbor. 34: 212 (1953).
Macaranga rufibarbis var. *campestris* Whitmore, Kew Bull., Addit. Ser. 8: 156 (1980).

Macaranga saccifera Pax, Bot. Jahrb. Syst. 19: 93 (1894). *Macaranga saccifera* var. *genuina* Pax & K.Hoffm. in H.G.A.Engler, Pflanzenr., IV, 147, VII: 312 (1914), nom. inval.
Gabon, Congo, Zaire. 23 CON GAB ZAI. Nanophan. or phan.
Macaranga saccifera var. *dentifera* Pax & K.Hoffm. in H.G.A.Engler, Pflanzenr., IV, 147, VII: 313 (1914).

Macaranga salicifolia Airy Shaw, Kew Bull. 23: 108 (1969).
New Guinea (Vogelkop Pen.). 42 NWG. Phan.

Macaranga salomonensis L.M.Perry, J. Arnold Arbor. 34: 210 (1953).
Papua New Guinea, Solomon Is. 42 NWG 60 SOL. Phan.

Macaranga sampsonii Hance, J. Bot. 9: 134 (1871).
S. China, Vietnam. 36 CHC CHS 41 VIE. Phan.

Macaranga sarcocarpa Airy Shaw, Kew Bull. 23: 88 (1969).
Borneo (SW. Sarawak). 42 BOR.

Macaranga schweinfurthii Pax, Bot. Jahrb. Syst. 19: 92 (1894).
Trop. Africa. 22 NGA 23 CMN CON ZAI 24 SUD 25 KEN TAN UGA 26 ANG ZAM. Phan.
Macaranga rosea Pax, Bot. Jahrb. Syst. 26: 328 (1899).
Macaranga lecomtei Beille, Bull. Soc. Bot. France 55(8): 78 (1908).
Macaranga calophylla Pax, Bot. Jahrb. Syst. 43: 221 (1909).

Macaranga secunda Müll.Arg. in A.P.de Candolle, Prodr. 15(2): 996 (1866).
Fiji. 60 FIJ. Phan.

Macaranga seemannii (Müll.Arg.) Müll.Arg. in A.P.de Candolle, Prodr. 15(2): 999 (1866).
Fiji, Tonga, Niue. 60 FIJ NUE TON. Phan.
* *Mappa seemannii* Müll.Arg., Flora 47: 468 (1864).

var. **capillata** A.C.Sm., J. Arnold Arbor. 33: 380 (1952).
Fiji, Niue. 60 FIJ NUE. Phan.

var. **deltoides** A.C.Sm., J. Arnold Arbor. 33: 380 (1952).
Fiji (W. Viti Levu). 60 FIJ. Nanophan. or phan.

var. **seemannii**
Fiji, Tonga, Niue. 60 FIJ NUE TON. Phan.

Macaranga semiglobosa J.J.Sm., Meded. Dept. Landb. Ned.-Indië 10: 501 (1910).
N. & E. Sumatera, Jawa. 42 JAW SUM.

Macaranga setosa Gage, Rec. Bot. Surv. India 9: 244 (1922).
Pen. Malaysia. 42 MLY.

Macaranga similis Pax & K.Hoffm. in H.G.A.Engler, Pflanzenr., IV, 147, XIV: 29 (1919).
New Guinea. 42 NWG. Phan.
Macaranga brassii Mansf., J. Arnold Arbor. 10: 78 (1929).

Macaranga sinensis (Baill.) Müll.Arg. in A.P.de Candolle, Prodr. 15(2): 1001 (1866).
S. China, Philippines. 36 CHS 42 PHI.
* *Mappa sinensis* Müll.Arg. in A.P.de Candolle, Prodr. 15(2): 1001 (1866).

Macaranga sphaerophylla Baker, J. Linn. Soc., Bot. 20: 257 (1883).
Madagascar. 29 MDG. Phan.

Macaranga spinosa Müll.Arg., Flora 47: 466 (1864).
Trop. Africa. 22 IVO LBR NGA 23 BUR GAB GGI CMN CON EQG ZAI 25 TAN UGA 26 ANG. Phan.
Macaranga pynaertii De Wild., Ann. Mus. Congo Belge, Bot., V, 2: 283 (1908).
Macaranga ledermanniana Pax, Bot. Jahrb. Syst. 43: 222 (1909).

Macaranga staudtii Pax, Bot. Jahrb. Syst. 23: 526 (1897).
Nigeria, Cameroon, Gabon, Cabinda. 22 NGA 23 CAB CMN GAB. Phan.

Macaranga stellimontium Whitmore, Kew Bull. 34: 604 (1980).
Papua New Guinea (Star Mts.). 42 NWG. Phan.

Macaranga stenophylla Pax & K.Hoffm. in H.G.A.Engler, Pflanzenr., IV, 147, VII: 371 (1914).
Papua New Guinea (Kani Mts.). 42 NWG. – Provisionally accepted.

Macaranga sterrophylla L.M.Perry, J. Arnold Arbor. 34: 231 (1953).
Papua New Guinea. 42 NWG. Phan.

Macaranga stipulosa Müll.Arg. in A.P.de Candolle, Prodr. 15(2): 1001 (1866).
Samoa. 60 SAM. Phan.

Macaranga strigosa Pax & K.Hoffm. in H.G.A.Engler, Pflanzenr., IV, 147, XIV: 30 (1919).
Papua New Guinea. 42 NWG. Nanophan. or phan.
Macaranga strigosa var. *carrii* L.M.Perry, J. Arnold Arbor. 34: 214 (1953).

Macaranga strigosissima Airy Shaw, Kew Bull. 19: 320 (1965).
Borneo (SW. & C. Sarawak). 42 BOR. Phan.

Macaranga subdentata Benth., Fl. Austral. 6: 145 (1873).
Queensland (Cape York Pen.). 50 QLD. Phan.

Macaranga subpeltata K.Schum. & Lauterb., Fl. Schutzgeb. Südsee: 400 (1900).
NE. Papua New Guinea. 42 NWG. Phan.

Macaranga suleensis Whitmore, Kew Bull. 34: 605 (1980).
Bismarck Archip. (New Britain: Mt. Sule). 42 BIS. Phan.

Macaranga sumatrana Müll.Arg. in A.P.de Candolle, Prodr. 15(2): 1003 (1866).
Sumatera (?). 42 SUM. – Provisionally accepted.

Macaranga suwo Whitmore, Kew Bull. 34: 605 (1980).
Irian Jaya. 42 NWG. Phan.

Macaranga sylvatica Elmer, Leafl. Philipp. Bot. 2: 431 (1908). *Macaranga ramiflora* var. *sylvatica* (Elmer) Pax & K.Hoffm. in H.G.A.Engler, Pflanzenr., IV, 147, VII: 324 (1914).
Philippines. 42 PHI.

Macaranga taitensis (Müll.Arg.) Müll.Arg. in A.P.de Candolle, Prodr. 15(2): 997 (1866).
Society Is. (Tahiti). 61 SCI. Nanophan. or phan.
* *Mappa taitensis* Müll.Arg., Linnaea 34: 197 (1865).

Macaranga tanarius (L.) Müll.Arg. in A.P.de Candolle, Prodr. 15(2): 997 (1866).
Trop. & Subtrop. Asia, N. & NE. Australia, SW. Pacific. 36 CHS 38 NNS TAI 41 AND NCB THA VIE 42 BIS BOR JAW LSI MOL NWG PHI SUL SUM 50 NSW NTA QLD 60 SOL VAN. Phan.
* *Ricinus tanarius* L., Herb. Amb.: 14 (1754).

Mappa tomentosa Blume, Bijdr.: 624 (1826). *Macaranga tomentosa* (Blume) Druce, Bot. Soc. Exch. Club Brit. Isles 4: 634 (1916 publ. 1917), nom. illeg.
Croton lacciferus Blanco, Fl. Filip.: 731 (1837).
Macaranga molliuscula Kurz, J. Asiat. Soc. Bengal, Pt. 2, Nat. Hist. 42(2): 245 (1873).
Macaranga vulcanica Elmer ex Merr., Enum. Philipp. Fl. Pl. 2: 443 (1923), pro syn.

Macaranga tchibangensis Pellegr., Bull. Mus. Natl. Hist. Nat. 34: 229 (1928).
Congo. 23 CON.

Macaranga tentaculata Airy Shaw, Kew Bull. 23: 102 (1969).
Borneo (Goodenough I.). 42 BOR. Phan.

Macaranga tenuifolia Müll.Arg. in A.P.de Candolle, Prodr. 15(2): 990 (1866).
Sumatera (Bangka). 42 SUM. – Probably identical with *M. triloba*.

Macaranga tessellata Gage, Nova Guinea 12: 481 (1917).
New Guinea. 42 NWG. Nanophan. or phan.

Macaranga teysmannii (Zoll.) Müll.Arg. in A.P.de Candolle, Prodr. 15(2): 999 (1866).
S. Sumatera. 42 SUM. – Provisionally accepted.
* *Mappa teysmannii* Zoll., Linnaea 29: 465 (1858).

Macaranga thompsonii Merr., Philipp. J. Sci., C 9: 102 (1914).
Marianas. 62 MRN.

Macaranga thorelii Gagnep., Bull. Soc. Bot. France 69: 704 (1922 publ. 1923).
Laos. 41 LAO.

Macaranga trachyphylla Airy Shaw, Kew Bull. 25: 534 (1971).
Borneo (Sarawak, Brunei). 42 BOR. Phan.

Macaranga trichanthera L.M.Perry, J. Arnold Arbor. 34: 243 (1953).
New Guinea. 42 NWG. Phan.

Macaranga trichocarpa (Rchb.f. & Zoll.) Müll.Arg. in A.P.de Candolle, Prodr. 15(2): 1003 (1866).
Andaman Is. ?, S. Burma, SE. Thailand, Pen. Malaysia, Vietnam, Sumatera (incl. Bangka), Borneo. 41 AND? BMA THA VIE 42 BOR MLY SUM. Nanophan. or phan.
* *Mappa trichocarpa* Rchb.f. & Zoll., Linnaea 28: 307 (1856).
Macaranga minutiflora Müll.Arg., Flora 47: 466 (1864).
Mappa borneensis Müll.Arg., Flora 47: 468 (1864). *Macaranga borneensis* (Müll.Arg.) Müll.Arg. in A.P.de Candolle, Prodr. 15(2): 1003 (1866).
Macaranga helferi Müll.Arg. in A.P.de Candolle, Prodr. 15(2): 1004 (1866).

Macaranga trigonostemonoides Croizat, J. Arnold Arbor. 23: 51 (1942).
Vietnam. 41 VIE.

Macaranga triloba (Thunb.) Müll.Arg. in A.P.de Candolle, Prodr. 15(2): 989 (1866).
S. Burma, S. Thailand, Pen. Malaysia, Sumatera, Jawa, Borneo, Philippines, Sulawesi (Sangihe I., Talaud I.) ? 41 BMA THA 42 BOR JAW MLY PHI SUL? SUM. Phan.
* *Ricinus trilobus* Thunb., Ricin.: 6 (1815).
Ricinus trilobus Reinw. ex Blume, Catalogus: 108 (1823), nom. nud.
Macaranga cornuta Müll.Arg. in A.P.de Candolle, Prodr. 15(2): 988 (1866).

Macaranga truncata Florence, Fl. Polynésie Fr. 1: 111 (1997).
Society Is. (Tahiti). 61 SCI. Phan.

Macaranga tsonane Whitmore, Kew Bull. 34: 606 (1980).
Papua New Guinea. 42 NWG. Phan.

Macaranga vanderystii De Wild., Pl. Bequaert. 3: 483 (1926).
Zaire. 23 ZAI.

Macaranga vedeliana (Baill.) Müll.Arg. in A.P.de Candolle, Prodr. 15(2): 1002 (1866).
New Caledonia (incl. Loyalty Is.). 60 NWC. Nanophan. or phan.
* *Acalypha vedeliana* Baill., Adansonia 2: 224 (1862).
Macaranga porrecta S.Moore, J. Linn. Soc., Bot. 45: 408 (1921).

Macaranga velutina (Rchb.f. & Zoll.) Müll.Arg. in A.P.de Candolle, Prodr. 15(2): 993 (1866).
S. Sumatera. 42 SUM. – Provisionally accepted.
* *Mappa velutina* Rchb.f. & Zoll., Linnaea 29: 468 (1858).

Macaranga venosa J.W.Moore, Bernice P. Bishop Mus. Bull. 226: 18 (1963).
Society Is. (Raiatea, Tahaa, Tahiti). 61 SCI. Nanophan. or phan.

Macaranga vermoesenii De Wild., Pl. Bequaert. 3: 484 (1926).
Zaire. 23 ZAI.

Macaranga versteeghii L.M.Perry, J. Arnold Arbor. 34: 242 (1953).
Irian Jaya. 42 NWG. Phan.

Macaranga vieillardii (Müll.Arg.) Müll.Arg. in A.P.de Candolle, Prodr. 15(2): 1007 (1866).
SE. New Caledonia. 60 NWC. Nanophan. or phan.
* *Mappa vieillardii* Müll.Arg., Linnaea 34: 199 (1865).
Cleidion tenuispica Schltr., Bot. Jahrb. Syst. 39: 150 (1906).

Macaranga villosula Pax & K.Hoffm. in H.G.A.Engler, Pflanzenr., IV, 147, XIV: 27 (1919).
NE. Papua New Guinea. 42 NWG. Phan.

Macaranga vitiensis Pax & K.Hoffm. in H.G.A.Engler, Pflanzenr., IV, 147, VII: 337 (1914).
Fiji. 60 FIJ. Nanophan. or phan.
Macaranga sanguinea Gillespie, Bernice P. Bishop Mus. Bull. 91: 17 (1932).

Macaranga warburgiana Pax & K.Hoffm. in H.G.A.Engler, Pflanzenr., IV, 147, VII: 347 (1914).
Papua New Guinea, Bismarck Archip. (New Ireland). 42 BIS NWG. Phan.
* *Macaranga cuspidata* Warb., Bot. Jahrb. Syst. 13: 351 (1891), nom. illeg.

Macaranga whitmorei Airy Shaw, Kew Bull. 23: 89 (1969).
Solomon Is., Santa Cruz Is. 60 SCZ SOL.

Macaranga winkleri Pax & K.Hoffm. in H.G.A.Engler, Pflanzenr., IV, 147, VII: 355 (1914).
Borneo. 42 BOR. Phan.

Macaranga winkleriella Whitmore, Kew Bull. 29: 449 (1974).
Borneo (Sarawak). 42 BOR. Nanophan.

Macaranga yakasii Airy Shaw, Kew Bull. 32: 410 (1978).
Bismarck Archip. (New Ireland). 42 BIS. Phan.

Synonyms:
Macaranga acuminata Pax & K.Hoffm. === **Macaranga hoffmannii** L.M.Perry
Macaranga acuminata Ridl. === **Macaranga densiflora** Warb.
Macaranga acuminata Warb. ex J.J.Sm. === **Macaranga densiflora** Warb.

Macaranga adenantha Gagnep. === **Macaranga indica** Wight
Macaranga amboinensis (Müll.Arg.) Müll.Arg. === **Macaranga involucrata** (Roxb.) Baill. var. **involucrata**
Macaranga andersonii Craib === **Macaranga kurzii** (Kuntze) Pax & K.Hoffm.
Macaranga angolensis var. *mollis* (Pax) Prain === **Macaranga angolensis** (Müll.Arg.) Müll.Arg.
Macaranga anjuanensis Leandri === ?
Macaranga ankafinensis Baill. === **Orfilea ankafinensis** (Baill.) Radcl.-Sm. & Govaerts
Macaranga apicifera Beille === **Macaranga heudelotii** Baill.
Macaranga apoensis Elmer === **Macaranga caudatifolia** Elmer
Macaranga asterolasia F.Muell. === **Macaranga involucrata** var. **mallotoides** (F.Muell.) L.M.Perry
Macaranga auriculata (Merr.) Airy Shaw === **Macaranga lowii** King ex Hook.f. var. **lowii**
Macaranga bachmannii Pax === **Macaranga capensis** (Baill.) Sim
Macaranga bartlettii Merr. === **Macaranga hullettii** King ex Hook.f. subsp. **hullettii**
Macaranga blumeana Müll.Arg. === **Macaranga rhizinoides** (Blume) Müll.Arg.
Macaranga borneensis (Müll.Arg.) Müll.Arg. === **Macaranga trichocarpa** (Rchb.f. & Zoll.) Müll.Arg.
Macaranga bracteata Merr. === **Macaranga andamanica** Kurz
Macaranga brandisii King ex Hook.f. === **Macaranga andamanica** Kurz
Macaranga brassii Mansf. === **Macaranga similis** Pax & K.Hoffm.
Macaranga brunneofloccosa var. *calvescens* Pax & K.Hoffm. === **Macaranga brunneofloccosa** Pax & K.Hoffm.
Macaranga caladiifolia var. *pilosula* Pax & K.Hoffm. === **Macaranga caladiifolia** Becc.
Macaranga caladiifolia var. *truncata* Pax & K.Hoffm. === **Macaranga caladiifolia** Becc.
Macaranga calcicola var. *calcifuga* Whitmore === **Macaranga bancana** (Miq.) Müll.Arg.
Macaranga calcifuga (Whitmore) R.I.Milne === **Macaranga bancana** (Miq.) Müll.Arg.
Macaranga calophylla Pax === **Macaranga schweinfurthii** Pax
Macaranga capensis var. *kilimandscharica* (Pax) Friis & M.G.Gilbert === **Macaranga kilimandscharica** Pax
Macaranga carrii var. *laevis* Whitmore === **Macaranga carrii** L.M.Perry
Macaranga carrii var. *leonardii* (L.M.Perry) Whitmore === **Macaranga carrii** L.M.Perry
Macaranga carrii var. *myolensis* Whitmore === **Macaranga carrii** L.M.Perry
Macaranga carrii var. *womersleyi* (L.M.Perry) Whitmore === **Macaranga carrii** L.M.Perry
Macaranga caryocarpa Airy Shaw === **Macaranga domatiosa** Airy Shaw
Macaranga chevalieri Beille === **Macaranga poggei** Pax
Macaranga chrysotricha var. *glaucescens* Mansf. === **Macaranga chrysotricha** K.Schum. & Lauterb.
Macaranga ciliata Bojer ex Baker === **Macaranga cupularis** Müll.Arg.
Macaranga cordifolia Boivin ex Baill. === **Macaranga bailloniana** Müll.Arg.
Macaranga cornuta Müll.Arg. === **Macaranga triloba** (Thunb.) Müll.Arg.
Macaranga coursi Leandri === ?
Macaranga crenata A.C.Sm. === **Macaranga graeffeana** var. **crenata** (A.C.Sm.) A.C.Sm.
Macaranga cuneata Elmer === **Macaranga caudatifolia** Elmer
Macaranga cuspidata Warb. === **Macaranga warburgiana** Pax & K.Hoffm.
Macaranga dalechampioides S.Moore === **Macaranga involucrata** (Roxb.) Baill. var. **involucrata**
Macaranga dallachyi (F.Muell.) F.Muell. ex Benth. === **Macaranga dallachyana** (Baill.) Airy Shaw
Macaranga danguyana Leandri === ?
Macaranga dawei Prain === **Myrica salicifolia** Hochst. ex A.Rich. (Myricaceae)
Macaranga depressa var. *genuina* Müll.Arg. === **Macaranga depressa** (Müll.Arg.) Müll.Arg.
Macaranga depressa var. *mollis* (Müll.Arg.) Müll.Arg. === **Macaranga indistincta** Whitmore
Macaranga divergens Müll.Arg. === **Macaranga depressa** f. **strigosa** Whitmore
Macaranga effusa Pax & K.Hoffm. === **Macaranga inermis** Pax & K.Hoffm.
Macaranga eglandulosa Baill. === **Macaranga bailloniana** Müll.Arg.
Macaranga eloba Pax & K.Hoffm. === **Macaranga caladiifolia** Becc.

Macaranga fimbriata S.Moore === **Macaranga polyadenia** Pax & K.Hoffm.
Macaranga fimbriata var. *doctersii* L.M.Perry === **Macaranga polyadenia** Pax & K.Hoffm.
Macaranga flexuosa Wight === **Macaranga indica** Wight
Macaranga formicarum Pax & K.Hoffm. === **Macaranga pruinosa** (Miq.) Müll.Arg.
Macaranga fulvescens Schltr. === **Macaranga corymbosa** (Müll.Arg.) Müll.Arg.
Macaranga gilletii De Wild. === **Macaranga poggei** Pax
Macaranga gmelinifolia King ex Hook.f. === **Macaranga pustulata** King ex Hook.f.
Macaranga gossypiifolia Pax & K.Hoffm. === **Macaranga hosei** King ex Hook.f.
Macaranga grandifolia Turrill === **Macaranga magna** Turrill
Macaranga griffithiana Müll.Arg. === **Macaranga motleyana** subsp. **griffithiana** (Müll.Arg.) Whitmore
Macaranga guignardii Beille === **Macaranga angolensis** (Müll.Arg.) Müll.Arg.
Macaranga gummiflua (Miq.) Müll.Arg. === **Macaranga denticulata** (Blume) Müll.Arg.
Macaranga haplostachya Pax & K.Hoffm. === **Macaranga glaberrima** (Hassk.) Airy Shaw var. **glaberrima**
Macaranga harveyana var. *glabrata* Pax & K.Hoffm. === **Macaranga harveyana** (Müll.Arg.) Müll.Arg.
Macaranga harveyana var. *puberula* Pax & K.Hoffm. === **Macaranga harveyana** (Müll.Arg.) Müll.Arg.
Macaranga havilandii var. *resecta* Airy Shaw === **Macaranga havilandii** Airy Shaw
Macaranga helferi Müll.Arg. === **Macaranga trichocarpa** (Rchb.f. & Zoll.) Müll.Arg.
Macaranga henricorum Hemsl. === **Macaranga denticulata** (Blume) Müll.Arg.
Macaranga heudelotii var. *nitida* Beille === **Macaranga barteri** Müll.Arg.
Macaranga hildebrandtii Baill. === **Macaranga boutonioides** Baill.
Macaranga hildebrandtii Pax & K.Hoffm. === **Macaranga oblongifolia** Baill.
Macaranga hispida var. *papuana* J.J.Sm. === **Macaranga papuana** (J.J.Sm.) Pax & K.Hoffm.
Macaranga humblotiana Baill. === **Macaranga boutonioides** Baill.
Macaranga hypoleuca var. *borneensis* Hutch. ex Gibbs === **Macaranga beccariana** Merr.
Macaranga incisa Gage === **Macaranga gigantea** (Rchb.f. & Zoll.) Müll.Arg.
Macaranga inermis var. *mallotiformis* (Pax & K.Hoffm.) L.M.Perry === **Macaranga inermis** Pax & K.Hoffm.
Macaranga inermis var. *penninervia* (Pax & K.Hoffm.) L.M.Perry === **Macaranga inermis** Pax & K.Hoffm.
Macaranga inermis var. *plurifoveata* L.M.Perry === **Macaranga inermis** Pax & K.Hoffm.
Macaranga inopinata Prain === **Macaranga capensis** (Baill.) Sim
Macaranga insignis Merr. === **Macaranga kingii** Hook.f.
Macaranga insularis Schltr. === **Macaranga corymbosa** (Müll.Arg.) Müll.Arg.
Macaranga involucrata var. *keyensis* Warb. === **Macaranga involucrata** (Roxb.) Baill. var. **involucrata**
Macaranga isadenia Pax & K.Hoffm. === **Macaranga punctata** K.Schum.
Macaranga javanica var. *montana* Müll.Arg. === **Macaranga heynei** I.M.Johnst.
Macaranga keyensis (Warb.) Pax & K.Hoffm. === **Macaranga involucrata** (Roxb.) Baill. var. **involucrata**
Macaranga kingii var. *platyphylla* Airy Shaw === **Macaranga kingii** Hook.f.
Macaranga lancifolia Pax === **Macaranga barteri** Müll.Arg.
Macaranga latifolia L.M.Perry === **Macaranga bifoveata** J.J.Sm.
Macaranga laurentii De Wild. === **Macaranga poggei** Pax
Macaranga lecomtei Beille === **Macaranga schweinfurthii** Pax
Macaranga ledermanniana Pax === **Macaranga spinosa** Müll.Arg.
Macaranga leightonii Whitmore === **Mallotus lackeyi** Elmer
Macaranga leonardii L.M.Perry === **Macaranga carrii** L.M.Perry
Macaranga leptostachya (Müll.Arg.) Müll.Arg. === **Cleidion leptostachyum** (Müll.Arg.) Pax & K.Hoffm.
Macaranga longispica S.Moore === **Macaranga alchorneoides** Pax & Lingelsh.
Macaranga macrophylla Müll.Arg. === **Endospermum macrophyllum** (Müll.Arg.) Pax & K.Hoffm.

Macaranga madagascariensis Steud. === ?
Macaranga maingayi Hook.f. === **Macaranga pruinosa** (Miq.) Müll.Arg.
Macaranga mallotiformis Pax & K.Hoffm. === **Macaranga inermis** Pax & K.Hoffm.
Macaranga mallotoides F.Muell. === **Macaranga involucrata** var. **mallotoides** (F.Muell.) L.M.Perry
Macaranga maluensis Pax & K.Hoffm. === **Macaranga punctata** K.Schum.
Macaranga maudsleyi Horne === **Trichospermum calyculatum** (Seem.) Burret (Tiliaceae)
Macaranga megalophylla (Müll.Arg.) Müll.Arg. === **Macaranga gigantea** (Rchb.f. & Zoll.) Müll.Arg.
Macaranga membranacea Kurz === **Macaranga kurzii** (Kuntze) Pax & K.Hoffm.
Macaranga merrilliana Pax & K.Hoffm. === **Macaranga montana** Merr.
Macaranga mildbraediana Pax === **Macaranga angolensis** (Müll.Arg.) Müll.Arg.
Macaranga mildbraediana Pax & K.Hoffm. === **Macaranga kilimandscharica** Pax
Macaranga minutiflora Müll.Arg. === **Macaranga trichocarpa** (Rchb.f. & Zoll.) Müll.Arg.
Macaranga mollis Pax === **Macaranga angolensis** (Müll.Arg.) Müll.Arg.
Macaranga molliuscula Kurz === **Macaranga tanarius** (L.) Müll.Arg.
Macaranga moluccana Radcl.-Sm. & Govaerts === **Macaranga cordifolia** (Roxb.) Müll.Arg.
Macaranga montana (Müll.Arg.) Pax & K.Hoffm. === **Macaranga heynei** I.M.Johnst.
Macaranga multiflora C.T.White === **Macaranga polyadenia** Pax & K.Hoffm.
Macaranga multiglandulosa Pax & K.Hoffm. === **Macaranga capensis** (Baill.) Sim
Macaranga myrmecophila Diels === **Macaranga caladiifolia** Becc.
Macaranga neomildbraediana Lebrun === **Macaranga kilimandscharica** Pax
Macaranga nyassae Pax & K.Hoffm. === **Macaranga kilimandscharica** Pax
Macaranga oreophila Pax & K.Hoffm. === **Macaranga corymbosa** (Müll.Arg.) Müll.Arg.
Macaranga ovalifolia Ridl. === **Macaranga papuana** var. **glabristipulata** Whitmore
Macaranga ovalifolia Pax & K.Hoffm. === **Macaranga punctata** K.Schum.
Macaranga ovata Boivin ex Baill. === **Macaranga boutonioides** Baill.
Macaranga pachyphylla Müll.Arg. === **Macaranga diepenhorstii** (Miq.) Müll.Arg.
Macaranga peltata Boivin ex Baill. === **Macaranga cuspidata** Boivin ex Baill.
Macaranga penninervia Pax & K.Hoffm. === **Macaranga inermis** Pax & K.Hoffm.
Macaranga perakensis Hook.f. === **Macaranga denticulata** (Blume) Müll.Arg.
Macaranga perrieri Leandri === ?
Macaranga platyphylla Baker ex Baill. === **Macaranga ferruginea** Baker
Macaranga pleiostemon f. *pubescens* L.M.Perry === **Macaranga pleiostemon** Pax & K.Hoffm.
Macaranga poggei var. *chevalieri* (Beille) Prain === **Macaranga poggei** Pax
Macaranga poilanei Gagnep. === **Macaranga lowii** King ex Hook.f. var. **lowii**
Macaranga populifolia (Miq.) Müll.Arg. === **Macaranga conifera** (Zoll.) Müll.Arg.
Macaranga porrecta S.Moore === **Macaranga vedeliana** (Baill.) Müll.Arg.
Macaranga porteana André === **Macaranga grandifolia** (Blanco) Merr.
Macaranga preussii Pax === **Macaranga occidentalis** (Müll.Arg.) Müll.Arg.
Macaranga pseudopeltata Pax & K.Hoffm. === **Macaranga punctata** K.Schum.
Macaranga pseudopruinosa Pax & K.Hoffm. === **Macaranga hosei** King ex Hook.f.
Macaranga punctata var. *whitei* L.M.Perry === **Macaranga punctata** K.Schum.
Macaranga pynaertii De Wild. === **Macaranga spinosa** Müll.Arg.
Macaranga quadriglandulosa var. *abbreviata* (J.J.Sm.) Pax & K.Hoffm. === **Macaranga quadriglandulosa** Warb.
Macaranga quadriglandulosa var. *digyna* Pax & K.Hoffm. === **Macaranga quadriglandulosa** Warb.
Macaranga quadriglandulosa var. *genuina* Pax & K.Hoffm. === **Macaranga quadriglandulosa** Warb.
Macaranga quadriglandulosa var. *variabilis* L.M.Perry === **Macaranga quadriglandulosa** Warb.
Macaranga quinquelobata Beille === **Macaranga heterophylla** (Müll.Arg.) Müll.Arg.
Macaranga ramiflora var. *sylvatica* (Elmer) Pax & K.Hoffm. === **Macaranga sylvatica** Elmer
Macaranga reineckei Pax === **Homalanthus acuminatus** (Müll.Arg.) Pax
Macaranga reticulata Baill. === **Macaranga obovata** Boivin ex Baill.
Macaranga riparia Engl. === **Macaranga aleuritoides** F.Muell.

Macaranga robiginosa Ridl. === **Macaranga heynei** I.M.Johnst.
Macaranga roemeri Pax & K.Hoffm. === **Macaranga platyclada** Pax & K.Hoffm.
Macaranga rosea Pax === **Macaranga schweinfurthii** Pax
Macaranga rosuliflora Croizat === **Macaranga andamanica** Kurz
Macaranga rottleroides Baill. === **Macaranga boutonioides** Baill.
Macaranga rotundifolia Elmer ex Pax & K.Hoffm. === **Macaranga hispida** (Blume) Müll.Arg.
Macaranga rowlandii Prain === **Macaranga barteri** Müll.Arg.
Macaranga roxburghii Wight === **Macaranga peltata** (Roxb.) Müll.Arg.
Macaranga rufibarbis var. *campestris* Whitmore === **Macaranga rufibarbis** Warb.
Macaranga rufibarbis var. *tenella* (Pax & K.Hoffm.) L.M.Perry === **Macaranga rufibarbis** Warb.
Macaranga rugosa Müll.Arg. === **Macaranga gigantea** (Rchb.f. & Zoll.) Müll.Arg.
Macaranga ruwenzorica Pax === **Macaranga capensis** (Baill.) Sim
Macaranga saccifera var. *dentifera* Pax & K.Hoffm. === **Macaranga saccifera** Pax
Macaranga saccifera var. *genuina* Pax & K.Hoffm. === **Macaranga saccifera** Pax
Macaranga sanguinea Gillespie === **Macaranga vitiensis** Pax & K.Hoffm.
Macaranga schleinitziana K.Schum. === **Macaranga involucrata** (Roxb.) Baill. var. **involucrata**
Macaranga schleinitziana var. *lobulata* Pax & K.Hoffm. === **Macaranga involucrata** (Roxb.) Baill. var. **involucrata**
Macaranga seta-felis Whitmore === **Macaranga involucrata** (Roxb.) Baill. var. **involucrata**
Macaranga sibuyanensis Elmer === **Macaranga caudatifolia** Elmer
Macaranga stricta (Rchb.f. & Zoll.) Müll.Arg. === **Mallotus philippensis** (Lam.) Müll.Arg.
Macaranga strigosa var. *carrii* L.M.Perry === **Macaranga strigosa** Pax & K.Hoffm.
Macaranga subfalcata (Rchb.f. & Zoll.) Müll.Arg. === **Macaranga glaberrima** (Hassk.) Airy Shaw var. **glaberrima**
Macaranga tamiana K.Schum. === **Cleidion spiciflorum** (Burm.f.) Merr. var. **spiciflorum**
Macaranga tanarius var. *abbreviata* J.J.Sm. === **Macaranga quadriglandulosa** Warb.
Macaranga tenella Pax & K.Hoffm. === **Macaranga rufibarbis** Warb.
Macaranga tenuiramea Pax & K.Hoffm. === **Macaranga caladiifolia** Becc.
Macaranga thonneri De Wild. === **Alchornea laxiflora** (Benth.) Pax & K.Hoffm.
Macaranga thouarsii Baill. === ?
Macaranga togoensis Pax === **Macaranga hurifolia** Beille
Macaranga tomentosa Wight === **Macaranga peltata** (Roxb.) Müll.Arg.
Macaranga tomentosa (Blume) Druce === **Macaranga tanarius** (L.) Müll.Arg.
Macaranga urophylla Pax & K.Hoffm. === **Macaranga dioica** (G.Forst.) Müll.Arg.
Macaranga usambarica Pax & K.Hoffm. === **Macaranga capensis** (Baill.) Sim
Macaranga utilis Elmer ex Merr. === **Macaranga bicolor** Müll.Arg.
Macaranga vulcanica Elmer ex Merr. === **Macaranga tanarius** (L.) Müll.Arg.
Macaranga wightiana Baill. === **Macaranga peltata** (Roxb.) Müll.Arg.
Macaranga womersleyi L.M.Perry === **Macaranga carrii** L.M.Perry
Macaranga zenkeri Pax === **Macaranga monandra** Müll.Arg.

Macraea

Synonyms:
Macraea Wight === **Phyllanthus** L.
Macraea gardneriana Wight === **Phyllanthus virgatus** var. **gardnerianus** (Wight) Govaerts & Radcl.-Sm.
Macraea myrtifolia Wight === **Phyllanthus myrtifolius** (Wight) Müll.Arg.
Macraea oblongifolia Wight === **Phyllanthus virgatus** G.Forst. var. **virgatus**
Macraea ovalifolia Wight === **Phyllanthus virgatus** G.Forst. var. **virgatus**
Macraea rheedii Wight === **Phyllanthus macraei** Müll.Arg.

Macrocroton

Synonyms:
Macrocroton Klotzsch === **Croton** L.

Maesobotrya

18 species, Africa (largely in W. & WC. Africa); shrubs or trees to 15(-20) m of closed or more open forest formations in which the inflorescences are sometimes turned into much-branched panicles ('balais de sorcère') possibly through parasitisation (Léonard, 1994). Some species have curious cresent-shaped stipules (*M. bertramiana*); and in others the inflorescences are in ramiflorous clusters recalling *Baccaurea*. A relationship also exists with the American *Richeria*, and Webster (Synopsis, 1994) has accordingly included all of them along with *Aporusa* and others in Antidesmeae subtribe Scapinae. [More recently R. Haegens, in his studies of *Baccaurea*, has confirmed a close relationship between it and *Maesobotrya*, with some species 'overlapping'.] (Phyllanthoideae)

Pax, F. & K. Hoffmann (1922). *Maesobotrya*. In A. Engler (ed.), Das Pflanzenreich, IV 147 XV (Euphorbiaceae-Phyllanthoideae-Phyllantheae): 17-26. Berlin. (Heft 81.) La/Ge. — 16 species, Africa.

- Léonard, J. (1994). Révision des espèces zaïroises du genre *Maesobotrya* Benth. (Euphorbiaceae). Bull. Jard. Bot. Natl. Belg. 63: 3-67, illus., maps. Fr. — Descriptive revision (11 species, one new) with key, synonymy, references and citations, types, vernacular names, indication of distribution, habitat and chorology, localities with exsiccatae, notes on uses, and sometimes detailed taxonomic commentary; index at end. The general part includes a key relating the genus to *Thecacoris* and *Antidesma* along with a discussion of the position of the genus in the family. [Most species here in sect. *Maesobotrya*; one in sect. *Staphyspora*. For additional coverage of the genus in Congo/Zaïre, see the author's account in *Flore d'Afrique Centrale: Euphorbiaceae*, 2: 46-79 (1995).]

Maesobotrya Benth., Hooker's Icon. Pl. 13: t. 1296 (1879).
Trop. Africa. 22 23 24 25 26. Nanophan. or phan.
Staphysora Pierre, Bull. Mens. Soc. Linn. Paris 2: 1233 (1896).

Maesobotrya barteri (Baill.) Hutch. in D.Oliver, Fl. Trop. Afr. 6(1): 669 (1912).
W. & WC. Trop. Africa. 22 GHA GUI IVO LBR NGA SIE 23 CMN EQG. Nanophan. or phan.
* *Pierardia barteri* Baill., Adansonia 4: 137 (1864). *Baccaurea barteri* (Baill.) Müll.Arg. in A.P.de Candolle, Prodr. 15(2): 464 (1866).

var. **barteri**
W. & WC. Trop. Africa. 22 NGA SIE 23 CMN EQG. Nanophan. or phan.
Maesobotrya brevispicata Pax in H.G.A.Engler & C.G.O.Drude, Veg. Erde 9, III(2): 15 (1921).
Maesobotrya dissitiflora Pax in H.G.A.Engler & C.G.O.Drude, Veg. Erde 9, III(2): 15 (1921).

var. **sparsiflora** (Scott-Elliot) Keay, Kew Bull. 10: 139 (1955).
W. Trop. Africa. 22 GHA GUI IVO LBR SIE. Phan.
* *Baccaurea sparsiflora* Scott-Elliot, J. Linn. Soc., Bot. 30: 97 (1894). *Maesobotrya sparsiflora* (Scott-Elliot) Hutch. in D.Oliver, Fl. Trop. Afr. 6(1): 665 (1912). *Maesobotrya floribunda* var. *sparsiflora* (Scott-Elliot) Pax & K.Hoffm. in H.G.A.Engler, Pflanzenr., IV, 147, XV: 20 (1922).
Baccaurea bonnetii Beille, Bull. Soc. Bot. France 55(8): 58 (1908).
Baccaurea caillei Beille, Bull. Soc. Bot. France 55(8): 61 (1908).
Baccaurea gagnepainii Beille, Bull. Soc. Bot. France 55(8): 59 (1908).
Baccaurea glaziovii Beille, Bull. Soc. Bot. France 55(8): 60 (1908).
Baccaurea longispicata Beille, Bull. Soc. Bot. France 55(8): 59 (1908).
Baccaurea poissonii Beille, Bull. Soc. Bot. France 55(8): 60 (1908).

Baccaurea cavalliensis Beille, Bull. Soc. Bot. France 57(8): 121 (1910).
Maesobotrya rufinervis Pierre ex Pax & K.Hoffm. in H.G.A.Engler, Pflanzenr., IV, 147, XV: 20 (1922).
Maesobotrya edulis Hutch. & Dalziel, Fl. W. Trop. Afr. 1: 284 (1928).

Maesobotrya bertramiana Büttner, Verh. Bot. Vereins Prov. Brandenburg 31: 93 (1889).
Gabon, Congo, Zaire. 23 CON GAB ZAI. Nanophan. or phan.
Maesobotrya sapinii De Wild., Ann. Mus. Congo Belge, Bot., V, 2: 268 (1908).
Maesobotrya sapinii var. *brevipetiolata* De Wild., Ann. Mus. Congo Belge, Bot., V, 2: 269 (1908).
Staphysora sapinii De Wild., Ann. Mus. Congo Belge, Bot., V, 2: 268 (1908), nom. nud.
Maesobotrya tsoukensis Pellegr., Mém. Soc. Linn. Normandie, Bot. 1: 64 (1928).

Maesobotrya bipindensis (Pax) Hutch. in D.Oliver, Fl. Trop. Afr. 6(1): 667 (1912).
Cameroon, Zaire. 23 CMN ZAI. Nanophan. or phan.
* *Baccaurea bipindensis* Pax, Bot. Jahrb. Syst. 34: 368 (1904).

Maesobotrya cordulata J.Léonard, Bull. Jard. Bot. État 17: 259 (1945).
Congo, Gabon, Zaire. 23 CON GAB ZAI. Nanophan. or phan.

Maesobotrya fallax Pax & K.Hoffm. in H.G.A.Engler, Pflanzenr., IV, 147, XV: 23 (1922).
Cameroon. 23 CMN.

Maesobotrya floribunda Benth., Hooker's Icon. Pl. 13: t. 1296 (1879). – FIGURE, p. 1109.
WC. Trop. Africa, Zambia. 23 CMN CON ZAI 26 ZAM. Nanophan. or phan.
Antidesma schweinfurthii Pax, Bot. Jahrb. Syst. 15: 530 (1893). *Maesobotrya floribunda* var. *schweinfurthii* (Pax) Pax & K.Hoffm. in H.G.A.Engler, Pflanzenr., IV, 147, XV: 20 (1922).
Maesobotrya hirtella Pax, Bot. Jahrb. Syst. 28: 21 (1899). *Maesobotrya floribunda* var. *hirtella* (Pax) Pax & K.Hoffm. in H.G.A.Engler, Pflanzenr., IV, 147, XV: 20 (1922).

Maesobotrya glabrata (Hutch.) Exell, Bull. Inst. Franç. Afrique Noire 21: 465 (1959).
São Tomé, Principe. 23 GGI.
* *Thecacoris glabrata* Hutch. in A.W.Exell, Cat. Vasc. Pl. S. Tome: 287 (1944).

Maesobotrya griffoniana (Baill.) Pierre ex Hutch. in D.Oliver, Fl. Trop. Afr. 6(1): 669 (1912).
Gabon. 23 GAB. Nanophan. or phan.
* *Pierardia griffoniana* Baill., Adansonia 4: 136 (1864). *Baccaurea griffoniana* (Baill.) Müll.Arg. in A.P.de Candolle, Prodr. 15(2): 464 (1866).

Maesobotrya klaineana (Pierre) J.Léonard, Bull. Jard. Bot. Belg. 63: 14 (1994).
Nigeria, WC. Trop. Africa. 22 NGA 23 CMN CON EQG GAB GGI ZAI. Nanophan. or phan.
Staphysora dusenii Pax, Bull. Mens. Soc. Linn. Paris 2: 1233 (1896). *Maesobotrya dusenii* (Pax) Pax, Bot. Jahrb. Syst. 23: 521 (1897).
* *Staphysora klaineana* Pierre, Exsicc. (L.Pierre): t. 6405 (1896).
Maesobotrya intermedia Pax & K.Hoffm. in H.G.A.Engler, Pflanzenr., IV, 147, XV: 19 (1922).

Maesobotrya longipes (Pax) Hutch. in D.Oliver, Fl. Trop. Afr. 6(1): 670 (1912).
WC. Trop. Africa. 23 CAF CMN GAB ZAI. Phan.
* *Antidesma longipes* Pax, Bot. Jahrb. Syst. 15: 529 (1893).
Staphysora albida Pierre ex Pax, Bull. Mens. Soc. Linn. Paris 2: 1233 (1896), nom. nud.
Maesobotrya longipes var. *albida* Pax & K.Hoffm. in H.G.A.Engler, Pflanzenr., IV, 147, XV: 25 (1922).
Maesobotrya longipes var. *lancifolia* Pax & K.Hoffm. in H.G.A.Engler, Pflanzenr., IV, 147, XV: 25 (1922).

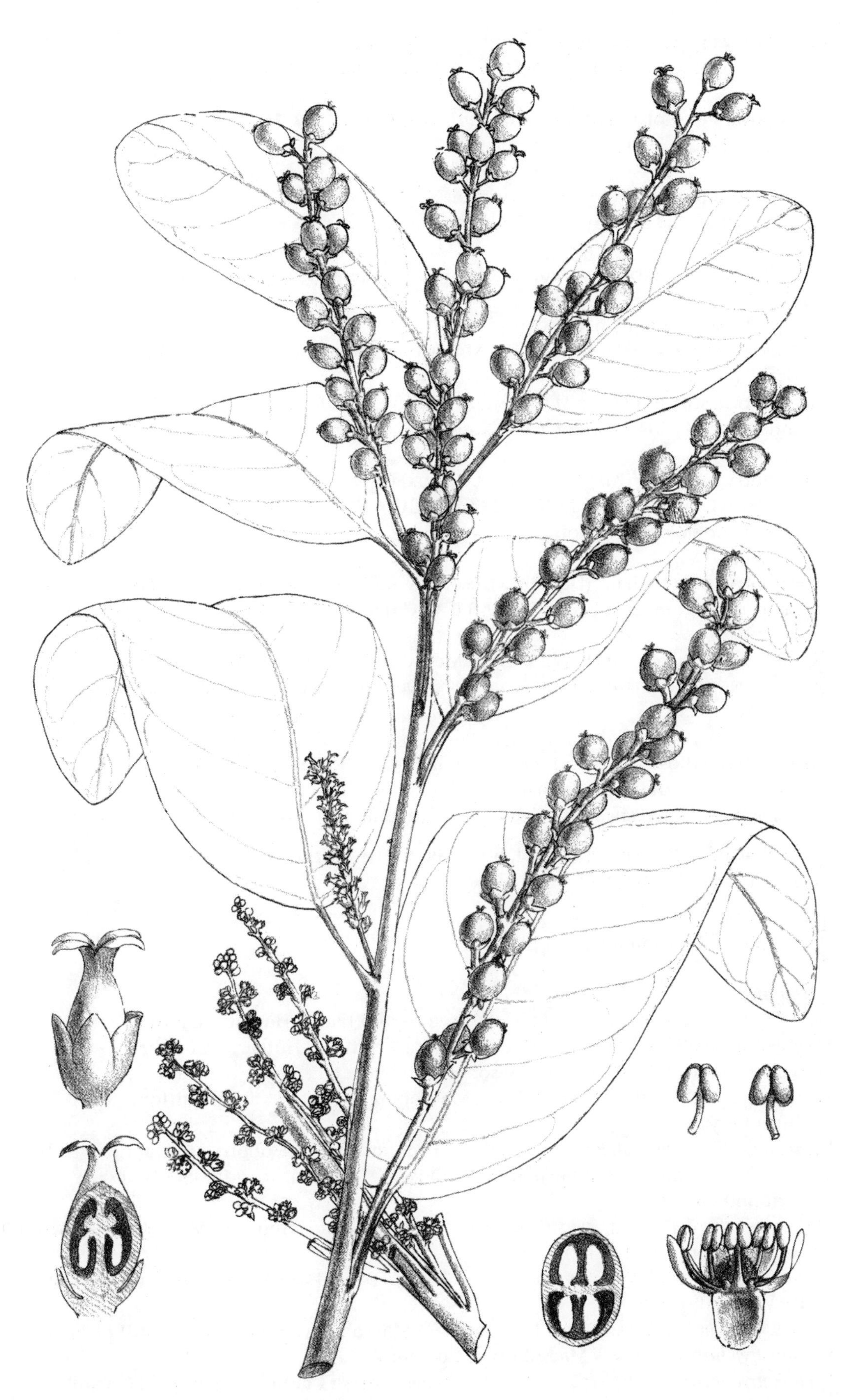

Maesobotrya floribunda Benth.
Artist: 'A.M.C.'
Ic. Pl. 13: pl. 1296 (1879)
KEW ILLUSTRATIONS COLLECTION

Maesobotrya pauciflora Pax, Bot. Jahrb. Syst. 33: 281 (1903).
WC. Trop. Africa. 23 CMN CON EQG ZAI? Nanophan.

Maesobotrya pierlotii J.Léonard, Bull. Jard. Bot. Belg. 63: 39 (1994).
E. Zaire. 23 ZAI. Phan.

Maesobotrya purseglovei Verdc., Kew Bull. 7: 357 (1952).
Zaire, Uganda. 23 ZAI 25 UGA. Nanophan. or phan.

Maesobotrya pynaertii (De Wild.) Pax in H.G.A.Engler & C.G.O.Drude, Veg. Erde 9, III(2): 15 (1921).
Zaire. 23 ZAI. Nanophan. or phan.
* *Baccaurea pynaertii* De Wild., Ann. Mus. Congo Belge, Bot., V, 3: 219 (1910).

Maesobotrya scariosa Pax in H.G.A.Engler & C.G.O.Drude, Veg. Erde 9, III(2): 15 (1921).
Cameroon. 23 CMN. Nanophan. - Close to *M. griffoniana*.

Maesobotrya staudtii (Pax) Hutch. in D.Oliver, Fl. Trop. Afr. 6(1): 668 (1912).
Nigeria, WC. Trop. Africa. 22 NGA 23 CMN CON CMN ZAI. Phan.
* *Baccaurea staudtii* Pax, Bot. Jahrb. Syst. 23: 521 (1897).

Maesobotrya vermeulenii (De Wild.) J.Léonard, Bull. Jard. Bot. Belg. 63: 51 (1994).
S. Nigeria to Zambia. 22 NGA 23 CAB CAF CMN CON GAB ZAI 26 ZAM. Nanophan. or phan.
* *Baccaurea vermeulenii* De Wild., Ann. Mus. Congo Belge, Bot., V, 3: 218 (1910).
Maesobotrya floribunda var. *vermeulenii* (De Wild.) J.Léonard, Bull. Jard. Bot. État 17: 258 (1945).

Maesobotrya villosa (J.Léonard) J.Léonard, Bull. Jard. Bot. Belg. 63: 60 (1994).
Zaire. 23 ZAI. Nanophan. or phan.
* *Maesobotrya floribunda* var. *villosa* J.Léonard, Bull. Jard. Bot. État 17: 257 (1945).

var. **lenifolia** J.Léonard, Bull. Jard. Bot. Belg. 63: 62 (1994).
Zaire. 23 ZAI. Nanophan. or phan.

var. **villosa**
Zaire. 23 ZAI. Nanophan. or phan.

Synonyms:
Maesobotrya brevispicata Pax === **Maesobotrya barteri** (Baill.) Hutch. var. **barteri**
Maesobotrya dissitiflora Pax === **Maesobotrya barteri** (Baill.) Hutch. var. **barteri**
Maesobotrya dusenii (Pax) Pax === **Maesobotrya klaineana** (Pierre) J.Léonard
Maesobotrya edulis Hutch. & Dalziel === **Maesobotrya barteri** var. **sparsiflora** (Scott-Elliot) Keay
Maesobotrya floribunda var. *hirtella* (Pax) Pax & K.Hoffm. === **Maesobotrya floribunda** Benth.
Maesobotrya floribunda var. *schweinfurthii* (Pax) Pax & K.Hoffm. === **Maesobotrya floribunda** Benth.
Maesobotrya floribunda var. *sparsiflora* (Scott-Elliot) Pax & K.Hoffm. === **Maesobotrya barteri** var. **sparsiflora** (Scott-Elliot) Keay
Maesobotrya floribunda var. *vermeulenii* (De Wild.) J.Léonard === **Maesobotrya vermeulenii** (De Wild.) J.Léonard
Maesobotrya floribunda var. *villosa* J.Léonard === **Maesobotrya villosa** (J.Léonard) J.Léonard
Maesobotrya hirtella Pax === **Maesobotrya floribunda** Benth.
Maesobotrya intermedia Pax & K.Hoffm. === **Maesobotrya klaineana** (Pierre) J.Léonard
Maesobotrya longipes var. *albida* Pax & K.Hoffm. === **Maesobotrya longipes** (Pax) Hutch.
Maesobotrya longipes var. *lancifolia* Pax & K.Hoffm. === **Maesobotrya longipes** (Pax) Hutch.
Maesobotrya oblonga Hutch. === **Antidesma oblongum** (Hutch.) Keay

Maesobotrya rufinervis Pierre ex Pax & K.Hoffm. === **Maesobotrya barteri** var. **sparsiflora** (Scott-Elliot) Keay
Maesobotrya sapinii De Wild. === **Maesobotrya bertramiana** Büttner
Maesobotrya sapinii var. *brevipetiolata* De Wild. === **Maesobotrya bertramiana** Büttner
Maesobotrya sparsiflora (Scott-Elliot) Hutch. === **Maesobotrya barteri** var. **sparsiflora** (Scott-Elliot) Keay
Maesobotrya stapfiana Beille === **Protomegabaria stapfiana** (Beille) Hutch.
Maesobotrya tsoukensis Pellegr. === **Maesobotrya bertramiana** Büttner

Mallotus

135 species, Africa (2), Madagascar (2, 1 shared with Africa), Asia, Malesia, Australia and the Pacific; here exclusive of *Deuteromallotus* but not *Coccoceras*. Shrubs or trees of forest and secondary formations, rather diverse in form but distinguished from *Macaranga* by usually stellate indumentum and the leaves sometimes being opposite as well as usually granular-glandulose. The anthers moreover are 2-celled rather than having 4 cells as in *Macaranga*. Pax & Hoffmann (1914) accepted 10 sections for the genus, the last three (*Hancea*, *Polyadenii* and *Axenfeldia*) with pinnate venation, the remainder tri- or palminerved. Sects. *Hancea* (Pax & Hoffmann no. 10) and *Diplochlamys* (no. 5) were moreover eglandulose. Airy-Shaw (1963) reduced *Coccoceras*; this is now sect. *Polyadenii*. Later (1968) Airy-Shaw combined sects. *Echinocroton* and *Plagianthera* (nos. 1-2) as sect. *Rottleropsis* and established sect. *Oliganthae*, while two others were for nomenclatural reasons renamed. There are now 11 sections, although this may change with realisation of a new revision for Malesia. The combined sect. *Rottleropsis* is the only one well represented west of South Asia. Anisophylly appears in sect. *Hancea*, each pair of leaves being unequal, as in *M. miquelianus*. A suggestion has been made that *Macaranga* sect. *Pseudo-rottlera* is transitional to *Mallotus*, although in the Webster system the two genera are in different subtribes. A study of sect. *Polyadenii* has been completed by Sarah Bollendorff (with 8 species, 4 in Malesia), while two other sections, *Hancea* and *Stylanthes* (Pax & Hoffmann sect. 4; species are characterised by a fenugreek smell), have been undertaken by Ferry Slik. Both of these are in connection with the Malesian Euphorbiaceae project. Slik has also undertaken studies showing that at least in East Kalimantan (Borneo) species of the genus, along with *Macaranga*, can be good indicators of forest succession and integrity. (Acalyphoideae)

Pax, F. (with K. Hoffmann) (1914). *Coccoceras*. In A. Engler (ed.), Das Pflanzenreich, IV 147 VII (Euphorbiaceae-Acalypheae-Mercurialinae): 209-211. Berlin. (Heft 63.) La/Ge. — 3 species, SE Asia, Malesia. [Now included in *Mallotus* as sect. *Polyadenii*.]

• Pax, F. (with K. Hoffmann) (1914). *Mallotus*. In A. Engler (ed.), Das Pflanzenreich, IV 147 VII (Euphorbiaceae-Acalypheae-Mercurialinae): 145-208. Berlin. (Heft 63.) La/Ge. — 101 species, of which nos. 94-101 doubtful; 10 sections recognized. [Some of the doubtful species are not *Mallotus*.] – Additions in ibid., XIV (Additamentum VI): 17-19 (1919).

Airy-Shaw, H. K. (1963). Notes on Malaysian and other Asiatic Euphorbiaceae, XXXI. Limestone association of *Mallotus wrayi*. Kew Bull. 16: 349. En. — Habitat note.

Airy-Shaw, H. K. (1963). Notes on Malaysian and other Asiatic Euphorbiaceae, XXXII. *Coccoceras* Miquel reduced to *Mallotus* Lour. Kew Bull. 16: 349-352. En. — New combinations, commentary and key.

Airy-Shaw, H. K. (1963). Notes on Malaysian and other Asiatic Euphorbiaceae, XXXIII. A misplaced Burmese *Mallotus*. Kew Bull. 16: 352. En. — Formerly in *Claoxylon*; localities with exsiccatae and commentary.

Airy-Shaw, H. K. (1966). Notes on Malaysian and other Asiatic Euphorbiaceae, LXIII. Notes on, and new species in, *Mallotus* Lour. Kew Bull. 20: 38-45. En. — Novelties and notes.

• Airy-Shaw, H. K. (1968). Notes on Malesian and other Asiatic Euphorbiaceae, XC. New or noteworthy species, transferences, etc., in *Mallotus* Lour. Kew Bull. 21: 379-400. En. — A substantial suite of additions and other changes, organized by sections; from all over SE Asia and Malesia. Particular features include the merger of sects. *Echinocroton* and *Plagianthera* as sect. *Rottleropsis*; the redesignation of sect. *Echinus* as sect. *Mallotus* and sect. *Philippinenses* as sect. *Rottlera* (for nomenclatural reasons); and a new sect. *Oliganthae* (carved out of sect. *Diplochlamys*), [Easily the largest regional contribution since Pax and Hoffmann's revision. No keys, however, are presented.]

Airy-Shaw, H. K. (1969). Notes on Malesian and other Asiatic Euphorbiaceae, CVIII. A new species of *Mallotus* Lour. from Borneo. Kew Bull. 23: 80-82. En. — Novelty and additional record; both species in sect. *Hancea*.

Airy-Shaw, H. K. (1971). Notes on Malesian and other Asiatic Euphorbiaceae, CXXXVIII. New or noteworthy species of *Mallotus* Lour. Kew Bull. 25: 526-528. En. — 2 species, one new; both in sect. *Rottleropsis*.

Airy-Shaw, H. K. (1972). Notes on Malesian and other Asiatic Euphorbiaceae, CLXII: A new *Mallotus* from Borneo. Kew Bull. 27: 86-87. En. —*M. beccarii* described. (Sect. *Hancea*).

Airy-Shaw, H. K. (1974). Notes on Malesian and other Asiatic Euphorbiaceae, CLXXIX. The female flower of *Mallotus macularis* Airy Shaw. Kew Bull. 29: 313. En. — Extension of knowledge of this New Guinean species (in sect. *Rottleropsis*).

Airy-Shaw, H. K. (1978). Notes on Malesian and other Asiatic Euphorbiaceae, CCXVI. *Mallotus* Lour. Kew Bull. 33: 62-66. En. — 4 spp (three new). in 3 sections, two from Malaya, two from New Guinea.

Airy-Shaw, H. K. (1978). Notes on Malesian and other Asiatic Euphorbiaceae, CXCIX. New or noteworthy species of *Mallotus* Lour. Kew Bull. 32: 400-407. En. — Descriptions of 5 new species; sections indicated.

Airy-Shaw, H. K. (1980). Notes on Euphorbiaceae from Indomalesia, Australia and the Pacific, CCXXXVII. *Mallotus* Lour. Kew Bull. 35: 393-395. En. — Description of *M. tenuipes* (sect. *Hancea*) from Borneo and the Moluccas; related to *M. penangensis*.

McPherson, G. (1995). On *Mallotus* and *Deuteromallotus* in Madagascar. Bull. Mus. Natl. Hist. Nat., IV, B (Adansonia, III), 17: 169-173, illus. En. — Reduction of *Deuteromallotus* to *Mallotus*; revival of *M. baillonianus* Muell. Arg. for *D. acuminatus*, transfer of *D. capuronii*, brief note on *M. oppositifolius* (also widespread in Africa), and description of a novelty, *M. spinulosus*, with figure and documentation; key to all four species now known from Madagascar (p. 172). [The reduction of *Deuteromallotus* is not universally accepted.]

Hussin, K. H., B. A. Washab & C. P. Teh (1996). Comparative leaf anatomical studies of some *Mallotus* Lour. (Euphorbiaceae) species. Bot. J. Linn. Soc. 122: 137-153, illus. En. — Comparative study of leaf blades and petioles of 16 species, all in Peninsular Malaysia; descriptions and comparative tables provided. [The authors indicate that their results do not show any clustering of characters useful for formation of broad groups; moreover, the leaf anatomy in many cases does not correlate with the sections established by Pax (as used by Ridley) and Airy-Shaw. The only real exception to the latter was the closeness of sects. *Echinocroton* (sic) and *Rottleropsis* (now united; see Airy-Shaw 1968).]

Slik, J. W. F. (1998a). Three new Malesian species of *Mallotus* section *Hancea* (Euphorbiaceae). Blumea 43: 225-232. En. — Descriptions of novelties.

Slik, J. W. F. (1998b). Keys to the taxa of *Macaranga* and *Mallotus* (Euphorbiaceae) of East Kalimantan (Indonesia). *Flora Males. Bull.* 12(4): 157-178. En. — Analytical key.

Mallotus Lour., Fl. Cochinch.: 635 (1790).

Trop. Africa to Fiji. 22 23 24 25 26 29 36 38 40 41 42 50 60.

Echinus Lour., Fl. Cochinch.: 633 (1790).
Rottlera Roxb., Pl. Coromandel 2: 36 (1802).
Adisca Blume, Bijdr.: 609 (1826).
Lasipana Raf., Sylva Tellur.: 21 (1838).
Adisa Steud., Nomencl. Bot., ed. 2, 1: 28 (1840).
Boutonia Bojer, Trav. Soc. Hist. Nat. Ile Maurice 1842-1846: 151 (1846).
Plagianthera Rchb.f. & Zoll., Acta Soc. Regiae Sci. Indo-Neerl. 1: 19 (1856).
Stylanthus Rchb. & Zoll., Linnaea 28: 312 (1856).
Hancea Seem., Bot. Voy. Herald: 409 (1857).
Axenfeldia Baill., Étude Euphorb.: 419 (1858).
Boutonia Bojer ex Baill., Étude Euphorb.: 400 (1858).
Coelodiscus Baill., Étude Euphorb.: 293 (1858).

Echinocroton F.Muell., Fragm. 1: 31 (1858).
Aconceveibum Miq., Fl. Ned. Ind. 1(2): 389 (1859).
Coccoceras Miq., Fl. Ned. Ind., Eerste Bijv.: 455 (1861).
Diplochlamys Müll.Arg., Flora 47: 539 (1864).

Mallotus actinoneurus Airy Shaw, Kew Bull. 33: 63 (1978).
Pen. Malaysia. 42 MLY.

Mallotus anisopodus (Gagnep.) Airy Shaw, Kew Bull. 16: 351 (1963).
C. Thailand, Vietnam. 41 THA VIE. Phan.
* *Coccoceras anisopodum* Gagnep., Bull. Soc. Bot. France 71: 1021 (1924 publ. 1925).

Mallotus anomalus Merr. & Chun, Sunyatsenia 5: 99 (1940).
Hainan. 36 CHH.

Mallotus apelta (Lour.) Müll.Arg., Linnaea 34: 189 (1865).
S. China (incl. Hong Kong), Hainan. 36 CHC CHH CHS. Nanophan. or phan.
* *Ricinus apelta* Lour., Fl. Cochinch.: 585 (1790).
Croton chinensis Geiseler, Croton. Monogr.: 24 (1807). *Mallotus apelta* var. *chinensis* (Geiseler) Pax & K.Hoffm. in H.G.A.Engler, Pflanzenr., IV, 147, VII: 171 (1914).
Mallotus tenuifolius Pax, Bot. Jahrb. Syst. 29: 429 (1900). *Mallotus apelta* var. *tenuifolius* (Pax) Pax in H.G.A.Engler, Pflanzenr., IV, 147, VII: 171 (1914).
Mallotus paxii Pamp., Nuovo Giorn. Bot. Ital. 17: 414 (1910).
Mallotus castanopsis F.P.Metcalf, Lingnan Sci. J. 10: 487 (1931). *Mallotus paxii* var. *castanopsis* (F.P.Metcalf) S.M.Hwang, Acta Phytotax. Sin. 23: 298 (1985).
Mallotus apelta var. *kwangsiensis* F.P.Metcalf, J. Arnold Arbor. 22: 204 (1941).

Mallotus atrovirens Müll.Arg., Linnaea 34: 195 (1865).
India. 40 IND.
Croton atrovirens Wall., Numer. List: 7771 (1847), nom. nud.

Mallotus attenuatus Airy Shaw, Kew Bull. 33: 63 (1978).
NE. Papua New Guinea. 42 NWG. Nanophan.

Mallotus aureopunctatus (Dalzell) Müll.Arg. in A.P.de Candolle, Prodr. 15(2): 973 (1866).
India. 40 IND. Nanophan. or phan.
* *Rottlera aureopunctata* Dalzell, Hooker's J. Bot. Kew Gard. Misc. 3: 122 (1851).
Mallotus lawii Müll.Arg., Linnaea 34: 192 (1865).

Mallotus barbatus Müll.Arg., Linnaea 34: 184 (1865).
India, S. China, Indo-China, Pen. Malaysia, Sumatera ?, Jawa, Maluku (Halamahera). 36 CHC CHS 40 IND 41 BMA THA VIE 42 JAW MLY MOL SUM? Nanophan. or phan.
Rottlera barbata Wall. ex Baill., Étude Euphorb.: 423 (1858), nom. nud.

var. **barbatus**
India, S. China, Indo-China, Pen. Malaysia, Sumatera ?, Jawa, Maluku (Halamahera). 36 CHC CHS 40 IND 41 BMA THA VIE 42 JAW MLY MOL SUM? Nanophan. or phan.
Mallotus barbatus var. *pedicellaris* Croizat, J. Arnold Arbor. 19: 135 (1938).
Mallotus barbatus var. *hubeiensis* S.M.Hwang, Acta Phytotax. Sin. 23: 296 (1985).

var. **croizatianus** (Metcalf) S.M.Hwang, Acta Phytotax. Sin. 23: 295 (1985).
China (Guangxi). 36 CHS. Nanophan. or phan.
* *Mallotus croizatianus* F.P.Metcalf, J. Arnold Arbor. 22: 204 (1941).

Mallotus beccarii Airy Shaw, Kew Bull. 27: 86 (1972).
Borneo (Sarawak). 42 BOR. Nanophan.

Mallotus beddomei Hook.f., Fl. Brit. India 5: 438 (1887).
S. India. 40 IND. Nanophan.

Mallotus blumeanus Müll.Arg., Linnaea 34: 195 (1865).
S. Sumatera, Jawa, Lesser Sunda Is. (Flores), Sulawesi. 42 JAW LSI SUL SUM. Phan.
* *Rottlera oppositifolia* Blume, Bijdr.: 608 (1826).

Mallotus brachythyrsus Merr., Sarawak Mus. J. 3: 526 (1928).
Borneo (SW. Sarawak). 42 BOR. Nanophan.

Mallotus brevipes Merr., Philipp. J. Sci., C 9: 487 (1914 publ. 1915).
Philippines. 42 PHI.

Mallotus brevipetiolatus Gage, Rec. Bot. Surv. India 9: 242 (1922).
S. Thailand, Pen. Malaysia. 41 THA 42 MLY. Phan.

Mallotus calocarpus Airy Shaw, Kew Bull. 21: 395 (1968).
SE. Thailand. 41 THA. Nanophan. or phan.

Mallotus cambodianus (Gagnep.) Airy Shaw, Kew Bull. 21: 381 (1968).
Cambodia. 41 CBD.
* *Coelodiscus cambodianus* Gagnep., Notul. Syst. (Paris) 4: 49 (1923).

Mallotus canii Thin, Euphorb. Vietnam: 46 (1995).
Vietnam. 41 VIE. Nanophan. or phan.

Mallotus cauliflorus Merr., Philipp. J. Sci., C 7: 399 (1912 publ. 1913).
Philippines. 42 PHI.

Mallotus chromocarpus Airy Shaw, Kew Bull. 32: 403 (1978).
Papua New Guinea. 42 NWG. Phan.

Mallotus chuyenii Thin, J. Biol. (Vietnam) 9: 38 (1987).
Vietnam. 41 VIE.

Mallotus claoxyloides (F.Muell.) Müll.Arg., Linnaea 34: 192 (1865).
Papua New Guinea, Queensland, New South Wales. 42 NWG 50 NSW QLD. Nanophan. or phan.
* *Echinocroton claoxyloides* F.Muell., Fragm. 1: 32 (1858).
Echinus claoxyloides var. *cordata* Baill., Adansonia 6: 315 (1866). *Mallotus claoxyloides* var. *cordatus* (Baill.) Airy Shaw, Muelleria 4: 232 (1980).
Echinus claoxyloides var. *ficifolia* Baill., Adansonia 6: 315 (1866). *Mallotus claoxyloides* var. *ficifolius* (Baill.) Benth., Fl. Austral. 6: 141 (1873). *Mallotus ficifolius* (Baill.) Pax & K.Hoffm. in H.G.A.Engler, Pflanzenr., IV, 147, VII: 151 (1914).
Mallotus cloaxyloides var. *macrophylla* Benth., Fl. Austral. 6: 141 (1873).
Mallotus claoxyloides var. *angustifolius* F.M.Bailey, Contr. Queensland Fl.: 18 (1891).
Mallotus claoxyloides f. *grossedentata* Domin in ?, : 888 (1927).

Mallotus clellandii Hook.f., Fl. Brit. India 5: 435 (1887).
Burma, N. Thailand, Vietnam. 41 BMA THA VIE. Nanophan.
Mallotus tristis Pax & K.Hoffm. in H.G.A.Engler, Pflanzenr., IV, 147, VII: 154 (1914).

Mallotus concinnus Airy Shaw, Kew Bull. 32: 406 (1978).
Pen. Malaysia. 42 MLY.

Mallotus confusus Merr., Philipp. J. Sci. 16: 559 (1920).
Philippines. 42 PHI.

Mallotus conspurcatus Croizat, J. Arnold Arbor. 21: 501 (1940).
China (Guangxi). 36 CHS.

Mallotus contubernalis Hance, J. Bot. 20: 293 (1882).
S. China. 36 CHC CHS. Nanophan. or phan.
Mallotus chrysocarpus Pamp., Nuovo Giorn. Bot. Ital. 17: 413 (1910). *Mallotus repandus* var. *chrysocarpus* (Pamp.) S.M.Hwang, Acta Phytotax. Sin. 23: 297 (1985).

Mallotus cordatifolius Slik, Blumea 43: 225 (1998).
Philippines (Samar). 42 PHI. Phan.

Mallotus cuneatus Ridl., J. Straits Branch Roy. Asiat. Soc. 59: 181 (1911). *Mallotus resinosus* var. *cuneatus* (Ridl.) N.P.Balakr. & Chakrab., Rheedea 1: 39 (1991).
Indo-China, Pen. Malaysia. 41 THA VIE 42 MLY. Nanophan.

Mallotus darbyshirei Airy Shaw, Kew Bull. 34: 597 (1980).
NE. Papua New Guinea. 42 NWG. Nanophan. or phan.

Mallotus decipiens Müll.Arg., Linnaea 34: 194 (1865).
Burma, Thailand. 41 BMA THA. Nanophan. or phan.

Mallotus didymochryseus Airy Shaw, Kew Bull. 20: 40 (1966).
Papua New Guinea, Northern Territory. 42 NWG 50 NTA. Phan.

Mallotus discolor F.Muell. ex Benth., Fl. Austral. 6: 143 (1873).
E. Queensland, NE. New South Wales. 50 NSW QLD. Phan.

Mallotus dispar (Blume) Müll.Arg. in A.P.de Candolle, Prodr. 15(2): 971 (1866).
Hainan, Trop. Asia. 36 CHH 38 TAI 41 THA VIE 42 BOR JAW LSI MOL PHI SUL SUM. Nanophan. or phan.
* *Rottlera dispar* Blume, Bijdr.: 608 (1826).
Claoxylon stipulosum Rchb.f. ex Zoll., Acta Soc. Regiae Sci. Indo-Neerl. 1: 20 (1856). *Mallotus stipulosus* (Rchb.f. ex Zoll.) Müll.Arg., Linnaea 34: 192 (1865).
Mallotus leucocalyx Müll.Arg. in A.P.de Candolle, Prodr. 15(2): 970 (1866).
Mallotus bracteatus Hook.f., Fl. Brit. India 5: 436 (1887).

Mallotus distans Müll.Arg., Linnaea 34: 194 (1865).
S. India. 40 IND. Nanophan.

Mallotus dunnii F.P.Metcalf, J. Arnold Arbor. 22: 205 (1941).
China (Fujian). 36 CHS.
Mallotus roxburghianus var. *glabra* Dunn, J. Linn. Soc., Bot. 38: 365 (1908).

Mallotus eberhardtii Gagnep., Notul. Syst. (Paris) 4: 52 (1923).
Vietnam. 41 VIE.

Mallotus echinatus Elmer, Leafl. Philipp. Bot. 3: 925 (1910).
Philippines, Maluku (Obi), New Guinea (incl. Admiralty Is.), Bismarck Archip. 42 BIS MOL NWG PHI. Phan.
Mallotus pseudopenangensis Pax & K.Hoffm. in H.G.A.Engler, Pflanzenr., IV, 147, VII: 203 (1914).
Mallotus papuanus var. *glabrescens* Pax & K.Hoffm. in H.G.A.Engler, Pflanzenr., IV, 147, XIV: 19 (1919).
Mallotus papuanus var. *intermedius* Pax & K.Hoffm. in H.G.A.Engler, Pflanzenr., IV, 147, XIV: 19 (1919).

Mallotus eriocarpus (Thwaites) Müll.Arg., Linnaea 34: 185 (1865).
Sri Lanka, Pen. Malaysia. 40 SRL 42 MLY. Phan.
* *Rottlera eriocarpa* Thwaites, Enum. Pl. Zeyl.: 273 (1861).

Mallotus esquirolii H.Lév., Repert. Spec. Nov. Regni Veg. 9: 327 (1911).
China (Yunnan, Guizhou). 36 CHC. Phan.

Mallotus eucaustus Airy Shaw, Kew Bull. 23: 80 (1969).
Borneo (Sarawak, Sabah, E. Kalimantan). 42 BOR. Phan.

Mallotus eximius Airy Shaw, Kew Bull. 32: 400 (1978).
Pen. Malaysia. 42 MLY.

Mallotus floribundus (Blume) Müll.Arg., Flora 47: 469 (1864).
Trop. Asia, SW. Pacific. 41 BMA NCB THA VIE 42 BIS BOR JAW MOL NWG PHI SUL SUM 60 SOL. Nanophan. or phan.
* *Adisca floribunda* Blume, Bijdr.: 610 (1826). *Mallotus floribundus* var. *genuinus* Pax & K.Hoffm. in H.G.A.Engler, Pflanzenr., IV, 147, VII: 174 (1914), nom. inval.

var. **cordifolius** Chakrab., J. Econ. Taxon. Bot. 6: 496 (1985).
Nicobar Is. 41 NCB.

var. **floribundus**
Trop. Asia, SW. Pacific. 41 BMA THA VIE 42 BIS BOR JAW MOL NWG PHI SUL SUM 60 SOL. Nanophan. or phan.
Mallotus amentiformis Müll.Arg., Flora 47: 468 (1864).
Ricinus floribundus Reinw. ex Müll.Arg. in A.P.de Candolle, Prodr. 15(2): 962 (1866).
Mallotus anamiticus Kuntze, Revis. Gen. Pl. 2: 608 (1891).
Mallotus floribundus var. *pilosus* Pax & K.Hoffm. in H.G.A.Engler, Pflanzenr., IV, 147, VII: 174 (1914).

Mallotus fuscescens (Thwaites) Müll.Arg., Linnaea 34: 195 (1865).
Sri Lanka. 40 SRL. Phan.
* *Rottlera fuscescens* Thwaites, Enum. Pl. Zeyl.: 273 (1861).

Mallotus garrettii Airy Shaw, Kew Bull. 21: 387 (1968).
N. Thailand, Laos. 41 LAO THA. Phan.

Mallotus glabriusculus (Kurz) Pax & K.Hoffm. in H.G.A.Engler, Pflanzenr., IV, 147, VII: 162 (1914).
Burma, Cambodia, Vietnam. 41 BMA CBD VIE.
* *Coelodiscus glabriusculus* Kurz, Forest Fl. Burma 2: 393 (1877).
Coelodiscus coudercii Gagnep., Notul. Syst. (Paris) 4: 49 (1923). *Mallotus coudercii* (Gagnep.) Airy Shaw, Kew Bull. 20: 42 (1966).

Mallotus grandistipularis Slik, Blumea 43: 227 (1998).
C. Sumatera. 42 SUM. Phan.

Mallotus griffithianus (Müll.Arg.) Hook.f., Fl. Brit. India 5: 433 (1887).
Pen. Malaysia, Borneo (Sarawak, Brunei, Sabah, E. Kalimantan). 42 BOR MLY. Phan.
* *Diplochlamys griffithianus* Müll.Arg., Flora 47: 539 (1864).
Mallotus impar Pax & K.Hoffm. in H.G.A.Engler, Pflanzenr., IV, 147, VII: 396 (1914).
Mallotus woodii Merr., Philipp. J. Sci., C 13: 81 (1918).

Mallotus grossedentatus Merr. & Chun, Sunyatsenia 5: 98 (1940).
Hainan. 36 CHH.

Mallotus hainanensis S.M.Hwang, Acta Phytotax. Sin. 23: 293 (1985).
Hainan. 36 CHH.

Mallotus hanheoensis Thin, Euphorb. Vietnam: 46 (1995).
Vietnam. 41 VIE. Nanophan.

Mallotus havilandii Airy Shaw, Kew Bull. 20: 39 (1966).
Borneo (SW. Sarawak). 42 BOR. Phan.

Mallotus hirsutus Elmer, Leafl. Philipp. Bot. 7: 2648 (1915).
Philippines (Mindanao). 42 PHI.

Mallotus hookerianus (Seem.) Müll.Arg., Linnaea 34: 193 (1865).
China (Hong Kong). 36 CHS. Nanophan.
* *Hancea hookeriana* Seem., Bot. Voy. Herald: 409 (1857).

Mallotus hymenophyllus Airy Shaw, Kew Bull. 21: 381 (1968).
S. Thailand. 41 THA. Nanophan.

Mallotus illudens Croizat, J. Arnold Arbor. 19: 146 (1938).
S. China. 36 CHC CHS. Phan.

Mallotus intercedens Pax & K.Hoffm. in H.G.A.Engler, Pflanzenr., IV, 147, VII: 179 (1914).
India. 40 IND. Nanophan. or phan.

Mallotus japonicus (L.f.) Müll.Arg., Linnaea 34: 189 (1865).
S. Japan, Nansei-shoto, Taiwan, S. China. 36 CHC CHS 38 JAP NNS TAI. Nanophan. or phan.
* *Croton japonicus* L.f., Suppl. Pl.: 422 (1782).
Croton acuminatus Lam., Encycl. 2: 207 (1786).
Mallotus oreophilus var. *ochraceoalbidus* Müll.Arg., Linnaea 34: 188 (1865). *Mallotus nepalensis* var. *ochraceoalbidus* (Müll.Arg.) Pax in H.G.A.Engler, Pflanzenr., IV, 147, VII: 166 (1914). *Mallotus japonicus* var. *ochraceoalbidus* (Müll.Arg.) S.M.Hwang, Acta Phytotax. Sin. 23: 298 (1985).

Mallotus khasianus Hook.f., Fl. Brit. India 5: 438 (1887).
Assam, Burma, Thailand. 40 ASS 41 BMA THA. Phan.
Mallotus filiformis Hook.f., Fl. Brit. India 5: 435 (1887).
Mallotus polyneurus Hook.f., Fl. Brit. India 5: 439 (1887).

Mallotus kingii Hook.f., Fl. Brit. India 5: 439 (1887).
S. Thailand, Pen. Malaysia. 41 THA 42 MLY. Nanophan. or phan.

Mallotus korthalsii Müll.Arg. in A.P.de Candolle, Prodr. 15(2): 976 (1866).
Pen. Malaysia, Borneo (Sarawak, Sabah, E. Kalimantan), Philippines, Sulawesi. 42 BOR MLY PHI SUL. Nanophan. or phan.
Mallotus smilaciformis Gage, Rec. Bot. Surv. India 9: 242 (1922).

Mallotus kweichowensis Lauener & W.T.Wang, Notes Roy. Bot. Gard. Edinburgh 38: 487 (1980).
China (Guizhou). 36 CHC.
* *Phytolacca esquirolii* H.Lév., Fl. Kouy-Tchéou: 313 (1915).

Mallotus lackeyi Elmer, Leafl. Philipp. Bot. 4: 1298 (1911).
N. Pen. Malaysia, Borneo (NE. Sarawak, Sabah, E. Kalimantan), S. Philippines. 42 BOR MLY PHI. Nanophan. or phan.
Mallotus sanchezii Merr., Philipp. J. Sci., C 7: 402 (1912 publ. 1913).
Macaranga leightonii Whitmore, Kew Bull. 39: 607 (1984).

Mallotus laevigatus (Müll.Arg.) Airy Shaw, Kew Bull. 19: 328 (1965).
Sumatera (Enggano), Borneo (Sarawak, Kalimantan), Jawa. 42 BOR JAW SUM. Phan.
* *Mallotus moritzianus* var. *laevigatus* Müll.Arg., Linnaea 34: 191 (1865), nom. inval.

Mallotus lanceolatus (Gagnep.) Airy Shaw, Kew Bull. 21: 380 (1968).
Thailand, Vietnam. 41 THA VIE. Nanophan. or phan.
* *Coelodiscus lanceolatus* Gagnep., Notul. Syst. (Paris) 4: 50 (1923).

Mallotus lappaceus Müll.Arg. in A.P.de Candolle, Prodr. 15(2): 957 (1866).
Burma. 41 BMA. Nanophan.
Croton laetus Wall., Numer. List: 7738 (1847), nom. nud.

Mallotus lauterbachianus (Pax & K.Hoffm.) Pax & K.Hoffm. in H.G.A.Engler, Pflanzenr., IV, 147, VII: 157 (1914).
New Guinea. 42 NWG. Nanophan. or phan.
* *Coelodiscus lauterbachianus* Pax & K.Hoffm., Repert. Spec. Nov. Regni Veg. 8: 481 (1910).
Mallotus macularis Airy Shaw, Kew Bull. 25: 526 (1971).

Mallotus leptostachyus Hook.f., Fl. Brit. India 5: 435 (1887).
S. Burma, S. Thailand. 41 BMA THA. Nanophan. or phan.

Mallotus leucocarpus (Kurz) Airy Shaw, Kew Bull. 16: 352 (1963).
Burma. 41 BMA.
* *Claoxylon leucocarpum* Kurz, J. Asiat. Soc. Bengal, Pt. 2, Nat. Hist. 42(2): 244 (1873).

Mallotus leucodermis Hook.f., Fl. Brit. India 5: 441 (1887).
Pen. Malaysia, Sumatera (Simeuluë), Borneo (C. Sarawak, Sabah), Solomon Is. 42 BOR MLY SUM 60 SOL. Phan.
Coccoceras muticum var. *pedicellatum* Hook.f., Fl. Brit. India 5: 424 (1887).
Mallotus leucodermis var. *puberulus* Airy Shaw, Kew Bull. 21: 398 (1968).

Mallotus leveillanus Fedde, Repert. Spec. Nov. Regni Veg. 10: 144 (1911).
SC. China. 36 CHC.
* *Mallotus esquirolii* H.Lév., Repert. Spec. Nov. Regni Veg. 9: 461 (1911), nom. illeg.

Mallotus lianus Croizat, J. Arnold Arbor. 19: 140 (1938).
SE. China (Guangdong, Fujian, Zhejiang). 36 CHS. Phan.

Mallotus longipes Müll.Arg., Linnaea 34: 193 (1865).
Burma. 41 BMA. Nanophan.
Coelodiscus hirsutulus Kurz, J. Asiat. Soc. Bengal, Pt. 2, Nat. Hist. 42(2): 243 (1873).
Mallotus hirsutulus (Kurz) Pax & K.Hoffm. in H.G.A.Engler, Pflanzenr., IV, 147, VII: 161 (1914).

Mallotus longistylus Merr., Philipp. J. Sci. 16: 560 (1920).
Philippines. 42 PHI.

Mallotus lotingensis F.P.Metcalf, J. Arnold Arbor. 22: 206 (1941).
China (Guangdong). 36 CHS.
* *Mallotus barbatus* var. *congestus* F.P.Metcalf, Lingnan Sci. J. 10: 487 (1931).

Mallotus luchenensis F.P.Metcalf, J. Arnold Arbor. 22: 206 (1941).
S. China, Vietnam. 36 CHS 41 VIE.

Mallotus lullulae Airy Shaw, Kew Bull. 21: 384 (1968).
New Guinea (Woodlark I.). 42 NWG. Nanophan. or phan.

Mallotus macrostachyus (Miq.) Müll.Arg. in A.P.de Candolle, Prodr. 15(2): 963 (1866).
S. Thailand, Pen. Malaysia, Sumatera (Bangka), Borneo. 41 THA 42 BOR MLY SUM. Nanophan. or phan.
* *Rottlera macrostachya* Miq., Fl. Ned. Ind., Eerste Bijv.: 454 (1861).
Mallotus insignis Müll.Arg., Linnaea 34: 193 (1865).

Mallotus metcalfianus Croizat, J. Arnold Arbor. 21: 501 (1940).
Vietnam, China (Guangxi). 36 CHS 41 VIE.

Mallotus micranthus Müll.Arg., Linnaea 34: 191 (1865).
Sri Lanka. 4O SRL. Nanophan. or phan.

Mallotus microcarpus Pax & K.Hoffm. in H.G.A.Engler, Pflanzenr., IV, 147, VII: 172 (1914).
S. China. 36 CHS. Nanophan.

Mallotus millietii H.Lév., Fl. Kouy-Tchéou: 165 (1914).
China (Guizhou). 36 CHC.

Mallotus miquelianus (Scheff.) Boerl., Handl. Fl. Ned. Ind. 3(1): 290 (1900).
S. Thailand to Maluku. 41 THA 42 BOR LSI MOL PHI SUM. Nanophan. or phan.
* *Rottlera miqueliana* Scheff., Ann. Mus. Bot. Lugduno-Batavi 4: 124 (1869).

var. **insularum** Airy Shaw, Kew Bull. 20: 40 (1966).
Maluku (Sula Is.), Lesser Sunda Is. (Tanimbar Is.). 42 LSI MOL.

var. **miquelianus**
S. Thailand, Sumatera, Borneo (Sarawak, Sabah, E. & SE. Kalimantan), Philippines. 41 THA 42 BOR PHI SUM. Nanophan. or phan.
Mallotus anisophyllus Hook.f., Fl. Brit. India 5: 436 (1887).

Mallotus mollissimus (Geiseler) Airy Shaw, Kew Bull. 26: 297 (1972).
Trop. Asia, NE. Queensland, SW. Pacific. 41 BMA THA VIE 42 BIS BOR JAW LSI MOL NWG PHI SUL SUM 50 QLD 60 SOL. Nanophan. or phan.
* *Croton mollissimus* Geiseler, Croton. Monogr.: 73 (1807). *Chrozophora mollissima* (Geiseler) A.Juss., Euphorb. Gen.: 28 (1824).
Croton ricinoides Pers., Syn. Pl. 2: 586 (1807). *Mallotus ricinoides* (Pers.) Müll.Arg., Linnaea 34: 187 (1865).
Adelia bernardia Blanco, Fl. Filip.: 814 (1837).
Adelia barbata Blanco, Fl. Filip., ed. 2: 561 (1845).
Mallotus pycnostachys F.Muell., Fragm. 4: 138 (1864).
Mallotus zippellii F.Muell., Fragm. 4: 139 (1864).

Mallotus monanthos Airy Shaw, Kew Bull. 32: 402 (1978).
Pen. Malaysia (Pahang). 42 MLY.

Mallotus montanus (Müll.Arg.) Airy Shaw, Kew Bull. 32: 78 (1977).
S. Thailand, Pen. Malaysia. 41 THA 42 MLY.
* *Coelodiscus montanus* Müll.Arg. in A.P.de Candolle, Prodr. 15(2): 759 (1866).

Mallotus muticus (Müll.Arg.) Airy Shaw, Kew Bull. 16: 351 (1963).
Pen. Malaysia, Borneo. 42 BOR MLY. Phan.
* *Coccoceras muticum* Müll.Arg., Flora 47: 470 (1864).
Mallotus borneensis Müll.Arg. in A.P.de Candolle, Prodr. 15(2): 980 (1866).
Coccoceras borneense J.J.Sm., Bull. Jard. Bot. Buitenzorg, III, 6: 94 (1924).

Mallotus myanmaricus P.T.Li, Guihaia 14: 131(1994).
Burma. 41 BMA.
* *Coelodiscus longipes* Kurz, J. Asiat. Soc. Bengal, Pt. 2, Nat. Hist. 42(2): 244 (1873). *Mallotus longipes* (Kurz) Pax & K.Hoffm. in H.G.A.Engler, Pflanzenr., IV, 147, VII: 162 (1914), nom. illeg.

Mallotus nanus Airy Shaw, Kew Bull. 21: 380 (1968).
Laos. 41 LAO.
* *Coelodiscus thorelii* Gagnep., Notul. Syst. (Paris) 4: 51 (1923).

Mallotus neocavaleriei H.Lév., Fl. Kouy-Tchéou: 165 (1914).
China (Guizhou). 36 CHC.

Mallotus nepalensis Müll.Arg., Linnaea 34: 188 (1865).
C. Himalaya to SC. China. 36 CHC 40 ASS EHM NEP 41 BMA. Nanophan. or phan.
Mallotus oreophilus var. *floccosus* Müll.Arg., Linnaea 34: 188 (1865). *Mallotus nepalensis* var. *floccosus* (Müll.Arg.) Pax in H.G.A.Engler, Pflanzenr., IV, 147, VII: 166 (1914). *Mallotus tenuifolius* var. *floccosus* (Müll.Arg.) Croizat, J. Arnold Arbor. 19: 138 (1938). *Mallotus japonicus* var. *floccosus* (Müll.Arg.) S.M.Hwang, Acta Phytotax. Sin. 23: 299 (1985).

Mallotus nesophilus Müll.Arg., Linnaea 34: 196 (1865).
N. Australia. 50 NTA QLD WAU. Nanophan. or phan.

Mallotus oblongifolius (Miq.) Müll.Arg., Linnaea 34: 192 (1865).
Trop. Asia, NE. Queensland. 41 AND BMA NCB THA VIE 42 BIS BOR JAW MOL NWG PHI SUL SUM 50 QLD. Nanophan. or phan.
* *Rottlera oblongifolia* Miq., Fl. Ned. Ind. 1(2): 396 (1859).

var. **oblongifolius**
Trop. Asia, NE. Queensland. 41 AND BMA NCB THA VIE 42 BIS BOR JAW MOL NWG PHI SUL SUM 50 QLD. Nanophan. or phan.
Mallotus furetianus Müll.Arg., Linnaea 34: 190 (1865).
Mallotus helferi Müll.Arg., Linnaea 34: 190 (1865). *Mallotus oblongifolius* var. *helferi* (Müll.Arg.) Pax & K.Hoffm. in H.G.A.Engler, Pflanzenr., IV, 147, VII: 194 (1914).
Mallotus lambertianus Müll.Arg., Linnaea 34: 190 (1865).
Mallotus porterianus Müll.Arg., Linnaea 34: 115 (1865).
Mallotus stylaris Müll.Arg. in A.P.de Candolle, Prodr. 15(2): 973 (1866).
Mallotus puberulus Hook.f., Fl. Brit. India 5: 435 (1887).
Mallotus columnaris Warb., Bot. Jahrb. Syst. 13: 349 (1891).
Mallotus odoratus Elmer, Leafl. Philipp. Bot. 4: 1299 (1911).
Mallotus alternifolius Merr., Philipp. J. Sci., C 7: 395 (1912 publ. 1913).
Mallotus camiguinensis Merr., Philipp. J. Sci., C 7: 397 (1912 publ. 1913).
Mallotus kietanus Rech., Denkschr. Kaiserl. Akad. Wiss., Wien. Math.-Naturwiss. Kl. 89: 568 (1913 publ. 1914).
Mallotus oblongifolius var. *siamensis* Pax & K.Hoffm. in H.G.A.Engler, Pflanzenr., IV, 147, VII: 194 (1914).
Mallotus batjanensis Pax & K.Hoffm. in H.G.A.Engler, Pflanzenr., IV, 147, XIV: 17 (1919).
Mallotus warburgianus Pax & K.Hoffm. in H.G.A.Engler, Pflanzenr., IV, 147, XIV: 18 (1919).
Mallotus maclurei Merr., Philipp. J. Sci. 21: 347 (1922).
Mallotus oblongifolius var. *rubriflorus* Chakrab., J. Econ. Taxon. Bot. 6: 496 (1985).

var. **villosulus** Pax & K.Hoffm. in H.G.A.Engler, Pflanzenr., IV, 147, VII: 194 (1914).
New Guinea, Bismarck Archip., NE. Queensland ? 42 BIS NWG 50 QLD? Nanophan. or phan.
Mallotus tenuispicus Pax & K.Hoffm. in H.G.A.Engler, Pflanzenr., IV, 147, VII: 201 (1914).
Mallotus peekelii Pax & K.Hoffm., Notizbl. Bot. Gart. Berlin-Dahlem 10: 384 (1928).

Mallotus oppositifolius (Geiseler) Müll.Arg., Linnaea 34: 194 (1865).
Trop. Africa, Madagascar. 22 BEN GAM GHA GUI IVO LBR NGA SEN SIE TOG 23 CAB CMN CON GAB ZAI 24 ETH SUD 25 KEN TAN 26 ANG MOZ ZAM ZIM 29 MDG. Nanophan. or phan.
* *Croton oppositifolius* Geiseler, Croton. Monogr.: 23 (1807). *Mallotus oppositifolius* var. *genuinus* Müll.Arg., Linnaea 34: 194 (1865), nom. inval.

f. **glabratus** (Müll.Arg.) Pax in H.G.A.Engler, Pflanzenr., IV, 147, VII: 160 (1914).
Trop. Africa, Madagascar. 22 BEN GAM GHA GUI IVO LBR NGA SEN SIE TOG 23 CAB CMN CON GAB ZAI 24 SUD 25 KEN TAN UGA 26 ANG MOZ ZAM ZIM 29 MDG. Nanophan. or phan.
* *Mallotus oppositifolius* var. *glabratus* Müll.Arg., Linnaea 24: 194 (1865).
Mallotus oppositifolius var. *integrifolius* Müll.Arg., Linnaea 34: 195 (1865).

var. **lindicus** (Radcl.-Sm.) Radcl.-Sm., Kew Bull. 39: 790 (1984).
S. Tanzania, Mozambique. 25 TAN 26 MOZ. Nanophan. or phan.
* *Mallotus oppositifolius* f. *lindicus* Radcl.-Sm., Kew Bull. 29: 437 (1974).

var. **oppositifolius**
Trop. Africa. 22 IVO NGA TOG 23 CMN CON ZAI 24 ETH 25 KEN TAN. Nanophan. or phan.
Acalypha dentata Schumach. & Thonn. in C.F.Schumacher, Beskr. Guin. Pl.: 410 (1827). *Ricinocarpus dentatus* (Schumach. & Thonn.) Kuntze, Revis. Gen. Pl. 2: 617 (1891). *Mallotus oppositifolius* f. *dentatus* (Schumach. & Thonn.) Pax & K.Hoffm. in H.G.A.Engler, Pflanzenr., IV, 147, VII: 159 (1914).
Claoxylon cordifolium Benth. in W.J.Hooker, Niger Fl.: 506 (1849).
Mallotus oppositifolius var. *pubescens* Pax, Bot. Jahrb. Syst. 23: 525 (1897). *Mallotus oppositifolius* f. *pubescens* (Pax) Pax in H.G.A.Engler, Pflanzenr., IV, 147, VII: 160 (1914).
Mallotus chevalieri Beille, Bull. Soc. Bot. France 55(8): 76 (1908).
Mallotus beillei A.Chev. ex Hutch. & Dalziel, Fl. W. Trop. Afr. 1: 307 (1928), nom. illeg.

f. **polycytotrichus** Radcl.-Sm., Kew Bull. 40: 658 (1985).
Zimbabwe. 26 ZIM. Nanophan. or phan.

Mallotus oreophilus Müll.Arg., Linnaea 34: 188 (1865).
Sikkim, SC. China. 36 CHC 40 EHM.

subsp. **latifolius** Boufford & T.S.Ying, J. Arnold Arbor. 71: 575 (1990).
China (Sichuan). 36 CHC.

subsp. **oreophilus**
Sikkim, SC. China. 36 CHC 40 EHM.

Mallotus pachypodus Pax & K.Hoffm. in H.G.A.Engler, Pflanzenr., IV, 147, VII: 196 (1914).
Burma. 41 BMA. Nanophan.

Mallotus palauensis Hosok., Trans. Nat. Hist. Soc. Taiwan 25: 25 (1935).
Caroline Is. (Palau). 62 CRL.

Mallotus pallidus (Airy Shaw) Airy Shaw, Kew Bull. 32: 78 (1977).
S. Thailand. 41 THA.
* *Mallotus philippensis* var. *pallidus* Airy Shaw, Kew Bull. 26: 300 (1972).

Mallotus paniculatus (Lam.) Müll.Arg., Linnaea 34: 189 (1865).
Trop. & Subtrop. Asia, N. Queensland. 36 CHS 38 JAP TAI 41 BMA THA VIE 42 BOR JAW MOL NWG PHI SUL SUM 50 QLD. Nanophan. or phan.
* *Croton paniculatus* Lam., Encycl. 2: 207 (1786).
Mallotus chinensis Lour., Fl. Cochinch.: 635 (1790).
Mallotus cochinchinensis Lour., Fl. Cochinch.: 635 (1790).

Trewia tricuspidata Willd., Sp. Pl. 4: 835 (1806).
Ricinus chinensis Thunb., Ricin.: 5 (1815).
Rottlera alba Roxb., Fl. Ind. ed. 1832, 3: 829 (1832). *Mallotus albus* (Roxb.) Müll.Arg., Linnaea 34: 188 (1865).
Croton appendiculatus Elmer, Leafl. Philipp. Bot. 1: 312 (1908).
Mallotus formosanus Hayata, J. Coll. Sci. Imp. Univ. Tokyo 30: 269 (1911).

Mallotus papuanus (J.J.Sm.) Pax & K.Hoffm. in H.G.A.Engler, Pflanzenr., IV, 147, VII: 202 (1914).
Irian Jaya. 42 NWG. Phan.
* *Mallotus hookerianus* var. *papuanus* J.J.Sm., Nova Guinea 8: 2648 (1915).

Mallotus peltatus (Geiseler) Müll.Arg., Linnaea 34: 187 (1865).
Andaman Is., Burma, Thailand, Pen. Malaysia, Vietnam, Sumatera (incl. Bangka), Jawa, Lesser Sunda Is. 41 AND BMA THA VIE 42 JAW LSI MLY SUM. Nanophan. or phan.
* *Aleurites peltata* Geiseler, Croton. Monogr.: 81 (1807).
Adisca acuminata Blume, Bijdr.: 610 (1826). *Mallotus acuminatus* (Blume) Müll.Arg., Linnaea 34: 117 (1865).
Rottlera longifolia Rchb.f. & Zoll., Acta Soc. Regiae Sci. Indo-Neerl. 1: 31 (1856). *Mallotus longifolius* (Rchb.f. & Zoll.) Müll.Arg. in A.P.de Candolle, Prodr. 15(2): 967 (1866).
Mallotus peltatus var. *rubriflorus* Chakrab., J. Econ. Taxon. Bot. 6: 497 (1985).

Mallotus penangensis Müll.Arg., Linnaea 34: 186 (1865).
Pen. Malaysia, Sumatera, Borneo (Sarawak, Brunei, Sabah, E. Kalimantan), Maluku, Philippines ? 42 BOR MLY MOL PHI? SUM. Nanophan. or phan.
Mallotus leptophyllus Pax & K.Hoffm. in H.G.A.Engler, Pflanzenr., IV, 147, VII: 203 (1914).
Mallotus sarawakensis Pax & K.Hoffm. in H.G.A.Engler, Pflanzenr., IV, 147, VII: 201 (1914).
Mallotus xylacanthus Pax & K.Hoffm. in H.G.A.Engler, Pflanzenr., IV, 147, VII: 203, 397 (1914).
Mallotus penangensis var. *grandifolia* Ridl. in ?, : 293 (1923).

Mallotus petanodon Airy Shaw, Kew Bull. 32: 401 (1978).
Borneo (Kalimantan). 42 BOR.

Mallotus philippensis (Lam.) Müll.Arg., Linnaea 34: 196 (1865).
Trop. & Subtrop. Asia, N. & E. Australia, Pacific. 36 CHS 38 TAI 40 ASS IND NEP PAK SRL 41 THA VIE 42 BIS BOR JAW LSI MLY MOL NWG PHI SUL SUM 50 NSW NTA QLD 60. Nanophan. or phan.
* *Croton philippensis* Lam., Encycl. 2: 206 (1786).
Croton punctatus Retz., Observ. Bot. 5: 30 (1789), nom. illeg.
Croton laurifolius Noronha, Verh. Batav. Genootsch. Kunsten 5(4): 13 (1790), nom. nud.
Croton coccineus Vahl, Symb. Bot. 2: 97 (1791), nom. illeg.
Croton montanus Willd., Sp. Pl. 4: 545 (1805).
Croton distans Wall., Numer. List: 7772A (1847), nom. nud.
Mappa stricta Rchb.f. & Zoll., Acta Soc. Regiae Sci. Indo-Neerl. 1: 31 (1856). *Macaranga stricta* (Rchb.f. & Zoll.) Müll.Arg. in A.P.de Candolle, Prodr. 15(2): 1004 (1866).
Mallotus reticulatus Dunn, J. Linn. Soc., Bot. 38: 365 (1908). *Mallotus philippensis* var. *reticulatus* (Dunn) F.P.Metcalf, J. Arnold Arbor. 22: 207 (1941).
Mallotus philippensis var. *mengliangensis* C.Y.Wu ex S.M.Hwang, Acta Phytotax. Sin. 23: 294 (1985).

Mallotus pierrei (Gagnep.) Airy Shaw, Kew Bull. 21: 380 (1968).
Thailand, Vietnam. 41 THA VIE. Nanophan. or phan.
* *Coelodiscus pierrei* Gagnep., Notul. Syst. (Paris) 4: 51 (1923).

Mallotus plicatus (Müll.Arg.) Airy Shaw, Kew Bull. 16: 352 (1963).
S. Burma, Pen. Malaysia. 41 BMA 42 MLY.
* *Coccoceras plicatum* Müll.Arg., Flora 47: 539 (1864). *Hymenocardia plicata* (Müll.Arg.) Kurz, Forest Fl. Burma 2: 395 (1877).
Mallotus eriocarpoides Müll.Arg., Linnaea 34: 185 (1865).
Mallotus wallichianus Müll.Arg., Linnaea 34: 196 (1865).

Mallotus poilanei Gagnep., Bull. Soc. Bot. France 72: 466 (1925).
Vietnam. 41 VIE.

Mallotus polyadenos F.Muell., Fragm. 6: 184 (1868).
New Guinea, Queensland (Cape York Pen.). 42 NWG 50 QLD. Nanophan. or phan.
Trewia glabrata Banks & Sol. ex Pax & K.Hoffm. in H.G.A.Engler, Pflanzenr., IV, 147, VII: 198 (1914).

Mallotus ponapensis Hosok., Trans. Nat. Hist. Soc. Taiwan 25: 26 (1935).
Caroline Is. (Ponape). 62 CRL.

Mallotus repandus (Willd.) Müll.Arg., Linnaea 34: 197 (1865).
Taiwan, Trop. Asia, NE. Queensland, New Caledonia. 38 TAI 40 IND SRL 41 THA 42 JAW LSI MLY NWG PHI SUL SUM 50 QLD 60 NWC. (Cl.) nanophan. or phan.
* *Croton repandus* Rottler, Ges. Naturf. Freunde Berlin Neue Schriften 4: 206 (1803).
Croton rhombifolius Willd., Sp. Pl. 4: 555 (1805).
Rottlera scabrifolia A.Juss., Euphorb. Gen.: 111 (1824). *Mallotus scabrifolius* (A.Juss.) Müll.Arg. in A.P.de Candolle, Prodr. 15(2): 699 (1866). *Mallotus repandus* var. *scabrifolius* (A.Juss.) Müll.Arg. in A.P.de Candolle, Prodr. 15(2): 982 (1866).
Rottlera scandens Span., Linnaea 15: 348 (1841). *Mallotus scandens* (Span.) Müll.Arg. in A.P.de Candolle, Prodr. 15(2): 982 (1866).
Croton bacciferus Wall., Numer. List: 7826 (1847), nom. nud.
Croton volubilis Llanos, Mem. Real. Acad. Ci. Exact. Madrid 4: 503 (1856).

Mallotus resinosus (Blanco) Merr., Sp. Blancoan.: 222 (1918).
Trop. Asia, Queensland (Cook). 40 IND SRL 41 AND NCB THA VIE 42 BOR JAW LSI MLY NWG PHI SUL 50 QLD. Nanophan. or phan.
* *Adelia resinosa* Blanco, Fl. Filip., ed. 2: 562 (1845).
Claoxylon muricatum Wight, Icon. Pl. Ind. Orient. 5: t. 1886 (1852). *Mallotus muricatus* (Wight) Müll.Arg., Linnaea 34: 191 (1865). *Mallotus muricatus* var. *genuinus* Pax & K.Hoffm. in H.G.A.Engler, Pflanzenr., IV, 147, VII: 191 (1914), nom. inval. *Mallotus resinosus* var. *muricatus* (Wight) N.P.Balakr. & Chakrab., Rheedea 1: 39 (1991).
Axenfeldia intermedia Baill., Étude Euphorb.: 419 (1858). *Mallotus intermedius* (Baill.) N.P.Balakr., Bull. Bot. Surv. India 10: 245 (1969).
Mallotus dispar var. *psiloneurus* Müll.Arg., Linnaea 34: 191 (1865).
Mallotus andamanicus Hook.f., Fl. Brit. India 5: 439 (1887).
Mallotus walkerae Hook.f., Fl. Brit. India 5: 437 (1887). *Mallotus muricatus* var. *walkerae* (Hook.f.) Pax & K.Hoffm. in H.G.A.Engler, Pflanzenr., IV, 147, VII: 190 (1914).
Mallotus sanguirensis Pax & K.Hoffm. in H.G.A.Engler, Pflanzenr., IV, 147, XIV: 18 (1919).
Mallotus subramanyamii J.L.Ellis, Bull. Bot. Surv. India 25: 199 (1983 publ. 1985). *Mallotus resinosus* var. *subramanyamii* (J.L.Ellis) Chakrab., J. Econ. Taxon. Bot. 6: 704 (1985).

Mallotus rhamnifolius (Willd.) Müll.Arg., Linnaea 34: 196 (1865).
S. India, Sri Lanka. 40 IND SRL. Nanophan. or phan.
Croton nervosus Rottler, Ges. Naturf. Freunde Berlin Neue Schriften 4: 190 (1803).
* *Croton rhamnifolius* Willd., Ges. Naturf. Freunde Berlin Neue Schriften 4: 190 (1803).
Croton reticulatus Willd., Sp. Pl. 4: 545 (1805).
Mallotus zeylanicus Müll.Arg., Linnaea 34: 195 (1865).
Mallotus rhamnifolius var. *ovatifolius* Hook.f., Fl. Brit. India 5: 440 (1887).

Mallotus roxburghianus Müll.Arg., Linnaea 34: 186 (1865).
Sikkim, Assam. 40 ASS EHM. Nanophan. or phan.

Mallotus rufidulus (Miq.) Müll.Arg. in A.P.de Candolle, Prodr. 15(2): 970 (1866).
Pen. Malaysia, Sumatera, Jawa, Lesser Sunda Is., Borneo (Sabah, Sarawak, E. Kalimantan), Philippines. 42 BOR JAW LSI MLY PHI SUL SUM. Nanophan. or phan.
* *Rottlera rufidula* Miq., Fl. Ned. Ind., Eerste Bijv.: 453 (1861).
Mallotus moritzianus var. *scaber* Müll.Arg., Linnaea 34: 190 (1865), nom. inval.
Mallotus zollingeri Müll.Arg., Linnaea 34: 193 (1865).
Mallotus moritzianus Müll.Arg. in A.P.de Candolle, Prodr. 15(2): 971 (1866).

Mallotus sathavensis Thin, Euphorb. Vietnam: 46 (1995).
Vietnam. 41 VIE. Nanophan.

Mallotus sphaerocarpus (Miq.) Müll.Arg. in A.P.de Candolle, Prodr. 15(2): 976 (1866).
N. Sumatera. 42 SUM. Phan.
* *Rottlera sphaerocarpa* Miq., Fl. Ned. Ind., Eerste Bijv.: 454 (1861).

Mallotus spinulosus McPherson, Bull. Mus. Natl. Hist. Nat., B, Adansonia 17: 170 (1995).
Madagascar. 29 MDG.

Mallotus spodocarpus Airy Shaw, Kew Bull. 21: 383 (1968).
Thailand, Vietnam. 41 THA VIE. Cham.

Mallotus stenanthus Müll.Arg., Linnaea 34: 191 (1865).
S. India. 40 IND. Phan.

Mallotus stewardii Merr. ex Metcalf, Lingnan Sci. J. 10: 488 (1931).
E. China. 36 CHS.

Mallotus stipularis Airy Shaw, Kew Bull. 21: 398 (1968).
S. Thailand, N. Sumatera, Borneo (Sarawak, Sabah). 41 THA 42 BOR SUM. Phan.

Mallotus subcuneatus (Gage) Airy Shaw, Kew Bull. 26: 304 (1972).
S. Thailand, Pen. Malaysia. 41 THA 42 MLY. Nanophan. or phan.
* *Coelodiscus subcuneatus* Gage, Rec. Bot. Surv. India 9: 240 (1922).

Mallotus subjaponicus (Croizat) Croizat, J. Arnold Arbor. 21: 502 (1940).
China (S. Anhui). 36 CHS. Nanophan.
* *Mallotus tenuifolius* var. *subjaponicus* Croizat, J. Arnold Arbor. 19: 138 (1938).

Mallotus subpeltatus (Blume) Müll.Arg., Linnaea 34: 189 (1865). – FIGURE, p. 1125.
S. Burma, Thailand, Pen. Malaysia, Sumatera, Jawa. 41 BMA THA 42 JAW MLY SUM. Nanophan. or phan.
* *Adisca subpeltata* Blume, Bijdr.: 610 (1826).

Mallotus subulatus Müll.Arg., Linnaea 34: 192 (1865).
W. & WC. Trop. Africa. 22 GHA IVO LBR NGA SIE 23 CMN EQG GAB GGI ZAI. Nanophan. or phan.
Mallotus buettneri Pax, Bot. Jahrb. Syst. 19: 89 (1894).

Mallotus sumatranus (Miq.) Airy Shaw, Kew Bull. 16: 351 (1963).
S. Sumatera, E. & SE. Kalimantan. 42 BOR SUM. Phan.
* *Coccoceras sumatranum* Miq., Fl. Ned. Ind., Eerste Bijv.: 181, 456 (1861).

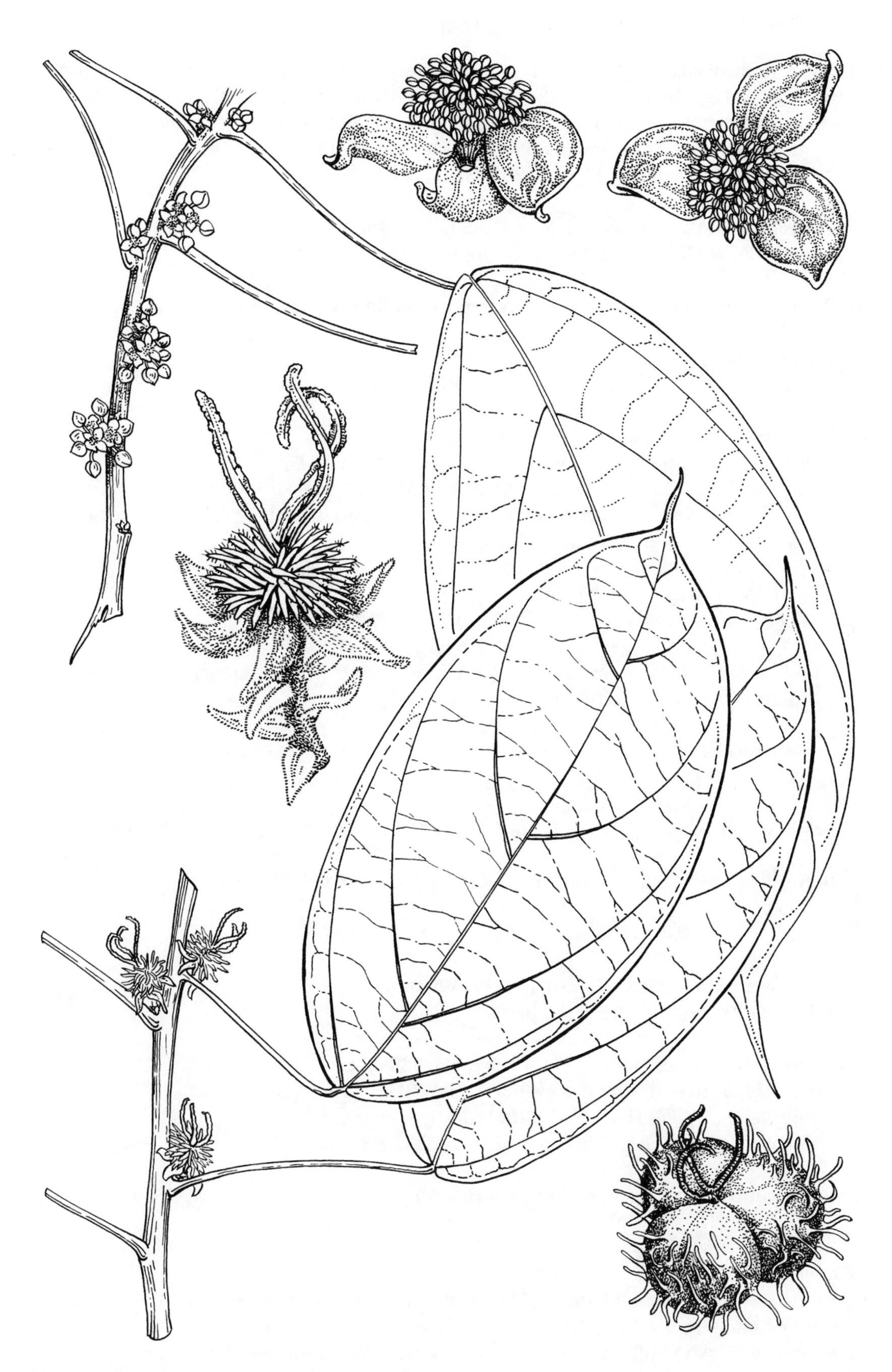

Mallotus subpeltatus (Blume) Müll. Arg.
Artist: Mary Grierson
Ic. Pl. 38(1): pl. 3715 (1974)
KEW ILLUSTRATIONS COLLECTION

Mallotus tenuipes Airy Shaw, Kew Bull. 35: 393 (1980).
Borneo, Maluku (Morotai). 42 BOR MOL. Phan.

Mallotus tetracoccus (Roxb.) Kurz, J. Asiat. Soc. Bengal, Pt. 2, Nat. Hist. 41(2): 245 (1873).
India, Sri Lanka, Bhutan, Assam, S. China (SE. Yunnan). 36 CHC 40 ASS EHM IND SRL.
Rottlera ferruginea Roxb., Fl. Ind. ed. 1832, 3: 828 (1832). *Mallotus ferrugineus* (Roxb.) Müll.Arg. in A.P.de Candolle, Prodr. 15(2): 982 (1866).
* *Rottlera tetracocca* Roxb., Fl. Ind. ed. 1832, 3: 826 (1832).

Mallotus thorelii Gagnep., Notul. Syst. (Paris) 4: 53 (1923).
Indo-China. 41 CBD LAO THA VIE. Nanophan. or phan.

Mallotus thunbergianus (Müll.Arg.) Pax & K.Hoffm. in H.G.A.Engler, Pflanzenr., IV, 147, VII: 162 (1914).
Sri Lanka. 40 SRL.
* *Coelodiscus thunbergianus* Müll.Arg. in A.P.de Candolle, Prodr. 15(2): 758 (1866).

Mallotus tiliifolius (Blume) Müll.Arg., Linnaea 34: 190 (1865).
Hainan, Taiwan, Trop. Asia, Queensland (Cook), Solomon Is., Fiji. 36 CHH 38 TAI 41 THA 42 BIS BOR JAW LSI MLY NWG MOL PHI SUL SUM 50 QLD 60 FIJ SOL. Nanophan. or phan.
Croton ovatus Noronha, Verh. Batav. Genootsch. Kunsten 5(4): 12 (1790), nom. nud.
* *Rottlera tiliifolia* Blume, Bijdr.: 607 (1826).
Adelia papillaris Blanco, Fl. Filip., ed. 2: 562 (1845). *Mallotus papillaris* (Blanco) Merr., Philipp. J. Sci., C 7: 238 (1912).
Croton aromaticus Miq., Fl. Ned. Ind. 1(2): 380 (1859), nom. illeg.
Mallotus playfairii Hemsl., J. Linn. Soc., Bot. 26: 441 (1894).
Croton enantiophyllus K.Schum. in K.M.Schumann & C.A.G.Lauterbach, Fl. Schutzgeb. Südsee, Nachtr.: 296 (1905).

Mallotus trinervius (K.Schum. & Lauterb.) Pax & K.Hoffm. in H.G.A.Engler, Pflanzenr., IV, 147, VII: 156 (1914).
New Guinea. 42 NWG. Phan.
* *Syndyophyllum trinervium* K.Schum. & Lauterb., Fl. Schutzgeb. Südsee: 405 (1900).

Mallotus ustulatus (Gagnep.) Airy Shaw, Kew Bull. 21: 381 (1968).
Cambodia. 41 CBD.
* *Coelodiscus ustulatus* Gagnep., Notul. Syst. (Paris) 4: 52 (1923).

Mallotus wenzelianus Slik, Blumea 43: 229 (1998).
Philippines (Mindanao). 42 PHI. Phan.

Mallotus wrayi King ex Hook.f., Fl. Brit. India 5: 433 (1887).
Pen. Malaysia, Sumatera, Borneo. 42 BOR MLY SUM. Nanophan. or phan.
Mallotus lanceifolius Hook.f., Fl. Brit. India 5: 434 (1887).
Mallotus caudatus Merr., Philipp. J. Sci., C 13: 83 (1918).

Mallotus yunnanensis Pax & K.Hoffm. in H.G.A.Engler, Pflanzenr., IV, 147, VII: 188 (1914).
S. China. 36 CHC. Nanophan.

Synonyms:
Mallotus acuminatus (Blume) Müll.Arg. === **Mallotus peltatus** (Geiseler) Müll.Arg.
Mallotus affinis Merr. === **Macaranga lowii** King ex Hook.f. var. **lowii**
Mallotus albus (Roxb.) Müll.Arg. === **Mallotus paniculatus** (Lam.) Müll.Arg.
Mallotus alternifolius Merr. === **Mallotus oblongifolius** (Miq.) Müll.Arg. var. **oblongifolius**
Mallotus amentiformis Müll.Arg. === **Mallotus floribundus** (Blume) Müll.Arg. var. **floribundus**
Mallotus anamiticus Kuntze === **Mallotus floribundus** (Blume) Müll.Arg. var. **floribundus**

Mallotus andamanicus Hook.f. === **Mallotus resinosus** (Blanco) Merr.
Mallotus angulatus (Miq.) Müll.Arg. === **Melanolepis multiglandulosa** (Reinw. ex Blume) Rchb. & Zoll. var. **multiglandulosa**
Mallotus angustifolius Benth. === **Rockinghamia angustifolia** (Benth.) Airy Shaw
Mallotus anisophyllus Hook.f. === **Mallotus miquelianus** (Scheff.) Boerl. var. **miquelianus**
Mallotus apelta var. *chinensis* (Geiseler) Pax & K.Hoffm. === **Mallotus apelta** (Lour.) Müll.Arg.
Mallotus apelta var. *kwangsiensis* F.P.Metcalf === **Mallotus apelta** (Lour.) Müll.Arg.
Mallotus apelta var. *tenuifolius* (Pax) Pax === **Mallotus apelta** (Lour.) Müll.Arg.
Mallotus arboreus Merr. === **Ptychopyxis arborea** (Merr.) Airy Shaw
Mallotus arboreus var. *platyphyllus* Merr. === **Ptychopyxis kingii** Ridl.
Mallotus auriculatus Merr. === **Macaranga lowii** King ex Hook.f. var. **lowii**
Mallotus baillonianus Müll.Arg. === **Deuteromallotus acuminatus** (Baill.) Pax & K.Hoffm.
Mallotus barbatus var. *congestus* F.P.Metcalf === **Mallotus lotingensis** F.P.Metcalf
Mallotus barbatus var. *hubeiensis* S.M.Hwang === **Mallotus barbatus** (Wall. ex Baill.) Müll.Arg. var. **barbatus**
Mallotus barbatus var. *pedicellaris* Croizat === **Mallotus barbatus** (Wall. ex Baill.) Müll.Arg. var. **barbatus**
Mallotus batjanensis Pax & K.Hoffm. === **Mallotus oblongifolius** (Miq.) Müll.Arg. var. **oblongifolius**
Mallotus beillei A.Chev. ex Hutch. & Dalziel === **Mallotus oppositifolius** (Geiseler) Müll.Arg. var. **oppositifolius**
Mallotus borneensis Müll.Arg. === **Mallotus muticus** (Müll.Arg.) Airy Shaw
Mallotus bracteatus Hook.f. === **Mallotus dispar** (Blume) Müll.Arg.
Mallotus brevipes Pax ex Engl. === **Acalypha neptunica** Müll.Arg. var. **neptunica**
Mallotus buettneri Pax === **Mallotus subulatus** Müll.Arg.
Mallotus calcosus (Miq.) Müll.Arg. === **Melanolepis multiglandulosa** (Reinw. ex Blume) Rchb. & Zoll. var. **multiglandulosa**
Mallotus calvus Pax & K.Hoffm. === **Spathiostemon javensis** Blume
Mallotus camiguinensis Merr. === **Mallotus oblongifolius** (Miq.) Müll.Arg. var. **oblongifolius**
Mallotus capensis (Baill.) Müll.Arg. === **Macaranga capensis** (Baill.) Sim
Mallotus capuronii (Leandri) McPherson === **Deuteromallotus capuronii** Leandri
Mallotus caput-medusae Hook.f. === **Ptychopyxis caput-medusae** (Hook.f.) Ridl.
Mallotus cardiophyllus Merr. === **Trewia nudiflora** L.
Mallotus castanopsis F.P.Metcalf === **Mallotus apelta** (Lour.) Müll.Arg.
Mallotus caudatus Merr. === **Mallotus wrayi** King ex Hook.f.
Mallotus cavaleriei H.Lév. === **Discocleidion rufescens** (Franch.) Pax & K.Hoffm.
Mallotus chevalieri Beille === **Mallotus oppositifolius** (Geiseler) Müll.Arg. var. **oppositifolius**
Mallotus chinensis Lour. === **Mallotus paniculatus** (Lam.) Müll.Arg.
Mallotus chrysanthus K.Schum. === **Ptychopyxis chrysantha** (K.Schum.) Airy Shaw
Mallotus chrysocarpus Pamp. === **Mallotus contubernalis** Hance
Mallotus claoxyloides var. *angustifolius* F.M.Bailey === **Mallotus claoxyloides** (F.Muell.) Müll.Arg.
Mallotus claoxyloides var. *cordatus* (Baill.) Airy Shaw === **Mallotus claoxyloides** (F.Muell.) Müll.Arg.
Mallotus claoxyloides var. *ficifolius* (Baill.) Benth. === **Mallotus claoxyloides** (F.Muell.) Müll.Arg.
Mallotus claoxyloides f. *grossedentata* Domin === **Mallotus claoxyloides** (F.Muell.) Müll.Arg.
Mallotus cloaxyloides var. *macrophylla* Benth. === **Mallotus claoxyloides** (F.Muell.) Müll.Arg.
Mallotus cochinchinensis Lour. === **Mallotus paniculatus** (Lam.) Müll.Arg.
Mallotus columnaris Warb. === **Mallotus oblongifolius** (Miq.) Müll.Arg. var. **oblongifolius**
Mallotus coudercii (Gagnep.) Airy Shaw === **Mallotus glabriusculus** (Kurz) Pax & K.Hoffm.
Mallotus croizatianus F.P.Metcalf === **Mallotus barbatus** var. **croizatianus** (Metcalf) S.M.Hwang
Mallotus cumingii Müll.Arg. === **Neotrewia cumingii** (Müll.Arg.) Pax & K.Hoffm.
Mallotus dallachyi F.Muell. === **Macaranga dallachyana** (Baill.) Airy Shaw
Mallotus derbyensis W.Fitzg. === **Grewia breviflora** Benth. (Tiliaceae)

Mallotus diadenus (Miq.) Müll.Arg. === **Endospermum diadenum** (Miq.) Airy Shaw
Mallotus dispar var. *psiloneurus* Müll.Arg. === **Mallotus resinosus** (Blanco) Merr.
Mallotus eglandulosus Elmer === **Spathiostemon javensis** Blume
Mallotus eriocarpoides Müll.Arg. === **Mallotus plicatus** (Müll.Arg.) Airy Shaw
Mallotus esquirolii H.Lév. === **Mallotus leveillanus** Fedde
Mallotus ferrugineus (Roxb.) Müll.Arg. === **Mallotus tetracoccus** (Roxb.) Kurz
Mallotus ficifolius (Baill.) Pax & K.Hoffm. === **Mallotus claoxyloides** (F.Muell.) Müll.Arg.
Mallotus filiformis Hook.f. === **Mallotus khasianus** Hook.f.
Mallotus floribundus var. *genuinus* Pax & K.Hoffm. === **Mallotus floribundus** (Blume) Müll.Arg.
Mallotus floribundus var. *pilosus* Pax & K.Hoffm. === **Mallotus floribundus** (Blume) Müll.Arg. var. **floribundus**
Mallotus formosanus Hayata === **Mallotus paniculatus** (Lam.) Müll.Arg.
Mallotus furetianus Müll.Arg. === **Mallotus oblongifolius** (Miq.) Müll.Arg. var. **oblongifolius**
Mallotus geloniifolius Müll.Arg. ex Pax & K.Hoffm. === **Cleidion spiciflorum** (Burm.f.) Merr. var. **spiciflorum**
Mallotus glaberrimus (Hassk.) Müll.Arg. === **Macaranga glaberrima** (Hassk.) Airy Shaw
Mallotus helferi Müll.Arg. === **Mallotus oblongifolius** (Miq.) Müll.Arg. var. **oblongifolius**
Mallotus hellwigianus K.Schum. === **Melanolepis multiglandulosa** (Reinw. ex Blume) Rchb. & Zoll. var. **multiglandulosa**
Mallotus henryi Pax & K.Hoffm. === **Macaranga henryi** (Pax & K.Hoffm.) Rehder
Mallotus hirsutulus (Kurz) Pax & K.Hoffm. === **Mallotus longipes** Müll.Arg.
Mallotus hollrungianus K.Schum. === **Melanolepis multiglandulosa** (Reinw. ex Blume) Rchb. & Zoll. var. **multiglandulosa**
Mallotus hookerianus var. *papuanus* J.J.Sm. === **Mallotus papuanus** (J.J.Sm.) Pax & K.Hoffm.
Mallotus impar Pax & K.Hoffm. === **Mallotus griffithianus** (Müll.Arg.) Hook.f.
Mallotus inamoenus F.Muell. ex Benth. === **Macaranga inamoena** F.Muell. ex Benth.
Mallotus insignis Müll.Arg. === **Mallotus macrostachyus** (Miq.) Müll.Arg.
Mallotus integrifolius (Willd.) Müll.Arg. === **Cordemoya integrifolia** (Willd.) Baill.
Mallotus intermedius (Baill.) N.P.Balakr. === **Mallotus resinosus** (Blanco) Merr.
Mallotus japonicus var. *floccosus* (Müll.Arg.) S.M.Hwang === **Mallotus nepalensis** Müll.Arg.
Mallotus japonicus var. *ochraceoalbidus* (Müll.Arg.) S.M.Hwang === **Mallotus japonicus** (L.f.) Müll.Arg.
Mallotus kietanus Rech. === **Mallotus oblongifolius** (Miq.) Müll.Arg. var. **oblongifolius**
Mallotus kunstleri King ex Hook.f. === **Chondrostylis kunstleri** (King ex Hook.f.) Airy Shaw
Mallotus kurzii Hook.f. === **Blumeodendron kurzii** (Hook.f.) J.J.Sm. ex Koord. & Valeton
Mallotus lambertianus Müll.Arg. === **Mallotus oblongifolius** (Miq.) Müll.Arg. var. **oblongifolius**
Mallotus lanceifolius Hook.f. === **Mallotus wrayi** King ex Hook.f.
Mallotus lawii Müll.Arg. === **Mallotus aureopunctatus** (Dalzell) Müll.Arg.
Mallotus leptophyllus Pax & K.Hoffm. === **Mallotus penangensis** Müll.Arg.
Mallotus leucocalyx Müll.Arg. === **Mallotus dispar** (Blume) Müll.Arg.
Mallotus leucodermis var. *puberulus* Airy Shaw === **Mallotus leucodermis** Hook.f.
Mallotus longifolius (Rchb.f. & Zoll.) Müll.Arg. === **Mallotus peltatus** (Geiseler) Müll.Arg.
Mallotus longipes (Kurz) Pax & K.Hoffm. === **Mallotus myanmaricus** P.T.Li
Mallotus maclurei Merr. === **Mallotus oblongifolius** (Miq.) Müll.Arg. var. **oblongifolius**
Mallotus macularis Airy Shaw === **Mallotus lauterbachianus** (Pax & K.Hoffm.) Pax & K.Hoffm.
Mallotus maingayi Durand & Jacks. === **Macaranga pruinosa** (Miq.) Müll.Arg.
Mallotus melleri Müll.Arg. === **Neoboutonia melleri** (Müll.Arg.) Prain
Mallotus minahassae Koord. === **Croton oblongus** Burm.f.
Mallotus moluccanus (L.) Müll.Arg. === **Aleurites moluccana** (L.) Willd.
Mallotus moluccanus var. *genuinus* Müll.Arg. === **Aleurites moluccana** (L.) Willd.
Mallotus moluccanus var. *glabratus* Müll.Arg. === **Melanolepis multiglandulosa** (Reinw. ex Blume) Rchb. & Zoll. var. **multiglandulosa**

Mallotus moluccanus var. *pendulus* Merr. === **Melanolepis multiglandulosa** var. **pendula** (Merr.) Merr.
Mallotus moritzianus Müll.Arg. === **Mallotus rufidulus** (Miq.) Müll.Arg.
Mallotus moritzianus var. *laevigatus* Müll.Arg. === **Mallotus laevigatus** (Müll.Arg.) Airy Shaw
Mallotus moritzianus var. *scaber* Müll.Arg. === **Mallotus rufidulus** (Miq.) Müll.Arg.
Mallotus multiglandulosus (Reinw. ex Blume) Hurus. === **Melanolepis multiglandulosa** (Reinw. ex Blume) Rchb. & Zoll.
Mallotus muricatus (Wight) Müll.Arg. === **Mallotus resinosus** (Blanco) Merr.
Mallotus muricatus var. *genuinus* Pax & K.Hoffm. === **Mallotus resinosus** (Blanco) Merr.
Mallotus muricatus var. *walkerae* (Hook.f.) Pax & K.Hoffm. === **Mallotus resinosus** (Blanco) Merr.
Mallotus nepalensis var. *floccosus* (Müll.Arg.) Pax === **Mallotus nepalensis** Müll.Arg.
Mallotus nepalensis var. *ochraceoalbidus* (Müll.Arg.) Pax === **Mallotus japonicus** (L.f.) Müll.Arg.
Mallotus nitidus Müll.Arg. === **Cleidion nitidum** (Müll.Arg.) Thwaites ex Kurz
Mallotus oblongifolius var. *helferi* (Müll.Arg.) Pax & K.Hoffm. === **Mallotus oblongifolius** (Miq.) Müll.Arg. var. **oblongifolius**
Mallotus oblongifolius var. *rubriflorus* Chakrab. === **Mallotus oblongifolius** (Miq.) Müll.Arg. var. **oblongifolius**
Mallotus oblongifolius var. *siamensis* Pax & K.Hoffm. === **Mallotus oblongifolius** (Miq.) Müll.Arg. var. **oblongifolius**
Mallotus odoratus Elmer === **Mallotus oblongifolius** (Miq.) Müll.Arg. var. **oblongifolius**
Mallotus oppositifolius f. *dentatus* (Schumach. & Thonn.) Pax & K.Hoffm. === **Mallotus oppositifolius** (Geiseler) Müll.Arg. var. **oppositifolius**
Mallotus oppositifolius var. *genuinus* Müll.Arg. === **Mallotus oppositifolius** (Geiseler) Müll.Arg.
Mallotus oppositifolius var. *glabratus* Müll.Arg. === **Mallotus oppositifolius** f. **glabratus** (Müll.Arg.) Pax
Mallotus oppositifolius var. *integrifolius* Müll.Arg. === **Mallotus oppositifolius** f. **glabratus** (Müll.Arg.) Pax
Mallotus oppositifolius f. *lindicus* Radcl.-Sm. === **Mallotus oppositifolius** var. **lindicus** (Radcl.-Sm.) Radcl.-Sm.
Mallotus oppositifolius f. *pubescens* (Pax) Pax === **Mallotus oppositifolius** (Geiseler) Müll.Arg. var. **oppositifolius**
Mallotus oppositifolius var. *pubescens* Pax === **Mallotus oppositifolius** (Geiseler) Müll.Arg. var. **oppositifolius**
Mallotus oreophilus var. *floccosus* Müll.Arg. === **Mallotus nepalensis** Müll.Arg.
Mallotus oreophilus var. *ochraceoalbidus* Müll.Arg. === **Mallotus japonicus** (L.f.) Müll.Arg.
Mallotus papillaris (Blanco) Merr. === **Mallotus tiliifolius** (Blume) Müll.Arg.
Mallotus papuanus var. *glabrescens* Pax & K.Hoffm. === **Mallotus echinatus** Elmer
Mallotus papuanus var. *intermedius* Pax & K.Hoffm. === **Mallotus echinatus** Elmer
Mallotus paxii Pamp. === **Mallotus apelta** (Lour.) Müll.Arg.
Mallotus paxii var. *castanopsis* (F.P.Metcalf) S.M.Hwang === **Mallotus apelta** (Lour.) Müll.Arg.
Mallotus peekelii Pax & K.Hoffm. === **Mallotus oblongifolius** var. **villosulus** Pax & K.Hoffm.
Mallotus peltatus var. *rubriflorus* Chakrab. === **Mallotus peltatus** (Geiseler) Müll.Arg.
Mallotus penangensis var. *grandifolia* Ridl. === **Mallotus penangensis** Müll.Arg.
Mallotus philippensis var. *mengliangensis* C.Y.Wu ex S.M.Hwang === **Mallotus philippensis** (Lam.) Müll.Arg.
Mallotus philippensis var. *pallidus* Airy Shaw === **Mallotus pallidus** (Airy Shaw) Airy Shaw
Mallotus philippensis var. *reticulatus* (Dunn) F.P.Metcalf === **Mallotus philippensis** (Lam.) Müll.Arg.
Mallotus playfairii Hemsl. === **Mallotus tiliifolius** (Blume) Müll.Arg.
Mallotus pleiogynus Pax & K.Hoffm. === **Octospermum pleiogynum** (Pax & K.Hoffm.) Airy Shaw
Mallotus polyneurus Hook.f. === **Mallotus khasianus** Hook.f.
Mallotus populifolius Hemsl. === **Macaranga hemsleyana** Pax & K.Hoffm.

Mallotus porterianus Müll.Arg. === **Mallotus oblongifolius** (Miq.) Müll.Arg. var. **oblongifolius**
Mallotus preussii Pax === **Tetracarpidium conophorum** (Müll.Arg.) Hutch. & Dalziel
Mallotus pseudopenangensis Pax & K.Hoffm. === **Mallotus echinatus** Elmer
Mallotus pseudoverticillatus Merr. === **Lasiococca comberi** Haines
Mallotus puberulus Hook.f. === **Mallotus oblongifolius** (Miq.) Müll.Arg. var. **oblongifolius**
Mallotus pycnostachys F.Muell. === **Mallotus mollissimus** (Geiseler) Airy Shaw
Mallotus ramosii Merr. === **Cleidion ramosii** (Merr.) Merr.
Mallotus repandus var. *chrysocarpus* (Pamp.) S.M.Hwang === **Mallotus contubernalis** Hance
Mallotus repandus var. *scabrifolius* (A.Juss.) Müll.Arg. === **Mallotus repandus** (Willd.) Müll.Arg.
Mallotus resinosus var. *cuneatus* (Ridl.) N.P.Balakr. & Chakrab. === **Mallotus cuneatus** Ridl.
Mallotus resinosus var. *muricatus* (Wight) N.P.Balakr. & Chakrab. === **Mallotus resinosus** (Blanco) Merr.
Mallotus resinosus var. *subramanyamii* (J.L.Ellis) Chakrab. === **Mallotus resinosus** (Blanco) Merr.
Mallotus reticulatus Dunn === **Mallotus philippensis** (Lam.) Müll.Arg.
Mallotus rhamnifolius var. *ovatifolius* Hook.f. === **Mallotus rhamnifolius** (Willd.) Müll.Arg.
Mallotus ricinoides (Pers.) Müll.Arg. === **Mallotus mollissimus** (Geiseler) Airy Shaw
Mallotus roxburghianus var. *glabra* Dunn === **Mallotus dunnii** F.P.Metcalf
Mallotus samarensis Merr. === **Cleidion ramosii** (Merr.) Merr.
Mallotus sanchezii Merr. === **Mallotus lackeyi** Elmer
Mallotus sanguirensis Pax & K.Hoffm. === **Mallotus resinosus** (Blanco) Merr.
Mallotus sarawakensis Pax & K.Hoffm. === **Mallotus penangensis** Müll.Arg.
Mallotus scabrifolius (A.Juss.) Müll.Arg. === **Mallotus repandus** (Willd.) Müll.Arg.
Mallotus scandens (Span.) Müll.Arg. === **Mallotus repandus** (Willd.) Müll.Arg.
Mallotus smilaciformis Gage === **Mallotus korthalsii** Müll.Arg.
Mallotus speciosus (Müll.Arg.) Pax & K.Hoffm. === **Sumbaviopsis albicans** (Blume) J.J.Sm.
Mallotus stipulosus (Rchb.f. ex Zoll.) Müll.Arg. === **Mallotus dispar** (Blume) Müll.Arg.
Mallotus stylaris Müll.Arg. === **Mallotus oblongifolius** (Miq.) Müll.Arg. var. **oblongifolius**
Mallotus subramanyamii J.L.Ellis === **Mallotus resinosus** (Blanco) Merr.
Mallotus tenuifolius Pax === **Mallotus apelta** (Lour.) Müll.Arg.
Mallotus tenuifolius var. *floccosus* (Müll.Arg.) Croizat === **Mallotus nepalensis** Müll.Arg.
Mallotus tenuifolius var. *subjaponicus* Croizat === **Mallotus subjaponicus** (Croizat) Croizat
Mallotus tenuispicus Pax & K.Hoffm. === **Mallotus oblongifolius** var. **villosulus** Pax & K.Hoffm.
Mallotus tokbrai (Blume) Müll.Arg. === **Blumeodendron tokbrai** (Blume) Kurz
Mallotus tristis Pax & K.Hoffm. === **Mallotus clellandii** Hook.f.
Mallotus tsiangii Merr. & Chun === **Macaranga lowii** King ex Hook.f. var. **lowii**
Mallotus vernicosus Hook.f. === **Blumeodendron tokbrai** (Blume) Kurz var. **tokbrai**
Mallotus vitifolius Kuntze === **Melanolepis vitifolia** (Kuntze) Gagnep.
Mallotus walkerae Hook.f. === **Mallotus resinosus** (Blanco) Merr.
Mallotus wallichianus Müll.Arg. === **Mallotus plicatus** (Müll.Arg.) Airy Shaw
Mallotus warburgianus Pax & K.Hoffm. === **Mallotus oblongifolius** (Miq.) Müll.Arg. var. **oblongifolius**
Mallotus woodii Merr. === **Mallotus griffithianus** (Müll.Arg.) Hook.f.
Mallotus xylacanthus Pax & K.Hoffm. === **Mallotus penangensis** Müll.Arg.
Mallotus yifengensis Hu & F.H.Chen === **Croton lachnocarpus** Benth.
Mallotus zeylanicus Müll.Arg. === **Mallotus rhamnifolius** (Willd.) Müll.Arg.
Mallotus zippellii F.Muell. === **Mallotus mollissimus** (Geiseler) Airy Shaw
Mallotus zollingeri Müll.Arg. === **Mallotus rufidulus** (Miq.) Müll.Arg.

Mancanilla

Synonyms:
Mancanilla Plum. ex Adans. === **Hippomane** L.

Mancinella

Synonyms:
Mancinella Tussac === **Hippomane** L.

Mandioca

Synonyms:
Mandioca Link === **Manihot** Mill.

Manihot

107 species, Americas (a few naturalised or adventive elsewhere); includes *Manihotoides*. Semiherbaceous, often tuberiferous, heliophilic subshrubs or woody shrubs or small trees to 10-12 m or clambering vines of predominately seasonal regions, some of them laticiferous; leaves simple or palmately lobed. The genus is most strongly represented in interior east-central Brazil (Goiás, Minas Gerais and Bahia); its entry into Middle America geologically is relatively late. *M. esculenta*, cassava or manioc, is now widely cultivated in warmer regions worldwide and has become a staple crop for millions of people; it has two wild subspecies, both widely ranging in South America (Allem 1994). The latex of *M. glaziovii* and related species was formerly (and still is here and there) tapped for rubber; before World War I attempts were made to establish plantations (Zimmermann 1913). The standard revision, covering most of the range, is that of Rogers & Appan (1973); 17 sections (exclusive of *Manihotoides*) are recognised. (Crotonoideae)

Pax, F. (1910). *Manihot*. In A. Engler (ed.), Das Pflanzenreich, IV 147 II (Euphorbiaceae-Adrianae): 21-99. Berlin. (Heft 44.) La/Ge. — 129 species, including one in the Addenda (p. 101); disposed in 11 sections (with 2 of uncertain position). [An additional species appears in Additamentum II (IV 147 III: 111 (1911), a further 7 species in ibid., VII (Additamentum V): 401-402 (1914) and more additions in ibid., XIV (Additamentum VI): 44-49 (1919). Now largely superseded.]

Zimmermann, A. (1913). Der *Manihot*-Kautschuk. xli, 342 pp., illus., map. Jena: Fischer. Ge. — Study of the biology and cultivation of *Manihot glaziovii* (Ceará rubber) and of its associated industry, with particular reference to German East Africa (now Tanzania). A botanical review of it and related rubber-yielding species of *Manihot* appears in the first two chapters (pp. 1-22), references to the cultivation and exploitation of the related species appear elsewhere in appropriate parts of the text. A distribution map of *M. glaziovii* and its relatives appears on p. 19.

Croizat, L. (1942). A study of *Manihot* in North America. J. Arnold Arbor. 23: 216-225, illus. En. — Brief introduction and survey of characters; novelties and notes; no keys. [Covers in effect Mexico and Central America.]

Rogers, D. J. & H. S. Fleming (1973). A monograph of *Manihot esculenta* - with an explanation of the taximetric methods used. Econ. Bot. 27: 1-113. En. — Part I encompasses an introduction and history of research on the genus (only one major study, by Pohl in the 1820s, based on much field experience); properties of *M. esculenta* (syn.: *M. utilissima*); review of taxonomic characters; methodology; summary. Part II is systematic, covering the species description along with cultivars and inculding evidence from chemical analysis; localities of collection of cultivars and list of references at end.

- Rogers, D. J. & S. G. Appan (1973). *Manihot, Manihotoides* (Euphorbiaceae). 272 pp., illus., maps (Flora Neotropica Monogr. 13). New York: New York Botanical Garden (for Organization Flora Neotropica). En. — Detailed revision (98 species of *Manihot*, 1 of *Manihotoides*) with keys, descriptions, synonymy, references, types, distribution, localities with exsiccatae, and commentary; references, synoptic list of taxa (pp. 255-256), list of specimens seen, index to botanical names, biographical data, and phenogram at end. The general part includes a historical survey, sections on geography and ecology and on speciation processes, and a review of characters (list, pp. 10-14). [*Manihotoides* not now accepted as distinct.]

Allem, A. C. (1994). The origin of *Manihot esculenta* Crantz (Euphorbiaceae). Genet. Resourc. Crop Evol. 41: 133-150, illus., maps. En. — Taxonomic investigation of *M. esculenta* and its putative wild relatives, with recognition of the latter as two subspecies (the cultivated plants comprising the nominate form); key and formal treatment (pp. 143-147), references (pp. 147-148) and list of exsiccatae seen (pp. 148-150). The wild populations are discussed along with their literature and a historical account of taxonomic investigations is also presented (priority of *M. esculenta* was first recognised in 1938).

Manihot Mill., Gard. Dict. Abr. ed. 4 (1754).
S. U.S.A., Mexico, S. America, naturalised elsewere. (22) (23) (24) (25) (40) (41) (42) (60) 76 77 79 80 81 82 83 84 85.
Janipha Kunth in F.W.H.von Humboldt, A.J.A.Bonpland & C.S.Kunth, Nov. Gen. Sp. 2: 106 (1817).
Mandioca Link, Handbuch 2: 436 (1831).
Hotnima A.Chev., J. Agric. Trop. 8: 111 (1908).
Manihotoides D.J.Rogers & Appan, Fl. Neotrop. Monogr. 13: 247 (1973).

Manihot acuminatissima Müll.Arg. in C.F.P.von Martius, Fl. Bras. 11(2): 455 (1874).
Brazil (Goiás, Bahia). 84 BZC BZE. Cham.

Manihot aesculifolia (Kunth) Pohl, Pl. Bras. Icon. Descr. 1: 55 (1827).
Mexico, C. America. 79 MXG MXN MXS MXT 80 BLZ COS ELS GUA HON NIC PAN. Tuber nanophan. or phan.
* *Jatropha aesculifolia* Kunth in von Humboldt, Bonpland & Kunth, Nov. Gen. Sp.: 85 (1817).
Manihot intermedia Weath., Proc. Amer. Acad. Arts 45: 427 (1910).
Manihot olfersiana Pax in H.G.A.Engler, Pflanzenr., IV, 147, I: 55 (1910).
Manihot gualanensis S.F.Blake, Contr. U. S. Natl. Herb. 24: 13 (1922).

Manihot alutacea D.J.Rogers & Appan, Fl. Neotrop. Monogr. 13: 130 (1973).
Brazil (Goiás). 84 BZC. Nanophan.

Manihot angustiloba (Torr.) Müll.Arg. in A.P.de Candolle, Prodr. 15(2): 1073 (1866).
Arizona, Mexico (Sonora, Baja California, Chihuahua, Sinaloa). 76 ARI 79 MXE MXN. Tuber nanophan.
* *Janipha manihot* var. *angustiloba* Torr. in W.E.Emory, Rep. U.S. Mex. Bound. Surv. 2: 199 (1859).
Manihot acutiloba Weath., Proc. Amer. Acad. Arts 45: 427 (1910).

Manihot anisophylla (Griseb.) Müll.Arg., J. Bot. 12: 230 (1874).
Argentina (Catamarca, Salta, Tucumán, Rioja, San Luís). 85 AGW. Tuber nanophan.
* *Janipha anisophylla* Griseb., Abh. Königl. Ges. Wiss. Göttingen 19: 95 (1874). *Manihot carthaginensis* var. *anisophylla* (Griseb.) Kuntze, Revis. Gen. Pl. 3(2): 288 (1898).

Manihot anomala Pohl, Pl. Bras. Icon. Descr. 1: 27 (1827). *Jatropha anomala* (Pohl) Steud., Nomencl. Bot., ed. 2, 1: 799 (1840).
S. Trop. America. 83 BOL PER 84 BZC BZL BZN 85 AGW PAR. Tuber nanophan.

subsp. **anomala**
Brazil (Mato Grosso, Goiás, Minas Gerais, São Paulo). 84 BZC BZL. Tuber nanophan.
Manihot caricifolia Pohl, Pl. Bras. Icon. Descr. 1: 40 (1827).
Manihot heterophylla Pohl, Pl. Bras. Icon. Descr. 1: 39 (1827).
Jatropha caricifolia Steud., Nomencl. Bot., ed. 2, 1: 799 (1840).
Jatropha heterophylla Steud., Nomencl. Bot., ed. 2, 1: 799 (1840).

subsp. **cujabensis** (Müll.Arg.) D.J.Rogers & Appan, Fl. Neotrop. Monogr. 13: 168 (1973).
Brazil (Mato Grosso). 84 BZC. Tuber nanophan.
* *Manihot cujabensis* Müll.Arg., Linnaea 34: 207 (1865).
Manihot alcicornis Klotzsch ex Pax in H.G.A.Engler, Pflanzenr., IV, 147, II: 84 (1910).

subsp. **glabrata** (Chodat & Hassl.) D.J.Rogers & Appan, Fl. Neotrop. Monogr. 13: 168 (1973).
Paraguay. 85 PAR. Tuber nanophan.
Manihot langsdorffii var. *glabra* Chodat & Hassl., Bull. Herb. Boissier, II, 5: 673 (1905).
* *Manihot pubescens* f. *glabrata* Chodat & Hassl., Bull. Herb. Boissier, II, 5: 672 (1905).
Manihot glabrata (Chodat & Hassl.) Pax & K.Hoffm. in H.G.A.Engler, Pflanzenr., IV, 147, II: 43 (1910).
Manihot klingensteinii Pax & K.Hoffm. in H.G.A.Engler, Pflanzenr., IV, 147, II: 66 (1910).
Manihot multiflora Pax & K.Hoffm. in H.G.A.Engler, Pflanzenr., IV, 147, II: 86 (1910).
Manihot pseudoheterophylla Pax & K.Hoffm. in H.G.A.Engler, Pflanzenr., IV, 147, II: 86 (1910).

subsp. **pavoniana** (Müll.Arg.) D.J.Rogers & Appan, Fl. Neotrop. Monogr. 13: 170 (1973).
SE. Peru, Bolivia, NW. Argentina (Jujuy, Salta, Catamarca, Tucumán). 83 BOL PER 85 AGW. Tuber nanophan.
* *Manihot pavoniana* Müll.Arg., Linnaea 34: 205 (1865).
Jatropha simayuca Ruiz & Pav. ex Pax in H.G.A.Engler, Pflanzenr., IV, 147, II: 65 (1910).
Manihot weberbaueri Pax & K.Hoffm. in H.G.A.Engler, Pflanzenr., IV, 147, XVI: 194 (1924).

subsp. **pubescens** (Pohl) D.J.Rogers & Appan, Fl. Neotrop. Monogr. 13: 167 (1973).
Brazil (Pará, Goiás, Brasília D.F., Minas Gerais). 84 BZC BZL BZN. Tuber nanophan.
* *Manihot pubescens* Pohl, Pl. Bras. Icon. Descr. 1: 50 (1827). *Jatropha pubescens* (Pohl) Steud., Nomencl. Bot., ed. 2, 1: 800 (1840).
Manihot warmingii Müll.Arg. in C.F.P.von Martius, Fl. Bras. 11(2): 481 (1874).

Manihot attenuata Müll.Arg. in C.F.P.von Martius, Fl. Bras. 11(2): 443 (1874).
Brazil (Goiás). 84 BZC. Hemicr. or cham.
Manihot brachystachys Pax & K.Hoffm., Verh. Bot. Vereins Prov. Brandenburg 50: 97 (1908 publ. 1910).

Manihot auriculata McVaugh, Brittonia 13: 190 (1961).
Mexico (Nayarit). 79 MXS. Tuber nanophan. or phan.

Manihot brachyandra Pax & K.Hoffm. in H.G.A.Engler, Pflanzenr., IV, 147, XVI: 196 (1924).
Brazil (Bahia: Rio de Contas). 84 BZE.

Manihot brachyloba Müll.Arg. in C.F.P.von Martius, Fl. Bras. 11(2): 451 (1874).
Costa Rica, Dominican Rep., Colombia, NW. Venezuela, Surinam, Guiana, Peru (Loreto), N. & W. Brazil. 80 COS 81 DOM 82 FRG SUR VEN 83 CLM PER 84 BZN. Cl. nanophan.
Manihot rusbyi Britton, Bull. Torrey Bot. Club 28: 302 (1901).
Manihot amazonica Ule, Verh. Bot. Vereins Prov. Brandenburg 50: 83 (1908 publ. 1909).

Manihot caerulescens Pohl, Pl. Bras. Icon. Descr. 1: 56 (1827). *Manihot caerulescens* var. *genuina* Müll.Arg. in A.P.de Candolle, Prodr. 15(2): 1070 (1866), nom. inval. *Jatropha coerulescens* (Pohl) Müll.Arg. in A.P.de Candolle, Prodr. 15(2): 1070 (1866).
Brazil, Paraguay. 84 BZC BZE BZL BZN 85 PAR. Nanophan. or phan.

subsp. **caerulescens**
Brazil (Pará, Maranhão, Piauí, Ceará, Pernambuco, Mato Grosso, Goiás, Bahia, Minas Gerais, Rio de Janeiro). 84 BZC BZE BZL BZN. Nanophan. or phan.
Jatropha coerulea Steud., Nomencl. Bot., ed. 2, 1: 799 (1840).
Manihot coerulea Steud., Nomencl. Bot., ed. 2, 1: 799 (1841).
Manihot caerulescens var. *pubescens* Müll.Arg. in A.P.de Candolle, Prodr. 15(2): 1070 (1866).

Manihot grandiflora Müll.Arg. in C.F.P.von Martius, Fl. Bras. 11(2): 471 (1874).
Manihot speciosa Müll.Arg. in C.F.P.von Martius, Fl. Bras. 11(2): 470 (1874).
Manihot piauhyensis Ule, Tropenpflanzer 11: 864 (1907).
Manihot riedeliana Klotzsch ex Pax in H.G.A.Engler, Pflanzenr., IV, 147, II: 34 (1910), nom. inval.
Manihot bahiensis Ule, Bot. Jahrb. Syst. 50(114): 4 (1914).
Manihot bahiensis var. *microsperma* Ule, Bot. Jahrb. Syst. 50(114): 5 (1914).
Manihot cuneata Ule, Bot. Jahrb. Syst. 50(114): 1 (1914).
Manihot discolor Ule, Bot. Jahrb. Syst. 50(114): 7 (1914).
Manihot ferruginea Ule, Bot. Jahrb. Syst. 50(114): 8 (1914).
Manihot harmsiana Ule, Bot. Jahrb. Syst. 50(114): 7 (1914).
Manihot labroyana Ule, Bot. Jahrb. Syst. 50(114): 6 (1914).
Manihot lyrata Ule, Bot. Jahrb. Syst. 50(114): 9 (1914).
Manihot microdendron Ule, Bot. Jahrb. Syst. 50(114): 4 (1914).
Manihot rotundata Ule, Bot. Jahrb. Syst. 50(114): 4 (1914).
Manihot toledii Labroy ex Ule, Bot. Jahrb. Syst. 50(114): 8 (1914).
Manihot trifoliata Ule, Bot. Jahrb. Syst. 50(114): 3 (1914).
Manihot trifoliata var. *platyphylla* Ule, Bot. Jahrb. Syst. 50(114): 3 (1914).
Manihot cearensis Pax & K.Hoffm. in H.G.A.Engler, Pflanzenr., IV, 147, XVI: 194 (1924).

subsp. **macrantha** (Pax & K.Hoffm.) D.J.Rogers & Appan, Fl. Neotrop. Monogr. 13: 245 (1973).
Paraguay (Sierra de Maracayu). 85 PAR. Nanophan. or phan.
Manihot speciosa Chodat & Hassl., Bull. Herb. Boissier, II, 5: 673 (1905).
* *Manihot macrantha* Pax & K.Hoffm. in H.G.A.Engler, Pflanzenr., IV, 147, II: 32 (1910).

subsp. **paraensis** (Müll.Arg.) D.J.Rogers & Appan, Fl. Neotrop. Monogr. 13: 245 (1973).
Brazil (Amapá, Pará). 84 BZN. Nanophan. or phan.
* *Manihot paraensis* Müll.Arg. in C.F.P.von Martius, Fl. Bras. 11(2): 470 (1874).

Manihot carthaginensis (Jacq.) Müll.Arg. in A.P.de Candolle, Prodr. 15(2): 1073 (1866).
Bonaire, Trinidad, Venezuela (incl. Patos I.), NE. Colombia. 81 NLA TRT VNA 82 VEN 83 CLM. Tuber nanophan. or phan.
* *Jatropha carthaginensis* Jacq., Enum. Syst. Pl.: 32 (1760).
Jatropha frutescens Loefl., Iter Hispan.: 397 (1766).
Jatropha janipha L., Mant. Pl. 1: 126 (1771). *Manihot janipha* (L.) Pohl, Pl. Bras. Icon. Descr. 1: 55 (1827).
Manihot pittieri Pax & K.Hoffm. in H.G.A.Engler, Pflanzenr., IV, 147, VII: 401 (1914).
Manihot meridensis Pittier, J. Wash. Acad. Sci. 19: 352 (1929).
Manihot remotiloba Pittier, J. Wash. Acad. Sci. 19: 352 (1929).

Manihot catingae Ule, Bot. Jahrb. Syst. 42: 221 (1908).
Brazil (Bahia). 84 BZE. Nanophan. or phan.

Manihot caudata Greenm., Proc. Amer. Acad. Arts 32: 82 (1903).
Mexico (Michoacán, Guanajuato, Zacatecas, Chihuahua). 79 MXE MXS. Phan.

Manihot cecropiifolia Pohl, Pl. Bras. Icon. Descr. 1: 49 (1827). *Jatropha cecropiifolia* (Pohl) Steud., Nomencl. Bot., ed. 2, 1: 799 (1840). *Manihot violacea* var. *cecropiifolia* (Pohl) Müll.Arg. in A.P.de Candolle, Prodr. 15(2): 1069 (1866).
Brazil (Goiás, Brasília D.F.). 84 BZC. Nanophan.

Manihot chlorosticta Standl. & Goldman, Contr. U. S. Natl. Herb. 13: 375 (1911).
Mexico (Baja California, Sinaloa, Jalisco, Colima, Michoacán, Guerrero). 79 MXN MXS. Cl. nanophan.
Manihot colimensis Croizat, J. Arnold Arbor. 23: 221 (1942).
Manihot mobilis Standl., Amer. Midl. Naturalist 36: 177 (1946).

Manihot compositifolia Allem, Revista Brasil. Biol. 49: 650 (1989 publ. 1990).
Brazil (Bahia). 84 BZE. Cl. nanophan.

Manihot condensata D.J.Rogers & Appan, Fl. Neotrop. Monogr. 13: 100 (1973).
Bolivia. 83 BOL. Nanophan. or phan.

Manihot corymbiflora Pax & K.Hoffm. in H.G.A.Engler, Pflanzenr., IV, 147, II: 80 (1910).
Brazil (Rio de Janeiro). 84 BZL. Nanophan. – Very close to *M. pilosa.*

Manihot crassisepala Pax & K.Hoffm. in H.G.A.Engler, Pflanzenr., IV, 147, II: 28 (1910).
Mexico (México State, Morelos). 79 MXC. Phan.

Manihot crotalariiformis Pohl, Pl. Bras. Icon. Descr. 1: 24 (1827). *Jatropha crotalariiformis* (Pohl) Steud., Nomencl. Bot., ed. 2, 1: 799 (1840).
Brazil (Minas Gerais). 84 BZL. Cham.

Manihot davisiae Croizat, J. Arnold Arbor. 23: 224 (1942).
Arizona, Mexico (Sonora, Chihuahua, Sinaloa). 76 ARI 79 MXE MXN. Nanophan.

Manihot diamantinensis Allem, Revista Brasil. Biol. 49: 658 (1989 publ. 1990).
Brazil (Bahia). 84 BZE. Cham.

Manihot dichotoma Ule, Tropenpflanzer 11: 863 (1907). *Manihot dichotoma* var. *genuina* Pax in H.G.A.Engler, Pflanzenr., IV, 147, II: 83 (1910), nom. inval.
Brazil (Pernambuco, Bahia). 84 BZE. Phan.
Manihot dichotoma var. *parvifolia* Ule, Notizbl. Bot. Gart. Berlin-Dahlem 5(41a): 20 (1908).
Manihot preciosa Schindler, Bull. Misc. Inform. Kew 1910: 96 (1910), nom. nud.

Manihot divergens Pohl, Pl. Bras. Icon. Descr. 1: 41 (1827). *Jatropha divergens* (Pohl) Steud., Nomencl. Bot., ed. 2, 1: 799 (1840). *Manihot violacea* var. *divergens* (Pohl) Müll.Arg. in A.P.de Candolle, Prodr. 15(2): 1069 (1866).
Brazil (Goiás, Brasília D.F., Minas Gerais). 84 BZC BZL. Nanophan.
Manihot arcuata Pohl, Pl. Bras. Icon. Descr. 1: 42 (1827). *Jatropha arcuata* (Pohl) Steud., Nomencl. Bot., ed. 2, 1: 799 (1840). *Manihot violacea* var. *arcuata* (Pohl) Müll.Arg. in A.P.de Candolle, Prodr. 15(2): 1069 (1866).

Manihot epruinosa Pax & K.Hoffm. in H.G.A.Engler, Pflanzenr., IV, 147, XVI: 196 (1924).
Brazil (Piauí, Ceará, Paraíba, Pernambuco, Bahia). 84 BZE. Tuber phan.
Manihot floribunda Pax & K.Hoffm. in H.G.A.Engler, Pflanzenr., IV, 147, XVI: 195 (1924).

Manihot esculenta Crantz, Inst. Rei Herb. 1: 167 (1766).
Mexico, Trop. America; widely cultivated therein and elsewhere in tropical regions. 79 MXC MXG MXS MXT 80 ALL 83 BOL CLM ECU PER 84 BZC BZE BZL BZN BZS. Tuber nanophan. or phan. – Tubers edible when the cyanide is removed; a major staple food (cassava, tapioca) in the tropics.
* *Jatropha manihot* L., Sp. Pl.: 1007 (1753). *Manihot manihot* (L.) H.Karst., Deut. Fl.: 588 (1882), nom. illeg. *Manihot manihot* (L.) Cockerell, Bull. Torrey Bot. Club 19: 95 (1892), nom. illeg.
Jatropha dulcis J.F.Gmel., Onomat. Bot. 5: 7 (1773). *Manihot dulcis* (J.F.Gmel.) Baill., Traité Bot. Méd. Phan. 2: 932 (1884).
Jatropha mitis Rottb., Acta Lit. Univ. Hafn. 1: 301 (1778).
Jatropha janipha Lour., Fl. Cochinch.: 585 (1790), nom. illeg.
Manihot aipi Pohl, Pl. Bras. Icon. Descr. 1: 29 (1827). *Manihot palmata* var. *aipi* (Pohl) Müll.Arg. in A.P.de Candolle, Prodr. 15(2): 1062 (1866). *Jatropha aipi* (Pohl) A.Moller, Phycomyc. & Ascomyc.: 92 (1901). *Manihot dulcis* var. *aipi* (Pohl) Pax in H.G.A.Engler, Pflanzenr., IV, 147, II: 71 (1910).

Manihot aipi var. *lanceolata* Pohl, Pl. Bras. Icon. Descr. 1: 31 (1827).
Manihot aipi var. *latifolia* Pohl, Pl. Bras. Icon. Descr. 1: 31 (1827).
Manihot aipi var. *lutescens* Pohl, Pl. Bras. Icon. Descr. 1: 31 (1827).
Manihot diffusa Pohl, Pl. Bras. Icon. Descr. 1: 55 (1827). *Jatropha diffusa* (Pohl) Steud., Nomencl. Bot., ed. 2, 1: 799 (1840). *Manihot palmata* var. *diffusa* (Pohl) Müll.Arg. in A.P.de Candolle, Prodr. 15(2): 1062 (1866). *Manihot dulcis* var. *diffusa* (Pohl) Pax in H.G.A.Engler, Pflanzenr., IV, 147, II: 71 (1910).
Manihot digitiformis Pohl, Pl. Bras. Icon. Descr. 1: 36 (1827). *Manihot palmata* var. *digitiformis* (Pohl) Müll.Arg. in A.P.de Candolle, Prodr. 15(2): 1063 (1833). *Jatropha digitiformis* (Pohl) Steud., Nomencl. Bot., ed. 2, 1: 799 (1840).
Manihot flabellifolia Pohl, Pl. Bras. Icon. Descr. 1: 35 (1827). *Jatropha flabellifolia* (Pohl) Steud., Nomencl. Bot., ed. 2, 1: 799 (1840). *Manihot palmata* var. *flabellifolia* (Pohl) Müll.Arg. in A.P.de Candolle, Prodr. 15(2): 1062 (1866). *Manihot dulcis* var. *flabellifolia* (Pohl) Pax in H.G.A.Engler, Pflanzenr., IV, 147, II: 72 (1910). *Manihot esculenta* subsp. *flabellifolia* (Pohl) Cif., Arch. Bot. (Forlì) 18: 31 (1942).
Manihot loureirii Pohl, Pl. Bras. Icon. Descr. 1: 55 (1827). *Jatropha loureirii* (Pohl) Steud., Nomencl. Bot., ed. 2, 1: 799 (1840).
Manihot utilissima Pohl, Pl. Bras. Icon. Descr. 1: 32 (1827).
Manihot utilissima var. *castellana* Pohl, Pl. Bras. Icon. Descr. 1: 34 (1827).
Manihot utilissima var. *sutinga* Pohl, Pl. Bras. Icon. Descr. 1: 34 (1827).
Manihot cannabina Sweet, Hort. Brit., ed. 2: 458 (1830).
Jatropha silvestris Vell., Fl. Flumin. 10: t. 83 (1831).
Jatropha stipulata Vell., Fl. Flumin. 10: t. 82 (1831).
Jatropha glauca A.Rich., Tent. Fl. Abyss. 2: 250 (1850), nom. illeg. *Jatropha lobata* var. *richardiana* Müll.Arg. in A.P.de Candolle, Prodr. 15(2): 1086 (1866).
Manihot edule A.Rich. in R.de la Sagra, Hist. Fis. Cuba, Bot. 2: 208 (1850).
Manihot aypi Spruce, J. Linn. Soc., Bot. 5: 10 (1861).
Manihot melanobasis Müll.Arg., Linnaea 34: 206 (1865).
Jatropha mitis Sessé & Moç., Pl. Nov. Hisp.: 167 (1890), nom. illeg.
Manihot cassava Cook & Collins, Contr. U. S. Natl. Herb. 8: 184 (1903), nom. nud.
Jatropha paniculata Ruiz & Pav. ex Pax in H.G.A.Engler, Pflanzenr., IV, 147, II: 71 (1910).
Manihot guyanensis Klotzsch ex Pax in H.G.A.Engler, Pflanzenr., IV, 147, II: 84 (1910), nom. nud.
Manihot sprucei Pax in H.G.A.Engler, Pflanzenr., IV, 147, I: 71 (1910).
Manihot flexuosa Pax & K.Hoffm. in H.G.A.Engler, Pflanzenr., IV, 147, XVI: 195 (1924).
Manihot esculenta var. *sprucei* Lanj., Euphorb. Surinam: 33 (1931).
Manihot esculenta subsp. *alboerecta* Cif., Arch. Bot. (Forlì) 18: 31 (1942), nom. illeg.
Manihot esculenta subsp. *diffusa* Cif., Arch. Bot. (Forlì) 18: 32 (1942), nom. illeg.
Manihot esculenta subsp. *grandifolia* Cif., Arch. Bot. (Forlì) 18: 31 (1942).
Manihot esculenta var. *argentea* Cif., Arch. Bot. (Forlì) 18: 31 (1942).
Manihot esculenta var. *coalescens* Cif., Arch. Bot. (Forlì) 18: 31 (1942).
Manihot esculenta var. *communis* Cif., Arch. Bot. (Forlì) 18: 32 (1942), nom. illeg.
Manihot esculenta var. *debilis* Cif., Arch. Bot. (Forlì) 18: 31 (1942).
Manihot esculenta var. *digitifolia* Cif., Arch. Bot. (Forlì) 18: 31 (1942).
Manihot esculenta var. *domingensis* Cif., Arch. Bot. (Forlì) 18: 32 (1942), nom. illeg.
Manihot esculenta var. *fertilis* Cif., Arch. Bot. (Forlì) 18: 32 (1942), nom. illeg.
Manihot esculenta var. *flavicaulis* Cif., Arch. Bot. (Forlì) 18: 31 (1942).
Manihot esculenta var. *fuscescens* Cif., Arch. Bot. (Forlì) 18: 31 (1942).
Manihot esculenta var. *hispaniolensis* Cif., Arch. Bot. (Forlì) 18: 31 (1942), nom. illeg.
Manihot esculenta var. *jamaicensis* Cif., Arch. Bot. (Forlì) 18: 32 (1942), nom. illeg.
Manihot esculenta var. *luteola* Cif., Arch. Bot. (Forlì) 18: 31 (1942), nom. illeg.
Manihot esculenta var. *mutabilis* Cif., Arch. Bot. (Forlì) 18: 32 (1942), nom. illeg.
Manihot esculenta var. *nodosa* Cif., Arch. Bot. (Forlì) 18: 31 (1942).
Manihot esculenta var. *pohlii* Cif., Arch. Bot. (Forlì) 18: 32 (1942), nom. illeg.

Manihot esculenta var. *ramosissima* Cif., Arch. Bot. (Forlì) 18: 31 (1942), nom. illeg.
Manihot esculenta var. *rufescens* Cif., Arch. Bot. (Forlì) 18: 32 (1942), nom. illeg.
Manihot esculenta var. *zimmermannii* Cif., Arch. Bot. (Forlì) 18: 32 (1942), nom. illeg.

Manihot falcata D.J.Rogers & Appan, Fl. Neotrop. Monogr. 13: 125 (1973).
Brazil (Goiás, Brasília D.F.). 84 BZC. Cham.

Manihot filamentosa Pittier, J. Wash. Acad. Sci. 20: 11 (1930).
Venezuela (Lara, Guarico). 82 VEN. Nanophan.

Manihot flemingiana D.J.Rogers & Appan, Fl. Neotrop. Monogr. 13: 143 (1973).
Brazil (Mato Grosso). 84 BZC. Nanophan.

Manihot foetida (Kunth) Pohl, Pl. Bras. Icon. Descr. 1: 55 (1827).
Mexico (México State). 79 MXC. Phan.
* *Janipha foetida* Kunth in F.W.H.von Humboldt, A.J.A.Bonpland & C.S.Kunth, Nov. Gen. Sp. 2: 106 (1817). *Jatropha foetida* (Kunth) Steud., Nomencl. Bot., ed. 2, 1: 799 (1840).

Manihot fruticulosa (Pax) D.J.Rogers & Appan, Fl. Neotrop. Monogr. 13: 149 (1973).
Brazil (Minas Gerais, Goiás, Brasília D.F.). 84 BZC BZL. Cham. or nanophan.
* *Manihot triphylla* var. *fruticulosa* Pax in H.G.A.Engler, Pflanzenr., IV, 147, II: 74 (1910).

Manihot gabrielensis Allem, Revista Brasil. Biol. 49: 653 (1989 publ. 1990).
Brazil (Goiás). 84 BZC. Nanophan.

Manihot glaziovii Müll.Arg. in C.F.P.von Martius, Fl. Bras. 11(2): 446 (1874).
Brazil (Ceará, Paraíba, Pernambuco, Bahia). 84 BZE. Phan. – Source of Ceará rubber & oilseeds; also used as vegetable greens.

Manihot gracilis Pohl, Pl. Bras. Icon. Descr. 1: 23 (1827). *Jatropha gracilis* (Pohl) Steud., Nomencl. Bot., ed. 2, 1: 799 (1840). *Manihot gracilis* var. *genuina* Müll.Arg. in A.P.de Candolle, Prodr. 15(2): 1065 (1866), nom. inval.
Brazil (Minas Gerais, Goiás, Brasília D.F., São Paulo), Paraguay. 84 BZC BZL 85 PAR. Cham.

subsp. **gracilis**
Brazil (Minas Gerais, Goiás, Brasília D.F., São Paulo), Paraguay. 84 BZC BZL 85 PAR. Cham.
Manihot pronifolia Pohl, Pl. Bras. Icon. Descr. 1: 24 (1827). *Jatropha pronifolia* (Pohl) Steud., Nomencl. Bot., ed. 2, 1: 800 (1840). *Manihot gracilis* var. *pronifolia* (Pohl) Müll.Arg. in A.P.de Candolle, Prodr. 15(2): 1065 (1866).
Manihot depauperata Pax & K.Hoffm. in H.G.A.Engler, Pflanzenr., IV, 147, II: 41 (1910).

subsp. **varians** (Pohl) D.J.Rogers & Appan, Fl. Neotrop. Monogr. 13: 160 (1973).
Brazil (Minas Gerais, Goiás, São Paulo). 84 BZC BZL. Cham.
* *Manihot varians* Pohl, Pl. Bras. Icon. Descr. 1: 53 (1827). *Jatropha varians* (Pohl) Steud., Nomencl. Bot., ed. 2, 1: 800 (1840).
Manihot pardina Müll.Arg. in C.F.P.von Martius, Fl. Bras. 11(2): 484 (1874).

Manihot grahamii Hook., Hooker's Icon. Pl. 11: t. 530 (1843).
SE. Brazil, N. Argentina, Paraguay, Uruguay. 84 BZC BZL BZS 85 AGE AGW PAR URU. Nanophan. or phan. – Ornamental.
Janipha loeflingii var. *multifida* Graham, Edinb. Philos. J. 29: 172 (1840). *Manihot palmata* var. *multifida* (Graham) Müll.Arg. in A.P.de Candolle, Prodr. 15(2): 1062 (1866). *Manihot loeflingii* var. *multifida* (Graham) Müll.Arg. in A.P.de Candolle, Prodr. 15(2): 1062 (1866). *Manihot dulcis* var. *multifida* (Graham) Pax in H.G.A.Engler, Pflanzenr., IV, 147, II: 72 (1910).
Manihot loeflingii Müll.Arg. in C.F.P.von Martius, Fl. Bras. 11(2): 460 (1874).
Manihot tweedieana Müll.Arg. in C.F.P.von Martius, Fl. Bras. 11(2): 450 (1874).

Manihot tweedieana f. *nana* Chodat & Hassl., Bull. Herb. Boissier, II, 5: 673 (1905).
Manihot tweedieana var. *lobata* Chodat & Hassl., Bull. Herb. Boissier, II, 5: 673 (1905).
Manihot lobata (Chodat & Hassl.) Pax in H.G.A.Engler, Pflanzenr., IV, 147, II: 82 (1910).
Manihot enneaphylla Pax & K.Hoffm. in H.G.A.Engler, Pflanzenr., IV, 147, XVI: 196 (1924).

Manihot guaranitica Chodat & Hassl., Bull. Herb. Boissier, II, 5: 671 (1905).
Bolivia, Paraguay, N. Argentina. 83 BOL 85 AGE AGW PAR. Nanophan.

subsp. **boliviana** (Pax & K.Hoffm.) D.J.Rogers & Appan, Fl. Neotrop. Monogr. 13: 112 (1973).
Bolivia. 83 BOL. Nanophan.
* *Manihot boliviana* Pax & K.Hoffm. in H.G.A.Engler, Pflanzenr., IV, 147, VII: 402 (1914).

subsp. **guaranitica**
SE. Bolivia, Paraguay, N. Argentina. 83 BOL 85 AGE AGW PAR. Nanophan.
Manihot fiebrigii Pax & K.Hoffm. in H.G.A.Engler, Pflanzenr., IV, 147, II: 75 (1910).
Manihot grandistipula Pax in H.G.A.Engler, Pflanzenr., IV, 147, II: 81 (1910).
Manihot recognita Pax in H.G.A.Engler, Pflanzenr., IV, 147, II: 91 (1910).
Manihot anisitsii Pax & K.Hoffm. in H.G.A.Engler, Pflanzenr., IV, 147, XVI: 197 (1924).

Manihot handroana Cruz, Brogantia 26: 318 (1967).
Brazil (Minas Gerais: Serra de Mantiqueira). 84 BZL. Nanophan.

Manihot hassleriana Chodat, Bull. Herb. Boissier, II, 5: 672 (1905).
Paraguay (near Carimbatay River). 85 PAR. Cham.

Manihot heptaphylla Ule, Tropenpflanzer 11: 863 (1907).
Brazil (Bahia). 84 BZE. Nanophan. or phan.

Manihot hilariana Baill., Adansonia 4: 282 (1864).
Brazil (Minas Gerais). 84 BZL. Cham.

Manihot hunzikeriana Mart.Crov., Bonplandia 1: 273 (1964).
Brazil (Rio Grande do Sul), Argentina (Misiones). 84 BZS 85 AGE. Cham.

Manihot inflata Müll.Arg. in C.F.P.von Martius, Fl. Bras. 11(2): 450 (1874).
Brazil (Paraná, São Paulo, Rio de Janeiro). 84 BZL BZS. Nanophan.
Manihot brasiliensis Klotzsch ex Pax, Verh. Bot. Vereins Prov. Brandenburg 50: 57 (1908 publ. 1910).

Manihot irwinii D.J.Rogers & Appan, Fl. Neotrop. Monogr. 13: 137 (1973).
Brazil (Goiás). 84 BZC. Nanophan.

Manihot jacobinensis Müll.Arg., Linnaea 34: 205 (1865). *Manihot violacea* subsp. *jacobinensis* (Müll.Arg.) Allem, Revista Brasil. Biol. 49: 25 (1989).
Brazil (Mato Grosso, Bahia). 84 BZC BZE. Nanophan.
Manihot occidentalis Müll.Arg. in C.F.P.von Martius, Fl. Bras. 11(2): 468 (1874).
Manihot rigidifolia Pax & K.Hoffm. in H.G.A.Engler, Pflanzenr., IV, 147, XVI: 194 (1924).

Manihot janiphoides Müll.Arg. in C.F.P.von Martius, Fl. Bras. 11(2): 480 (1874).
Brazil (Minas Gerais, São Paulo). 84 BZL. Nanophan.

Manihot jolyana Cruz, Brogantia 24: 360 (1965).
Brazil (São Paulo). 84 BZL. Nanophan.

Manihot leptophylla Pax & K.Hoffm. in H.G.A.Engler, Pflanzenr., IV, 147, II: 57 (1910).
Ecuador, Peru, Brazil (Amazonas, Para, Pernambuco, Acre). 83 ECU PER 84 BZE BZN. Cl. nanophan.
Manihot palmata var. *ferruginea* Müll.Arg. in A.P.de Candolle, Prodr. 15(2): 1063 (1866).
Manihot dulcis var. *ferruginea* (Müll.Arg.) Pax in H.G.A.Engler, Pflanzenr., IV, 147, II: 71 (1910).

Manihot longipetiolata Pohl, Pl. Bras. Icon. Descr. 1: 25 (1827). *Jatropha longipetiolata* (Pohl) Steud., Nomencl. Bot., ed. 2, 1: 799 (1840).
Brazil (Goiás, Minas Gerais). 84 BZE BZL. Cham.

Manihot maguireana D.J.Rogers & Appan, Fl. Neotrop. Monogr. 13: 161 (1973).
Venezuela (Bolívar). 82 VEN. Nanophan.

Manihot maracasensis Ule, Bot. Jahrb. Syst. 42: 221 (1908).
Brazil (Bahia). 84 BZE. Phan.
Manihot maracasensis var. *vestita* Pax & K.Hoffm. in H.G.A.Engler, Pflanzenr., IV, 147, XVI: 194 (1924).

Manihot marajoara Huber, Bol. Mus. Goeldi Paraense Hist. Nat. Ethnogr. 5: 120 (1908).
Brazil (Pará, Amapá). 84 BZN. Nanophan.

Manihot membranacea Pax & K.Hoffm. in H.G.A.Engler, Pflanzenr., IV, 147, III: 111 (1911).
Brazil (Mato Grosso). 84 BZC. – Provisionally accepted.

Manihot michaelis McVaugh, Brittonia 13: 190 (1961).
Mexico (Colima, Jalisco). 79 MXS. Phan.

Manihot mirabilis Pax in H.G.A.Engler, Pflanzenr., IV, 147, II: 91 (1910).
Paraguay (Sierra de Amabay). 85 PAR. Nanophan.

Manihot mossamedensis Taub., Bot. Jahrb. Syst. 21: 442 (1896).
Brazil (Goiás). 84 BZC. Nanophan.

Manihot nana Müll.Arg. in C.F.P.von Martius, Fl. Bras. 11(2): 448 (1874).
Brazil (Goiás, Minas Gerais). 84 BZC BZL. Hemicr. or cham.

Manihot neusana Nassar, Ci. & Cult. 38: 340 (1986).
Brazil (Paraná). 84 BZS.

Manihot nogueirae Allem, Revista Brasil. Biol. 49: 656 (1989 publ. 1990).
Brazil (Brasília D.F.). 84 BZC. Hemicr.

Manihot oaxacana D.J.Rogers & Appan, Fl. Neotrop. Monogr. 13: 46 (1973).
Mexico (Oaxaca). 79 MXS. Nanophan. or phan.

Manihot obovata J.Jiménez Ram., Anales Inst. Biol. Univ. Nac. Auton. Mexico, Bot. 60: 52 (1990).
Mexico (Guerrero). 79 MXS. Nanophan.

Manihot oligantha Pax & K.Hoffm. in H.G.A.Engler, Pflanzenr., IV, 147, II: 53 (1910).
Brazil (Goiás). 84 BZC. Hemicr. or cham.

Manihot orbicularis Pohl, Pl. Bras. Icon. Descr. 1: 20 (1827). *Jatropha orbicularis* (Pohl) Steud., Nomencl. Bot., ed. 2, 1: 799 (1840).
Brazil (Goiás). 84 BZC. Cham.

Manihot palmata Müll.Arg. in A.P.de Candolle, Prodr. 15(2): 1062 (1866).
Brazil (Rio de Janeiro). 84 BZL. Nanophan.
Jatropha manihot Vell., Fl. Flumin. 10: 80 (1831).
* *Jatropha palmata* Vell., Fl. Flumin. 10: 81 (1831), nom. illeg. *Manihot palmata* var. *genuina* Müll.Arg. in A.P.de Candolle, Prodr. 15(2): 1062 (1866), nom. inval. *Manihot palmata* (Vell.) Pax in H.G.A.Engler, Pflanzenr., IV, 147, II: 55 (1910), nom. illeg.
Manihot palmata var. *leptopoda* Müll.Arg. in C.F.P.von Martius, Fl. Bras. 11(2): 459 (1874). *Manihot dulcis* var. *leptopoda* (Müll.Arg.) Pax in H.G.A.Engler, Pflanzenr., IV, 147, II: 72 (1910). *Manihot leptopoda* (Müll.Arg.) D.J.Rogers & Appan, Fl. Neotrop. Monogr. 13: 96 (1973).

Manihot pauciflora Brandegee, Univ. Calif. Publ. Bot. 4: 89 (1910).
Mexico (Puebla, Oaxaca). 79 MXC MXS. Phan.

Manihot paviifolia Pohl, Pl. Bras. Icon. Descr. 1: 52 (1827). *Jatropha paviifolia* (Pohl) Steud., Nomencl. Bot., ed. 2, 1: 799 (1840). *Manihot pentaphylla* var. *paviifolia* (Pohl) Müll.Arg. in A.P.de Candolle, Prodr. 15(2): 1071 (1866).
Brazil (Goiás). 84 BZC. Nanophan.

Manihot peltata Pohl, Pl. Bras. Icon. Descr. 1: 18 (1827). *Jatropha peltata* (Pohl) Steud., Nomencl. Bot., ed. 2, 1: 800 (1840), nom. illeg.
Brazil (Goiás). 84 BZC. Cham. or nanophan.

Manihot pentaphylla Pohl, Pl. Bras. Icon. Descr. 1: 53 (1827). *Jatropha pentaphylla* (Pohl) Steud., Nomencl. Bot., ed. 2, 1: 800 (1840). *Manihot pentaphylla* var. *genuina* Müll.Arg. in A.P.de Candolle, Prodr. 15(2): 1071 (1866), nom. inval.
Brazil, Paraguay. 84 BZC BZL BZN 85 PAR. Nanophan.

subsp. **graminifolia** (Chodat & Hassl.) D.J.Rogers & Appan, Fl. Neotrop. Monogr. 13: 155 (1973).
Paraguay (Sierra de Maracayu). 85 PAR. Nanophan.
* *Manihot graminifolia* Chodat & Hassl., Bull. Herb. Boissier, II, 5: 671 (1905).

subsp. **pentaphylla**
Brazil (Goiás). 84 BZC. Nanophan.
Manihot uleana Pax & K.Hoffm. in H.G.A.Engler, Pflanzenr., IV, 147, II: 37 (1910).

subsp. **rigidula** (Müll.Arg.) D.J.Rogers & Appan, Fl. Neotrop. Monogr. 13: 153 (1973).
Brazil (Minas Gerais, Goiás). 84 BZC BZL. Nanophan.
Manihot conulifera Müll.Arg. in C.F.P.von Martius, Fl. Bras. 11(2): 474 (1874).
* *Manihot rigidula* Müll.Arg. in C.F.P.von Martius, Fl. Bras. 11(2): 474 (1874).

subsp. **tenuifolia** (Pohl) D.J.Rogers & Appan, Fl. Neotrop. Monogr. 13: 153 (1973).
Brazil (Pará, Goiás). 84 BZC BZN. Nanophan.
Manihot tenerrima Pohl, Pl. Bras. Icon. Descr. 1: 39 (1827). *Jatropha tenerrima* (Pohl) Steud., Nomencl. Bot., ed. 2, 1: 800 (1840). *Manihot gracilis* var. *tenerrima* (Pohl) Müll.Arg. in A.P.de Candolle, Prodr. 15(2): 1066 (1866). *Manihot esculenta* var. *tenerrima* (Pohl) Cif., Arch. Bot. (Forlì) 18: 31 (1942).
* *Manihot tenuifolia* Pohl, Pl. Bras. Icon. Descr. 1: 38 (1827). *Jatropha tenuifolia* (Pohl) Steud., Nomencl. Bot., ed. 2, 1: 800 (1840). *Manihot gracilis* var. *tenuifolia* (Pohl) Müll.Arg. in A.P.de Candolle, Prodr. 15(2): 1066 (1866).
Manihot reflexa Klotzsch ex Pax in H.G.A.Engler, Pflanzenr., IV, 147, II: 78 (1910).

Manihot peruviana Müll.Arg., Linnaea 34: 206 (1865). *Manihot heterandra* Ule, Verh. Bot. Vereins Prov. Brandenburg 50: 84 (1908 publ. 1909), nom. illeg. *Manihot esculenta* subsp. *peruviana* (Müll.Arg.) Allan, Gen. Res. Cap. Ev. 41: 146 (1994).
Peru (Loreto, San Martin). 83 PER. Cl. nanophan.

Manihot pilosa Pohl, Pl. Bras. Icon. Descr. 1: 55 (1827). *Jatropha pilosa* (Pohl) Steud., Nomencl. Bot., ed. 2, 1: 800 (1840).
Brazil (Minas Gerais, São Paulo, Rio de Janeiro). 84 BZL. Nanophan. or phan.
Manihot hemigynandra Müll.Arg. in C.F.P.von Martius, Fl. Bras. 11(2): 454 (1874).
Manihot hemitrichandra Müll.Arg. in C.F.P.von Martius, Fl. Bras. 11(2): 454 (1874).
Manihot langsdorffii Müll.Arg. in C.F.P.von Martius, Fl. Bras. 11(2): 455 (1874).
Manihot pedicellaris Müll.Arg. in C.F.P.von Martius, Fl. Bras. 11(2): 453 (1874).
Manihot brevipedicellata Pax & K.Hoffm., Verh. Bot. Vereins Prov. Brandenburg 50: 63 (1908 publ. 1910).
Manihot meyeriana Klotzsch ex Pax in H.G.A.Engler, Pflanzenr., IV, 147, II: 65 (1910).
Manihot tubuliflora Pax & K.Hoffm. in H.G.A.Engler, Pflanzenr., IV, 147, II: 61 (1910).

Manihot pohliana Müll.Arg. in C.F.P.von Martius, Fl. Bras. 11(2): 464 (1874).
Brazil (Bahia). 84 BZE. – Provisionally accepted.
Manihot johannis Pax in H.G.A.Engler, Pflanzenr., IV, 147, II: 78 (1910).

Manihot pohlii Wawra, Flora 47: 252 (1864).
Brazil (Espírito Santo, Rio de Janeiro). 84 BZL. Nanophan.
Manihot tripartita var. *quinqueloba* Pax & K.Hoffm. in H.G.A.Engler, Pflanzenr., IV, 147, XVI: 194 (1924).

Manihot populifolia Pax in H.G.A.Engler, Pflanzenr., IV, 147, II: 93 (1910).
Paraguay. 85 PAR. Cham.
Manihot cordifolia Pax in H.G.A.Engler, Pflanzenr., IV, 147, II: 94 (1910).

Manihot pringlei S.Watson, Proc. Amer. Acad. Arts 26: 148 (1891).
Mexico (Tamaulipas, San Luis Potosí). 79 MXE. Tuber nanophan.

Manihot procumbens Müll.Arg., Linnaea 34: 206 (1865). *Manihot procumbens* var. *genuina* Pax in H.G.A.Engler, Pflanzenr., IV, 147, II: 40 (1910), nom. inval.
Brazil (Minas Gerais, São Paulo), Paraguay. 84 BZL 85 PAR. Cham.
Manihot elegans Müll.Arg. in C.F.P.von Martius, Fl. Bras. 11(2): 485 (1874).
Manihot procumbens var. *grandifolia* Chodat & Hassl., Bull. Herb. Boissier, II, 5: 673 (1905).
Manihot affinis Pax & K.Hoffm. in H.G.A.Engler, Pflanzenr., IV, 147, II: 48 (1910).
Manihot meeboldii Pax & K.Hoffm. in H.G.A.Engler, Pflanzenr., IV, 147, II: 47 (1910).
Manihot sellowiana Klotzsch ex Pax in H.G.A.Engler, Pflanzenr., IV, 147, II: 40 (1910).

Manihot pruinosa Pohl, Pl. Bras. Icon. Descr. 1: 28 (1827). *Jatropha pruinosa* (Pohl) Steud., Nomencl. Bot., ed. 2, 1: 800 (1840). *Manihot pruinosa* var. *genuina* Müll.Arg. in A.P.de Candolle, Prodr. 15(2): 1061 (1866), nom. inval.
Brazil (Mato Grosso, Goiás). 84 BZC. Nanophan.
Manihot pruinosa var. *pumila* Müll.Arg. in A.P.de Candolle, Prodr. 15(2): 1061 (1866).
Manihot burchellii Müll.Arg. in C.F.P.von Martius, Fl. Bras. 11(2): 457 (1874).
Manihot pseudopruinosa Pax & K.Hoffm. in H.G.A.Engler, Pflanzenr., IV, 147, II: 61 (1910).

Manihot pseudoglaziovii Pax & K.Hoffm. in H.G.A.Engler, Pflanzenr., IV, 147, XVI: 196 (1924).
Brazil (Ceará, Pernambuco, Rio Grande do Norte, Paraíba). 84 BZE. Phan.

Manihot purpureocostata Pohl, Pl. Bras. Icon. Descr. 1: 19 (1827). *Jatropha purpureocostata* (Pohl) Steud., Nomencl. Bot., ed. 2, 1: 800 (1840).
Brazil (Goiás: near Cavalcante). 84 BZC. Cham.

Manihot pusilla Pohl, Pl. Bras. Icon. Descr. 1: 36 (1827). *Jatropha pusilla* (Pohl) Steud., Nomencl. Bot., ed. 2, 1: 800 (1840). *Manihot palmata* var. *pusilla* (Pohl) Müll.Arg. in A.P.de Candolle, Prodr. 15(2): 1063 (1866).
Brazil (Goiás: Serra de Cristaes). 84 BZC. Hemicr.

Manihot quinquefolia Pohl, Pl. Bras. Icon. Descr. 1: 56 (1827). *Jatropha quinquefolia* (Pohl) Steud., Nomencl. Bot., ed. 2, 1: 800 (1840).
Brazil (C. Bahia). 84 BZE. Nanophan.
Jatropha quinqueformis Steud., Nomencl. Bot., ed. 2, 1: 800 (1840), orth. var.
Manihot quinqueformis Steud., Nomencl. Bot., ed. 2, 1: 800 (1840), orth. var.

Manihot quinqueloba Pohl, Pl. Bras. Icon. Descr. 1: 21 (1827). *Jatropha quinqueloba* (Pohl) Steud., Nomencl. Bot., ed. 2, 1: 800 (1840), nom. illeg.
Brazil (Mato Grosso, Goiás). 84 BZC. Nanophan.
Manihot subquinqueloba Müll.Arg. in C.F.P.von Martius, Fl. Bras. 11(2): 446 (1874).
Manihot polyantha Pax & K.Hoffm. in H.G.A.Engler, Pflanzenr., IV, 147, II: 88 (1910).

Manihot quinquepartita Huber ex D.J.Rogers & Appan, Fl. Neotrop. Monogr. 13: 196 (1973).
Brazil (Amapá, Pará, Maranhão, Mato Grosso). 84 BZC BZE BZN. Cl. nanophan.

Manihot reniformis Pohl, Pl. Bras. Icon. Descr. 1: 56 (1827). *Jatropha reniformis* (Pohl) Steud., Nomencl. Bot., ed. 2, 1: 800 (1840).
Brazil (Bahia). 84 BZE. Cham.

Manihot reptans Pax in H.G.A.Engler, Pflanzenr., IV, 147, II: 30 (1910).
Brazil (Goiás, Minas Gerais). 84 BZC BZL. Cham.

Manihot rhomboidea Müll.Arg., Linnaea 34: 205 (1865).
Mexico, C. America. 79 MXC MXE MXS MXT 80 ELS GUA HON NIC. Tuber cham. or nanophan.

subsp. **microcarpa** (Müll.Arg.) D.J.Rogers & Appan, Fl. Neotrop. Monogr. 13: 60 (1973).
C. & S. Mexico, Guatemala, El Salvador, Honduras, Nicaragua. 79 MXC MXS MXT 80 ELS GUA HON NIC. Tuber cham. or nanophan.
* *Manihot microcarpa* Müll.Arg., Flora 55: 42 (1872).
Manihot parvicocca Croizat, J. Arnold Arbor. 23: 219 (1942).

subsp. **rhomboidea**
Mexico. 79 MXC MXE MXS. Tuber cham. or nanophan.
Manihot mexicana I.M.Johnst., Contr. Gray Herb. 68: 90 (1923).
Manihot ludibunda Croizat, J. Arnold Arbor. 23: 219 (1942).

Manihot rubricaulis I.M.Johnst., Contr. Gray Herb. 68: 90 (1923).
Mexico (Sinaloa, Durango). 79 MXE MXN. Nanophan. or phan.

subsp. **isoloba** (Standl.) D.J.Rogers & Appan, Fl. Neotrop. Monogr. 13: 46 (1973).
Mexico (Sinaloa, Sonora, Chihuahua). 79 MXE MXN. Nanophan. or phan.
* *Manihot isoloba* Standl., Publ. Field Mus. Nat. Hist., Bot. Ser. 17: 197 (1937).

subsp. **rubricaulis**
Mexico (Sinaloa, Durango). 79 MXE MXN. Nanophan. or phan.

Manihot sagittatopartita Pohl, Pl. Bras. Icon. Descr. 1: 22 (1827). *Jatropha sagittatopartita* (Pohl) Steud., Nomencl. Bot., ed. 2, 1: 800 (1840).
Brazil (Bahia, Goiás, Minas Gerais). 84 BZC BZE BZL. Cham.

Manihot salicifolia Pohl, Pl. Bras. Icon. Descr. 1: 18 (1827). *Jatropha salicifolia* (Pohl) Steud., Nomencl. Bot., ed. 2, 1: 800 (1840).
Brazil (Mato Grosso, Goiás). 84 BZC. Hemicr. or cham.
Manihot riedeliana Müll.Arg. in C.F.P.von Martius, Fl. Bras. 11(2): 443 (1874).
Manihot mattogrossensis Pax & K.Hoffm. in H.G.A.Engler, Pflanzenr., IV, 147, XVI: 197 (1924).

Manihot sparsifolia Pohl, Pl. Bras. Icon. Descr. 1: 26 (1827). *Jatropha sparsifolia* (Pohl) Steud., Nomencl. Bot., ed. 2, 1: 800 (1840).
Brazil (Goiás). 84 BZC. Cham. or nanophan.
Manihot amaroleitensis Baill., Adansonia 4: 281 (1864).

Manihot stipularis Pax in H.G.A.Engler, Pflanzenr., IV, 147, II: 50 (1910).
Brazil (Goiás, Brasília D.F.). 84 BZC. Cham.

Manihot stricta Baill., Adansonia 4: 282 (1864).
Peru, Brazil (Mato Grosso, Goiás). 83 PER 84 BZC. Cham. or nanophan.
Manihot linearifolia Müll.Arg., Flora 55: 43 (1872).

Manihot subspicata D.J.Rogers & Appan, Fl. Neotrop. Monogr. 13: 62 (1973).
Mexico (Coahuila, Nuevo León, Tamaulipas). 79 MXE. Tuber nanophan.

Manihot surinamensis D.J.Rogers & Appan, Fl. Neotrop. Monogr. 13: 80 (1973).
Venezuela (Amazonas), Guianas. 82 FRG GUY SUR VEN. Nanophan.

Manihot tenella Müll.Arg. in C.F.P.von Martius, Fl. Bras. 11(2): 484 (1874).
Brazil (São Paulo), Paraguay. 84 BZL 85 PAR. Nanophan.

Manihot tomatophylla Standl., Amer. Midl. Naturalist 36: 178 (1946).
Mexico (Michoacán: Apatzingan Reg.). 79 MXS. Phan.

Manihot tomentosa Pohl, Pl. Bras. Icon. Descr. 1: 50 (1827). *Jatropha tomentosa* (Pohl) Steud., Nomencl. Bot., ed. 2, 1: 800 (1840).
Brazil (Goiás, Minas Gerais). 84 BZC BZL. Hemicr.

subsp. **araliifolia** (Pax) D.J.Rogers & Appan, Fl. Neotrop. Monogr. 13: 215 (1973).
Brazil (Goiás). 84 BZC. Hemicr.
* *Manihot araliifolia* Pax, Verh. Bot. Vereins Prov. Brandenburg 50: 26 (1908 publ. 1910).
Manihot canastrana Glaz., Bull. Soc. Bot. France 59(3): 628 (1912 publ. 1913), nom. nud.

subsp. **tomentosa**
Brazil (Goiás, Minas Gerais). 84 BZC BZL. Hemicr.

Manihot triloba (Sessé ex Cerv.) McVaugh ex Miranda, Bol. Soc. Bot. México 29: 38 (1965).
Mexico (Colima, Michoacán, Morelos, Guerrero, México State, Puebla, Oaxaca, Chiapas). 79 MXC MXE MXS MXT. Phan.
* *Jatropha triloba* Sessé ex Cerv., Supl. Gaz. Lit. Mexico: 4 (2 July 1794).

Manihot tripartita (Spreng.) Müll.Arg. in A.P.de Candolle, Prodr. 15(2): 1068 (1866).
Brazil, Paraguay. 84 BZC BZE BZL BZN 85 PAR. Cham. or nanophan.
* *Jatropha tripartita* Spreng., Syst. Veg. 3: 76 (1826). *Manihot tripartita* var. *genuina* Müll.Arg. in A.P.de Candolle, Prodr. 15(2): 1068 (1866), nom. inval.

subsp. **humilis** (Müll.Arg.) D.J.Rogers & Appan, Fl. Neotrop. Monogr. 13: 236 (1973).
Brazil (Goiás, Minas Gerais, São Paulo). 84 BZC BZL. Cham. or nanophan.
* *Manihot humilis* Müll.Arg. in C.F.P.von Martius, Fl. Bras. 11(2): 448 (1874).
Manihot tripartita var. *glabra* Müll.Arg. in C.F.P.von Martius, Fl. Bras. 11(2): 478 (1874).

subsp. **indivisa** Allem, Revista Brasil. Biol. 49: 13 (1989).
Brazil (Mato Grosso). 84 BZC. Nanophan.

subsp. **laciniosa** (Pohl) D.J.Rogers & Appan, Fl. Neotrop. Monogr. 13: 238 (1973).
Brazil (Bahia, Goiás, Minas Gerais, Rio de Janeiro). 84 BZC BZE BZL. Cham. or nanophan.

* *Manihot laciniosa* Pohl, Pl. Bras. Icon. Descr. 1: 54 (1827). *Jatropha lanciniosa* (Pohl) Steud., Nomencl. Bot., ed. 2, 1: 799 (1840). *Jatropha laciniosa* (Pohl) Steud., Nomencl. Bot., ed. 2, 1: 799 (1840). *Manihot sinuata* var. *laciniosa* (Pohl) Müll.Arg. in A.P.de Candolle, Prodr. 15(2): 1075 (1866). *Manihot laciniosa* var. *genuina* Müll.Arg. in C.F.P.von Martius, Fl. Bras. 11(2): 483 (1874), nom. inval.
Manihot intercedens Müll.Arg. in C.F.P.von Martius, Fl. Bras. 11(2): 483 (1874).
Manihot laciniosa var. *lanata* Müll.Arg. in C.F.P.von Martius, Fl. Bras. 11(2): 483 (1874).
Manihot lagoensis Müll.Arg. in C.F.P.von Martius, Fl. Bras. 11(2): 475 (1874).
Manihot tripartita var. *subintegra* Müll.Arg. in C.F.P.von Martius, Fl. Bras. 11(2): 478 (1874).

subsp. **tripartita**
Brazil (Pará, Mato Grosso, Bahia, Goiás, Minas Gerais, São Paulo, Rio de Janeiro), Paraguay. 84 BZC BZE BZL BZN 85 PAR. Cham. or nanophan.
Manihot cajaniformis Pohl, Pl. Bras. Icon. Descr. 1: 45 (1827). *Jatropha cajaniformis* (Pohl) Steud., Nomencl. Bot., ed. 2, 1: 799 (1840). *Manihot tripartita* var. *cajaniformis* (Pohl) Müll.Arg. in A.P.de Candolle, Prodr. 15(2): 1068 (1866).
Manihot cleomifolia Pohl, Pl. Bras. Icon. Descr. 1: 51 (1827). *Jatropha cleomifolia* (Pohl) Steud., Nomencl. Bot., ed. 2, 1: 799 (1840).
Manihot dalechampiiformis Pohl, Pl. Bras. Icon. Descr. 1: 51 (1827). *Jatropha dalechampiiformis* (Pohl) Steud., Nomencl. Bot., ed. 2, 1: 799 (1840). *Manihot tripartita* var. *dalechampiiformis* (Pohl) Pax in H.G.A.Engler, Pflanzenr., IV, 147, II: 38 (1910).
Manihot porrecta Pohl, Pl. Bras. Icon. Descr. 1: 46 (1827). *Jatropha porrecta* (Pohl) Steud., Nomencl. Bot., ed. 2, 1: 800 (1840). *Manihot tripartita* var. *porrecta* (Pohl) Müll.Arg. in A.P.de Candolle, Prodr. 15(2): 1068 (1866).
Manihot sinuata Pohl, Pl. Bras. Icon. Descr. 1: 18 (1827). *Jatropha sinuata* (Pohl) Steud., Nomencl. Bot., ed. 2, 1: 800 (1840). *Manihot sinuata* var. *genuina* Müll.Arg. in A.P.de Candolle, Prodr. 15(2): 1074 (1866), nom. inval.
Manihot tomentella Pohl, Pl. Bras. Icon. Descr. 1: 45 (1827). *Jatropha tomentella* (Pohl) Steud., Nomencl. Bot., ed. 2, 1: 800 (1840).
Manihot tripartita var. *glauca* Müll.Arg. in A.P.de Candolle, Prodr. 15(2): 1069 (1866).
Manihot tripartita var. *lanceolata* Müll.Arg. in A.P.de Candolle, Prodr. 15(2): 1068 (1866).
Manihot tripartita var. *seminuda* Müll.Arg. in C.F.P.von Martius, Fl. Bras. 11(2): 477 (1874).
Manihot tripartita var. *apaensis* Chodat & Hassl., Bull. Herb. Boissier, II, 5: 651 (1905).
Manihot consanguinea Klotzsch ex Pax in H.G.A.Engler, Pflanzenr., IV, 147, II: 38 (1910), nom. nud.

subsp. **vestita** (S.Moore) D.J.Rogers & Appan, Fl. Neotrop. Monogr. 13: 236 (1973).
Brazil (Mato Grosso: Serra de Chapada). 84 BZC. Cham. or nanophan.
Manihot tripartita var. *vestita* S.Moore, Trans. Linn. Soc. London, Bot. 4: 466 (1895).
Manihot trichandra Pax & K.Hoffm. in H.G.A.Engler, Pflanzenr., IV, 147, II: 39 (1910).

Manihot triphylla Pohl, Pl. Bras. Icon. Descr. 1: 37 (1827). *Jatropha triphylla* (Pohl) Steud., Nomencl. Bot., ed. 2, 1: 800 (1840). *Manihot gracilis* var. *triphylla* (Pohl) Müll.Arg. in A.P.de Candolle, Prodr. 15(2): 1066 (1866). *Manihot triphylla* var. *genuina* Pax in H.G.A.Engler, Pflanzenr., IV, 147, II: 74 (1910), nom. inval.
Brazil (Minas Gerais, Goiás, Brasília D.F.). 84 BZC BZL. Nanophan.
Manihot angustifrons Müll.Arg. in C.F.P.von Martius, Fl. Bras. 11(2): 461 (1874).
Manihot stenophylla Pax & K.Hoffm. in H.G.A.Engler, Pflanzenr., IV, 147, II: 73 (1910).

Manihot tristis Müll.Arg. in C.F.P.von Martius, Fl. Bras. 11(2): 449 (1874).
N. South America, N. Brazil. 82 FRG GUY SUR VEN 84 BZN. Nanophan.
Manihot orinocensis Croizat, J. Arnold Arbor. 24: 169 (1943).

subsp. **saxicola** (Lanj.) D.J.Rogers & Appan, Fl. Neotrop. Monogr. 13: 84 (1973).
Guyana, Guiana, Surinam, Brazil (Amapá). 82 FRG GUY SUR 84 BZN. Nanophan.
* *Manihot saxicola* Lanj., Meded. Bot. Mus. Herb. Rijks Univ. Utrecht 67: 544 (1939).

subsp. **surumuensis** (Ule) D.J.Rogers & Appan, Fl. Neotrop. Monogr. 13: 86 (1973).
Brazil (Roraima). 84 BZN. Nanophan.
* *Manihot surumuensis* Ule, Bot. Jahrb. Syst. 50(114): 12 (1914).

subsp. **tristis**
Venezuela (Bolívar, Amazonas). 82 VEN. Nanophan.

Manihot variifolia Pax & K.Hoffm. in H.G.A.Engler, Pflanzenr., IV, 147, II: 85 (1910).
Paraguay (Sierra de Amabay), Brazil (Goiás: Campo Duro) ? 84 BZC? 85 PAR. Nanophan.
Manihot katharinae Pax in H.G.A.Engler, Pflanzenr., IV, 147, II: 87 (1910).

Manihot violacea Pohl, Pl. Bras. Icon. Descr. 1: 43 (1827). *Jatropha violacea* (Pohl) Steud., Nomencl. Bot., ed. 2, 1: 800 (1840). *Manihot violacea* var. *genuina* Müll.Arg. in A.P.de Candolle, Prodr. 15(2): 1070 (1866), nom. inval.
Brazil (Mato Grosso, Brasília D.F., Goiás, Minas Gerais). 84 BZC BZL. Cham. or nanophan.

subsp. **recurvata** D.J.Rogers & Appan, Fl. Neotrop. Monogr. 13: 134 (1973).
Brazil (Goiás, Brasília D.F.). 84 BZC. Cham. or nanophan.

subsp. **violacea**
Brazil (Mato Grosso, Brasília D.F., Goiás, Minas Gerais). 84 BZC BZL. Cham. or nanophan.

Manihot walkerae Croizat, Bull. Torrey Bot. Club 69: 452 (1942).
Texas, Mexico (Tamaulipas). 77 TEX 79 MXE. Tuber cham.

Manihot websteri D.J.Rogers & Appan, Fl. Neotrop. Monogr. 13: 72 (1973).
Mexico (Puebla: SE. of Izucar de Matamoros). 79 MXC. Phan.

Manihot weddelliana Baill., Adansonia 4: 281 (1864).
Brazil (Goiás). 84 BZC. Cham.

Manihot xavantinensis D.J.Rogers & Appan, Fl. Neotrop. Monogr. 13: 124 (1973). *Manihot tripartita* subsp. *xavantinensis* (D.J.Rogers & Appan) Allem, Revista Brasil. Biol. 49: 13 (1989).
Brazil (Mato Grosso). 84 BZC. Nanophan.

Manihot zehntneri Ule, Bot. Jahrb. Syst. 50(114): 10 (1914).
Brazil (SC. Bahia). 84 BZE. Nanophan.

Synonyms:
Manihot acutiloba Weath. === **Manihot angustiloba** (Torr.) Müll.Arg.
Manihot affinis Pax & K.Hoffm. === **Manihot procumbens** Müll.Arg.
Manihot aipi Pohl === **Manihot esculenta** Crantz
Manihot aipi var. *lanceolata* Pohl === **Manihot esculenta** Crantz
Manihot aipi var. *latifolia* Pohl === **Manihot esculenta** Crantz
Manihot aipi var. *lutescens* Pohl === **Manihot esculenta** Crantz
Manihot alcicornis Klotzsch ex Pax === **Manihot anomala** subsp. **cujabensis** (Müll.Arg.) D.J.Rogers & Appan
Manihot amaroleitensis Baill. === **Manihot sparsifolia** Pohl
Manihot amazonica Ule === **Manihot brachyloba** Müll.Arg.
Manihot angustifrons Müll.Arg. === **Manihot triphylla** Pohl
Manihot anisitsii Pax & K.Hoffm. === **Manihot guaranitica** Chodat & Hassl. subsp. **guaranitica**
Manihot araliifolia Pax === **Manihot tomentosa** subsp. **araliifolia** (Pax) D.J.Rogers & Appan
Manihot arcuata Pohl === **Manihot divergens** Pohl
Manihot aypi Spruce === **Manihot esculenta** Crantz
Manihot bahiensis Ule === **Manihot caerulescens** Pohl subsp. **caerulescens**
Manihot bahiensis var. *microsperma* Ule === **Manihot caerulescens** Pohl subsp. **caerulescens**

Manihot berroana Beauverd === ?
Manihot boliviana Pax & K.Hoffm. === **Manihot guaranitica** subsp. **boliviana** (Pax & K.Hoffm.) D.J.Rogers & Appan
Manihot brachystachys Pax & K.Hoffm. === **Manihot attenuata** Müll.Arg.
Manihot brasiliensis Klotzsch ex Pax === **Manihot inflata** Müll.Arg.
Manihot brevipedicellata Pax & K.Hoffm. === **Manihot pilosa** Pohl
Manihot burchellii Müll.Arg. === **Manihot pruinosa** Pohl
Manihot caerulescens var. *genuina* Müll.Arg. === **Manihot caerulescens** Pohl
Manihot caerulescens var. *pubescens* Müll.Arg. === **Manihot caerulescens** Pohl subsp. **caerulescens**
Manihot cajaniformis Pohl === **Manihot tripartita** (Spreng.) Müll.Arg. subsp. **tripartita**
Manihot canastrana Glaz. === **Manihot tomentosa** subsp. **araliifolia** (Pax) D.J.Rogers & Appan
Manihot cannabina Sweet === **Manihot esculenta** Crantz
Manihot caricifolia Pohl === **Manihot anomala** Pohl subsp. **anomala**
Manihot carthaginensis var. *anisophylla* (Griseb.) Kuntze === **Manihot anisophylla** (Griseb.) Müll.Arg.
Manihot cassava Cook & Collins === **Manihot esculenta** Crantz
Manihot cearensis Pax & K.Hoffm. === **Manihot caerulescens** Pohl subsp. **caerulescens**
Manihot cleomifolia Pohl === **Manihot tripartita** (Spreng.) Müll.Arg. subsp. **tripartita**
Manihot coerulea Steud. === **Manihot caerulescens** Pohl subsp. **caerulescens**
Manihot colimensis Croizat === **Manihot chlorosticta** Stand. & Goldman
Manihot consanguinea Klotzsch ex Pax === **Manihot tripartita** (Spreng.) Müll.Arg. subsp. **tripartita**
Manihot conulifera Müll.Arg. === **Manihot pentaphylla** subsp. **rigidula** (Müll.Arg.) D.J.Rogers & Appan
Manihot cordifolia Pax === **Manihot populifolia** Pax
Manihot cujabensis Müll.Arg. === **Manihot anomala** subsp. **cujabensis** (Müll.Arg.) D.J.Rogers & Appan
Manihot cuneata Ule === **Manihot caerulescens** Pohl subsp. **caerulescens**
Manihot curcas (L.) Crantz === **Jatropha curcas** L.
Manihot dalechampiiformis Pohl === **Manihot tripartita** (Spreng.) Müll.Arg. subsp. **tripartita**
Manihot depauperata Pax & K.Hoffm. === **Manihot gracilis** Pohl subsp. **gracilis**
Manihot dichotoma var. *genuina* Pax === **Manihot dichotoma** Ule
Manihot dichotoma var. *parvifolia* Ule === **Manihot dichotoma** Ule
Manihot diffusa Pohl === **Manihot esculenta** Crantz
Manihot digitata Sweet === ?
Manihot digitiformis Pohl === **Manihot esculenta** Crantz
Manihot discolor Ule === **Manihot caerulescens** Pohl subsp. **caerulescens**
Manihot diversifolia (Steud.) Sweet === ?
Manihot dulcis (J.F.Gmel.) Baill. === **Manihot esculenta** Crantz
Manihot dulcis var. *aipi* (Pohl) Pax === **Manihot esculenta** Crantz
Manihot dulcis var. *diffusa* (Pohl) Pax === **Manihot esculenta** Crantz
Manihot dulcis var. *ferruginea* (Müll.Arg.) Pax === **Manihot leptophylla** Pax & K.Hoffm.
Manihot dulcis var. *flabellifolia* (Pohl) Pax === **Manihot esculenta** Crantz
Manihot dulcis var. *leptopoda* (Müll.Arg.) Pax === **Manihot palmata** Müll.Arg.
Manihot dulcis var. *multifida* (Graham) Pax === **Manihot grahamii** Hook.
Manihot edule A.Rich. === **Manihot esculenta** Crantz
Manihot elegans Müll.Arg. === **Manihot procumbens** Müll.Arg.
Manihot enneaphylla Pax & K.Hoffm. === **Manihot grahamii** Hook.
Manihot esculenta subsp. *alboerecta* Cif. === **Manihot esculenta** Crantz
Manihot esculenta var. *argentea* Cif. === **Manihot esculenta** Crantz
Manihot esculenta var. *coalescens* Cif. === **Manihot esculenta** Crantz
Manihot esculenta var. *communis* Cif. === **Manihot esculenta** Crantz
Manihot esculenta var. *debilis* Cif. === **Manihot esculenta** Crantz
Manihot esculenta subsp. *diffusa* Cif. === **Manihot esculenta** Crantz
Manihot esculenta var. *digitifolia* Cif. === **Manihot esculenta** Crantz

Manihot esculenta var. *domingensis* Cif. === **Manihot esculenta** Crantz
Manihot esculenta var. *fertilis* Cif. === **Manihot esculenta** Crantz
Manihot esculenta subsp. *flabellifolia* (Pohl) Cif. === **Manihot esculenta** Crantz
Manihot esculenta var. *flavicaulis* Cif. === **Manihot esculenta** Crantz
Manihot esculenta var. *fuscescens* Cif. === **Manihot esculenta** Crantz
Manihot esculenta subsp. *grandifolia* Cif. === **Manihot esculenta** Crantz
Manihot esculenta var. *hispaniolensis* Cif. === **Manihot esculenta** Crantz
Manihot esculenta var. *jamaicensis* Cif. === **Manihot esculenta** Crantz
Manihot esculenta var. *luteola* Cif. === **Manihot esculenta** Crantz
Manihot esculenta var. *mutabilis* Cif. === **Manihot esculenta** Crantz
Manihot esculenta var. *nodosa* Cif. === **Manihot esculenta** Crantz
Manihot esculenta subsp. *peruviana* (Müll.Arg.) Allan === **Manihot peruviana** Müll.Arg.
Manihot esculenta var. *pohlii* Cif. === **Manihot esculenta** Crantz
Manihot esculenta var. *ramosissima* Cif. === **Manihot esculenta** Crantz
Manihot esculenta var. *rufescens* Cif. === **Manihot esculenta** Crantz
Manihot esculenta var. *sprucei* Lanj. === **Manihot esculenta** Crantz
Manihot esculenta var. *tenerrima* (Pohl) Cif. === **Manihot pentaphylla** subsp. **tenuifolia** (Pohl) D.J.Rogers & Appan
Manihot esculenta var. *zimmermannii* Cif. === **Manihot esculenta** Crantz
Manihot ferruginea Ule === **Manihot caerulescens** Pohl subsp. **caerulescens**
Manihot fiebrigii Pax & K.Hoffm. === **Manihot guaranitica** Chodat & Hassl. subsp. **guaranitica**
Manihot flabellifolia Pohl === **Manihot esculenta** Crantz
Manihot flexuosa Pax & K.Hoffm. === **Manihot esculenta** Crantz
Manihot floribunda Pax & K.Hoffm. === **Manihot epruinosa** Pax & K.Hoffm.
Manihot glabrata (Chodat & Hassl.) Pax & K.Hoffm. === **Manihot anomala** subsp. **glabrata** (Chodat & Hassl.) D.J.Rogers & Appan
Manihot gossypiifolia (L.) Crantz === **Jatropha gossypiifolia** L.
Manihot gracilis var. *genuina* Müll.Arg. === **Manihot gracilis** Pohl
Manihot gracilis var. *pronifolia* (Pohl) Müll.Arg. === **Manihot gracilis** Pohl subsp. **gracilis**
Manihot gracilis var. *tenerrima* (Pohl) Müll.Arg. === **Manihot pentaphylla** subsp. **tenuifolia** (Pohl) D.J.Rogers & Appan
Manihot gracilis var. *tenuifolia* (Pohl) Müll.Arg. === **Manihot pentaphylla** subsp. **tenuifolia** (Pohl) D.J.Rogers & Appan
Manihot gracilis var. *triphylla* (Pohl) Müll.Arg. === **Manihot triphylla** Pohl
Manihot graminifolia Chodat & Hassl. === **Manihot pentaphylla** subsp. **graminifolia** (Chodat & Hassl.) D.J.Rogers & Appan
Manihot grandiflora Müll.Arg. === **Manihot caerulescens** Pohl subsp. **caerulescens**
Manihot grandistipula Pax === **Manihot guaranitica** Chodat & Hassl. subsp. **guaranitica**
Manihot gualanensis S.F.Blake === **Manihot aesculifolia** (Kunth) Pohl
Manihot guyanensis Klotzsch ex Pax === **Manihot esculenta** Crantz
Manihot harmsiana Ule === **Manihot caerulescens** Pohl subsp. **caerulescens**
Manihot hemigynandra Müll.Arg. === **Manihot pilosa** Pohl
Manihot hemitrichandra Müll.Arg. === **Manihot pilosa** Pohl
Manihot herbacea (L.) Crantz === **Cnidoscolus urens** (L.) Arthur var. **urens**
Manihot heterandra Ule === **Manihot peruviana** Müll.Arg.
Manihot heterophylla Pohl === **Manihot anomala** Pohl subsp. **anomala**
Manihot humilis Müll.Arg. === **Manihot tripartita** subsp. **humilis** (Müll.Arg.) D.J.Rogers & Appan
Manihot intercedens Müll.Arg. === **Manihot tripartita** subsp. **laciniosa** (Pohl) D.J.Rogers & Appan
Manihot intermedia Weath. === **Manihot aesculifolia** (Kunth) Pohl
Manihot isoloba Standl. === **Manihot rubricaulis** subsp. **isoloba** (Standl.) D.J.Rogers & Appan
Manihot janipha (L.) Pohl === **Manihot carthaginensis** (Jacq.) Müll.Arg.
Manihot japonica Semler ===**?**
Manihot johannis Pax === **Manihot pohliana** Müll.Arg.
Manihot katharinae Pax === **Manihot variifolia** Pax & K.Hoffm.

Manihot klingensteinii Pax & K.Hoffm. === **Manihot anomala** subsp. **glabrata** (Chodat & Hassl.) D.J.Rogers & Appan
Manihot labroyana Ule === **Manihot caerulescens** Pohl subsp. **caerulescens**
Manihot laciniosa Pohl === **Manihot tripartita** subsp. **laciniosa** (Pohl) D.J.Rogers & Appan
Manihot laciniosa var. *genuina* Müll.Arg. === **Manihot tripartita** subsp. **laciniosa** (Pohl) D.J.Rogers & Appan
Manihot laciniosa var. *lanata* Müll.Arg. === **Manihot tripartita** subsp. **laciniosa** (Pohl) D.J.Rogers & Appan
Manihot lagoensis Müll.Arg. === **Manihot tripartita** subsp. **laciniosa** (Pohl) D.J.Rogers & Appan
Manihot langsdorffii Müll.Arg. === **Manihot pilosa** Pohl
Manihot langsdorffii var. *glabra* Chodat & Hassl. === **Manihot anomala** subsp. **glabrata** (Chodat & Hassl.) D.J.Rogers & Appan
Manihot leptopoda (Müll.Arg.) D.J.Rogers & Appan === **Manihot palmata** Müll.Arg.
Manihot linearifolia Müll.Arg. === **Manihot stricta** Baill.
Manihot lobata (Chodat & Hassl.) Pax === **Manihot grahamii** Hook.
Manihot loeflingii Müll.Arg. === **Manihot grahamii** Hook.
Manihot loeflingii var. *multifida* (Graham) Müll.Arg. === **Manihot grahamii** Hook.
Manihot loureirii Pohl === **Manihot esculenta** Crantz
Manihot ludibunda Croizat === **Manihot rhomboidea** Müll.Arg. subsp. **rhomboidea**
Manihot lyrata Ule === **Manihot caerulescens** Pohl subsp. **caerulescens**
Manihot macrantha Pax & K.Hoffm. === **Manihot caerulescens** subsp. **macrantha** (Pax & K.Hoffm.) D.J.Rogers & Appan
Manihot manihot (L.) Cockerell === **Manihot esculenta** Crantz
Manihot manihot (L.) H.Karst. === **Manihot esculenta** Crantz
Manihot maracasensis var. *vestita* Pax & K.Hoffm. === **Manihot maracasensis** Ule
Manihot mattogrossensis Pax & K.Hoffm. === **Manihot salicifolia** Pohl
Manihot meeboldii Pax & K.Hoffm. === **Manihot procumbens** Müll.Arg.
Manihot melanobasis Müll.Arg. === **Manihot esculenta** Crantz
Manihot meridensis Pittier === **Manihot carthaginensis** (Jacq.) Müll.Arg.
Manihot mexicana I.M.Johnst. === **Manihot rhomboidea** Müll.Arg. subsp. **rhomboidea**
Manihot meyeriana Klotzsch ex Pax === **Manihot pilosa** Pohl
Manihot microcarpa Müll.Arg. === **Manihot rhomboidea** subsp. **microcarpa** (Müll.Arg.) D.J.Rogers & Appan
Manihot microdendron Ule === **Manihot caerulescens** Pohl subsp. **caerulescens**
Manihot mobilis Standl. === **Manihot chlorosticta** Stand. & Goldman
Manihot moluccana (L.) Crantz === **Aleurites moluccana** (L.) Willd.
Manihot multifida (L.) Crantz === **Jatropha multifida** L.
Manihot multiflora Pax & K.Hoffm. === **Manihot anomala** subsp. **glabrata** (Chodat & Hassl.) D.J.Rogers & Appan
Manihot neoglaziovii Pax & K.Hoffm. ex Luetzelb. === ?
Manihot occidentalis Müll.Arg. === **Manihot jacobinensis** Müll.Arg.
Manihot olfersiana Pax === **Manihot aesculifolia** (Kunth) Pohl
Manihot orinocensis Croizat === **Manihot tristis** Müll.Arg.
Manihot palmata (Vell.) Pax === **Manihot palmata** Müll.Arg.
Manihot palmata var. *aipi* (Pohl) Müll.Arg. === **Manihot esculenta** Crantz
Manihot palmata var. *diffusa* (Pohl) Müll.Arg. === **Manihot esculenta** Crantz
Manihot palmata var. *digitiformis* (Pohl) Müll.Arg. === **Manihot esculenta** Crantz
Manihot palmata var. *ferruginea* Müll.Arg. === **Manihot leptophylla** Pax & K.Hoffm.
Manihot palmata var. *flabellifolia* (Pohl) Müll.Arg. === **Manihot esculenta** Crantz
Manihot palmata var. *genuina* Müll.Arg. === **Manihot palmata** Müll.Arg.
Manihot palmata var. *leptopoda* Müll.Arg. === **Manihot palmata** Müll.Arg.
Manihot palmata var. *multifida* (Graham) Müll.Arg. === **Manihot grahamii** Hook.
Manihot palmata var. *pusilla* (Pohl) Müll.Arg. === **Manihot pusilla** Pohl
Manihot paraensis Müll.Arg. === **Manihot caerulescens** subsp. **paraensis** (Müll.Arg.) D.J.Rogers & Appan
Manihot pardina Müll.Arg. === **Manihot gracilis** subsp. **varians** (Pohl) D.J.Rogers & Appan

Manihot parvicocca Croizat === **Manihot rhomboidea** subsp. **microcarpa** (Müll.Arg.) D.J.Rogers & Appan
Manihot pavoniana Müll.Arg. === **Manihot anomala** subsp. **pavoniana** (Müll.Arg.) D.J.Rogers & Appan
Manihot pedicellaris Müll.Arg. === **Manihot pilosa** Pohl
Manihot pentaphylla var. *genuina* Müll.Arg. === **Manihot pentaphylla** Pohl
Manihot pentaphylla var. *paviifolia* (Pohl) Müll.Arg. === **Manihot paviifolia** Pohl
Manihot piauhyensis Ule === **Manihot caerulescens** Pohl subsp. **caerulescens**
Manihot pittieri Pax & K.Hoffm. === **Manihot carthaginensis** (Jacq.) Müll.Arg.
Manihot polyantha Pax & K.Hoffm. === **Manihot quinqueloba** Pohl
Manihot porrecta Pohl === **Manihot tripartita** (Spreng.) Müll.Arg. subsp. **tripartita**
Manihot preciosa Schindler === **Manihot dichotoma** Ule
Manihot procumbens var. *genuina* Pax === **Manihot procumbens** Müll.Arg.
Manihot procumbens var. *grandifolia* Chodat & Hassl. === **Manihot procumbens** Müll.Arg.
Manihot pronifolia Pohl === **Manihot gracilis** Pohl subsp. **gracilis**
Manihot pruinosa var. *genuina* Müll.Arg. === **Manihot pruinosa** Pohl
Manihot pruinosa var. *pumila* Müll.Arg. === **Manihot pruinosa** Pohl
Manihot pseudoheterophylla Pax & K.Hoffm. === **Manihot anomala** subsp. **glabrata** (Chodat & Hassl.) D.J.Rogers & Appan
Manihot pseudopruinosa Pax & K.Hoffm. === **Manihot pruinosa** Pohl
Manihot pubescens Pohl === **Manihot anomala** subsp. **pubescens** (Pohl) D.J.Rogers & Appan
Manihot pubescens f. *glabrata* Chodat & Hassl. === **Manihot anomala** subsp. **glabrata** (Chodat & Hassl.) D.J.Rogers & Appan
Manihot quinqueformis Steud. === **Manihot quinquefolia** Pohl
Manihot recognita Pax === **Manihot guaranitica** Chodat & Hassl. subsp. **guaranitica**
Manihot reflexa Klotzsch ex Pax === **Manihot pentaphylla** subsp. **tenuifolia** (Pohl) D.J.Rogers & Appan
Manihot remotiloba Pittier === **Manihot carthaginensis** (Jacq.) Müll.Arg.
Manihot riedeliana Klotzsch ex Pax === **Manihot caerulescens** Pohl subsp. **caerulescens**
Manihot riedeliana Müll.Arg. === **Manihot salicifolia** Pohl
Manihot rigidifolia Pax & K.Hoffm. === **Manihot jacobinensis** Müll.Arg.
Manihot rigidula Müll.Arg. === **Manihot pentaphylla** subsp. **rigidula** (Müll.Arg.) D.J.Rogers & Appan
Manihot rotundata Ule === **Manihot caerulescens** Pohl subsp. **caerulescens**
Manihot rusbyi Britton === **Manihot brachyloba** Müll.Arg.
Manihot saxicola Lanj. === **Manihot tristis** subsp. **saxicola** (Lanj.) D.J.Rogers & Appan
Manihot sellowiana Klotzsch ex Pax === **Manihot procumbens** Müll.Arg.
Manihot sinuata Pohl === **Manihot tripartita** (Spreng.) Müll.Arg. subsp. **tripartita**
Manihot sinuata var. *genuina* Müll.Arg. === **Manihot tripartita** (Spreng.) Müll.Arg. subsp. **tripartita**
Manihot sinuata var. *laciniosa* (Pohl) Müll.Arg. === **Manihot tripartita** subsp. **laciniosa** (Pohl) D.J.Rogers & Appan
Manihot speciosa Müll.Arg. === **Manihot caerulescens** Pohl subsp. **caerulescens**
Manihot speciosa Chodat & Hassl. === **Manihot caerulescens** subsp. **macrantha** (Pax & K.Hoffm.) D.J.Rogers & Appan
Manihot spinosissima Mill. ex Steud. === ?
Manihot sprucei Pax === **Manihot esculenta** Crantz
Manihot stenophylla Pax & K.Hoffm. === **Manihot triphylla** Pohl
Manihot subquinqueloba Müll.Arg. === **Manihot quinqueloba** Pohl
Manihot surumuensis Ule === **Manihot tristis** subsp. **surumuensis** (Ule) D.J.Rogers & Appan
Manihot teissonnieri A.Chev. === ?
Manihot tenerrima Pohl === **Manihot pentaphylla** subsp. **tenuifolia** (Pohl) D.J.Rogers & Appan
Manihot tenuifolia Pohl === **Manihot pentaphylla** subsp. **tenuifolia** (Pohl) D.J.Rogers & Appan
Manihot toledii Labroy ex Ule === **Manihot caerulescens** Pohl subsp. **caerulescens**
Manihot tomentella Pohl === **Manihot tripartita** (Spreng.) Müll.Arg. subsp. **tripartita**

Manihot trichandra Pax & K.Hoffm. === **Manihot tripartita** subsp. **vestita** (S.Moore) D.J.Rogers & Appan
Manihot trifoliata Ule === **Manihot caerulescens** Pohl subsp. **caerulescens**
Manihot trifoliata var. *platyphylla* Ule === **Manihot caerulescens** Pohl subsp. **caerulescens**
Manihot tripartita var. *apaensis* Chodat & Hassl. === **Manihot tripartita** (Spreng.) Müll.Arg. subsp. **tripartita**
Manihot tripartita var. *cajaniformis* (Pohl) Müll.Arg. === **Manihot tripartita** (Spreng.) Müll.Arg. subsp. **tripartita**
Manihot tripartita var. *dalechampiiformis* (Pohl) Pax === **Manihot tripartita** (Spreng.) Müll.Arg. subsp. **tripartita**
Manihot tripartita var. *genuina* Müll.Arg. === **Manihot tripartita** (Spreng.) Müll.Arg.
Manihot tripartita var. *glabra* Müll.Arg. === **Manihot tripartita** subsp. **humilis** (Müll.Arg.) D.J.Rogers & Appan
Manihot tripartita var. *glauca* Müll.Arg. === **Manihot tripartita** (Spreng.) Müll.Arg. subsp. **tripartita**
Manihot tripartita var. *lanceolata* Müll.Arg. === **Manihot tripartita** (Spreng.) Müll.Arg. subsp. **tripartita**
Manihot tripartita var. *porrecta* (Pohl) Müll.Arg. === **Manihot tripartita** (Spreng.) Müll.Arg. subsp. **tripartita**
Manihot tripartita var. *quinqueloba* Pax & K.Hoffm. === **Manihot pohlii** Wawra
Manihot tripartita var. *seminuda* Müll.Arg. === **Manihot tripartita** (Spreng.) Müll.Arg. subsp. **tripartita**
Manihot tripartita var. *subintegra* Müll.Arg. === **Manihot tripartita** subsp. **laciniosa** (Pohl) D.J.Rogers & Appan
Manihot tripartita var. *vestita* S.Moore === **Manihot tripartita** subsp. **vestita** (S.Moore) D.J.Rogers & Appan
Manihot tripartita subsp. *xavantinensis* (D.J.Rogers & Appan) Allem === **Manihot xavantinensis** D.J.Rogers & Appan
Manihot triphylla var. *fruticulosa* Pax === **Manihot fruticulosa** (Pax) D.J.Rogers & Appan
Manihot triphylla var. *genuina* Pax === **Manihot triphylla** Pohl
Manihot tubuliflora Pax & K.Hoffm. === **Manihot pilosa** Pohl
Manihot tweedieana Müll.Arg. === **Manihot grahamii** Hook.
Manihot tweedieana var. *lobata* Chodat & Hassl. === **Manihot grahamii** Hook.
Manihot tweedieana f. *nana* Chodat & Hassl. === **Manihot grahamii** Hook.
Manihot uleana Pax & K.Hoffm. === **Manihot pentaphylla** Pohl subsp. **pentaphylla**
Manihot urens (L.) Crantz === **Cnidoscolus urens** (L.) Arthur
Manihot utilissima Pohl === **Manihot esculenta** Crantz
Manihot utilissima var. *castellana* Pohl === **Manihot esculenta** Crantz
Manihot utilissima var. *sutinga* Pohl === **Manihot esculenta** Crantz
Manihot varians Pohl === **Manihot gracilis** subsp. **varians** (Pohl) D.J.Rogers & Appan
Manihot violacea var. *arcuata* (Pohl) Müll.Arg. === **Manihot divergens** Pohl
Manihot violacea var. *cecropiifolia* (Pohl) Müll.Arg. === **Manihot cecropiifolia** Pohl
Manihot violacea var. *divergens* (Pohl) Müll.Arg. === **Manihot divergens** Pohl
Manihot violacea var. *genuina* Müll.Arg. === **Manihot violacea** Pohl
Manihot violacea subsp. *jacobinensis* (Müll.Arg.) Allem === **Manihot jacobinensis** Müll.Arg.
Manihot warmingii Müll.Arg. === **Manihot anomala** subsp. **pubescens** (Pohl) D.J.Rogers & Appan
Manihot weberbaueri Pax & K.Hoffm. === **Manihot anomala** subsp. **pavoniana** (Müll.Arg.) D.J.Rogers & Appan

Manihotoides

Erected for *Manihot pauciflora* on the basis of a 1-2-flowered inflorescence (vs. many-flowered in *Manihot*) and other unique features including leaf-bearing short shoots; here treated as *Manihot*.

Synonyms:
Manihotoides D.J.Rogers & Appan === **Manihot** Mill.

Manniophyton

1 species, W. and C. Africa (Sierra Leone to Principe and Angola); woody lianas of primary, secondary, and gallery forest to 30(-40) m, the stems with irritating sap. A good flora treatment is that in *Fl. Congo Belge* 8(1): 171-174 (1962). Webster (Synopsis, 1994) has included this genus in his Aleuritideae subtribe Crotonogyninae but this seems questionable; the leaves are palmately veined as in subtribe Neoboutoninae. (Crotonoideae)

Pax, F. (with K. Hoffmann) (1912). *Manniophyton.* In A. Engler (ed.), Das Pflanzenreich, IV 147 VI (Euphorbiaceae-Acalypheae-Chrozophorinae): 120-123. Berlin. (Heft 57.) La/Ge. — 1 species, W and C Africa.

Léonard, J. (1955). Notulae systematicae, XVIII. Euphorbiaceae africanae novae. Bull. Jard. Bot. État 25: 281-301, illus. Fr. —*Manniophyton*, pp. 290-291; treatment of *M. fulvum* with reduction of *M. africanum.*

Manniophyton Müll.Arg., Flora 47: 530 (1864).
W. Trop. Africa to Angola. 22 23 26.

Manniophyton africanum Müll.Arg., Flora 47: 531 (1864). – FIGURE, p. 1152.
W. Trop. Africa to Angola. 22 GHA IVO LBR NGA SIE 23 CMN CON EQG GAB GGI ZAI 26 ANG. (Cl.) nanophan.
Manniophyton fulvum Müll.Arg., J. Bot. 2: 332 (1864). *Manniophyton africanum* var. *fulvum* (Müll.Arg.) Hutch. in D.Oliver, Fl. Trop. Afr. 6(1): 819 (1912).
Manniophyton chevalieri Beille, Bull. Soc. Bot. France 55(8): 74 (1908).
Manniophyton wildemanii Beille, Bull. Soc. Bot. France 57(8): 124 (1910).
Manniophyton tricuspe Pierre ex A.Chev., Rev. Bot. Appl. Agric. Trop. 20: 564 (1940).

Synonyms:
Manniophyton africanum var. *fulvum* (Müll.Arg.) Hutch. === **Manniophyton africanum** Müll.Arg.
Manniophyton angustifolium Baill. === **Crotonogyne parvifolia** Prain
Manniophyton chevalieri Beille === **Manniophyton africanum** Müll.Arg.
Manniophyton fulvum Müll.Arg. === **Manniophyton africanum** Müll.Arg.
Manniophyton tricuspe Pierre ex A.Chev. === **Manniophyton africanum** Müll.Arg.
Manniophyton wildemanii Beille === **Manniophyton africanum** Müll.Arg.

Mappa

Synonyms:
Mappa A.Juss. === **Macaranga** Thouars
Mappa amboinensis Müll.Arg. === **Macaranga involucrata** (Roxb.) Baill. var. **involucrata**
Mappa angolensis Müll.Arg. === **Macaranga angolensis** (Müll.Arg.) Müll.Arg.
Mappa borneensis Müll.Arg. === **Macaranga trichocarpa** (Rchb.f. & Zoll.) Müll.Arg.
Mappa capensis Baill. === **Macaranga capensis** (Baill.) Sim
Mappa chatiniana Baill. === **Macaranga chatiniana** (Baill.) Müll.Arg.
Mappa cissifolia Zoll. & Rchb.f. === **Macaranga cissifolia** (Zoll. & Rchb.f.) Müll.Arg.
Mappa conifera Zoll. === **Macaranga conifera** (Zoll.) Müll.Arg.
Mappa corymbosa Müll.Arg. === **Macaranga corymbosa** (Müll.Arg.) Müll.Arg.
Mappa costulata Miq. === **Baccaurea costulata** (Miq.) Müll.Arg.
Mappa cumingii Baill. === **Macaranga cumingii** (Baill.) Müll.Arg.
Mappa cuneifolia Zoll. === **Macaranga cuneifolia** (Zoll.) Müll.Arg.
Mappa denticulata Blume === **Macaranga denticulata** (Blume) Müll.Arg.

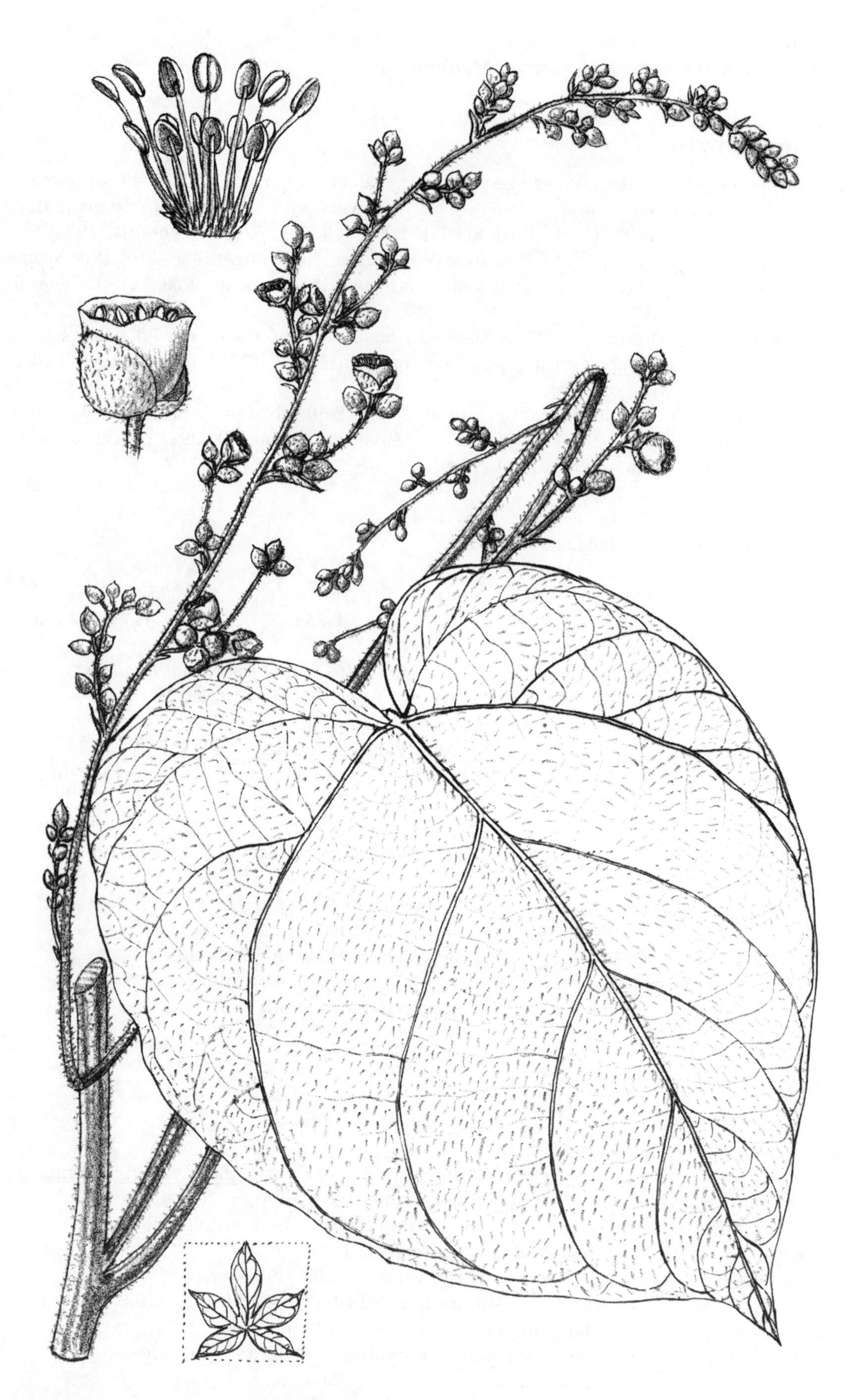

Manniophyton africanum Müll. Arg.
Artist: 'A.M.C.'
Ic. Pl. 13: pl. 1267 (1879)
KEW ILLUSTRATIONS COLLECTION

Mappa diepenhorstii Miq. === **Macaranga diepenhorstii** (Miq.) Müll.Arg.
Mappa fastuosa Linden === **Homalanthus fastuosus** (Linden) Villar
Mappa gigantea Zoll. === **Macaranga gigantea** (Rchb.f. & Zoll.) Müll.Arg.
Mappa glabra Juss. === **Macaranga glabra** (Juss.) Pax & K.Hoffm.
Mappa gummiflua Miq. === **Macaranga denticulata** (Blume) Müll.Arg.
Mappa harveyana Müll.Arg. === **Macaranga harveyana** (Müll.Arg.) Müll.Arg.
Mappa heterophylla Müll.Arg. === **Macaranga heterophylla** (Müll.Arg.) Müll.Arg.
Mappa hispida Blume === **Macaranga hispida** (Blume) Müll.Arg.
Mappa hypoleuca Rchb.f. & Zoll. === **Macaranga hypoleuca** (Rchb.f. & Zoll.) Müll.Arg.
Mappa javanica Blume === **Macaranga javanica** (Blume) Müll.Arg.
Mappa leptostachya Müll.Arg. === **Cleidion leptostachyum** (Müll.Arg.) Pax & K.Hoffm.
Mappa megalophylla Müll.Arg. === **Macaranga gigantea** (Rchb.f. & Zoll.) Müll.Arg.
Mappa motleyana Müll.Arg. === **Macaranga motleyana** (Müll.Arg.) Müll.Arg.
Mappa occidentalis Müll.Arg. === **Macaranga occidentalis** (Müll.Arg.) Müll.Arg.
Mappa pruinosa Miq. === **Macaranga pruinosa** (Miq.) Müll.Arg.
Mappa seemannii Müll.Arg. === **Macaranga seemannii** (Müll.Arg.) Müll.Arg.
Mappa sinensis Müll.Arg. === **Macaranga sinensis** (Baill.) Müll.Arg.
Mappa stricta Rchb.f. & Zoll. === **Mallotus philippensis** (Lam.) Müll.Arg.
Mappa taitensis Müll.Arg. === **Macaranga taitensis** (Müll.Arg.) Müll.Arg.
Mappa teysmannii Zoll. === **Macaranga teysmannii** (Zoll.) Müll.Arg.
Mappa tomentosa Blume === **Macaranga tanarius** (L.) Müll.Arg.
Mappa trichocarpa Rchb.f. & Zoll. === **Macaranga trichocarpa** (Rchb.f. & Zoll.) Müll.Arg.
Mappa velutina Rchb.f. & Zoll. === **Macaranga velutina** (Rchb.f. & Zoll.) Müll.Arg.
Mappa vieillardii Müll.Arg. === **Macaranga vieillardii** (Müll.Arg.) Müll.Arg.

Maprounea

4 species, S. America (2), Africa (2); small or large glabrous shrubs or trees to as much as 25 m, in forest, woodland or savanna (the latter usual for *M. africana*). The seeds are in the family distinguished by an unusually large caruncule (aril) and the staminate spikes are capitulate (well-illustrated in Senna, 1984: fig. 22). The pinnately veined, entire leaves are reminiscent of some species of *Sapium*. Opinions with respect to the number of species in the Americas vary. Allen (1976) accepted only one, the widely-ranging and variable *M. guyanensis*, while Senna (1984), arguing for the worth of the number of rows of palisade tissue in cross-sections of leaves and the distribution of extrafloral nectaries, accepted four taxa (two species, one with two additional varieties). A more definitive judgement must await comparative studies in both Africa and the Americas. (Euphorbioideae (except Euphorbieae))*

Pax, F. (with K. Hoffmann) (1912). *Maprounea*. In A. Engler (ed.), Das Pflanzenreich, IV 147 V (Euphorbiaceae-Hippomaneae): 174-180. Berlin. (Heft 52.) La/Ge. — 4 species, 2 in S. America, 2 in Africa, one of the latter with five non-nominate varieties.

Jablonski, E. (1967). *Maprounea*. Euphorbiaceae, Guayana Highland (Mem. New York Bot. Gard. 17(1)): 179-180. New York. En. — 1 species, *M. guyanensis*.

Allem, A. C. (1976). Uma espécie única de *Maprounea* (Euphorbiaceae) na América do Sul. Acta Amaz., 6: 417-422. Pt. — Reduction of *M. brasiliensis* to *Maprounea guianensis*.

Senna, L. Mendonça de (1984). *Maprounea* Aubl. (Euphorbiaceae). Considerações taxinômicas e anatômicas das espécies sul-americanas. Rodriguesia 36(61): 51-77, illus. Pt. — Detailed review of history and characters, with particular attention to morphology, anatomy and palynology; copiously illustrated descriptive treatment of 2 species (one with two additional varieties) with key (p. 59), descriptions, synonymy, references and citations, types, vernacular names, indication of distribution, localities with exsiccatae, and ecological and other notes; general discussion and conclusions; lists of collections and literature seen at end. The illustrations include (fig. 27) a scatter diagram of the four taxa based on leaf and inflorescence measurements.

* Esser (Novon 9: 32-35. 1999) has recognised a third South American species, *M. amazonica*, based on *M. guianensis* var. *obtusata*.

Maprounea Aubl., Hist. Pl. Guiane 2: 895 (1775).
Trop. America, Trop. Africa. 22 23 25 26 81 82 83 84.
Aegopricum L., Pl. Surin.: 15 (1775).

Maprounea africana Müll.Arg. in A.P.de Candolle, Prodr. 15(2): 1191 (1866).
– FIGURE, p. 1155.
Trop. & S. Africa. 22 NGA 23 CAF CMN CON GAB ZAI 25 TAN 25 ANG MLW MOZ ZAM ZIM 27 NAM. Nanophan. or phan.

var. **africana**
Trop. & S. Africa. 22 NGA 23 CAF CMN CON GAB ZAI 25 TAN 25 ANG MLW MOZ ZAM ZIM 27 NAM. Nanophan. or phan.
Maprounea obtusa Pax, Bot. Jahrb. Syst. 19: 116 (1894). *Maprounea africana* var. *obtusa* (Pax) Pax & K.Hoffm. in H.G.A.Engler, Pflanzenr., IV, 147, V: 180 (1912).
Maprounea vaccinioides Pax, Bot. Jahrb. Syst. 19: 116 (1894). *Maprounea africana* var. *benguelensis* Pax & K.Hoffm. in H.G.A.Engler, Pflanzenr., IV, 147, V: 180 (1912).
Maprounea africana var. *cinnamomea* Pax & K.Hoffm. in H.G.A.Engler, Pflanzenr., IV, 147, V: 179 (1912).
Maprounea africana var. *leucosperma* Pax & K.Hoffm. in H.G.A.Engler, Pflanzenr., IV, 147, V: 179 (1912).
Maprounea africana var. *orientalis* Pax & K.Hoffm. in H.G.A.Engler, Pflanzenr., IV, 147, V: 179 (1912).

var. **gracilis** Pax & K.Hoffm. in H.G.A.Engler, Pflanzenr., IV, 147, V: 179 (1912).
Maprounea gracilis (Pax & K.Hoffm.) Dewèvre ex Prain in D.Oliver, Fl. Trop. Afr. 6(1): 1003 (1913).
Cameroon, Congo. 23 CMN CON. Nanophan.

Maprounea brasiliensis A.St.-Hil., Pl. Usuel. Bras.: t. 65 (1828).
WC. & E. Brazil. 84 BZC BZE BZL. Nanophan.

Maprounea guianensis Aubl., Hist. Pl. Guiane 2: 895 (1775). *Stillingia guianensis* (Aubl.) Baill., Adansonia 5: 332 (1865). *Maprounea guianensis* var. *genuina* Müll.Arg. in A.P.de Candolle, Prodr. 15(2): 1191 (1866), nom. inval. *Excoecaria guianensis* (Aubl.) Baill., Hist. Pl. 5: 133 (1874).
Trop. America. 80 PAN 81 TRT 82 FRG GUY SUR VEN 83 BOL CLM PER 84 BZC BZE BZL BZN BZS. Nanophan. or phan.
Maprounea guianensis var. *nervosa* Müll.Arg., Linnaea 32: 115 (1863).
Maprounea guianensis var. *obtusata* Müll.Arg., Linnaea 32: 115 (1863).
Stillingia hilariana Baill., Adansonia 5: 332 (1865).
Maprounea guianensis var. *undulata* Müll.Arg. in C.F.P.von Martius, Fl. Bras. 11(2): 544 (1874).

Maprounea membranacea Pax & K.Hoffm. in H.G.A.Engler, Pflanzenr., IV, 147, V: 178 (1912).
W. & WC. Trop. Africa. 22 NGA 23 CMN CON EQG GAB ZAI. Phan.
Maprounea bridelioides Pierre ex Prain in D.Oliver, Fl. Trop. Afr. 6(1): 1003 (1913), nom. illeg.

Synonyms:
Maprounea africana var. *benguelensis* Pax & K.Hoffm. === **Maprounea africana** Müll.Arg. var. **africana**
Maprounea africana var. *cinnamomea* Pax & K.Hoffm. === **Maprounea africana** Müll.Arg. var. **africana**
Maprounea africana var. *leucosperma* Pax & K.Hoffm. === **Maprounea africana** Müll.Arg. var. **africana**
Maprounea africana var. *obtusa* (Pax) Pax & K.Hoffm. === **Maprounea africana** Müll.Arg. var. **africana**

Maprounea africana Müll. Arg.
Artist: Christine Grey-Wilson
Fl. Trop. East Africa, Euphorbiaceae 1: 396, fig. 75 (1987)
KEW ILLUSTRATIONS COLLECTION

Maprounea africana var. *orientalis* Pax & K.Hoffm. === **Maprounea africana** Müll.Arg. var. **africana**
Maprounea bridelioides Pierre ex Prain === **Maprounea membranacea** Pax & K.Hoffm.
Maprounea glauca Desv. === **Mabea taquari** Aubl.
Maprounea gracilis (Pax & K.Hoffm.) Dewèvre ex Prain === **Maprounea africana** var. **gracilis** Pax & K.Hoffm.
Maprounea guianensis var. *genuina* Müll.Arg. === **Maprounea guianensis** Aubl.
Maprounea guianensis var. *nervosa* Müll.Arg. === **Maprounea guianensis** Aubl.
Maprounea guianensis var. *obtusata* Müll.Arg. === **Maprounea guianensis** Aubl.
Maprounea guianensis var. *undulata* Müll.Arg. === **Maprounea guianensis** Aubl.
Maprounea obtusa Pax === **Maprounea africana** Müll.Arg. var. **africana**
Maprounea vaccinioides Pax === **Maprounea africana** Müll.Arg. var. **africana**

Mareya

4 species, West and West Central Tropical Africa, *M. brevipes* extending east to Uganda. Shrubs or small to medium trees (? rarely to as much as 30 m) in forest understorey; leaves with more or less translucent dots. The maps in Léonard (1996), while accounting for three of the four known species, cover only the distribution in Congo (Zaïre). The genus is considered to be related to *Claoxylon* (cf. Webster, Synopsis, 1994) but Léonard indicates that there is some resemblance to species of *Alchornea*. (Acalyphoideae)

Pax, F. & K. Hoffmann (1919). *Mareya*. In A. Engler (ed.), Das Pflanzenreich, IV 147 XIV (Euphorbiaceae-Additamentum VI): 11-12. Berlin. (Heft 68.) La/Ge. — Treatment of genus with 2 species and one additional variety.

- Léonard, J. (1996). Révision des espèces zaïroises des genres *Mareya* Baill. et *Mareyopsis* Pax & K. Hoffm. (Euphorbiaceae). Bull. Jard. Bot. Natl. Belg. 65: 3-22, illus., maps. Fr. — Includes treatment of *Mareya* with 3 species in Congo/Zaïre; key, synonymy, references and citations, types, localities with exsiccatae, indication of habitat and chorology, and notes on uses and biology. [For related flora treatment, see *Flore d'Afrique centrale, Euphorbiaceae* 3: 18, 20-28 (1996).]

Mareya Baill., Adansonia 1: 73 (1860).
W., WC. & E. Trop. Africa. 22 23 25.

Mareya acuminata Prain, Bull. Misc. Inform. Kew 1912: 103 (1912).
Gabon. 23 GAB. Nanophan. or phan.

Mareya brevipes Pax, Bot. Jahrb. Syst. 28: 24 (1899). – FIGURE, p. 1157.
WC. Trop. Africa to Uganda. 23 CAF CMN CON EQG GAB ZAI 25 UGA. Nanophan. or phan.

Mareya congolensis (J.Léonard) J.Léonard, Bull. Jard. Bot. Belg. 65: 12 (1996).
C. Zaire. 23 ZAI. Nanophan. or phan.
Mareya micrantha subsp. *congolensis* J.Léonard, Bull. Jard. Bot. Etat 25: 292 (1955).

Mareya micrantha (Benth.) Müll.Arg. in A.P.de Candolle, Prodr. 15(2): 792 (1866).
W. & WC. Trop. Africa. 22 GHA GNB GUI IVO LBR NGA SIE TOG 23 CAB CMN CON EQG GAB GGI ZAI. Nanophan. or phan.
Acalypha leonensis Benth. in W.J.Hooker, Niger Fl.: 504 (1849). *Mareya micrantha* var. *leonensis* (Benth.) Müll.Arg. in A.P.de Candolle, Prodr. 15(2): 796 (1866). *Mareya leonensis* (Benth.) Baill., Hist. Pl. 5: 213 (1874). *Mareya spicata* var. *leonensis* (Benth.) Pax & K.Hoffm. in H.G.A.Engler, Pflanzenr., IV, 147, XIV: 12 (1919).
* *Acalypha micrantha* Benth. in W.J.Hooker, Niger Fl.: 505 (1849). *Mareya spicata* var. *micrantha* (Benth.) Pax & K.Hoffm. in H.G.A.Engler, Pflanzenr., IV, 147, XIV: 12 (1919).
Mareya spicata Baill., Adansonia 1: 74 (1860).
Mareya micrantha var. *nitida* Beille, Bull. Soc. Bot. France 55(8): 82 (1908).

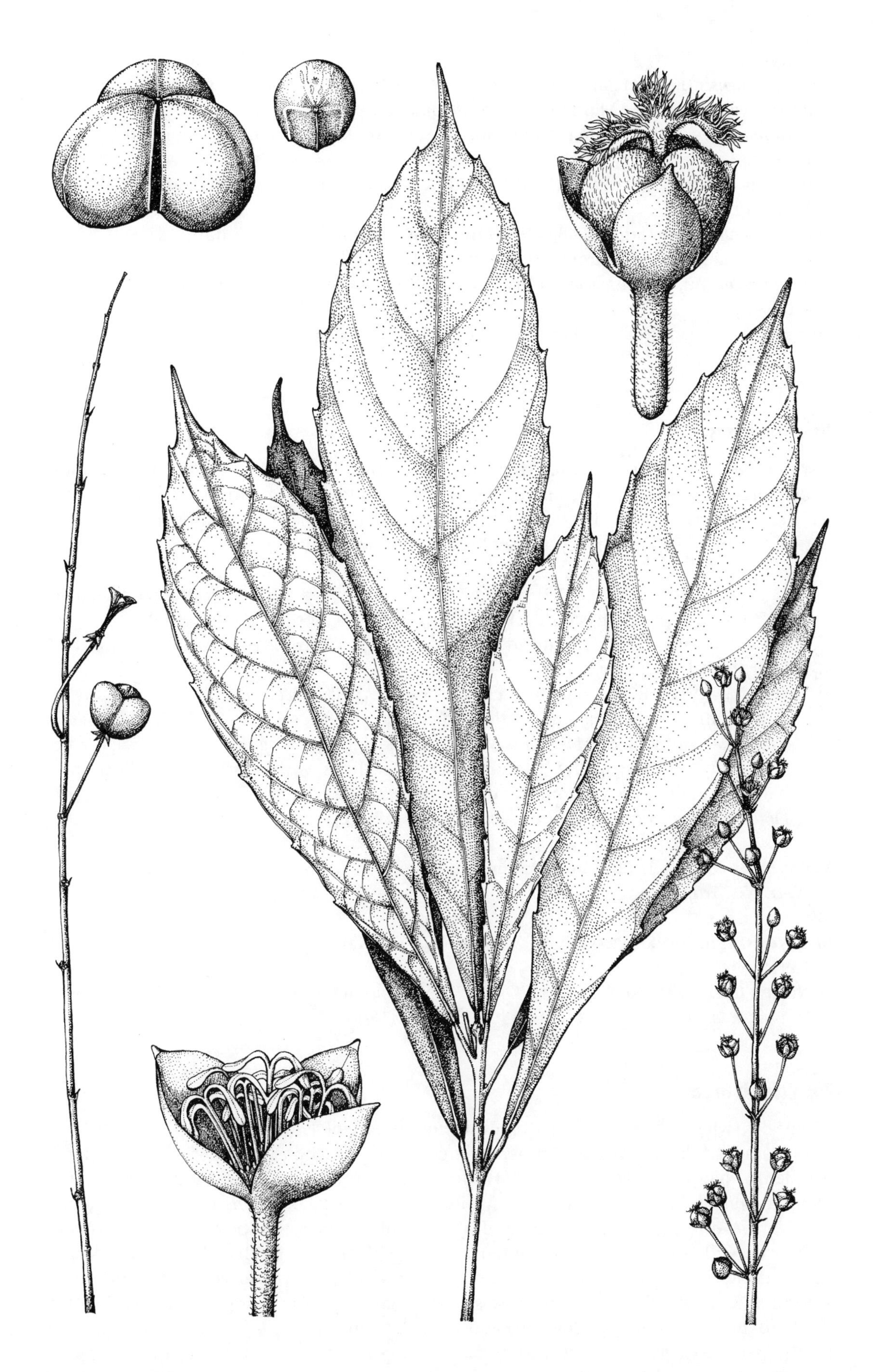

Mareya brevipes Pax
Artist: Judy C. Dunkley
Fl. Trop. East Africa, Euphorbiaceae, 1: 217, fig. 41 (1987)
KEW ILLUSTRATIONS COLLECTION

Synonyms:
Mareya leonensis (Benth.) Baill. === **Mareya micrantha** (Benth.) Müll.Arg.
Mareya longifolia Pax === **Mareyopsis longifolia** (Pax) Pax & K.Hoffm.
Mareya micrantha subsp. *congolensis* J.Léonard === **Mareya congolensis** (J.Léonard) J.Léonard
Mareya micrantha var. *leonensis* (Benth.) Müll.Arg. === **Mareya micrantha** (Benth.) Müll.Arg.
Mareya micrantha var. *nitida* Beille === **Mareya micrantha** (Benth.) Müll.Arg.
Mareya spicata Baill. === **Mareya micrantha** (Benth.) Müll.Arg.
Mareya spicata var. *leonensis* (Benth.) Pax & K.Hoffm. === **Mareya micrantha** (Benth.) Müll.Arg.
Mareya spicata var. *micrantha* (Benth.) Pax & K.Hoffm. === **Mareya micrantha** (Benth.) Müll.Arg.

Mareyopsis

1 species, West and West Central Tropical Africa; shrubs or small trees (to 10 m) of forest understorey. The opinion of Léonard (1996) is here followed in contrast to Webster (1994; see **General**) who united the genus with *Mareya*. A second species is known from Gabon but, according to Léonard, not yet described. (Acalyphoideae)

Pax, F. & K. Hoffmann (1919). *Mareyopsis*. In A. Engler (ed.), Das Pflanzenreich, IV 147 XIV (Euphorbiaceae-Additamentum VI): 13. Berlin. (Heft 68.) La/Ge. — A segregate from *Mareya*, reduced to that genus by Webster but not by Brummitt or Lebrun and Stork.

- Léonard, J. (1996). Révision des espèces zaïroises des genres *Mareya* Baill. et *Mareyopsis* Pax & K. Hoffm. (Euphorbiaceae). Bull. Jard. Bot. Natl. Belg. 65: 3-22, illus., maps. Fr. — Includes treatment of *Mareyopsis* with a single species, *M. longifolia*, in Congo/Zaïre; synonymy, references and citations, type, localities with exsiccatae, indication of habitat and chorology, and notes on uses and biology. [For related flora treatment, see *Flore d'Afrique centrale, Euphorbiaceae* 3: 16-19 (1996).]

Mareyopsis Pax & K.Hoffm. in H.G.A.Engler, Pflanzenr., IV, 147, XIV: 13 (1919).
W. & WC. Trop. Africa. 23.

Mareyopsis longifolia (Pax) Pax & K.Hoffm. in H.G.A.Engler, Pflanzenr., IV, 147, XIV: 13 (1919).
Nigeria to Zaire. 22 NGA 23 CMN GAB ZAI. Nanophan. or phan.
* *Mareya longifolia* Pax, Bot. Jahrb. Syst. 38: 283 (1906).

Margaritaria

13 species, widely distributed in tropics; formerly part of *Phyllanthus* with some still not recognising its distinctiveness. Shrubs or trees with distichously arranged foliage (rarely scandent), mainly of secondary growth and edges and more usually found in areas of seasonal climate. Some species are very widely distributed (*M. nobilis* in the Americas, *M. discoidea* in Africa, *M. indica* in Asia and Malesia) and it may well be that the genus is an early yet still successful development from within the Phyllantheae. The largest concentration of species is in Madagascar. A full revision was produced by Webster (1979) who took as a basis the common expression of a growth form, 'Margaritaria' (a variant of the Roux model of Hallé & Oldeman) and – for Phyllanthoideae – the unique morphology of the seed coat (with bony endotesta and fleshy exotesta). Growth of all axes is monopodial and indefinite, the axes themselves persisting. Webster suggests that the genus is more closely related to *Flueggea* than to *Phyllanthus*. (Phyllanthoideae)

Pax, F. & K. Hoffmann (1931). *Phyllanthus*. In A. Engler (ed.), Die natürlichen Pflanzenfamilien, 2. Aufl., 19c: 60-66. Leipzig. En. — Synopsis with description of genus. [*Margaritaria* covered in sect. *Cicca* (pp. 62-63). No Pflanzenreich revision.]

Airy-Shaw, H. K. (1963). Notes on Malaysian and other Asiatic Euphorbiaceae, XXI. *Prosorus indicus* in Borneo and other regions. Kew Bull. 16: 342-343. En. — Range extension. [Genus now part of *Margaritaria*.]

Airy-Shaw, H. K. (1966). Notes on Malaysian and other Asiatic Euphorbiaceae, LXIX. New combinations in *Margaritaria* L. f. Kew Bull. 20: 386-387. En. — Transfer of 3 spp. of *Prosorus*, now reduced.

Jablonski, E. (1967). *Margaritaria*. Euphorbiaceae, Guayana Highland (Mem. New York Bot. Gard. 17(1)): 118. New York. En. — 1 species, *M. nobilis*, a 'wide' in the Americas.

• Webster, G. L. (1979). A revision of *Margaritaria* (Euphorbiaceae). J. Arnold Arbor. 60: 403-444, illus., maps. En. — Critical revision (14 species) with key, descriptions, synonymy, references, types, localities with exsiccatae, indication of distribution and habitat, and commentary; list of collections seen at end but no separate index. The general part includes a world map of the distribution and a review of characters as well as past research. [The genus is considered closer to *Flueggea* than *Phyllanthus* within which it was long included.]

Radcliffe-Smith, A. (1981). Notes on African Euphorbiaceae, XI. *Margaritaria discoidea*, a re-appraisal. Kew Bull. 36: 219-221. En. — Synopsis of *M. discoidea* and its subdivisions (here treated as varieties), with key, synonymy, indication of distribution and commentary. [Precursory to *Flora of Tropical East Africa*.]

Brunel, J. F. (1987). *Margaritaria*. In *idem*, Sur le genre *Phyllanthus* L. et quelques genres voisins de la tribu des Phyllantheae Dumort. Strasbourg. [See also *Phyllanthus*.] Fr. — Description of genus, with illustrations of key characters.

Margaritaria L.f., Suppl. Pl.: 66 (1782).
Trop. & Subtrop. 22 23 24 25 26 27 29 36 38 40 41 42 50 79 80 81 82 83 84.
Nanophan. or phan.
Prosorus Dalzell, Hooker's J. Bot. Kew Gard. Misc. 4: 345 (1852).
Zygospermum Thwaites ex Baill., Étude Euphorb.: 620 (1858).
Wurtzia Baill., Adansonia 1: 186 (1861).
Calococcus Kurz ex Teijsm. & Binn., Natuurw. Tijdschr. Ned.-Indië 26: 48 (1864).

Margaritaria anomala (Baill.) Fosberg, Kew Bull. 33: 185 (1978). – FIGURE, p. 1160.
Aldabra (incl. Astove I., Cosmoledo I.), Comoros, Madagascar, Mauritius. 29 ALD COM MAU MDG. Nanophan. or phan.
Anisonema eglandulosum Decne., Herb. Timor: 154 (1834). *Kirganelia eglandulosa* Baill., Étude Euphorb.: 614 (1858), nom. nud.
Securinega hysterantha Bojer, Hortus Maurit.: 278 (1837), nom. nud.
* *Cicca anomala* Baill., Étude Euphorb.: 619 (1858). *Phyllanthus anomalus* (Baill.) Müll.Arg., Linnaea 32: 52 (1863).
Flueggea eglandulosa Baill., Étude Euphorb.: 593 (1858). *Phyllanthus eglandulosus* (Baill.) Leandri, Cat. Pl. Madag. Euphorbiac.: 23 (1935).
Flueggea major Baill., Étude Euphorb.: 593 (1858).
Phyllanthus hysteranthus Müll.Arg., Linnaea 32: 52 (1863).
Phyllanthus erythroxyloides Müll.Arg. in A.P.de Candolle, Prodr. 15(2): 418 (1866). *Phyllanthus anomalus* subsp. *erythroxyloides* (Müll.Arg.) Leandri, Notul. Syst. (Paris) 6: 191 (1938).
Phyllanthus cheloniphorbe Hutch., Bull. Misc. Inform. Kew 1918: 204 (1918). *Margaritaria anomala* var. *cheloniphorbe* (Hutch.) Fosberg, Kew Bull. 33: 185 (1978).

Margaritaria cyanosperma (Gaertn.) Airy Shaw, Kew Bull. 20: 387 (1966).
Sri Lanka. 40 SRL. Phan.
* *Croton cyanospermus* Gaertn., Fruct. Sem. Pl. 2: 120 (1790). *Cicca gaertneriana* Baill., Étude Euphorb.: 619 (1858). *Phyllanthus cyanospermus* (Gaertn.) Müll.Arg., Linnaea 32: 51 (1863).

Margaritaria anomala (Baill.) Fosberg
Artist: Ann Davies
Fl. Mascareignes, 160. Euphorbiacées: 31, pl. 4 (1982)
KEW ILLUSTRATIONS COLLECTION

Prosorus gaertneri Thwaites, Hooker's J. Bot. Kew Gard. Misc. 8: 272 (1856).
Zygospermum zeylanicum Thwaites ex Baill., Étude Euphorb.: 620 (1858).
Prosorus cyanospermus (Gaertn.) Thwaites, Enum. Pl. Zeyl.: 281 (1861).

Margaritaria decaryana (Leandri) G.L.Webster, J. Arnold Arbor. 60: 433 (1979).
S. Madagascar. 29 MDG. Nanophan. or phan.
* *Phyllanthus decaryanus* Leandri, Notul. Syst. (Paris) 6: 198 (1938).
Phyllanthus decaryanus var. *manambia* Leandri, Mém. Inst. Sci. Madagascar, Sér. B, Biol. Vég. 8: 228 (1957).

Margaritaria discoidea (Baill.) G.L.Webster, J. Arnold Arbor. 48: 311 (1967).
Trop. & S. Africa. 22 BEN GAM GHA GNB GUI IVO LBR MLI NGA SEN SIE TOG 23 BUR CAF CMN GAB GGI ZAI 24 ETH SUD 25 KEN TAN UGA 26 ANG MLW MOZ ZAM ZIM 27 BOT NAM NAT SWZ TVL. Nanophan. or phan.
* *Cicca discoidea* Baill., Adansonia 1: 85 (1860). *Phyllanthus discoideus* (Baill.) Müll.Arg., Linnaea 32: 51 (1863).

var. **discoidea**
Trop. & S. Africa. 22 BEN GAM GHA GNB GUI IVO LBR MLI NGA SEN SIE TOG 23 BUR CAF CMN GAB GGI ZAI 24 ETH SUD 25 KEN TAN UGA 26 ANG MLW MOZ ZAM ZIM 27 BOT NAT SWZ TVL. Nanophan. or phan.

var. **fagifolia** (Pax) Radcl.-Sm., Kew Bull. 36: 220 (1981).
Trop. & S. Africa. 22 BEN GHA GNB GUI IVO MLI NGA TOG 23 BUR CAF CMN GAB GGI ZAI 24 ETH SUD 25 KEN TAN UGA 26 ANG MLI MOZ ZAM ZIM 27 BOT NAT SWZ TVL. Nanophan. or phan.
* *Flueggea fagifolia* Pax in H.G.A.Engler, Pflanzenw. Ost-Afrikas C: 236 (1895).

var. **nitida** (Pax) Radcl.-Sm., Kew Bull. 36: 221 (1981).
Ethiopia to KwaZulu-Natal. 23 BUR 24 ETH 25 KEN TAN UGA 26 MAL MOZ ZAM ZIM 27 BOT NAM NAT SWZ TVL. Nanophan. or phan.
* *Flueggea nitida* Pax, Bot. Jahrb. Syst. 19: 76 (1894). *Margaritaria discoidea* subsp. *nitida* (Pax) G.L.Webster, J. Arnold Arbor. 60: 418 (1979).
Phyllanthus amapondensis Sim, Forest Fl. Cape: 325 (1907).
Phyllanthus flacourtioides Hutch., Bull. Misc. Inform. Kew 1915: 48 (1915).

var. **triplosphaera** Radcl.-Sm., Kew Bull. 30: 680 (1975 publ. 1976).
Zaire, Burundi, Kenya, Tanzania, Malawi, Mozambique, Zambia. 23 BUR ZAI 25 KEN TAN 26 MLW MOZ ZAM. Nanophan. or phan.
* *Flueggea obovata* Baill., Adansonia 2: 41 (1861). *Margaritaria obovata* (Baill.) G.L.Webster, J. Arnold Arbor. 60: 434 (1979).
Securinega baillonana Müll.Arg. in A.P.de Candolle, Prodr. 15(2): 451 (1866). *Acidoton baillonianus* (Müll.Arg.) Kuntze, Revis. Gen. Pl. 2: 592 (1891). *Flueggea baillonana* (Müll.Arg.) Pax, Bot. Jahrb. Syst. 19: 76 (1894).
Margaritaria discoidea f. *glabra* Radcl.-Sm., Kew Bull. 39: 790 (1984).

Margaritaria dubium-traceyi Airy Shaw & B.Hyland, Kew Bull. 31: 357 (1976).
N. Queensland. 50 QLD. Nanophan. or phan.

Margaritaria hispidula G.L.Webster, J. Arnold Arbor. 60: 432 (1979).
Madagascar (Tulear). 29 MDG. Phan.

Margaritaria hotteana (Urb. & Ekman) G.L.Webster, J. Arnold Arbor. 38: 66 (1957).
S. Haiti. 81 HAI. Cl. nanophan.
* *Phyllanthus hotteanus* Urb. & Ekman, Ark. Bot. 22A(8): 61 (1929).

Margaritaria indica (Dalzell) Airy Shaw, Kew Bull. 20: 387 (1966).
Trop. & Subtrop. Asia. 36 CHS 38 TAI 40 IND SRL 41 BMA MLY THA VIE 42 BOR JAW MOL NWG PHI SUM. Phan.
Bridelia berryana Wall., Numer. List: 7876, 7960 (1847), nom. inval.
* *Prosorus indicus* Dalzell, Hooker's J. Bot. Kew Gard. Misc. 4: 346 (1852). *Phyllanthus indicus* (Dalzell) Müll.Arg., Linnaea 32: 52 (1863).
Phyllanthus stocksii Müll.Arg., Linnaea 32: 51 (1863).
Calococcus sundaicus Kurz ex Teijsm. & Binn., Natuurw. Tijdschr. Ned.-Indië 26: 48 (1864). *Phyllanthus sundaicus* (Kurz ex Teijsm. & Binn.) Müll.Arg. in A.P.de Candolle, Prodr. 15(2): 1272 (1866).
Cicca arborea C.Wright ex Griseb., Nachr. Königl. Ges. Wiss. Georg-Augusts-Univ. 1: 166 (1865).
Phyllanthus indicus f. *vestita* J.J.Sm. in S.H.Koorders & T.Valeton, Bijdr. Boomsoort. Java 12: 87 (1910).

Margaritaria luzoniensis (Merr.) Airy Shaw, Kew Bull. 20: 387 (1966).
Philippines (Luzon). 42 PHI. Phan.
* *Phyllanthus luzoniensis* Merr., Philipp. J. Sci., C 7: 404 (1912 publ. 1913). *Prosorus luzoniensis* (Merr.) Airy Shaw, Kew Bull. 16: 343 (1963).

Margaritaria nobilis L.f., Suppl. Pl.: 428 (1782). *Phyllanthus nobilis* (L.f.) Müll.Arg. in A.P.de Candolle, Prodr. 15(2): 414 (1866).
Mexico, Trop. America. 79 MXC MXE MXS MXT 80 BLZ COS ELS GUA HON NIC PAN 81 CUB DOM HAI JAM LEE PUE TRT VNA WIN 82 FRG GUY SUR VEN 83 ALL 84 ALL 85 AGE PAR. Nanophan. or phan.
Margaritaria alternifolia L., Pl. Surin.: 16 (1775), nom. illeg.
Cicca antillana A.Juss., Euphorb. Gen.: 108 (1824). *Phyllanthus antillanus* (A.Juss.) Müll.Arg., Linnaea 32: 51 (1863). *Phyllanthus nobilis* var. *antillanus* (A.Juss.) Müll.Arg. in A.P.de Candolle, Prodr. 15(2): 415 (1866). *Margaritaria nobilis* var. *antillana* (A.Juss.) Stehlé & Quetin, Cat. Fl. Guad. 2: 47 (1937).
Cicca surinamensis Miq., Linnaea 21: 479 (1848). *Phyllanthus antillanus* var. *concolor* Müll.Arg., Linnaea 32: 51 (1863).
Cicca pavoniana Baill., Étude Euphorb.: 618 (1858). *Phyllanthus nobilis* var. *pavonianus* (Baill.) Müll.Arg. in A.P.de Candolle, Prodr. 15(2): 415 (1866).
Cicca sinica Baill., Étude Euphorb.: 618 (1858), nom. nud. *Phyllanthus sinicus* Müll.Arg., Linnaea 32: 50 (1863).
Margaritaria adelioides Rich. ex Baill., Étude Euphorb.: 618 (1858), nom. nud.
Cicca antillana var. *pedicellaris* Griseb., Pl. Wright. 8: 158 (1860).
Phyllanthus antillanus var. *pedicellaris* Müll.Arg., Linnaea 32: 51 (1863).
Bradleia sinica Müll.Arg. in A.P.de Candolle, Prodr. 15(2): 414 (1866), nom. illeg.
Phyllanthus nobilis var. *brasiliensis* Müll.Arg. in A.P.de Candolle, Prodr. 15(2): 415 (1866).
Phyllanthus nobilis var. *guyanensis* Müll.Arg. in A.P.de Candolle, Prodr. 15(2): 414 (1866).
Phyllanthus nobilis var. *peruvianus* Müll.Arg. in A.P.de Candolle, Prodr. 15(2): 414 (1866).
Phyllanthus nobilis var. *riedelianus* Müll.Arg. in A.P.de Candolle, Prodr. 15(2): 415 (1866).
Phyllanthus nobilis var. *martii* Müll.Arg. in C.F.P.von Martius, Fl. Bras. 11(2): 70 (1873).
Phyllanthus nobilis var. *panamensis* Müll.Arg. in C.F.P.von Martius, Fl. Bras. 11(2): 70 (1873).
Phyllanthus heteromorphus Rusby, Descr. S. Amer. Pl.: 42 (1920).
Phyllanthus nobilis var. *hypomalacus* Standl., Publ. Carnegie Inst. Wash. 461: 68 (1935). *Phyllanthus antillanus* var. *hypomalacus* (Standl.) Lundell, Phytologia 1: 337 (1939).

Margaritaria rhomboidalis (Baill.) G.L.Webster, J. Arnold Arbor. 60: 434 (1979).
Madagascar. 29 MDG. Nanophan. or phan.
* *Cicca rhomboidalis* Baill., Adansonia 2: 51 (1861). *Phyllanthus rhomboidalis* (Baill.) Müll.Arg., Linnaea 32: 52 (1863).

Margaritaria scandens (Wright ex Griseb.) G.L.Webster, J. Arnold Arbor. 38: 66 (1957).
Bahamas, E. Cuba (incl. I. de la Juventud). 81 BAH CUB. Cl. nanophan. or phan.
Cicca antillana var. *glaucescens* Griseb., Mem. Amer. Acad. Arts, n.s., 8: 157 (1860).
Phyllanthus antillanus var. *glaucescens* (Griseb.) Müll.Arg., Linnaea 32: 51 (1863).
* *Cicca scandens* C.Wright ex Griseb., Nachr. Königl. Ges. Wiss. Georg-Augusts-Univ. 1: 165 (1865). *Phyllanthus scandens* (C.Wright ex Griseb.) Müll.Arg. in A.P.de Candolle, Prodr. 15(2): 415 (1866).
Phyllanthus bahamensis Urb., Symb. Antill. 3: 289 (1902). *Margaritaria bahamensis* (Urb.) Britton & Millsp., Bahama Fl.: 220 (1920).

Margaritaria tetracocca (Baill.) G.L.Webster, J. Arnold Arbor. 38: 66 (1957).
E. Cuba, Haiti. 81 CUB HAI. Nanophan.
Cicca antillana var. *virens* Griseb., Mem. Amer. Acad. Arts, n.s., 8: 158 (1860).
Phyllanthus antillanus var. *virens* (Griseb.) Müll.Arg., Linnaea 32: 51 (1863). *Cicca virens* (Griseb.) C.Wright ex Griseb., Nachr. Königl. Ges. Wiss. Georg-Augusts-Univ. 1: 166 (1865). *Phyllanthus virens* (Griseb.) Müll.Arg. in A.P.de Candolle, Prodr. 15(2): 415 (1866).
* *Wurtzia tetracocca* Baill., Adansonia 1: 187 (1861).

Synonyms:
Margaritaria adelioides Rich. ex Baill. === **Margaritaria nobilis** L.f.
Margaritaria alternifolia L. === **Margaritaria nobilis** L.f.
Margaritaria anomala var. *cheloniphorbe* (Hutch.) Fosberg === **Margaritaria anomala** (Baill.) Fosberg
Margaritaria bahamensis (Urb.) Britton & Millsp. === **Margaritaria scandens** (Wright ex Griseb.) G.L.Webster
Margaritaria discoidea f. *glabra* Radcl.-Sm. === **Margaritaria discoidea** var. **triplosphaera** Radcl.-Sm.
Margaritaria discoidea subsp. *nitida* (Pax) G.L.Webster === **Margaritaria discoidea** var. **nitida** (Pax) Radcl.-Sm.
Margaritaria nobilis var. *antillana* (A.Juss.) Stehlé & Quetin === **Margaritaria nobilis** L.f.
Margaritaria obovata (Baill.) G.L.Webster === **Margaritaria discoidea** var. **triplosphaera** Radcl.-Sm.
Margaritaria oppositifolia L. === ? (Combretaceae)

Martretia

1 species, W. & WC. tropical Africa; well-branched shrubs or small to medium trees of low-lying forest land or along riverbanks with distichously arranged, finely-veined, thick leaves. Subfamily position apparently uncertain, with claims being made for both Phyllanthoideae and Acalyphoideae. An association with *Hymenocardia* has also been suggested. Léonard (1989), in advocating Phyllanthoideae, proposed for it a distinct tribe Martretieae following a proposal of Köhler; this, however, was not adopted by Webster (Synopsis, 1994). For the writer of these lines, the arrangement of the fruits on short axillary racemes recalls *Antidesma*. (Phyllanthoideae (?))

Pax, F. & K. Hoffmann (1922). *Martretia.* In A. Engler (ed.), Das Pflanzenreich, IV 147 XV (Euphorbiaceae-Phyllanthoideae-Phyllantheae): 79-80. Berlin. (Heft 81.) La/Ge. — 1 species, Africa.

• Léonard, J. (1989). Révision du genre africain *Martretia* Beille (Euphorbiaceae) et la nouvelle tribu des Martretieae. Bull. Jard. Bot. Natl. Belg. 59: 319-332, illus., map. Fr. — Includes review of *Martretia*, its position in the family, and a proposal for its placement in a distinct tribe of Phyllanthoideae (protologue, p. 326). Full documentation for the single species is also given, including localities with exsiccatae from throughout the range. [For an associated flora account, see *Flore d'Afrique Centrale, Euphorbiaceae* 2: 12-16 (1995).]

Martretia Beille, Compt. Rend. Hebd. Séances Acad. Sci. 145: 1294 (1907).
W. & WC. Trop. Africa. 22 23.

Martretia quadricornis Beille, Compt. Rend. Hebd. Séances Acad. Sci. 145: 64 (1908).
Sierra Leone, Ivory Coast, Ghana, Nigeria, Central African Rep., Zaire. 22 GHA IVO NGA SIE 23 CAF ZAI. Phan. – In swamp forests.

Maschalanthus

Synonyms:
Maschalanthus Nutt. === **Andrachne** L.
Maschalanthus obovatus Nutt. === **Andrachne phyllanthoides** (Nutt.) Müll.Arg.
Maschalanthus polygonoides Nutt. === **Andrachne phyllanthoides** (Nutt.) Müll.Arg.

Mazinna

Synonyms:
Mazinna Spach === **Jatropha** L.

Meborea

1 species, Guianas; originally described from French Guiana but also covered in *Euphorbiaceae of Suriname* (Lanjouw, 1931, with figure; see **Americas**). It remains imperfectly known and was by Webster (1994, Synopsis) not assigned to a tribe. Mueller in the nineteenth century mistakenly made *M. guianensis* (along with *P. attenuatus* Miq.) a basis for his *Phyllanthus guianensis* but its correct identity was clarified by Lanjouw (1931: 9-10). It was nonetheless overlooked in Lemée's *Flore de la Guyane Française*, while in the *Checklist of the Plants of the Guianas* (2nd edn.; Boggan et al., 1997) it has been treated as an unnamed species of *Phyllanthus* (p. 103). The leaves are distichously arranged but the fruits are borne at the ends of axillary shoots from which staminate flowers appear to have fallen. (Phyllanthoideae)

Lanjouw, J. (1931). The Euphorbiaceae of Suriname. See **Americas**. — A photograph (pl. 1) of a type specimen, collected by Aublet and now in the Natural History Museum, London, is included.

Meborea Aubl., Hist. Pl. Guiane 2: 825 (1775).
Guiana. 82.

Meborea guianensis Aubl., Hist. Pl. Guiane 2: 825 (1775). *Phyllanthus guianensis* (Aubl.) Müll.Arg. in A.P.de Candolle, Prodr. 15(2): 376 (1866), nom. illeg. *Phyllanthus protoguayanensis* Herter, Revista Sudamer. Bot. 6: 93 (1939).
Guiana. 82 FRG. – Known only from the type.

Mecostylis

Synonyms:
Mecostylis Kurz ex Teijsm. & Binn. === **Macaranga** Thouars

Medea

Synonyms:
Medea Klotzsch === **Croton** L.
Medea timandroides Didr. === **Croton timandroides** (Didr.) Müll.Arg.

Medusea

An early segregate of *Euphorbia*.

Synonyms:
Medusea Haw. === **Euphorbia** L.
Medusea tessellata Haw. === **Euphorbia caput-medusae** L.

Megabaria

Synonyms:
Megabaria Pierre ex Hutch. === **Spondianthus** Engl.
Megabaria klaineana Pierre ex Pax === **Protomegabaria macrophylla** (Pax) Hutch.
Megabaria macrophylla (Pax) Pierre ex A.Chev. === **Protomegabaria macrophylla** (Pax) Hutch.
Megabaria obovata Pierre ex Hutch. === **Protomegabaria stapfiana** (Beille) Hutch.
Megabaria trillesii Pierre ex Hutch. === **Spondianthus preussii** Engl. var. **preussii**
Megabaria ugandensis Hutch. === **Spondianthus preussii** var. **glaber** (Engl.) Engl.

Megalostylis

Synonyms:
Megalostylis S.Moore === **Dalechampia** L.
Megalostylis poeppigii S.Moore === **Dalechampia micrantha** Poepp. & Endl.

Megistostigma

5 species, China and SE Asia to Borneo and the Philippines; small, slender woody climbers or twiners with stinging hairs and triveined leaves of open places or scrub, similar (and closely related) to *Cnesmone*. (Acalyphoideae)

Hooker, J. D. (1887). *Megistostigma malaccense*. Ic. Pl. 16: pl. 1592. En. — Plant portrait with associated description and commentary. [Now correctly *M. glabratum*.]

Pax, F. & K. Hoffmann (1919). *Clavistylus*. In A. Engler (ed.), Das Pflanzenreich, IV 147 IX (Euphorbiaceae-Acalypheae-Plukenetiinae): 104. Berlin. (Heft 68.) La/Ge. — 1 species, W Malesia; now in *Megistostigma* as *M. peltatum*.

Pax, F. & K. Hoffmann (1919). *Sphaerostylis*. In A. Engler (ed.), Das Pflanzenreich, IV 147 IX (Euphorbiaceae-Acalypheae-Plukenetiinae): 106-107. Berlin. (Heft 68.) La/Ge. — 2 species, Madagascar (1), W Malesia (1). [*S. malaccensis* now in *Megistostigma*.]

Airy-Shaw, H. K. (1969). Notes on Malesian and other Asiatic Euphorbiaceae, CXII. Notes on the subtribe Plukenetiinae Pax. Kew Bull. 23: 114-121. En. —*Megistostigma*, pp. 119-121; includes key to all species (formerly also in *Clavistylus* and *Sphaerostylis*).

Megistostigma Hook.f., Hooker's Icon. Pl. 16: 1592 (1887).
S. China, Indo-China, Malesia. 36 41 42.
Clavistylus J.J.Sm. ex Koord. & Valeton, Meded. Dept. Landb. Ned.-Indië 10: 517 (1910).

Megistostigma burmanicum (Kurz) Airy Shaw, Kew Bull. 23: 119 (1969).
Burma, Thailand, Pen. Malaysia (Perlis), S. China ? 36 CHC? 41 BMA MLY THA.
Cl. cham.
* *Tragia burmanica* Kurz, J. Asiat. Soc. Bengal, Pt. 2, Nat. Hist. 42(2): 244 (1873).

Megistostigma cordatum Merr., Philipp. J. Sci. 16: 563 (1920). *Sphaerostylis cordata* (Merr.) Pax & K.Hoffm. in H.G.A.Engler, Nat. Pflanzenfam. ed. 2, 19c: 148 (1931).
Sumatera , Borneo (Sabah), Philippines (Samar). 42 BOR PHI SUM. Cl. nanophan.

Megistostigma glabratum (Kurz) Govaerts in R.Govaerts, D.G.Frodin & A.Radcliffe-Smith, World Checklist Bibliogr. Euphorbiaceae: 1166 (2000).
Pen. Malaysia (incl. Anamba Is.), N. Sumatera. 42 MLY SUM. Cl. nanophan.
* *Cnesmone glabrata* Kurz, Flora 58: 31 (1875). *Sphaerostylis glabrata* (Kurz) Merr., Pap. Michigan Acad. Sci. 24: 78 (1938 publ. 1939).
Megistostigma malaccense Hook.f., Hooker's Icon. Pl. 16: 1592 (1887). *Sphaerostylis malaccensis* (Hook.f.) Pax & K.Hoffm. in H.G.A.Engler, Pflanzenr., IV, 147, IX: 107 (1919).

Megistostigma peltatum (J.J.Sm.) Croizat, J. Arnold Arbor. 22: 426 (1941).
Sumatera (Siberut, Natuna I.), Jawa. 42 JAW SUM. Cl. nanophan.
* *Clavistylus peltatus* J.J.Sm., Meded. Dept. Landb. Ned.-Indië 10: 519 (1910).

Megistostigma yunnanense Croizat, J. Arnold Arbor. 22: 426 (1941).
SC. China. 36 CHC. Cl. nanophan.

Synonyms:
Megistostigma malaccense Hook.f. === **Megistostigma glabratum** (Kurz) Govaerts

Meialisa

Synonyms:
Meialisa Raf. === **Adriana** Gaudich.

Meineckia

28 species, Middle and South America, Africa, Socotra, Madagascar (10), Arabian Peninsula, Sri Lanka and India (S. India, Assam and Arunachal Pradesh); woody herbs, subshrubs, shrubs or small trees to 5 m with distichously arranged leaves in seasonally dry areas, sometimes in rocky places. Includes *Peltandra*, *Cluytiandra* (following Webster 1965), *Zimmermannia* (following Brunel, 1987 and Radcliffe-Smith, 1997) and *Zimmermaniopsis* (following Radcliffe-Smith, 1997). The genus has in turn been included in *Securinega* (Pax & Hoffmann, 1931) and also in *Flueggea* but in 1965 Webster resurrected the genus. Save for the species referred to *Cluytiandra* it was, as a close ally of or even included in *Phyllanthus*, never revised for *Pflanzenreich*. A key to the presently recognised four sections appears in Radcliffe-Smith (1997). (Phyllanthoideae)

Pax, F. & K. Hoffmann (1922). *Cluytiandra*. In A. Engler (ed.), Das Pflanzenreich, IV 147 XV (Euphorbiaceae-Phyllanthoideae-Phyllantheae): 209-211. Berlin. (Heft 81.) La/Ge. — 4 species; Africa. [Now included with *Meineckia* following revision by Webster (1965), a genus otherwise treated by Pax and Hoffmann only in 1931.]

Pax, F. & K. Hoffmann (1931). *Securinega*. In A. Engler (ed.), Die natürlichen Pflanzenfamilien, 2. Aufl., 19c: 60. Leipzig. Ge. — Synopsis with description of genus. [Not revised for Pflanzenreich. *Meineckia* here incorporated in sect. *Flueggea*. Some later authors have united it with a 'stand-alone' *Flueggea*, including Webster (1984) and Webster (1994; see **General**).]

Pax, F. & K. Hoffmann (1931). *Zimmermannia*. In A. Engler (ed.), Die natürlichen Pflanzenfamilien, 2. Aufl., 19c: 59-60. Leipzig. Ge. — Synopsis with description of genus; 1 species, *Z. capillipes*; E Africa. [Not revised for Pflanzenreich.]

Verdcourt, B. (1954). Revision of the genus *Zimmermannia* Pax. Kew Bull. 9: 38-40. (Notes from the East African Herbarium, II.) En. — Synoptic treatment with key, description of a novelty, synonymy, localities with exsiccatae, and notes. [One of the four species here was left undescribed for want of sufficient material.]

Leandri, J. (1958). *Cluytiandra*. Fl. Madag. Comores 111 (Euphorbiacées), I: 138-143. Paris. Fr. — Flora treatment (6 species); key. [Included with *Meineckia* by Webster (1965; Synopsis, 1994).]

- Webster, G. L. (1965). A revision of the genus *Meineckia* (Euphorbiaceae). Acta Bot. Neerl. 14: 323-365, illus., maps. En. — Revision (19 species) with keys, descriptions, synonymy, references and citations, types, distribution, localities with exsiccatae, and commentary; literature and list of specimens seen at end but no separate index. The general part includes a review of characters along with the history of the genus, its relationships with respect to other Phyllantheae, and its geography (with maps). [Includes *Cluytiandra* but not *Zimmermannia*. The genus is considered to have a centre of origin in Africa.]

Radcliffe-Smith, A. (1981). Notes on African Euphorbiaceae, X. *Zimmermannia*. Kew Bull. 36: 127-128. En. — Precursory to *Flora of Tropical East Africa*.

Brunel, J. F. (1987). *Meineckia*. In *idem*, Sur le genre *Phyllanthus* L. et quelques genres voisins de la tribu des Phyllantheae Dumort. Strasbourg. [See also *Phyllanthus*.] Fr. — Description of genus, with synopsis of the species and illustrations of key characters. [*Zimmermannia* is here synonymized, a position also adopted for the present work.]

Radcliffe-Smith, A. & M. M. Harley (1990). Notes on African Euphorbiaceae, XXI. *Aerisilvaea* and *Zimmermanniopsis*, two new phyllanthoid genera for the flora of Tanzania. Kew Bull. 45: 147-156. En. — Includes protologue of *Zimmermanniopsis* (a segregate of *Zimmermannia*); 1 species (also new). [Now reduced to *Meineckia*.]

Radcliffe-Smith, A. (1997). Notes on African and Madagascan Euphorbiaceae. Kew Bull. 52: 171-176. En. — Pp. 173-176, 'The status of *Zimmermanniopsis* and *Zimmermania*', incorporates material from Brunel (1987) and with it reduction of both genera to *Meineckia*. A synoptic key to four sections recognised for the enlarged genus is furnished.

Meineckia Baill., Étude Euphorb.: 586 (1858).
Trop. Africa to Assam. 23 24 25 26 29 35 40. Cham. or nanophan.
Peltandra Wight, Icon. Pl. Ind. Orient. 5(2): 24 (1852).
Cluytiandra Müll.Arg., J. Bot. 2: 328 (1864).
Zimmermannia Pax, Bot. Jahrb. Syst. 45: 235 (1910).
Neopeltandra Gamble, Fl. Madras: 1285 (1925).
Zimmermanniopsis Radcl.-Sm., Kew Bull. 45: 152 (1990).

Meineckia acuminata (Verdc.) Brunel ex Radcl.-Sm., Kew Bull. 52: 174 (1997).
Tanzania (Morogoro). 25 TAN. Nanophan. or phan.
* *Zimmermannia acuminata* Verdc., Kew Bull. 9: 39 (1954).

Meineckia baronii (Hutch.) G.L.Webster, Acta Bot. Neerl. 14: 356 (1965).
NW. Madagascar. 29 MDG. Nanophan.
* *Cluytiandra baronii* Hutch., Bull. Misc. Inform. Kew 1918: 205 (1918).

Meineckia bartlettii (Standl.) G.L.Webster, Acta Bot. Neerl. 14: 346 (1965).
Mexico (Chiapas), Belize. 79 MXT 80 BLZ. Nanophan.
* *Phyllanthus bartlettii* Standl., Publ. Carnegie Inst. Wash. 461: 68 (1935).

Meineckia calycina G.L.Webster, Acta Bot. Neerl. 14: 354 (1965).
S. India. 40 IND. Nanophan.

Meineckia capillipes (Blake) G.L.Webster, Acta Bot. Neerl. 14: 345 (1965).
Guatemala, Honduras. 80 GUA HON. Nanophan.
* *Phyllanthus capillipes* S.F.Blake, Contr. U. S. Natl. Herb. 24: 10 (1922).

Meineckia decaryi (Leandri) Brunel ex Radcl.-Sm., Kew Bull. 52: 173 (1997).
Madagascar. 29 MDG. Nanophan.
* *Cluytiandra decaryi* Leandri, Notul. Syst. (Paris) 7: 193 (1939). *Zimmermannia decaryi* (Leandri) G.L.Webster, Acta Bot. Neerl. 14: 363 (1965).

var. **decaryi**
Madagascar. 29 MDG. Nanophan.

var. **occidentalis** (Leandri) Radcl.-Sm., Kew Bull. 52: 174 (1997).
Madagascar. 29 MDG.
* *Cluytiandra decaryi* var. *occidentalis* Leandri, Notul. Syst. (Paris) 7: 193 (1939).
Zimmermannia decaryi var. *occidentalis* (Leandri) Poole, Kew Bull. 36: 129 (1981).

Meineckia filipes (Balf.f.) G.L.Webster, Acta Bot. Neerl. 14: 344 (1965).
Socotra. 24 SOC.
* *Phyllanthus filipes* Balf.f., Proc. Roy. Soc. Edinburgh 12: 94 (1884).

Meineckia fruticans (Pax) G.L.Webster, Acta Bot. Neerl. 14: 350 (1965).
Kenya, Tanzania. 25 KEN TAN. Nanophan.
* *Cluytiandra fruticans* Pax, Bot. Jahrb. Syst. 33: 276 (1903).

var. **engleri** (Pax) G.L.Webster, Acta Bot. Neerl. 14: 351 (1965).
Kenya, Tanzania. 25 KEN TAN. Nanophan.
* *Cluytiandra engleri* Pax, Bot. Jahrb. Syst. 34: 368 (1904).

var. **fruticans**
Kenya, Tanzania. 25 KEN TAN. Nanophan.

Meineckia grandiflora (Verdc.) Brunel ex Radcl.-Sm., Kew Bull. 52: 174 (1997).
Tanzania (Ulanga, Lindi). 25 TAN. Nanophan. or phan.
* *Zimmermannia grandiflora* Verdc., Kew Bull. 9: 40 (1954).

Meineckia humbertii G.L.Webster, Acta Bot. Neerl. 14: 360 (1965).
SW. Madagascar. 29 MDG. Nanophan.

Meineckia leandrii G.L.Webster, Acta Bot. Neerl. 14: 357 (1965).
WC. Madagascar. 29 MDG. Nanophan.

Meineckia longipes (Wight) G.L.Webster, Acta Bot. Neerl. 14: 352 (1965).
S. India. 40 IND. Nanophan.
* *Peltandra longipes* Wight, Icon. Pl. Ind. Orient. 5: t. 1891 (1852). *Phyllanthus longipes* (Wight) Müll.Arg., Linnaea 32: 11 (1863).

Meineckia macropus (Hook.f.) G.L.Webster, Acta Bot. Neerl. 14: 352 (1965).
Arunachal Pradesh (Mishmi Hills). 40 EHM. Nanophan.
* *Phyllanthus macropus* Hook.f., Fl. Brit. India 5: 287 (1887).

Meineckia madagascariensis (Leandri) G.L.Webster, Acta Bot. Neerl. 14: 358 (1965).
E. Madagascar. 29 MDG. Nanophan.
* *Cluytiandra madagascariensis* Leandri ex Humbert, Notul. Syst. (Paris) 7: 193 (1939).

Meineckia neogranatensis (Müll.Arg.) G.L.Webster, Acta Bot. Neerl. 14: 347 (1965).
C. Colombia, Brazil. 83 CLM 84 BZE BZL. Nanophan.
* *Phyllanthus neogranatensis* Müll.Arg., Linnaea 32: 10 (1863).

subsp. **gardneri** G.L.Webster, Acta Bot. Neerl. 14: 349 (1965).
Brazil (Piauí). 84 BZE. Nanophan.

subsp. **hilariana** (Baill.) G.L.Webster, Acta Bot. Neerl. 14: 349 (1965).
Brazil (Minas Gerais). 84 BZL. Nanophan.
* *Flueggea hilariana* Baill., Adansonia 5: 346 (1865). *Securinega hilariana* (Baill.) Müll.Arg. in A.P.de Candolle, Prodr. 15(2): 1273 (1866).
Acidoton hilarianus (Müll.Arg.) Kuntze, Revis. Gen. Pl. 2: 592 (1891).

subsp. **neogranatensis**
C. Colombia. 83 CLM. Nanophan.

Meineckia nguruensis (Radcl.-Sm.) Brunel ex Radcl.-Sm., Kew Bull. 52: 174 (1997).
Tanzania (Nguru Mts.). 25 TAN. Phan.
* *Zimmermannia nguruensis* Radcl.-Sm., Kew Bull. 36: 127 (1981).

Meineckia orientalis (Leandri) G.L.Webster, Acta Bot. Neerl. 14: 361 (1965).
EC. Madagascar. 29 MDG. Nanophan.
* *Cluytiandra orientalis* Leandri, Notul. Syst. (Paris) 7: 193 (1939).

Meineckia ovata (E.A.Bruce) Brunel ex Radcl.-Sm., Kew Bull. 52: 174 (1997).
Kenya (Teita Mts.). 25 KEN. Nanophan. or phan.
* *Zimmermannia ovata* E.A.Bruce, Bull. Misc. Inform. Kew 1933: 468 (1933).

Meineckia parvifolia (Wight) G.L.Webster, Acta Bot. Neerl. 14: 342 (1965).
S. India, Sri Lanka. 40 IND SRL. Nanophan.
* *Peltandra parvifolia* Wight, Icon. Pl. Ind. Orient. 5: t. 1892 (1852).
Peltandra flexuosa Thwaites, Enum. Pl. Zeyl.: 281 (1861).
Phyllanthus peltandrus Müll.Arg., Linnaea 32: 11 (1863).
Phyllanthus suberosus Wight ex Müll.Arg., Linnaea 32: 11 (1863).
Phyllanthus thwaitesianus Müll.Arg., Linnaea 32: 11 (1863).
Andrachne fruticosa B.Heyne ex Hook.f., Fl. Brit. India 5: 287 (1887), nom. illeg.

Meineckia paxii Brunel ex Radcl.-Sm., Kew Bull. 52: 174 (1997).
Tanzania (Lushoto). 25 TAN. Nanophan. or phan.
* *Zimmermannia capillipes* Pax, Bot. Jahrb. Syst. 45: 235 (1910).

Meineckia peltata (Hutch.) G.L.Webster, Acta Bot. Neerl. 14: 356 (1965).
C. & NE. Madagascar. 29 MDG. Nanophan.
* *Cluytiandra peltata* Hutch., Bull. Misc. Inform. Kew 1918: 204 (1918).

Meineckia phyllanthoides Baill., Étude Euphorb.: 587 (1858). *Flueggea meineckia* Müll.Arg., Linnaea 34: 76 (1865). *Securinega phyllanthoides* (Baill.) Müll.Arg. in A.P.de Candolle, Prodr. 15(2): 448 (1866).
Ethiopia to Angola, Arabian Pen. 23 ZAI 24 ETH SOM 25 KEN TAN UGA 26 ANG 35 OMA YEM. Nanophan.

subsp. **capillariformis** (Vatke & Pax ex Pax) G.L.Webster, Acta Bot. Neerl. 14: 340 (1965).
Zaire (Kivu), Kenya, Tanzania, Uganda. 23 ZAI 25 KEN TAN UGA. Nanophan.
* *Phyllanthus capillariformis* Vatke & Pax ex Pax, Bot. Jahrb. Syst. 15: 523 (1893).
Cluytiandra capillariformis (Vatke & Pax ex Pax) Pax & K.Hoffm. in H.G.A.Engler & C.G.O.Drude, Veg. Erde 9, III(2): 31 (1921).
Phyllanthus capilliformis Pax & Vatke ex Engl., Abh. Königl. Akad. Wiss. Berlin 1894: 20 (1894).

subsp. **phyllanthoides**
Yemen, S. Oman. 35 OMA YEM. Nanophan.

subsp. **somalensis** (Pax) G.L.Webster, Acta Bot. Neerl. 14: 340 (1965).
S. & E. Ethiopia, Somalia, Kenya, Uganda. 24 ETH SOM 25 KEN UGA. Nanophan.
* *Cluytiandra somalensis* Pax, Bot. Jahrb. Syst. 33: 277 (1903).

subsp. **trichopoda** (Müll.Arg.) G.L.Webster, Acta Bot. Neerl. 14: 341 (1965).
NW. Angola. 26 ANG. Nanophan.
* *Cluytiandra trichopoda* Müll.Arg., J. Bot. 2: 328 (1864).

Meineckia pubiflora G.L.Webster, Acta Bot. Neerl. 14: 360 (1965).
SW. Madagascar. 29 MDG. Nanophan. or phan.

Meineckia stipularis (Radcl.-Sm.) Brunel ex Radcl.-Sm., Kew Bull. 52: 175 (1997).
Tanzania. 25 TAN. Nanophan. or phan.
* *Zimmermannia stipularis* Radcl.-Sm., Kew Bull. 36: 128 (1981).

Meineckia trichogynis (Baill.) G.L.Webster, Acta Bot. Neerl. 14: 355 (1965).
Madagascar. 29 MDG. Nanophan.
* *Flueggea trichogynis* Baill., Étude Euphorb.: 593 (1858). *Securinega trichogynis* (Baill.) Müll.Arg. in A.P.de Candolle, Prodr. 15(2): 447 (1866). *Acidoton trichogynus* (Baill.) Kuntze, Revis. Gen. Pl. 2: 592 (1891).

Meineckia uzungwaensis (Radcl.-Sm.) Radcl.-Sm., Kew Bull. 52: 175 (1997).
Tanzania (Mufindi). 25 TAN. Nanophan.
* *Zimmermanniopsis uzungwaensis* Radcl.-Sm., Kew Bull. 45: 154 (1990).

Meineckia vestita G.L.Webster, Acta Bot. Neerl. 14: 343 (1965).
C. Tanzania. 25 TAN. Cham.

Meineckia websteri Brunel & J.P.Roux, Bull. Mus. Natl. Hist. Nat., B, Adansonia 4: 83 (1982).
Madagascar. 29 MDG. Nanophan.

Melanolepis

2 species, Nansei-shoto, Taiwan, SE. Asia, Marianas Is., Malesia east to the Bismarck Archipelago. *M. multiglandulosa* is a widespread small to medium-sized soft-wooded tree (rarely to as much as 25 m) of regrowth and light forest with often lobed leaves and terminal, branched inflorescences; all non-woody parts are stellate-hairy. The male flowers have up to 200 or more stamens. In the past, the genus was viewed as closely related to *Mallotus* (e.g. by Pax & Hoffmann) but according to Webster (Synopsis, 1994) the two are now in different tribes, *Melanolepis* being in the Chrozophoreae on account of petals being present. It also contrasts with both *Macaranga* and *Mallotus* in the presence of a disk in female flowers. No modern summary of the botany and ecology of the genus is available; it has, however, been under study in Leiden as part of Peter van Welzen's survey of Malesian *Chrozophora* and allies and for a *Flora Malesiana* account. (Acalyphoideae)

Pax, F. (with K. Hoffmann) (1914). *Melanolepis*. In A. Engler (ed.), Das Pflanzenreich, IV 147 VII (Euphorbiaceae-Acalypheae-Mercurialinae): 142-144. Berlin. (Heft 63.) La/Ge. — Monotypic, SE Asia and Taiwan through Malesia to Papuasia (including the Bismarck Archipelago) and in the Marianas.

Melanolepis Rchb. & Zoll., Acta Soc. Regiae Sci. Indo-Neerl. 1: 22 (1856).
Indo-China to W. Pacific. 38 41 42 60 62.

Melanolepis multiglandulosa (Reinw. ex Blume) Rchb. & Zoll., Linnaea 28: 324 (1856).
Trop. & E. Asia, Solomon Is., Marianas (Guam). 38 NNS TAI 41 THA 42 BIS BOR JAW LSI MLY MOL NWG PHI SUL SUM 60 SOL 62 MRN. Phan.
* *Croton multiglandulosus* Reinw. ex Blume, Catalogus: 105 (1823). *Mallotus multiglandulosus* (Reinw. ex Blume) Hurus., J. Fac. Sci. Univ. Tokyo, Sect. 3, Bot. 6: 308 (1954).

var. **multiglandulosa**
Trop. & E. Asia, Solomon Is., Guam. 38 NNS TAI 41 THA 42 BIS BOR JAW LSI MLY MOL NWG PHI SUL SUM 60 SOL 62 MRN. Phan.
Ricinus dioicus Wall. ex Roxb., Fl. Ind. ed. 1832, 3: 690 (1832).
Melanolepis angulata Miq., Fl. Ned. Ind., Eerste Bijv.: 455 (1861). *Mallotus angulatus* (Miq.) Müll.Arg. in A.P.de Candolle, Prodr. 15(2): 958 (1866).

Melanolepis calcosa Miq., Fl. Ned. Ind., Eerste Bijv.: 399 (1861). *Mallotus calcosus* (Miq.) Müll.Arg. in A.P.de Candolle, Prodr. 15(2): 958 (1866).
Mallotus moluccanus var. *glabratus* Müll.Arg., Linnaea 34: 186 (1865).
Mallotus hellwigianus K.Schum. in K.M.Schumann & U.M.Hollrung, Fl. Kais. Wilh. Land: 79 (1889).
Mallotus hollrungianus K.Schum. in K.M.Schumann & U.M.Hollrung, Fl. Kais. Wilh. Land: 79 (1889).
Melanolepis moluccana Pax & K.Hoffm. in H.G.A.Engler, Pflanzenr., IV, 147, VII: 142 (1914).

var. **pendula** (Merr.) Merr. in Enum. Philip. Fl. Pl. 2: 432 (1923).
Philippines. 42 PHI. Phan.
Mallotus moluccanus var. *pendulus* Merr., Philipp. J. Sci., C 7: 401 (1912 publ. 1913). *Melanolepis moluccana* var. *pendula* (Merr.) Pax & K.Hoffm. in H.G.A.Engler, Pflanzenr., IV, 147, VII: 144 (1914).

Melanolepis vitifolia (Kuntze) Gagnep., Bull. Soc. Bot. France 71: 1026 (1924 publ. 1925).
Cambodia, Vietnam. 41 CBD VIE. Phan.
* *Mallotus vitifolius* Kuntze, Revis. Gen. Pl. 2: 608 (1891).

Synonyms:
Melanolepis angulata Miq. === **Melanolepis multiglandulosa** (Reinw. ex Blume) Rchb. & Zoll. var. **multiglandulosa**
Melanolepis calcosa Miq. === **Melanolepis multiglandulosa** (Reinw. ex Blume) Rchb. & Zoll. var. **multiglandulosa**
Melanolepis diadena Miq. === **Endospermum diadenum** (Miq.) Airy Shaw
Melanolepis moluccana Pax & K.Hoffm. === **Melanolepis multiglandulosa** (Reinw. ex Blume) Rchb. & Zoll. var. **multiglandulosa**
Melanolepis moluccana var. *pendula* (Merr.) Pax & K.Hoffm. === **Melanolepis multiglandulosa** var. **pendula** (Merr.) Merr.

Melanthes

An orthographic variant of *Melanthesa*.

Melanthesa

Synonyms:
Melanthesa Blume === **Breynia** J.R.Forst. & G.Forst.
Melanthesa acuminata Müll.Arg. === **Breynia racemosa** (Blume) Müll.Arg. var. **racemosa**
Melanthesa anceps (Vahl) Miq. === **Phyllanthus virgatus** G.Forst. var. **virgatus**
Melanthesa canescens Zipp. ex Span. === ?
Melanthesa cernua (Poir.) Decne. === **Breynia cernua** (Poir.) Müll.Arg.
Melanthesa cernua var. *acutifolia* Müll.Arg. === **Breynia cernua** (Poir.) Müll.Arg.
Melanthesa chinensis Blume === **Breynia fruticosa** (L.) Hook.f.
Melanthesa glaucescens Miq. === ?
Melanthesa microphylla Kurz ex Teijsm. & Binn. === **Breynia microphylla** (Kurz ex Teijsm. & Binn.) Müll.Arg.
Melanthesa neocaledonica Baill. === **Breynia disticha** J.R.Forst. & G.Forst.
Melanthesa neocaledonica var. *forsteri* Müll.Arg. === **Breynia disticha** J.R.Forst. & G.Forst.
Melanthesa obliqua Wight === **Breynia retusa** (Dennst.) Alston
Melanthesa oblongifolia Oken === **Phyllanthus reticulatus** Poir. var. **reticulatus**
Melanthesa ovalifolia Kostel. === **Breynia vitis-idaea** (Burm.f.) C.E.C.Fisch.
Melanthesa racemosa Blume === **Breynia racemosa** (Blume) Müll.Arg.
Melanthesa racemosa var. *pubescens* Müll.Arg. === **Breynia discigera** Müll.Arg.
Melanthesa reclinata (Roxb.) Müll.Arg. === **Breynia reclinata** (Roxb.) Hook.f.

Melanthesa retusa (Dennst.) Kostel. === **Breynia retusa** (Dennst.) Alston
Melanthesa rhamnoides Decne. === **Breynia oblongifolia** Müll.Arg. var. **oblongifolia**
Melanthesa rhamnoides (Retz.) Blume === **Breynia vitis-idaea** (Burm.f.) C.E.C.Fisch.
Melanthesa rhamnoides var. *hypoglauca* Müll.Arg. === **Breynia racemosa** (Blume) Müll.Arg. var. **racemosa**
Melanthesa rhamnoides var. *oblongifolia* Müll.Arg. === **Breynia oblongifolia** (Müll.Arg.) Müll.Arg.
Melanthesa rhamnoides var. *pubescens* Müll.Arg. === **Breynia discigera** Müll.Arg.
Melanthesa rubra Blume === **Breynia cernua** (Poir.) Müll.Arg.
Melanthesa rupestris Miq. === **Phyllanthus virgatus** G.Forst. var. **virgatus**
Melanthesa turbinata (K.D.Koenig ex Roxb.) Oken === **Breynia retusa** (Dennst.) Alston
Melanthesa virgata Blume === **Breynia virgata** (Blume) Müll.Arg.

Melanthesopsis

Synonyms:
Melanthesopsis Müll.Arg. === **Breynia** J.R.Forst. & G.Forst.
Melanthesopsis fruticosa Müll.Arg. === **Breynia racemosa** (Blume) Müll.Arg. var. **racemosa**
Melanthesopsis lucens Müll.Arg. === **Breynia racemosa** (Blume) Müll.Arg. var. **racemosa**
Melanthesopsis patens (Roxb.) Müll.Arg. === **Breynia retusa** (Dennst.) Alston
Melanthesopsis patens var. *oblongifolia* Müll.Arg. === **Breynia retusa** (Dennst.) Alston
Melanthesopsis patens var. *turbinata* (Wight) Müll.Arg. === **Breynia retusa** (Dennst.) Alston
Melanthesopsis patens var. *vulgaris* Müll.Arg. === **Breynia retusa** (Dennst.) Alston
Melanthesopsis variabilis Müll.Arg. === **Breynia retusa** (Dennst.) Alston

Menarda

Synonyms:
Menarda Comm. ex A.Juss. === **Phyllanthus** L.
Menarda capillaris Baill. === **Phyllanthus nummulariifolius** Poir. subsp. **nummulariifolius**
Menarda cryptophila Comm. ex A.Juss. === **Phyllanthus cryptophilus** (Comm. ex A.Juss.) Müll.Arg.
Menarda goudotiana Baill. === **Phyllanthus goudotianus** (Baill.) Müll.Arg.
Menarda linifolia Baill. === **Phyllanthus pentandrus** Schumach. & Thonn.
Menarda nummulariifolia (Poir.) Baill. === **Phyllanthus nummulariifolius** Poir.
Menarda pulchella Baill. === **Phyllanthus multiflorus** Poir.

Mercadoa

Synonyms:
Mercadoa Náves === **Doryxylon** Zoll.

Mercurialis

8 species, Macaronesia, Mediterranean, Europe, E. and SE. Asia (*M. annua* now adventive or naturalised in southern Africa and elsewhere); annual or perennial herbs with opposite leaves, sometimes basally woody and yielding a purple dye. The genus is most closely related to the southern African genera *Leidesia* and *Seidelia*. The E. and SE. Asian *M. leiocarpa* is argued by Pax (1914) to be more closely related to central European than the Mediterranean species. This alliance of perennials was very fully studied by Krähenbühl & Küpfer (1995) without, however, presentation of a formal taxonomic treatment. An extensive earlier literature on the genus, much of it populational, biological and biosystematic, is listed by Pax & Hoffmann in their 1931 *Pflanzenfamilien* family survey; another list appears under the generic heading in the *Nachtrag* (1968) to Hegi's *Illustrierte Flora von Mitteleuropa*, vol. 5(1)(1925) (see **Eurasia**). (Acalyphoideae)

Pax, F. (with K. Hoffmann) (1914). *Mercurialis*. In A. Engler (ed.), Das Pflanzenreich, IV 147 VII (Euphorbiaceae-Acalypheae-Mercurialinae): 271-282. Berlin. (Heft 63.) La/Ge. — 8 species, Macaronesia and Mediterranean to N Europe and in E and SE Asia. 2 additional hybrids. [Additions in *ibid.*, Additamentum V: 397, and in Additamentum VI: 22 (1919).]

Durand, B. (1963). Le complexe *Mercurialis annua* L. s.l. – une étude biosystématique. Ann. Sci. Nat. Bot., XII, 4: 579-623, 625-736, illus., maps. Fr. — Extremely detailed biosystematic and evolutionary study of infraspecific variation, including ploidy levels and directions; resumé and conclusions in chapter 6 (pp. 724-732); no formal systematic treatment or index although populational distribution maps are presented.

Krähenbühl, M. & P. Küpfer (1995). Le genre *Mercurialis* (Euphorbiaceae): cytogéographie et évolution du complexe polyploïde des *M. perennis* L., *M. ovata* Sternb. et Hoppe et *M. leiocarpa* Sieb. et Zucc. Candollea 50: 411-430, illus., maps. Fr. — Biosystematic study of 3 species, with numerous chromosome counts covering a wide range of populations; suggestions regarding the origin of the observed chromosomal polymorphism; crossing diagrams and distribution maps; references at end. No formal taxonomic treatment or key is presented. [The presumed least advanced diploids are relictual and limited to areas of high biological diversity (southern Balkans and Yunnan).]

Mercurialis L., Sp. Pl.: 1035 (1753).
Macaronesia, Europe, Medit., Temp. Asia. 10 11 12 13 14 20 21 (27) (28) 34 35 36 38 40 41 (50) (60) (81).
Cynocrambe Hill, Brit. Herb.: 483 (1756).
Discoplis Raf., New Fl. 4: 9 (1838).
Synema Dulac, Fl. Hautes-Pyrénées: 154 (1867).

Mercurialis annua L., Sp. Pl.: 1035 (1753). *Mercurialis annua* var. *genuina* Müll.Arg. in A.P.de Candolle, Prodr. 15(2): 797 (1866), nom. inval. *Mercurialis annua* var. *typica* Fiori in A.Fiori, A.Béguinot & G.Paoletti, Fl. Italia 2: 291 (1901), nom. inval. *Mercurialis annua* subvar. *euannua* Litard. in J.I.Briquet, Prodr. Fl. Corse 2, 2: 66 (1935), nom. inval.
Europe, Macaronesia, N. Africa, W. Asia, Arabian Pen. 10 GRB 11 AUT BGM CZE den GER HUN POL SWI 12 BAL COR FRA SAR SPA 13 ALB BUL GRC ITA KRI ROM SIC YUG 14 BLR KRY RUC UKR 20 ALG EGY LBY MOR TUN 21 AZO CNY MDR (27) (28) 34 CYP IRN IRQ LBS SIN TUR 35 SAU. Ther.
Mercurialis annua var. *serratifolia* Ball in ?, . *Mercurialis serratifolia* (Ball) Pau, Bol. Soc. Esp. Hist. Nat. 21: 280 (1921).
Mercurialis ambigua L.f., Dec. Pl. Horti Upsal. 1: 15 (1762). *Mercurialis annua* lusus *ambigua* (L.f.) Müll.Arg. in A.P.de Candolle, Prodr. 15(2): 797 (1866). *Mercurialis annua* subsp. *ambigua* (L.f.) Arcang., Comp. Fl. Ital.: 622 (1882). *Mercurialis annua* subsp. *ambigua* (L.f.) Maire, Mém. Soc. Sci. Nat. Maroc 7: 178 (1924).
Mercurialis ciliata J.Presl & C.Presl, Delic. Prag.: 56 (1822). *Mercurialis annua* f. *ciliata* (J.Presl & C.Presl) Pax & K.Hoffm. in H.G.A.Engler, Pflanzenr., IV, 147, VII: 274 (1914).
Mercurialis annua var. *ambigua* Duby, Bot. Gall., ed. 2, 1: 417 (1830).
Mercurialis annua var. *angustifolia* Gaudin, Fl. Helv. 6: 296 (1830).
Mercurialis annua var. *capillacea* Guépin, Fl. Maine et Loire, ed. 3: 401 (1845).
Mercurialis ladanum Hartm., Bot. Not. 1849: 67 (1849).
Mercurialis annua f. *cordata* Wirtg., Flora 33: 82 (1850).
Mercurialis annua f. *cuneatolanceolata* Wirtg., Flora 33: 82 (1850).
Mercurialis annua f. *lanceolata* Wirtg., Flora 33: 82 (1850).
Mercurialis annua f. *ovata* Wirtg., Flora 33: 82 (1850).
Mercurialis annua var. *dioica* Moris, Fl. Sardoa 3: 478 (1859).
Mercurialis annua var. *monoica* Moris, Fl. Sardoa 3: 478 (1859). *Mercurialis monoica* (Moris) B.M.Durand, Ann. Sci. Nat., Bot. Biol. Vég., XII, 4: 727 (1963).
Mercurialis huetii Hanry, Billotia 1: 21 (1864). *Mercurialis annua* var. *huetii* (Hanry) Müll.Arg. in A.P.de Candolle, Prodr. 15(2): 798 (1866). *Mercurialis annua* subsp. *huetii* (Hanry) Lange in M.Willkom & J.M.C.Lange, Prodr. Fl. Hispan. 3: 509 (1877).

Mercurialis annua f. *huetii* (Hanry) Pax & K.Hoffm. in H.G.A.Engler, Pflanzenr., IV, 147, VII: 275 (1914).
Mercurialis annua var. *laciniata* Müll.Arg. in A.P.de Candolle, Prodr. 15(2): 797 (1866).
Mercurialis annua var. *transsylvanica* Schur, Enum. Pl. Transsilv.: 600 (1866).
Mercurialis annua var. *camberiensis* Chabert, Bull. Soc. Bot. France 38: 300 (1881).
Mercurialis pinnatifida Sennen, Bol. Soc. Aragonesa Ci. Nat. 8: 146 (1909).
Mercurialis annua var. *variegata* Löhr, Bot. Zeitung (Berlin) 2: 62 (1910).
Mercurialis annua subvar. *serrata* Litard. in J.I.Briquet, Prodr. Fl. Corse 2: 2 (1935).

Mercurialis corsica Coss. & Kralik, Notes Pl. Crit.: 63 (1850).
Corse, Sardegna. 12 COR SAR. Cham.
Mercurialis elliptica Duby, Bot. Gall., ed. 2, 1: 417 (1830), nom. illeg.

Mercurialis elliptica Lam., Encycl. 4: 119 (1797).
C. & S. Portugal, S. Spain, Morocco. 12 POR SPA 20 MOR. Cham.

Mercurialis leiocarpa Siebold & Zucc., Fl. Jap. Fam. Nat. 1: 37 (1845).
Nepal, Bhutan, N. Assam, N. Thailand, SW. & C. China, Korea, Taiwan, C. & S. Japan. 36 CHC CHN 38 JAP KOR TAI 40 ASS EHM NEP 41 THA. Hemicr.
Mercurialis transmorrisonensis Hayata, Icon. Pl. Formosan. 5: 199 (1915).
Mercurialis leiocarpa var. *trichocarpa* W.T.Wang, Acta Bot. Yunnan. 10: 39 (1988).

Mercurialis × longifolia Lam., Encycl. 4: 119 (1797). M. annua × M. tomentosa.
SW. Europe. 12 FRA POR SPA.
Mercurialis tomentosa var. *pubescens* Loscos & Pardo, Ser. Inconf. Pl. Aragon.: 97 (1863).
Mercurialis × bichei Magnier, Scrin. Fl. Select. 6: 118 (1887).
Mercurialis × bichei nothovar. *malinvaldi* (Sennen) Rouy in G.Rouy & J.Foucaud, Fl. France 12: 135 (1910).
Mercurialis × malinvaldi Sennen, Bol. Soc. Aragonesa Ci. Nat. 10: 171 (1911). *Mercurialis × longifolia* nothosubsp. *malinvaldi* (Sennen) Jauzein, Monde Pl. 85(439): 28 (1990).

Mercurialis ovata Sternb. & Hoppe, Denkschr. Bayer. Bot. Ges. Regensburg 1: 170 (1815). *Mercurialis perennis* var. *ovata* (Sternb. & Hoppe) Müll.Arg. in A.P.de Candolle, Prodr. 15(2): 796 (1866). *Mercurialis perennis* subsp. *ovata* (Sternb. & Hoppe) Celak., Prodr. Fl. Böhmen 2: 125 (1871). *Mercurialis ovata* f. *genuina* Pax & K.Hoffm. in H.G.A.Engler, Pflanzenr., IV, 147, VII: 279 (1914), nom. inval.
C., E. & SE. Europe, Turkey, W. Syria, Caucasus. 11 AUT CZE GER HUN POL SWI 13 ALB BUL GRC ITA ROM 14 KRY RUS UKR 33 NCS TCS 34 LBS TUR. Hemicr.
Mercurialis livida Port. ex Baumg., Enum. Stirp. Transsilv. 3: 344 (1846).
Mercurialis ovata f. *croatica* Degen, Magyar Bot. Lapok 4: 250 (1905).

Mercurialis × paxii Graebn. in P.F.A.Ascherson & P.Graebner, Syn. Mitteleur. Fl. 7: 408 (1917). M. ovata × M. perennis.
C. & SE. Europe. 11 AUT GER 13 ROM 14 UKR. Hemicr.

Mercurialis perennis L., Sp. Pl.: 1037 (1753). *Mercurialis perennis* f. *genuina* Müll.Arg. in A.P.de Candolle, Prodr. 15(2): 796 (1866), nom. inval.
Europe, Medit. to N. Iran. 10 DEN FIN GBR ire NOR SWE 11 AUT BGM CZE GER HUN NET POL SWI 12 COR FRA POR SPA 13 ALB BUL GRC ITA ROM SIC TUE YUG 14 BLR BLT KRY RUC RUE RUS UKR 20 ALG 33 NCS TCS 34 CYP IRN TUR. Hemicr.
Mercurialis perennis var. *longistipes* Borbás in ?, . *Mercurialis longistipes* (Borbás) Baksay, Ann. Hist.-Nat. Mus. Natl. Hung. 49: 170 (1957).
Mercurialis cynocrambe Scop., Fl. Carniol., ed. 2, 1: 266 (1772).
Mercurialis nemoralis Salisb., Prodr. Stirp. Chap. Allerton: 390 (1796).
Mercurialis sylvatica Hoppe, Flora 1: 472 (1818).
Mercurialis longifolia Host, Fl. Austriac. 2: 666 (1831).

Mercurialis perennis var. *brachyphylla* Willk., Flora 35: 309 (1852).
Mercurialis alpina Schur, Enum. Pl. Transsilv.: 600 (1866).
Mercurialis perennis var. *subalpina* Schur, Enum. Pl. Transsilv.: 600 (1866).
Mercurialis perennis f. *saxicola* Beck, Ann. K. K. Naturhist. Hofmus. 2: 107 (1887).
Mercurialis perennis f. *glabra* Beck, Fl. Nieder-Österreich 1: 554 (1890).
Mercurialis sylvestris Bubani, Fl. Pyren. 1: 89 (1897).
Mercurialis perennis f. *robusta* Gross, Mitt. Bot. Vereins Kreis Freiburg: 79 (1906).

Mercurialis reverchonii Rouy, Naturaliste 1887: 199 (1887).
SW. Spain, Morocco. 12 SPA 20 MOR. Cham.

Mercurialis tomentosa L., Sp. Pl.: 1035 (1753).
SW. Europe. 12 BAL FRA POR SPA. Cham.
Mercurialis sericea Salisb., Prodr. Stirp. Chap. Allerton: 390 (1796).

Synonyms:
Mercurialis abyssinica Hochst. ex Pax & K.Hoffn. === **Micrococca mercurialis** (L.) Benth.
Mercurialis acanthocarpa H.Lév. & Vaniot === **Speranskia cantonensis** (Hance) Pax & K.Hoffm.
Mercurialis afra L. === **Centella villosa** L. (Apiaceae)
Mercurialis alpina Schur === **Mercurialis perennis** L.
Mercurialis alternifolia Lam. === **Micrococca mercurialis** (L.) Benth.
Mercurialis alternifolia Hochst. ex Baill. === **Acalypha hochstetteriana** Müll.Arg.
Mercurialis ambigua L.f. === **Mercurialis annua** L.
Mercurialis androgyna Steud. === **Leidesia procumbens** (L.) Prain
Mercurialis angustifolia (Müll.Arg.) Baill. === **Claoxylon angustifolium** Müll.Arg.
Mercurialis annua lusus *ambigua* (L.f.) Müll.Arg. === **Mercurialis annua** L.
Mercurialis annua var. *ambigua* Duby === **Mercurialis annua** L.
Mercurialis annua subsp. *ambigua* (L.f.) Arcang. === **Mercurialis annua** L.
Mercurialis annua subsp. *ambigua* (L.f.) Maire === **Mercurialis annua** L.
Mercurialis annua var. *angustifolia* Gaudin === **Mercurialis annua** L.
Mercurialis annua var. *camberiensis* Chabert === **Mercurialis annua** L.
Mercurialis annua var. *capillacea* Guépin === **Mercurialis annua** L.
Mercurialis annua f. *ciliata* (J.Presl & C.Presl) Pax & K.Hoffm. === **Mercurialis annua** L.
Mercurialis annua f. *cordata* Wirtg. === **Mercurialis annua** L.
Mercurialis annua f. *cuneatolanceolata* Wirtg. === **Mercurialis annua** L.
Mercurialis annua var. *dioica* Moris === **Mercurialis annua** L.
Mercurialis annua subvar. *euannua* Litard. === **Mercurialis annua** L.
Mercurialis annua var. *genuina* Müll.Arg. === **Mercurialis annua** L.
Mercurialis annua subsp. *huetii* (Hanry) Lange === **Mercurialis annua** L.
Mercurialis annua var. *huetii* (Hanry) Müll.Arg. === **Mercurialis annua** L.
Mercurialis annua f. *huetii* (Hanry) Pax & K.Hoffm. === **Mercurialis annua** L.
Mercurialis annua var. *laciniata* Müll.Arg. === **Mercurialis annua** L.
Mercurialis annua f. *lanceolata* Wirtg. === **Mercurialis annua** L.
Mercurialis annua var. *monoica* Moris === **Mercurialis annua** L.
Mercurialis annua f. *ovata* Wirtg. === **Mercurialis annua** L.
Mercurialis annua subvar. *serrata* Litard. === **Mercurialis annua** L.
Mercurialis annua var. *serratifolia* Ball === **Mercurialis annua** L.
Mercurialis annua var. *transsylvanica* Schur === **Mercurialis annua** L.
Mercurialis annua var. *typica* Fiori === **Mercurialis annua** L.
Mercurialis annua var. *variegata* Löhr === **Mercurialis annua** L.
Mercurialis australis Baill. === **Claoxylon australe** Baill. ex Müll.Arg.
Mercurialis × *bichei* Magnier === **Mercurialis** × **longifolia** Lam.
Mercurialis × *bichei* nothovar. *malinvaldi* (Sennen) Rouy === **Mercurialis** × **longifolia** Lam.
Mercurialis bupleuroides Meisn. === **Adenocline pauciflora** var. **bupleuroides** (Meisn.) Müll.Arg.

Mercurialis caffra Meisn. === **Adenocline acuta** (Thunb.) Baill.
Mercurialis capensis (L.f.) Spreng. ex Sond. === **Leidesia procumbens** (L.) Prain
Mercurialis capensis Sond. === **Leidesia procumbens** (L.) Prain
Mercurialis ciliata J.Presl & C.Presl === **Mercurialis annua** L.
Mercurialis cucullata Dinter ex Pax === **Acalypha segetalis** Müll.Arg.
Mercurialis cynocrambe Scop. === **Mercurialis perennis** L.
Mercurialis dregeana Buchinger ex Krauss === **Adenocline acuta** (Thunb.) Baill.
Mercurialis elliptica Duby === **Mercurialis corsica** Coss. & Kralik
Mercurialis glabra M.E.Jones === ?
Mercurialis huetii Hanry === **Mercurialis annua** L.
Mercurialis indica Lour. === **Claoxylon hainanense** Pax & K.Hoffm.
Mercurialis ladanum Hartm. === **Mercurialis annua** L.
Mercurialis leiocarpa var. *trichocarpa* W.T.Wang === **Mercurialis leiocarpa** Siebold & Zucc.
Mercurialis livida Port. ex Baumg. === **Mercurialis ovata** Sternb. & Hoppe
Mercurialis longifolia Host === **Mercurialis perennis** L.
Mercurialis × *longifolia* nothosubsp. *malinvaldi* (Sennen) Jauzein === **Mercurialis** × **longifolia** Lam.
Mercurialis longistipes (Borbás) Baksay === **Mercurialis perennis** L.
Mercurialis × *malinvaldi* Sennen === **Mercurialis** × **longifolia** Lam.
Mercurialis monoica (Moris) B.M.Durand === **Mercurialis annua** L.
Mercurialis nemoralis Salisb. === **Mercurialis perennis** L.
Mercurialis ovata f. *croatica* Degen === **Mercurialis ovata** Sternb. & Hoppe
Mercurialis ovata f. *genuina* Pax & K.Hoffm. === **Mercurialis ovata** Sternb. & Hoppe
Mercurialis pauciflora Baill. === **Adenocline pauciflora** Turcz. var. **pauciflora**
Mercurialis perennis var. *brachyphylla* Willk. === **Mercurialis perennis** L.
Mercurialis perennis f. *genuina* Müll.Arg. === **Mercurialis perennis** L.
Mercurialis perennis f. *glabra* Beck === **Mercurialis perennis** L.
Mercurialis perennis var. *longistipes* Borbás === **Mercurialis perennis** L.
Mercurialis perennis var. *ovata* (Sternb. & Hoppe) Müll.Arg. === **Mercurialis ovata** Sternb. & Hoppe
Mercurialis perennis subsp. *ovata* (Sternb. & Hoppe) Celak. === **Mercurialis ovata** Sternb. & Hoppe
Mercurialis perennis f. *robusta* Gross === **Mercurialis perennis** L.
Mercurialis perennis f. *saxicola* Beck === **Mercurialis perennis** L.
Mercurialis perennis var. *subalpina* Schur === **Mercurialis perennis** L.
Mercurialis pinnatifida Sennen === **Mercurialis annua** L.
Mercurialis procumbens L. === **Leidesia procumbens** (L.) Prain
Mercurialis × *pubescens* Samp. === **M. elliptica** × **M. tomentosa**
Mercurialis pumila Sond. === **Seidelia triandra** (E.Mey.) Pax
Mercurialis sericea Salisb. === **Mercurialis tomentosa** L.
Mercurialis serrata Meisn. === **Adenocline pauciflora** var. **ovalifolia** (Turcz.) Müll.Arg.
Mercurialis serratifolia (Ball) Pau === **Mercurialis annua** L.
Mercurialis subcordata Buchinger ex Krauss === **Adenocline acuta** (Thunb.) Baill.
Mercurialis sylvatica Hoppe === **Mercurialis perennis** L.
Mercurialis sylvestris Bubani === **Mercurialis perennis** L.
Mercurialis tarraconensis Sennen === ?
Mercurialis × *taurica* Juz. === ?
Mercurialis tenella Meisn. === **Adenocline pauciflora** var. **tenella** (Meisn.) Müll.Arg.
Mercurialis tenerifolia Baill. === **Claoxylon tenerifolium** (Baill.) F.Muell.
Mercurialis tomentosa var. *pubescens* Loscos & Pardo === **Mercurialis** × **longifolia** Lam.
Mercurialis transmorrisonensis Hayata === **Mercurialis leiocarpa** Siebold & Zucc.
Mercurialis triandra E.Mey. === **Seidelia triandra** (E.Mey.) Pax
Mercurialis tricocca Eckl. & Zeyh. ex Krauss === **Leidesia procumbens** (L.) Prain
Mercurialis tricocca E.Mey. ex Sond. === **Adenocline violifolia** (Kunze) Prain
Mercurialis violifolia Kunze === **Adenocline violifolia** (Kunze) Prain
Mercurialis zeyheri Kunze === **Adenocline pauciflora** var. **transiens** Müll.Arg.

Mercuriastrum

Synonyms:
Mercuriastrum Heist. ex Fabr. === **Acalypha** L.

Merleta

Synonyms:
Merleta Raf. === **Croton** L.

Mesandrinia

Synonyms:
Mesandrinia Raf. === **Jatropha** L.

Meterana

Synonyms:
Meterana Raf. === **Caperonia** A.St.-Hil.

Mettenia

Reduced to *Chaetocarpus* since 1992.

Synonyms:
Mettenia Griseb. === **Chaetocarpus** Thwaites
Mettenia acutifolia Britton & P.Wilson === **Chaetocarpus acutifolius** (Britton & P.Wilson) Borhidi
Mettenia cordifolia Urb. === **Chaetocarpus cordifolius** (Urb.) Borhidi
Mettenia humilis Ekman ex Urb. === **Chaetocarpus cubensis** Fawc. & Rendle
Mettenia lepidota Urb. === **Chaetocarpus globosus** (Sw.) Fawc. & Rendle subsp. **globosus**
Mettenia oblongata Alain === **Chaetocarpus globosus** subsp. **oblongatus** (Alain) Borhidi

Micrandra

10 species, S. America. Laticiferous small to medium forest trees closely related to *Hevea* and in the past sometimes united with it but in the first instance differing by virtue of having simple leaves. The shoot flushes are, however, very similar in general appearance and, as in *Hevea*, the inflorescences are terminal, developing above the leaves. Most species are in the north of the continent but *M. elata* extends to Peru and southern Brazil. *Cunuria* is here included (following Schultes, 1952); this position has been supported by Jablonski (1967) and in Brummitt (1992) but not by Webster who maintains that genus as distinct. A modern summary treatment taking into account the many observations of Schultes (and other ethnobiologists) would be desirable. The name is conserved against *Micrandra* R.Br. (Crotonoideae)

Pax, F. (1910). *Cunuria*. In A. Engler (ed.), Das Pflanzenreich, IV 147 [I] (Euphorbiaceae-Jatropheae): 16-17. Berlin. (Heft 42.) La/Ge. — 2 species, S. Venezuela (T.F. Amazonas).

Pax, F. (1910). *Micrandra*. In A. Engler (ed.), Das Pflanzenreich, IV 147 [I] (Euphorbiaceae-Jatropheae): 18-21. Berlin. (Heft 42.) La/Ge. — 5 species, S. America (Brazil); *Cunuria* not included.

Baldwin, J. T. & R. E. Schultes (1947). A conspectus of the genus *Cunuria*. Bot. Mus. Leafl. 12: 325-351, 6 pls., map. En. — Treatment of 5 species and some infraspecific taxa with key, descriptions, references and citations, localities with exsiccatae, and extensive discussion. [Schultes later combined the genus with *Micrandra*, but this has not been accepted in all quarters.]

Schultes, R. E. (1952). Studies in the genus *Micrandra*, I. The relationship of the genus *Cunuria* to *Micrandra*. Bot. Mus. Leafl. 15: 201-221, 10 pls. (incl. map). En. —*Cunuria* here reduced (arguments, p. 202); individual species discussed and two novelties described.

Jablonski, E. (1967). *Micrandra*. Euphorbiaceae, Guayana Highland (Mem. New York Bot. Gard. 17(1)): 157-161. New York. En. — 7 species, with inclusion of those in *Cunuria*; none new.

Schultes, R. E. (1979(1980)). Studies in the genus *Micrandra*, II. Miscellaneous taxonomic and economic notes. Bot. Mus. Leafl. 27: 93-111, 3 pls. En. — On utilisation of *Micrandra* for rubber, with investigation of past reports, the taxonomy of various species with notes and new records, and a list of vernacular names.

Micrandra Benth., Hooker's J. Bot. Kew Gard. Misc. 6: 371 (1854), nom. & typ. cons.
S. Trop. America. 82 83 84.
Pogonophyllum Didr., Vidensk. Meddel. Dansk Naturhist. Foren. Kjøbenhavn 1857: 144 (1857).
Clusiophyllum Müll.Arg., Flora 47: 518 (1864).
Cunuria Baill., Adansonia 4: 287 (1864).

Micrandra australis (R.E.Schult.) R.E.Schult., Bot. Mus. Leafl. 15: 202 (1952).
Brazil (Amazonas). 84 BZN. Phan.
* *Cunuria australis* R.E.Schult., Bot. Mus. Leafl. 12: 333 (1947).

Micrandra elata (Didr.) Müll.Arg., Linnaea 34: 142 (1865).
Brazil (Bahia, Minas Gerais, São Paulo), Colombia (Santander), Guiana, Surinam, Peru (Loreto). 82 FRG SUR 83 CLM PER 84 BZE BZL. Phan.
* *Pogonophyllum elatum* Didr., Vidensk. Meddel. Dansk Naturhist. Foren. Kjøbenhavn 1857: 145 (1857).
Micrandra bracteosa Müll.Arg. in C.F.P.von Martius, Fl. Bras. 11(2): 290 (1873). *Cunuria bracteosa* (Müll.Arg.) Ducke, Notizbl. Bot. Gart. Berlin-Dahlem 11: 586 (1932). *Cunuria spruceana* var. *bracteosa* (Müll.Arg.) Baldwin & R.E.Schult., Bot. Mus. Leafl. 12: 345 (1947).
Micrandra glaziovii Pax in H.G.A.Engler, Pflanzenr., IV, 147, I: 20 (1910).
Micrandra brownsbergensis Lanj., Euphorb. Surinam: 34 (1931).
Micrandra santanderensis Croizat, J. Arnold Arbor. 24: 169 (1943).

Micrandra glabra (R.E.Schult.) R.E.Schult., Bot. Mus. Leafl. 15: 203 (1952).
Guyana, Surinam, S. Venezuela. 82 GUY SUR VEN. Phan.
* *Cunuria glabra* R.E.Schult., Bot. Mus. Leafl. 12: 339 (1947).

Micrandra gleasoniana (Croizat) R.E.Schult., Bot. Mus. Leafl. 15: 203 (1952).
Guyana. 82 GUY. Phan.
* *Cunuria gleasoniana* Croizat, Bull. Torrey Bot. Club 67: 289 (1940).

Micrandra heterophylla Poiss., Bull. Mus. Hist. Nat. (Paris) 8: 561 (1902).
Venezuela. 82 VEN. Phan. – Provisionally accepted.

Micrandra lopezii R.E.Schult., Bot. Mus. Leafl. 15: 204 (1952).
Brazil (Amazonas). 84 BZN. Phan.
Micrandra lopezii f. *anteridifera* R.E.Schult., Bot. Mus. Leafl. 15: 210 (1952).
Micrandra lopezii var. *microcarpa* R.E.Schult., Bot. Mus. Leafl. 27: 104 (1979).

Micrandra rossiana R.E.Schult., Bot. Mus. Leafl. 15: 211 (1952).
Colombia, S. Venezuela, Guyana, Brazil (Amazonas). 82 GUY VEN 83 CLM 84 BZN. Phan.

Micrandra siphonioides Benth., Hooker's J. Bot. Kew Gard. Misc. 6: 371 (1854). *Micrandra siphonioides* var. *genuina* Müll.Arg. in A.P.de Candolle, Prodr. 15(2): 710 (1866), nom. inval.

Brazil (Amazonas), Colombia, S. Venezuela, Guiana. 82 FRG VEN 83 CLM 84 BZN. Phan. – A minor source of rubber.

Micrandra minor Benth., Hooker's J. Bot. Kew Gard. Misc. 6: 372 (1854). *Micrandra siphonioides* var. *minor* (Benth.) Müll.Arg. in A.P.de Candolle, Prodr. 15(2): 710 (1866).

Micrandra major Baill., Adansonia 4: 286 (1864). *Micrandra siphonioides* var. *major* (Baill.) Müll.Arg. in A.P.de Candolle, Prodr. 15(2): 709 (1866).

Micrandra spruceana (Baill.) R.E.Schult., Bot. Mus. Leafl. 15: 217 (1952).

Brazil (Amazonas), Venezuela (Amazonas), Colombia, Peru. 82 VEN 83 CLM PER 84 BZN. Phan.

* *Cunuria spruceana* Baill., Adansonia 4: 288 (1864).

Pogonophora cunurii Baill., Adansonia 4: 288 (1864).

Micrandra cunurii Baill. ex Müll.Arg. in A.P.de Candolle, Prodr. 15(2): 1123 (1866).

Micrandra sprucei (Müll.Arg.) R.E.Schult., Bot. Mus. Leafl. 15: 218 (1952).

Brazil (Amazonas), Venezuela (Amazonas), Colombia. 82 VEN 83 CLM 84 BZN. Phan.

* *Clusiophyllum sprucei* Müll.Arg., Flora 47: 519 (1864).

Cunuria crassipes Müll.Arg. in C.F.P.von Martius, Fl. Bras. 11(2): 510 (1874).

Synonyms:

Micrandra bracteosa Müll.Arg. === **Micrandra elata** (Didr.) Müll.Arg.
Micrandra brownsbergensis Lanj. === **Micrandra elata** (Didr.) Müll.Arg.
Micrandra cunurii Baill. ex Müll.Arg. === **Micrandra spruceana** (Baill.) R.E.Schult.
Micrandra glaziovii Pax === **Micrandra elata** (Didr.) Müll.Arg.
Micrandra lopezii f. *anteridifera* R.E.Schult. === **Micrandra lopezii** R.E.Schult.
Micrandra lopezii var. *microcarpa* R.E.Schult. === **Micrandra lopezii** R.E.Schult.
Micrandra major Baill. === **Micrandra siphonioides** Benth.
Micrandra minor Benth. === **Micrandra siphonioides** Benth.
Micrandra santanderensis Croizat === **Micrandra elata** (Didr.) Müll.Arg.
Micrandra scleroxylon W.A.Rodrigues === **Micrandropsis scleroxylon** (W.A.Rodrigues) W.A.Rodrigues
Micrandra siphonioides var. *genuina* Müll.Arg. === **Micrandra siphonioides** Benth.
Micrandra siphonioides var. *major* (Baill.) Müll.Arg. === **Micrandra siphonioides** Benth.
Micrandra siphonioides var. *minor* (Benth.) Müll.Arg. === **Micrandra siphonioides** Benth.

Micrandra

of R. Brown; rejected against *Micrandra* Benth.

Synonyms:

Micrandra R.Br. === **Hevea** Aubl.
Micrandra ternata R.Br. === **Hevea spruceana** (Benth.) Müll.Arg.

Micrandropsis

1 species, Northern S. America (Brazil); a segregate of *Micrandra* (including *Cunuria*) but differing in the presence of stellate instead of simple hairs and, in addition, linear anthers. The general appearance of shoot flushes is similar to that of *Micrandra* and *Vaupesia*. (Crotonoideae)

Rodrigues, W. (1973). *Micrandropsis*, novo gênero de Euphorbiaceae da Amazônia. Acta Amaz. 3(2): 5-6. Pt. — Protologue of genus; 1 species, *M. scleroxylon*, transferred from *Micrandra scleroxylon* (see ibid., 1(3): 3-8 (1971)).

Micrandropsis W.A.Rodrigues, Acta Amazon. 3: 5 (1973).
N. Brazil. 84.

Micrandropsis scleroxylon (W.A.Rodrigues) W.A.Rodrigues, Acta Amazon. 3: 5 (1973).
Brazil (Amazonas). 84 BZN.
* *Micrandra scleroxylon* W.A.Rodrigues, Acta Amazon. 1: 3 (1971).

Micranthea

Synonyms:
Micranthea A.Juss. === **Micrantheum** Desf.
Micranthea hexandra (Hook.f.) Walp. === **Micrantheum hexandrum** Hook.f.

Micrantheum

4 species, Australia (including Tasmania). The presence of narrow cotyledons in these microphyllous herbs or undershrubs resulted it being placed in a subdivision known as 'Stenolobae' by Mueller, in this respect followed by Pax; however, like other Australian genera of that group it appears instead to be a specialised derivative, in this case (like the related *Pseudanthus*) of the Oldfieldioideae (following work by Webster, Hayden and Levin & Simpson; see **General** and **Special**). The seeds appear to be at least sometimes dispersed by ants. A key to all species appears in Orchard (1991). (Oldfieldioideae)

Gruening, G. (1913). *Micrantheum*. In A. Engler (ed.), Das Pflanzenreich, IV 147 [Stenolobieae] (Euphorbiaceae-Porantheroideae et Ricinocarpoideae): 21-25. Berlin. (Heft 58.) La/Ge. — 3 species, Australia (1 also in Tasmania).
Anonymous (1971). The west coast *Micrantheum*. Launceston Nat. 5(2): 3. En. — Naturalist's report of plants distinct from *M. hexandrum*.
Orchard, A. E. (1991). A new species of *Micrantheum* (Euphorbiaceae) from Tasmania. In M. R. Banks et al., *Aspects of Tasmanian botany – a tribute to Winifred Curtis*: 59-64. Hobart: Royal Society of Tasmania. En. — Description of a novelty, *M. serpentinum*; a key to all species is also included along with an illustration and notes on *M. hexandrum*.

Micrantheum Desf., Mém. Mus. Hist. Nat. 4: 253 (1818).
Australia. 50. Nanophan.
Micranthea A.Juss., Euphorb. Gen.: 24 (1824).
Caletia Baill., Étude Euphorb.: 553 (1858).
Allenia Ewart, Proc. Roy. Soc. Victoria 22: 7 (1909).

Micrantheum demissum F.Muell., Victorian Naturalist 7: 67 (1890).
South Australia. 50 SOA. Nanophan.
Allenia blackiana Ewart, Jean White & Rees, Proc. Roy. Soc. Victoria 22: 8 (1909).
Allenia blackiana var. *microphylla* Ewart, Jean White & Rees, Proc. Roy. Soc. Victoria 22: 8 (1909). *Micrantheum demissum* var. *microphyllum* (Ewart, Jean White & Rees) Grüning in H.G.A.Engler, Pflanzenr., IV, 147: 25 (1913).
Micrantheum demissum var. *typicum* Grüning in H.G.A.Engler, Pflanzenr., IV, 147: 25 (1913), nom. inval.

Micrantheum ericoides Desf., Mém. Mus. Hist. Nat. 4: 53 (1818). *Caletia ericodes* (Desf.) Kuntze, Revis. Gen. Pl. 2: 595 (1891). *Micrantheum ericoides* var. *genuinum* Grüning in H.G.A.Engler, Pflanzenr., IV, 147: 24 (1913), nom. inval.
New South Wales, Queensland. 50 NSW QLD. Nanophan.
Phyllanthus lhotzkyanus Hochst. ex Steud., Nomencl. Bot., ed. 2, 2: 327 (1841).
Micrantheum boroniaceum F.Muell., Fragm. 1: 32 (1858).

Micrantheum ericoides var. *intermedium* Grüning in H.G.A.Engler, Pflanzenr., IV, 147: 25 (1913).
Micrantheum ericoides var. *juniperinum* Grüning in H.G.A.Engler, Pflanzenr., IV, 147: 25 (1913).

Micrantheum hexandrum Hook.f., London J. Bot. 6: 283 (1847). *Micranthea hexandra* (Hook.f.) Walp., Ann. Bot. Syst. 1: 630 (1849). *Caletia hexandra* (Hook.f.) Müll.Arg. in A.P.de Candolle, Prodr. 15(2): 194 (1866).
Tasmania, E. Victoria, E. New South Wales, SE. Queensland. 50 NSW QLD TAS. Nanophan.
Caletia micrantheoides Baill., Étude Euphorb.: 554 (1858), nom. illeg.
Phyllanthus boroniacus F.Muell. ex Grüning in H.G.A.Engler, Pflanzenr., IV, 147: 23 (1913), pro syn.

Micrantheum serpentinum Orchard in M.R.Banks & al., Aspects of Tasmanian Botany: 60 (1991).
NW. Tasmania. 50 TAS. Cham. or nanophan.

Synonyms:
Micrantheum boroniaceum F.Muell. === **Micrantheum ericoides** Desf.
Micrantheum demissum var. *microphyllum* (Ewart, Jean White & Rees) Grüning === **Micrantheum demissum** F.Muell.
Micrantheum demissum var. *typicum* Grüning === **Micrantheum demissum** F.Muell.
Micrantheum ericoides var. *genuinum* Grüning === **Micrantheum ericoides** Desf.
Micrantheum ericoides var. *intermedium* Grüning === **Micrantheum ericoides** Desf.
Micrantheum ericoides var. *juniperinum* Grüning === **Micrantheum ericoides** Desf.
Micrantheum inversum Pancher ex Baill. === **Phyllanthus faguetii** Baill. var. **faguetii**
Micrantheum nervosum Pancher ex Guillaumin === **Phyllanthus mitchellii** Benth.
Micrantheum tatei J.M.Black === **Phyllanthus tatei** F.Muell.
Micrantheum triandrum Hook. === **Phyllanthus mitchellii** Benth.

Micrococca

12 species, tropics of Old World (tropical and southern Africa, Madagascar, Arabian Peninsula, India and Malesia); allied to *Claoxylon*. Annual or perennial herbs or shrubs, less often small trees. *M. mercurialis* is very widely distributed, covering most if not all of the range, and has become a garden weed beyond. No full revision has been published since 1914. It is an open question whether or not *M. mercurialis* is properly included with the rest of the species; moreover, *M. johorica* is also rather distinctive. (Acalyphoideae)

Prain, D. (1911). The genera *Erythrococca* and *Micrococca*. Ann. Bot. 25: 575-638. En. — *Micrococca*, pp. 628-632; synopsis of 8 species with key, brief descriptions, synonymy, references, and indication of distribution; biogeographic summary and notes on uses at end. The general part includes an extensive historical review.
Pax, F. (with K. Hoffmann) (1914). *Micrococca*. In A. Engler (ed.), Das Pflanzenreich, IV 147 VII (Euphorbiaceae-Acalypheae-Mercurialinae): 131-137. Berlin. (Heft 63.) La/Ge. — 10 species, Africa and tropical Asia.
Airy-Shaw, H. K. (1971). Notes on Malesian and other Asiatic Euphorbiaceae, CXXXVII. Two new species of *Micrococca* Benth. from Malaya. Kew Bull. 25: 524-526. En. — Descriptions of *MM. malaccensis* and *johorica*; both from Peninsular Malaysia.

Micrococca Benth. in W.J.Hooker, Niger Fl.: 503 (1849).
Trop. & S. Africa, Madagascar, Arabian Pen., Trop. Asia. 22 23 24 25 26 27 29 35 40 41 42.

Micrococca beddomei (Hook.f.) Prain, Ann. Bot. (Usteri) 25: 630 (1911).
S. India (Kerala, Tamil Nadu). 40 IND. Nanophan.
* *Claoxylon beddomei* Hook.f., Fl. Brit. India 5: 413 (1887).

Micrococca capensis (Baill.) Prain, Ann. Bot. (Usteri) 25: 630 (1911).
Mozambique, KwaZulu-Natal. 26 MOZ 27 NAT. Nanophan. or phan.
* *Claoxylon capense* Baill., Étude Euphorb.: 493 (1858).

Micrococca holstii (Pax) Prain, Ann. Bot. (Usteri) 25: 630 (1911).
C. Kenya, E. Tanzania. 25 KEN TAN. Nanophan.
* *Claoxylon holstii* Pax, Bot. Jahrb. Syst. 34: 372 (1904).

Micrococca humblotiana (Baill.) Prain, Ann. Bot. (Usteri) 25: 630 (1911).
Comoros. 29 COM. Nanophan.
* *Claoxylon humblotianum* Baill., Bull. Mens. Soc. Linn. Paris 2: 996 (1892).

Micrococca johorica Airy Shaw, Kew Bull. 25: 525 (1971).
Pen. Malaysia. 42 MLY.

Micrococca lancifolia Prain, Bull. Misc. Inform. Kew 1912: 282 (1912). *Claoxylon lancifolium* (Prain) Leandri, Notul. Syst. (Paris) 9: 174 (1941).
N. Madagascar. 29 MDG.

Micrococca malaccensis Airy Shaw, Kew Bull. 25: 524 (1971).
Pen. Malaysia. 42 MLY.

Micrococca mercurialis (L.) Benth. in W.J.Hooker, Niger Fl.: 503 (1849).
Trop. & S. Africa, Madagascar, Arabian Pen., India to Pen. Malaysia. 22 GHA GUI IVO LBR NGA SEN SIE TOG 23 CMN CON EQG ZAI 24 ETH SUD 25 KEN TAN UGA 26 ANG MLW MOZ ZAM ZIM 27 BOT 29 MDG 35 YEM 40 IND KUM LDV SRL 41 BMA THA 42 MLY. Ther. or cham.
* *Tragia mercurialis* L., Sp. Pl.: 980 (1753). *Microstachys mercurialis* (L.) Dalzell & Gibson, Bombay Fl.: 227 (1861). *Claoxylon mercurialis* (L.) Thwaites, Enum. Pl. Zeyl.: 271 (1861).
Mercurialis alternifolia Lam., Encycl. 4: 120 (1797).
Mercurialis abyssinica Hochst. ex Pax & K.Hoffn. in H.G.A.Engler, Pflanzenr., IV, 147, VII: 133 (1914).

Micrococca oligandra (Müll.Arg.) Prain, Ann. Bot. (Usteri) 25: 629 (1911).
Sri Lanka, S. India. 40 IND SRL. Nanophan.
Claoxylon longifolium Baill., Étude Euphorb.: 493 (1858), nom. illeg.
* *Claoxylon oligandrum* Müll.Arg., Linnaea 34: 164 (1865). *Micrococca oligandra* var. *genuina* Pax & K.Hoffm. in H.G.A.Engler, Pflanzenr., IV, 147, VII: 133 (1914), nom. inval.
Micrococca oligandra var. *pubescens* Pax & K.Hoffm. in H.G.A.Engler, Pflanzenr., IV, 147, VII: 133 (1914).

Micrococca scariosa Prain, Bull. Misc. Inform. Kew 1912: 192 (1912).
SE. Kenya, E. Tanzania (incl. Zanzibar), Mozambique. 25 KEN TAN 26 MOZ. Nanophan. or phan.

Micrococca volkensii (Pax) Prain, Ann. Bot. (Usteri) 25: 631 (1911).
N. & E. Tanzania. 25 TAN. Cham. or nanophan.
* *Claoxylon volkensii* Pax in H.G.A.Engler, Pflanzenw. Ost-Afrikas C: 238 (1895).

Micrococca wightii (Hook.f.) Prain, Ann. Bot. (Usteri) 25: 630 (1911).
S. India (Tamil Nadu). 40 IND. Nanophan.
* *Claoxylon wightii* Hook.f., Fl. Brit. India 5: 413 (1887). *Micrococca wightii* var. *genuina* Pax & K.Hoffm. in H.G.A.Engler, Pflanzenr., IV, 147, VII: 133 (1914), nom. inval.

var. **angustata** (S.R.M.Susila Rani & N.P.Balakr.) Radcl.-Sm. & Govaerts, Kew Bull. 52: 480 (1997).

India (Tamil Nadu). 40 IND. Nanophan.
* *Claoxylon wightii* var. *angustatum* S.R.M.Susila Rani & N.P.Balakr., J. Econ. Taxon. Bot. 16: 736 (1992).

var. **glabrata** (S.R.M.Susila Rani & N.P.Balakr.) Radcl.-Sm. & Govaerts, Kew Bull. 52: 480 (1997).
India (Tamil Nadu). 40 IND. Nanophan.
Claoxylon wightii var. *glabratum* S.R.M.Susila Rani & N.P.Balakr., J. Econ. Taxon. Bot. 16: 735 (1992).

var. **hirsuta** (Hook.f.) Prain, Ann. Bot. (Usteri) 25: 630 (1911).
S. India. 40 IND. Nanophan.
* *Claoxylon hirsutum* Hook.f., Fl. Brit. India 5: 413 (1887). *Claoxylon wightii* var. *hirsutum* (Hook.f.) S.R.M.Susila Rani & Balakr., J. Econ. Taxon. Bot. 16: 735 (1992).

var. **wightii**
India (Tamil Nadu). 40 IND. Nanophan.

Synonyms:
Micrococca berberidea (Prain) E.Phillips === **Erythrococca berberidea** Prain
Micrococca natalensis (Prain) E.Phillips === **Erythrococca natalensis** Prain
Micrococca oligandra var. *genuina* Pax & K.Hoffm. === **Micrococca oligandra** (Müll.Arg.) Prain
Micrococca oligandra var. *pubescens* Pax & K.Hoffm. === **Micrococca oligandra** (Müll.Arg.) Prain
Micrococca wightii var. *genuina* Pax & K.Hoffm. === **Micrococca wightii** (Hook.f.) Prain

Microelus

Synonyms:
Microelus Wight & Arn. === **Bischofia** Blume
Microelus roeperianus (Decne.) Wight & Arn. === **Bischofia javanica** Blume

Micropetalum

An ms. name of Poiteau proposed by Baillon; now part of *Amanoa*. [Not in *Index Kewensis*.]

Microsepala

Synonyms:
Microsepala Miq. === **Baccaurea** Lour.
Microsepala acuminata Miq. === **Baccaurea javanica** (Blume) Müll.Arg.

Microstachys

15 species, tropics and subtropics (mainly in S. America); small shrubs or, in the case of *MM. chamaelea* and *corniculata*, widely distributed annual or perennial herbs respectively in the Eastern and Western Hemisphere. Some, at least in Brazil, have lignotuberous rootstocks and are characteristic of the cerrados and other open formations of the Planalto. A similar feature has been recorded for what have respectively been known as *Sapium acetosella* from southern Zaïre, Zambia and Angola and *S. faradianense* from the Sahel, both also characteristic of open land. Many if not all have relatively small, narrow, distichously arranged leaves. *M. chamaelea* is found in grasslands and enters cultivated land. In the nineteenth century and in the more recent past the genus has been included within *Sebastiania*. The numerous formal varieties erected by Mueller within most of the species and largely maintained by Pax are here reduced without transfer. In reinstating the genus, Esser (1994; see **Euphorbioideae (except Euphorbieae)**) suggests it to be inclusive of *Sebastiania* sects. *Microstachys*, *Elachocroton* and *Microstachyopsis* (nos. 1, 2 and 5 in Pax &

Hoffmann, 1931) and to have as many as 17 species (at present somewhat fewer, all in the Americas save for, as already noted, 2 from Africa and the eastern Old World *M. chamaelea*). The only recent revision of any part of the genus as thus conceived covers the Brazilian species of sect. *Elachocroton* (Oliveira, 1988, in *Sebastiania*). Since then, the two above mentioned African species have been transferred along with those studied by Oliviera (*SS. ditassoides* and *revoluta*) and five others, preparatory to a full revision (see Esser, 1998). (Euphorbiaceae (except Euphorbieae))

Pax, F. (with K. Hoffmann) (1912). *Sapium*. In A. Engler (ed.), Das Pflanzenreich, IV 147 V (Euphorbiaceae-Hippomaneae): 199-258. Berlin. (Heft 52.) La/Ge. — No. 84 (pp. 247), *S. faradianense*, a shrub from the Sahel in Africa, is referable to *Microstachys*. [*S. acetosella* Milne-Redh. is closely related.]

Pax, F. (with K. Hoffmann) (1912). *Sebastiania*. In A. Engler (ed.), Das Pflanzenreich, IV 147 V (Euphorbiaceae-Hippomaneae): 88-153. Berlin. (Heft 52.) La/Ge. — Sect. *Microstachys* (pp. 91-114), part of sect. *Elachocroton* (pp. 114-117, without *S. stipulacea*), and sect. *Microstachyopsis* (p. 118) are now part of *Microstachys*.

Milne-Redhead, E. (1933). *Sapium acetosella*. Ic. Pl. 32: pl. 3199. En. — Plant portrait with description (here as a novelty) and commentary. [Related to *S. faradianense* and, like it, with a lignotuberous rootstock. Now transferred to *Microstachys*.]

Steenis, C. G. G. J. van (1948). Provisional note on the genus *Sebastiania* in Malaysia. Bull. Bot. Gard. Buitenzorg, III, 17: 409-410. En. — Includes coverage of the coastal *S. chamaelea* (now referable to *Microstachys*).

Léonard, J. (1961). Notulae systematicae XXXII. Observations sur des espèces africaines de *Clutia*, *Ricinodendron* et *Sapium* (Euphorbiacées). Bull. Jard. Bot. État 31: 391-406. Fr. —*Sapium* sens. lat., pp. 401-406; treatment of *S. leonardii-crispi* (a novelty) and *S. acetosella*, without key. [The former is now *Duvigneaudia leonardii-crispi* and the latter is likely to be transferred to *Microstachys*.]

Jablonski, E. (1967). *Sebastiania*. Euphorbiaceae, Guayana Highland (Mem. New York Bot. Gard. 17(1)): 179. New York. En. — 1 species, *S. corniculata*. [This species is now in *Microstachys*.]

Oliveira, A. Souza de (1988). Taxinomia das espécies do gênero *Sebastiania* secção *Elachocroton* (Baill.) Pax (Imphorbiaceae)[sic!] ocorrentes no Brasil. Arq. Jard. Bot. Rio de Janeiro 27: 3-65, illus., map. Pt. — Detailed, well illustrated revision of two species (one with two additional varieties) with key, brief descriptions, synonymy, localities with exsiccatae, and extensive commentary; general discussion covering the nature and value of characters including those indicative of xeromorphic trends; list of collections seen, literature, distribution map and graphs and illustrations at end. The revision is preceded by a character review and a history of studies in *Sebastiania*. [Both species have since been transferred to *Microstachys*.]

Esser, H.-J. (1998). New combinations in *Microstachys* (Euphorbiaceae). Kew Bull. 53: 955-960. En. — Short history of genus (united with *Sebastiania* by Müller in 1866, an opinion generally supported until 1994); scope of the redelimited genus and its differences from *Sebastiania* and related hippomanoid genera; 9 new combinations; exclusion of *Sebastiania panamensis* (not referable either to that genus or to *Microstachys*).

Microstachys A.Juss., Euphorb. Gen.: 48 (1824).
Trop. & Subtrop. 22 23 36 40 41 42 50 60 79 80 81 82 83 84 85.
Cnemidostachys Mart., Nov. Gen. Sp. Pl. 1: 66 (1824).
Elachocroton F.Muell., Hooker's J. Bot. Kew Gard. Misc. 9: 17 (1857).
Tragiopsis H.Karst., Wochenschr. Gärtnerei Pflanzenk. 2: 5 (1859).

Microstachys acetosella (Milne-Redh.) Esser, Kew Bull. 53: 958 (1998 publ. 1999).
Zaire, Angola, Zambia. 23 ZAI 26 ANG ZAM. – FIGURE, p. 1185.
* *Sapium acetosella* Milne-Redh., Hooker's Icon. Pl. 32: t. 3199 (1933). *Sebastiania acetosella* (Milne-Redh.) Kruijt, Biblioth. Bot. 146: 83 (1996).
Sapium acetosella var. *elatius* Radcl.-Sm., Kew Bull. 39: 794 (1984).

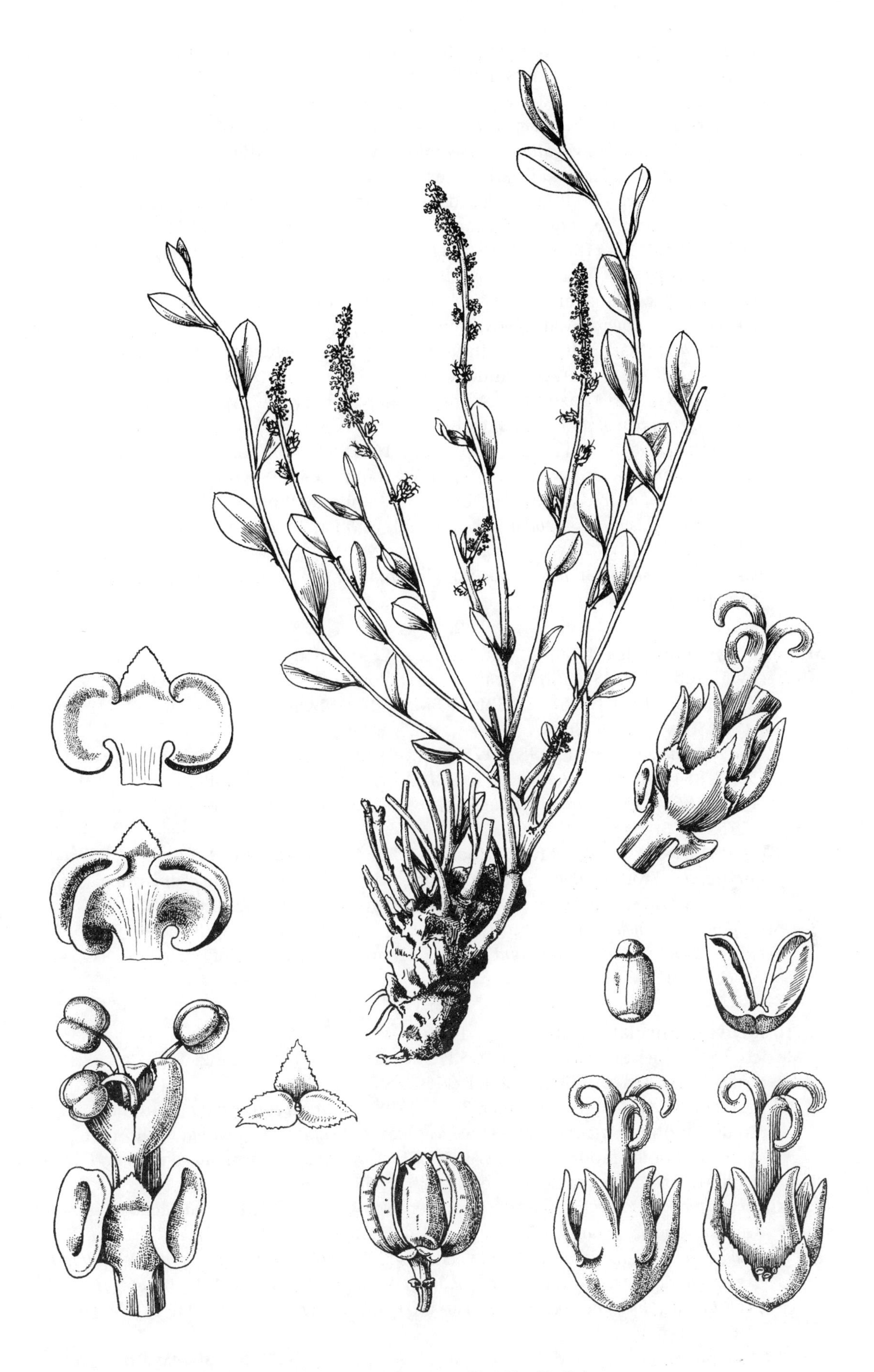

Microstachys acetosella (Milne-Redh.) Esser (as *Sapium acetosella* Milne-Redh.)
Artist: Stella Ross-Craig
Ic. Pl. 32: pl. 3199 (1933)
KEW ILLUSTRATIONS COLLECTION

Microstachys bidentata (Mart. & Zucc.) Esser, Kew Bull. 53: 958 (1998 publ. 1999).

S. Trop. America. 82 SUR 84 BZC BZE BZL BZN.

* *Cnemidostachys bidentata* Mart. & Zucc., Flora 7(1) Beibl. 4: 137 (1824); Nov. Gen. Sp. Pl. 1: 69, pl. 43 (1824). *Microstachys virgata* Müll.Arg., Linnaea 32: 94 (1863), nom. superfl. *Microstachys virgata* var. *bidentata* (Mart. & Zucc.) Müll.Arg., Linnaea 32: 94 (1863). *Stillingia bidentata* (Mart. & Zucc.) Baill., Adansonia 5: 324 (1865). *Sebastiania virgata* (Müll.Arg.) Müll.Arg. in A.P.de Candolle, Prodr. 15(2): 1173 (1866), nom. illeg. *Sebastiania virgata* var. *bidentata* (Mart. & Zucc.) Müll.Arg. in A.P.de Candolle, Prodr. 15(2): 1174 (1866). *Sebastiania bidentata* (Mart. & Zucc.) Pax in H.G.A.Engler, Pflanzenr., IV, 147, V: 113 (1912). *Sebastiania bidentata* var. *genuina* Pax in H.G.A.Engler, Pflanzenr., IV, 147, V: 114 (1912), nom. inval.

Cnemidostachys scoparia Mart., Nov. Gen. Sp. Pl. 1: 70, pl. 44 (1824). *Microstachys virgata* var. *scoparia* (Mart.) Müll.Arg., Linnaea 32: 94 (1863). *Sebastiania virgata* var. *scoparia* (Mart.) Müll.Arg. in A.P. de Candolle, Prodr. 15(2): 1174 (1866). *Sebastiania bidentata* var. *scoparia* (Mart.) Pax in H.G.A.Engler, Pflanzenr., IV, 147, V: 114 (1912).

Cnemidostachys bidentata Klotzsch ex Baill., Étude Euphorb.: 516 (1858), nom. nud. *Stillingia bidentata* Baill., Étude Euphorb.: 516 (1858), nom. nud.

Microstachys virgata var. *odontococca* Müll.Arg. Linnaea 32: 94 (1863). *Sebastiania virgata* var. *odontococca* Müll.Arg. in A.P. de Candolle, Prodr. 15(2): 1173 (1866). *Sebastiania bidentata* var. *odontococca* (Müll.Arg.) Pax in H.G.A.Engler, Pflanzenr., IV, 147, V: 13 (1912).

Sebastiania bidentata var. *pilgeri* Pax & K.Hoffm. in H.G.A.Engler, Pflanzenr., IV, 147, V: 113 (1912).

Microstachys chamaelea (L.) Müll.Arg., Linnaea 32: 95 (1863).

W. & WC. Trop. Africa, S. China, Trop. Asia to Solomon Is. 22 BEN GHA NGA TOG 23 CAF CMN 36 CHC CHS 40 IND SRL 41 BMA THA 42 BOR JAW MLY SUM 50 NTA QLD WAU 60 SOL. Ther. or cham.

* *Tragia chamaelea* L., Sp. Pl.: 981 (1753). *Excoecaria chamelaea* (L.) Baill., Adansonia 4: 323 (1864). *Stillingia chamaelea* (L.) Müll.Arg. in A.P.de Candolle, Prodr. 15(2): 1175 (1866). *Sebastiania chamaelea* (L.) Müll.Arg. in A.P.de Candolle, Prodr. 15(2): 1175 (1866).

Elachocroton asperococcum F.Muell., Hooker's J. Bot. Kew Gard. Misc. 9: 17 (1857). *Stillingia asperococca* (F.Muell.) Baill., Étude Euphorb.: 517 (1858). *Sebastiania chamaelea* var. *asperococca* (F.Muell.) Pax in H.G.A.Engler, Pflanzenr., IV, 147, V: 116 (1912).

Sebastiania chamaelea var. *chariensis* Beille, Bull. Soc. Bot. France 57(8): 128 (1910).

Sebastiania chamaelea var. *africana* Pax & K.Hoffm. in H.G.A.Engler, Pflanzenr., IV, 147, V: 117 (1912).

Microstachys corniculata (Vahl) Griseb., Fl. Brit. West Ind.: 49 (1864).

Mexico, Trop. America. 79 MXN 80 COS PAN 81 CUB DOM HAI LEE PUE TRT 82 FRG GUY SUR VEN 83 CLM 84 BZC BZE BZN BZS 85 PAR. Ther.

* *Tragia corniculata* Vahl, Eclog. Amer. 2: 55 (1798). *Stillingia corniculata* (Vahl) Baill., Étude Euphorb., Atlas: 17 (1858). *Sebastiania corniculata* var. *genuina* Müll.Arg. in A.P.de Candolle, Prodr. 15(2): 1173 (1866), nom. inval. *Sebastiania corniculata* (Vahl) Müll.Arg. in A.P.de Candolle, Prodr. 15(2): 1168 (1866).

Cnemidostachys acalyphoides Mart., Nov. Gen. Sp. Pl. 1: 71 (1824). *Sebastiania corniculata* var. *acalyphoides* (Mart.) Müll.Arg. in A.P.de Candolle, Prodr. 15(2): 1172 (1866).

Cnemidostachys glandulosa Mart., Nov. Gen. Sp. Pl. 1: 71 (1824). *Sebastiania corniculata* f. *glandulosa* (Mart.) Müll.Arg. in A.P.de Candolle, Prodr. 15(2): 1168 (1866). *Sebastiania glandulosa* f. *calvescens* Pax in H.G.A.Engler, Pflanzenr., IV, 147, V: 101 (1912), nom. inval.

Cnemidostachys longifolia Mart., Nov. Gen. Sp. Pl. 1: 71 (1824). *Sebastiania corniculata* f. *longifolia* (Mart.) Müll.Arg. in A.P.de Candolle, Prodr. 15(2): 1171 (1866). *Sebastiania corniculata* var. *longifolia* (Mart.) Müll.Arg. in C.F.P.von Martius, Fl. Bras. 11(2): 556 (1874). *Sebastiania salicifolia* var. *longifolia* (Mart.) Pax in H.G.A.Engler, Pflanzenr., IV, 147, V: 104 (1912).

Cnemidostachys prostrata Mart., Nov. Gen. Sp. Pl. 1: 70 (1824). *Stillingia prostrata* (Mart.) Baill., Adansonia 5: 324 (1865). *Sebastiania corniculata* var. *prostrata* (Mart.) Müll.Arg. in A.P.de Candolle, Prodr. 15(2): 1172 (1866).

Cnemidostachys salicifolia Mart., Nov. Gen. Sp. Pl. 1: 70 (1824). *Sebastiania corniculata* f. *salicifolia* (Mart.) Müll.Arg. in A.P.de Candolle, Prodr. 15(2): 1171 (1866). *Sebastiania corniculata* var. *salicifolia* (Mart.) Müll.Arg. in C.F.P.von Martius, Fl. Bras. 11(2): 556 (1874). *Sebastiania salicifolia* (Mart.) Pax in H.G.A.Engler, Pflanzenr., IV, 147, V: 103 (1912).

Cnemidostachys tragioides Mart., Nov. Gen. Sp. Pl. 1: 70 (1824). *Sebastiania corniculata* var. *tragioides* (Mart.) Pax in H.G.A.Engler, Pflanzenr., IV, 147, V: 98 (1912), nom. illeg.

Microstachys bicornis A.Juss., Euphorb. Gen.: 49 (1824).

Tragia bicornis Vahl ex A.Juss., Euphorb. Gen.: 49 (1824).

Tragia pilosa Vell., Fl. Flumin. 10: 9 (1831).

Cnemidostachys patula Mart., Flora 24(2): 8 (1841). *Sebastiania corniculata* var. *patula* (Mart.) Müll.Arg. in A.P.de Candolle, Prodr. 15(2): 1172 (1866). *Sebastiania hispida* var. *patula* (Mart.) Pax in H.G.A.Engler, Pflanzenr., IV, 147, V: 111 (1912).

Microstachys guianensis Klotzsch, Hooker's J. Bot. Kew Gard. Misc. 2: 45 (1843). *Sebastiania corniculata* var. *guianensis* (Klotzsch) Pax in H.G.A.Engler, Pflanzenr., IV, 147, V: 99 (1912).

Microstachys vahlii A.Rich. in R.de la Sagra, Hist. Fis. Cuba, Bot. 2: 202 (1850).

Microstachys crotonoides Klotzsch ex Benth., Hooker's J. Bot. Kew Gard. Misc. 6: 325 (1854).

Microstachys micrantha Benth., Hooker's J. Bot. Kew Gard. Misc. 6: 326 (1854). *Sebastiania corniculata* var. *micrantha* (Benth.) Müll.Arg. in A.P.de Candolle, Prodr. 15(2): 1173 (1866). *Sebastiania micrantha* (Benth.) Lanj., Recueil Trav. Bot. Néerl. 36: 704 (1939 publ.1940).

Stillingia campestris Baill., Étude Euphorb.: 516 (1858). *Sebastiania corniculata* var. *campestris* (Baill.) Müll.Arg. in A.P.de Candolle, Prodr. 15(2): 1169 (1866). *Sebastiania glandulosa* var. *campestris* (Baill.) Pax in H.G.A.Engler, Pflanzenr., IV, 147, V: 101 (1912).

Stillingia sellowiana Baill., Étude Euphorb.: 516 (1858), nom. inval.

Stillingia velutina Baill., Étude Euphorb.: 516 (1858).

Cnemidostachys dubia Wawra, Oesterr. Bot. Z. 12: 241 (1862). *Cnemidostachys patula* f. *dubia* (Wawra) Wawra, Bot. Ergebn.: 35 (1866).

Microstachys fruticulosa H.Karst., Fl. Columb. 2: 33 (1862).

Microstachys polymorpha Müll.Arg., Linnaea 32: 91 (1863).

Sebastiania corniculata f. *crotonoides* Müll.Arg. in A.P.de Candolle, Prodr. 15(2): 1171 (1866).

Sebastiania corniculata f. *discopoda* Müll.Arg. in A.P.de Candolle, Prodr. 15(2): 1170 (1866).

Sebastiania corniculata f. *fallax* Müll.Arg. in A.P.de Candolle, Prodr. 15(2): 1170 (1866). *Sebastiania corniculata* var. *fallax* (Müll.Arg.) Müll.Arg. in C.F.P.von Martius, Fl. Bras. 11(2): 553 (1874). *Sebastiania glandulosa* var. *fallax* (Müll.Arg.) Pax in H.G.A.Engler, Pflanzenr., IV, 147, V: 102 (1912).

Sebastiania corniculata f. *glabrata* Müll.Arg. in A.P.de Candolle, Prodr. 15(2): 1170 (1866).

Sebastiania corniculata f. *olfersiana* Müll.Arg. in A.P.de Candolle, Prodr. 15(2): 1170 (1866). *Sebastiania glandulosa* f. *olfersiana* (Müll.Arg.) Pax in H.G.A.Engler, Pflanzenr., IV, 147, V: 102 (1912).

Sebastiania corniculata f. *ovata* Müll.Arg. in A.P.de Candolle, Prodr. 15(2): 1169 (1866). *Sebastiania glandulosa* f. *ovata* (Müll.Arg.) Pax in H.G.A.Engler, Pflanzenr., IV, 147, V: 102 (1912).

Sebastiania corniculata f. *velutina* Müll.Arg. in A.P.de Candolle, Prodr. 15(2): 1168 (1866). *Sebastiania glandulosa* f. *velutina* (Müll.Arg.) Pax in H.G.A.Engler, Pflanzenr., IV, 147, V: 101 (1912).

Sebastiania corniculata var. *blepharophylla* Müll.Arg. in A.P.de Candolle, Prodr. 15(2): 1170 (1866).

Sebastiania corniculata var. *egensis* Müll.Arg. in A.P.de Candolle, Prodr. 15(2): 562 (1866).
Sebastiania corniculata var. *fischeri* Müll.Arg. in A.P.de Candolle, Prodr. 15(2): 1169 (1866). *Sebastiania salicifolia* var. *fischeri* (Müll.Arg.) Pax in H.G.A.Engler, Pflanzenr., IV, 147, V: 104 (1912).
Sebastiania corniculata var. *glabrata* (Mart.) Müll.Arg. in A.P.de Candolle, Prodr. 15(2): 1172 (1866).
Sebastiania corniculata var. *obtusifolia* Müll.Arg. in A.P.de Candolle, Prodr. 15(2): 1168 (1866). *Sebastiania glandulosa* var. *obtusifolia* (Müll.Arg.) Pax in H.G.A.Engler, Pflanzenr., IV, 147, V: 100 (1912).
Sebastiania corniculata var. *parvifolia* Müll.Arg. in A.P.de Candolle, Prodr. 15(2): 1169 (1866). *Sebastiania glandulosa* var. *parvifolia* (Müll.Arg.) Pax in H.G.A.Engler, Pflanzenr., IV, 147, V: 102 (1912).
Sebastiania corniculata var. *poeppingii* Müll.Arg. in A.P.de Candolle, Prodr. 15(2): 1173 (1866).
Sebastiania corniculata var. *rufescens* Müll.Arg. in A.P.de Candolle, Prodr. 15(2): 1170 (1866).
Sebastiania corniculata var. *scabra* Müll.Arg. in A.P.de Candolle, Prodr. 15(2): 1168 (1866). *Sebastiania glandulosa* var. *scabra* (Müll.Arg.) Pax in H.G.A.Engler, Pflanzenr., IV, 147, V: 102 (1912).
Sebastiania corniculata var. *sellowiana* Müll.Arg. in A.P.de Candolle, Prodr. 15(2): 1169 (1866). *Sebastiania glandulosa* var. *sellowiana* (Müll.Arg.) Pax in H.G.A.Engler, Pflanzenr., IV, 147, V: 102 (1912).
Sebastiania corniculata var. *villaricensis* Müll.Arg. in A.P.de Candolle, Prodr. 15(2): 1170 (1866).
Sebastiania corniculata var. *heterophylla* Müll.Arg. in C.F.P.von Martius, Fl. Bras. 11(2): 563 (1874).
Sebastiania corniculata var. *hispida* (Mart.) Müll.Arg. in C.F.P.von Martius, Fl. Bras. 11(2): 560 (1874).
Sebastiania corniculata var. *leptoclada* Müll.Arg. in C.F.P.von Martius, Fl. Bras. 11(2): 554 (1874). *Sebastiania salicifolia* var. *leptoclada* (Müll.Arg.) Pax in H.G.A.Engler, Pflanzenr., IV, 147, V: 104 (1912).
Sebastiania corniculata var. *lurida* Müll.Arg. in C.F.P.von Martius, Fl. Bras. 11(2): 554 (1874).
Sebastiania corniculata var. *microdendron* Müll.Arg. in C.F.P.von Martius, Fl. Bras. 11(2): 553 (1874). *Sebastiania glandulosa* var. *microdendron* (Müll.Arg.) Pax in H.G.A.Engler, Pflanzenr., IV, 147, V: 102 (1912).
Sebastiania corniculata var. *petiolaris* Müll.Arg. in C.F.P.von Martius, Fl. Bras. 11(2): 561 (1874).
Sebastiania corniculata var. *pohlii* Müll.Arg. in C.F.P.von Martius, Fl. Bras. 11(2): 553 (1874). *Sebastiania glandulosa* var. *pohlii* (Müll.Arg.) Pax in H.G.A.Engler, Pflanzenr., IV, 147, V: 103 (1912).
Sebastiania corniculata var. *potamophila* Müll.Arg. in C.F.P.von Martius, Fl. Bras. 11(2): 562 (1874).
Sebastiania corniculata var. *psilophylla* Müll.Arg. in C.F.P.von Martius, Fl. Bras. 11(2): 552 (1874). *Sebastiania glandulosa* var. *psilophylla* (Müll.Arg.) Pax in H.G.A.Engler, Pflanzenr., IV, 147, V: 103 (1912).
Sebastiania corniculata var. *transiens* Müll.Arg. in C.F.P.von Martius, Fl. Bras. 11(2): 553 (1874). *Sebastiania glandulosa* var. *transiens* (Müll.Arg.) Pax in H.G.A.Engler, Pflanzenr., IV, 147, V: 102 (1912).
Sebastiania corniculata var. *hassleriana* Chodat, Bull. Herb. Boissier, II, 1: 398 (1901).
Sebastiania mexicana Brandegee, Zoe 5: 205 (1905).
Sebastiania glandulosa Pax in H.G.A.Engler, Pflanzenr., IV, 147, V: 100 (1912), nom. inval.
Sebastiania linearifolia Lanj., Recueil Trav. Bot. Néerl. 36: 701 (1939 publ. 1940).

Microstachys dalzielii (Hutch.) Esser, Kew Bull. 53: 958 (1998 publ. 1999).
W. Trop. Africa. 22 GHA NGA.
* *Sapium dalzielii* Hutch., Bul. Misc. Inform. Kew 1917: 234 (1917).

Microstachys daphnoides (Mart.) Müll.Arg., Linnaea 32: 91 (1863).
Brazil (Bahia, Minas Gerais, Goiás). 84 BZC BZE BZL. Nanophan.
* *Cnemidostachys daphnoides* Mart., Nov. Gen. Sp. Pl. 1: 71 (1824). *Sebastiania daphnoides* var. *genuina* Müll.Arg. in A.P.de Candolle, Prodr. 15(2): 1168 (1866), nom. inval. *Sebastiania daphnoides* (Mart.) Müll.Arg. in A.P.de Candolle, Prodr. 15(2): 1167 (1866). *Sebastiania myrtilloides* var. *daphnoides* (Mart.) Pax in H.G.A.Engler, Pflanzenr., IV, 147, II: 94 (1910).
Cnemidostachys myrtilloides Mart., Nov. Gen. Sp. Pl. 1: 67 (1824). *Stillingia myrtilloides* (Mart.) Baill., Adansonia 5: 323 (1865). *Sebastiania daphnoides* var. *myrtilloides* (Mart.) Müll.Arg. in A.P.de Candolle, Prodr. 15(2): 1167 (1866). *Sebastiania myrtilloides* var. *martiana* Pax in H.G.A.Engler, Pflanzenr., IV, 147, V: 94 (1912), nom. inval. *Sebastiania myrtilloides* (Mart.) Pax in H.G.A.Engler, Pflanzenr., IV, 147, V: 93 (1912).
Cnemidostachys oleoides Mart., Nov. Gen. Sp. Pl. 1: 71 (1824). *Sebastiania daphnoides* var. *oleoides* (Mart.) Müll.Arg. in A.P.de Candolle, Prodr. 15(2): 1168 (1866). *Sebastiania oleoides* (Mart.) Müll.Arg. in C.F.P.von Martius, Fl. Bras. 11(2): 548 (1874).
Microstachys daphnoides var. *incana* Müll.Arg., Linnaea 32: 91 (1863). *Sebastiania daphnoides* var. *incana* (Müll.Arg.) Müll.Arg. in A.P.de Candolle, Prodr. 15(2): 1168 (1866). *Sebastiania myrtilloides* var. *incana* (Müll.Arg.) Pax in H.G.A.Engler, Pflanzenr., IV, 147, V: 94 (1912).
Sebastiania daphnoides var. *intermedia* Müll.Arg. in C.F.P.von Martius, Fl. Bras. 11(2): 547 (1874). *Sebastiania myrtilloides* var. *intermedia* (Müll.Arg.) Pax in H.G.A.Engler, Pflanzenr., IV, 147, V: 94 (1912).
Sebastiania daphnoides var. *major* Müll.Arg. in C.F.P.von Martius, Fl. Bras. 11(2): 547 (1874). *Sebastiania myrtilloides* var. *major* (Müll.Arg.) Pax in H.G.A.Engler, Pflanzenr., IV, 147, V: 94 (1912).

Microstachys ditassoides (Didr.) Esser, Kew Bull. 53: 958 (1998) publ. 1999).
Brazil (Goiás, Minas Gerais). 84 BZC BZL. Hemicr.
* *Cnemidostachys ditassoides* Didr., Vidensk. Meddel. Dansk Naturhist. Foren. Kjøbenhavn 1853: 88 (1853). *Sebastiania ditassoides* var. *genuina* Müll.Arg. in A.P.de Candolle, Prodr. 15(2): 1174 (1866), nom. inval. *Sebastiania ditassoides* (Didr.) Müll.Arg. in A.P.de Candolle, Prodr. 15(2): 1174 (1866).
Microstachys sessilifolia Müll.Arg., Linnaea 32: 95 (1863).
Microstachys sessilifolia f. *apiculata* Müll.Arg., Linnaea 32: 95 (1863).
Microstachys sessilifolia f. *hastata* Müll.Arg., Linnaea 32: 95 (1863).
Microstachys sessilifolia var. *glabrata* Müll.Arg., Linnaea 32: 95 (1863). *Sebastiania ditassoides* var. *glabrata* (Müll.Arg.) Müll.Arg. in A.P.de Candolle, Prodr. 15(2): 1174 (1866).
Microstachys sessilifolia var. *ledifolia* Müll.Arg., Linnaea 32: 95 (1863). *Sebastiania ditassoides* var. *ledifolia* (Müll.Arg.) Müll.Arg. in A.P.de Candolle, Prodr. 15(2): 1175 (1866).
Microstachys sessilifolia var. *parvifolia* Müll.Arg., Linnaea 32: 95 (1863). *Sebastiania ditassoides* var. *parvifolia* (Müll.Arg.) Müll.Arg. in A.P.de Candolle, Prodr. 15(2): 1175 (1866).
Microstachys sessilifolia var. *vellerifolia* Müll.Arg., Linnaea 32: 95 (1863). *Sebastiania ditassoides* var. *vellerifolia* (Müll.Arg.) Pax in H.G.A.Engler, Pflanzenr., IV, 147, V: 114 (1912).
Stillingia hastata Klotzsch ex Baill., Adansonia 5: 325 (1865).
Sebastiania ditassoides f. *apiculata* Müll.Arg. in A.P.de Candolle, Prodr. 15(2): 1174 (1866).
Sebastiania ditassoides f. *hastata* Müll.Arg. in A.P.de Candolle, Prodr. 15(2): 1174 (1866).
Sebastiania ditassoides var. *discolor* Pax & K.Hoffm. in H.G.A.Engler, Pflanzenr., IV, 147, V: 115 (1912).

Microstachys faradianensis (Beille) Esser, Kew Bull. 53: 959 (1998 publ. 1999).
W. Trop. Africa. 22 GHA GUI NGA SEN. Cham.
* *Excoecaria faradianensis* Beille, Bull. Soc. Bot. France 57(8): 128 (1910). *Sebastiania faradianensis* (Beille) Kruijt, Biblioth. Bot. 146: 88 (1996).
Sapium faradianense (Beille) Pax in H.G.A.Engler, Pflanzenr., IV, 147, V: 247 (1912).

Microstachys heterodoxa (Müll.Arg.) Esser, Kew Bull. 53: 959 (1998 publ. 1999).
Brazil (Bahia). 84 BZE.
* *Stillingia heterodoxa* Müll.Arg., Linnaea 32: 89 (1863). *Sebastiania heterodoxa* (Müll.Arg.) Benth. in G.Bentham & J.D.Hooker, Gen. Pl. 3: 334 (1880).

Microstachys hispida (Mart.) Govaerts in R.Govaerts, D.G.Frodin & A.Radcliffe-Smith, World Checklist Bibliogr. Euphorbiaceae: 1190 (2000).
Brazil, Bolivia, Paraguay. 83 BOL 84 BZC BZE BZL BZN BZS 85 PAR. Cham.
Cnemidostachys crotonoides Mart., Nov. Gen. Sp. Pl. 1: 71 (1824). *Stillingia crotonoides* (Mart.) Baill., Étude Euphorb.: 516 (1858). *Sebastiania hispida* var. *crotonoides* (Mart.) Pax in H.G.A.Engler, Pflanzenr., IV, 147, V: 110 (1912).
* *Cnemidostachys hispida* Mart., Nov. Gen. Sp. Pl. 1: 71 (1824). *Sebastiania corniculata* var. *hispida* (Mart.) Müll.Arg. in C.F.P.von Martius, Fl. Bras. 11(2): 560 (1874). *Sebastiania hispida* var. *euhispida* Pax in H.G.A.Engler, Pflanzenr., IV, 147, V: 109 (1912), nom. inval. *Sebastiania hispida* (Mart.) Pax in H.G.A.Engler, Pflanzenr., IV, 147, V: 105 (1912).
Sebastiania corniculata var. *klotzschiana* Müll.Arg. in A.P.de Candolle, Prodr. 15(2): 1169 (1866). *Sebastiania hispida* var. *klotzschiana* (Müll.Arg.) Pax in H.G.A.Engler, Pflanzenr., IV, 147, V: 106 (1912).
Sebastiania corniculata var. *leucoblepharis* Müll.Arg. in A.P.de Candolle, Prodr. 15(2): 1172 (1866). *Sebastiania hispida* var. *leucoblepharis* (Müll.Arg.) Pax in H.G.A.Engler, Pflanzenr., IV, 147, V: 110 (1912).
Sebastiania corniculata var. *macrophylla* Müll.Arg. in A.P.de Candolle, Prodr. 15(2): 1170 (1866). *Sebastiania hispida* var. *macrophylla* (Müll.Arg.) Pax in H.G.A.Engler, Pflanzenr., IV, 147, V: 108 (1912).
Sebastiania corniculata var. *major* Müll.Arg. in A.P.de Candolle, Prodr. 15(2): 1171 (1866). *Sebastiania hispida* var. *major* (Müll.Arg.) Pax in H.G.A.Engler, Pflanzenr., IV, 147, V: 113 (1912).
Sebastiania corniculata var. *tomentosa* Müll.Arg. in A.P.de Candolle, Prodr. 15(2): 1171 (1866). *Sebastiania hispida* var. *tomentosa* (Müll.Arg.) Pax in H.G.A.Engler, Pflanzenr., IV, 147, V: 109 (1912).
Sebastiania corniculata var. *ferruginea* Müll.Arg. in C.F.P.von Martius, Fl. Bras. 11(2): 107 (1873). *Sebastiania hispida* var. *ferruginea* (Müll.Arg.) Pax in H.G.A.Engler, Pflanzenr., IV, 147, V: 107 (1912).
Sebastiania corniculata var. *megapontica* Müll.Arg. in C.F.P.von Martius, Fl. Bras. 11(2): 108 (1873). *Sebastiania hispida* var. *megapontica* (Müll.Arg.) Pax in H.G.A.Engler, Pflanzenr., IV, 147, V: 108 (1912).
Sebastiania corniculata var. *incana* Müll.Arg. in C.F.P.von Martius, Fl. Bras. 11(2): 557 (1874). *Sebastiania hispida* var. *incana* (Müll.Arg.) Pax in H.G.A.Engler, Pflanzenr., IV, 147, V: 107 (1912).
Sebastiania corniculata var. *intercedens* Müll.Arg. in C.F.P.von Martius, Fl. Bras. 11(2): 556 (1874). *Sebastiania hispida* var. *intercedens* (Müll.Arg.) Pax in H.G.A.Engler, Pflanzenr., IV, 147, V: 111 (1912).
Sebastiania corniculata var. *laeta* Müll.Arg. in C.F.P.von Martius, Fl. Bras. 11(2): 559 (1874). *Sebastiania hispida* var. *laeta* (Müll.Arg.) Pax in H.G.A.Engler, Pflanzenr., IV, 147, V: 109 (1912).
Sebastiania corniculata var. *lagoensis* Müll.Arg. in C.F.P.von Martius, Fl. Bras. 11(2): 557 (1874). *Sebastiania hispida* var. *lagoensis* (Müll.Arg.) Pax in H.G.A.Engler, Pflanzenr., IV, 147, V: 107 (1912).

Sebastiania corniculata var. *mansoana* Müll.Arg. in C.F.P.von Martius, Fl. Bras. 11(2): 557 (1874). *Sebastiania hispida* var. *mansoana* (Müll.Arg.) Pax in H.G.A.Engler, Pflanzenr., IV, 147, V: 110 (1912).
Sebastiania corniculata var. *occidentalis* Müll.Arg. in C.F.P.von Martius, Fl. Bras. 11(2): 557 (1874). *Sebastiania hispida* var. *occidentalis* (Müll.Arg.) Pax in H.G.A.Engler, Pflanzenr., IV, 147, V: 107 (1912).
Sebastiania corniculata var. *oligophylla* Müll.Arg. in C.F.P.von Martius, Fl. Bras. 11(2): 556 (1874). *Sebastiania hispida* var. *oligophylla* (Müll.Arg.) Pax in H.G.A.Engler, Pflanzenr., IV, 147, V: 107 (1912).
Sebastiania corniculata var. *purpurella* Müll.Arg. in C.F.P.von Martius, Fl. Bras. 11(2): 558 (1874). *Sebastiania hispida* var. *purpurella* (Müll.Arg.) Pax in H.G.A.Engler, Pflanzenr., IV, 147, V: 110 (1912).
Sebastiania corniculata var. *regnellii* Müll.Arg. in C.F.P.von Martius, Fl. Bras. 11(2): 558 (1874). *Sebastiania hispida* var. *regnellii* (Müll.Arg.) Pax in H.G.A.Engler, Pflanzenr., IV, 147, V: 106 (1912).
Sebastiania corniculata var. *riedelii* Müll.Arg. in C.F.P.von Martius, Fl. Bras. 11(2): 557 (1874). *Sebastiania hispida* var. *riedelii* (Müll.Arg.) Pax in H.G.A.Engler, Pflanzenr., IV, 147, V: 110 (1912).
Sebastiania corniculata var. *schuechiana* Müll.Arg. in C.F.P.von Martius, Fl. Bras. 11(2): 559 (1874). *Sebastiania hispida* var. *schuechiana* (Müll.Arg.) Pax in H.G.A.Engler, Pflanzenr., IV, 147, V: 111 (1912).
Sebastiania corniculata var. *sclerophylla* Müll.Arg. in C.F.P.von Martius, Fl. Bras. 11(2): 560 (1874). *Sebastiania hispida* var. *sclerophylla* (Müll.Arg.) Pax in H.G.A.Engler, Pflanzenr., IV, 147, V: 113 (1912).
Sebastiania corniculata var. *speciosa* Müll.Arg. in C.F.P.von Martius, Fl. Bras. 11(2): 560 (1874). *Sebastiania hispida* var. *speciosa* (Müll.Arg.) Pax in H.G.A.Engler, Pflanzenr., IV, 147, V: 111 (1912).
Sebastiania corniculata var. *subglabrata* Müll.Arg. in C.F.P.von Martius, Fl. Bras. 11(2): 561 (1874). *Sebastiania hispida* var. *subglabrata* (Müll.Arg.) Pax in H.G.A.Engler, Pflanzenr., IV, 147, V: 113 (1912).
Sebastiania corniculata var. *weddelliana* Müll.Arg. in C.F.P.von Martius, Fl. Bras. 11(2): 558 (1874). *Sebastiania hispida* var. *weddeliana* (Müll.Arg.) Pax in H.G.A.Engler, Pflanzenr., IV, 147, V: 109 (1912).
Sebastiania corniculata var. *paraguayensis* Chodat, Bull. Herb. Boissier, II, 1: 398 (1901). *Sebastiania hispida* var. *paraguayensis* (Chodat) Pax in H.G.A.Engler, Pflanzenr., IV, 147, V: 111 (1912).
Sebastiania hispida f. *brevipila* Pax in H.G.A.Engler, Pflanzenr., IV, 147, V: 109 (1912).
Sebastiania hispida f. *glabrescens* Pax in H.G.A.Engler, Pflanzenr., IV, 147, V: 109 (1912).
Sebastiania hispida f. *villosula* Pax in H.G.A.Engler, Pflanzenr., IV, 147, V: 109 (1912).
Sebastiania hispida var. *ambigua* Pax & K.Hoffm. in H.G.A.Engler, Pflanzenr., IV, 147, V: 110 (1912).
Sebastiania hispida var. *aspera* Pax & K.Hoffm. in H.G.A.Engler, Pflanzenr., IV, 147, V: 109 (1912).
Sebastiania hispida var. *graciliramea* Pax & K.Hoffm. in H.G.A.Engler, Pflanzenr., IV, 147, V: 107 (1912).
Sebastiania hispida var. *scandens* Pax & K.Hoffm. in H.G.A.Engler, Pflanzenr., IV, 147, V: 111 (1912).
Sebastiania hispida var. *stenophylla* Pax & K.Hoffm. in H.G.A.Engler, Pflanzenr., IV, 147, V: 107 (1912).
Sebastiania hispida var. *subpatula* Pax & K.Hoffm. in H.G.A.Engler, Pflanzenr., IV, 147, V: 111 (1912).

Microstachys marginata (Mart.) Klotzsch ex Müll.Arg., Linnaea 32: 90 (1863).
Brazil (Minas Gerais). 84 BZL. Nanophan.

Cnemidostachys coriacea Mart., Nov. Gen. Sp. Pl. 1: 68 (1824). *Stillingia coriacea* (Mart.) Baill., Adansonia 5: 328 (1865). *Sebastiania marginata* var. *coriacea* (Mart.) Pax in H.G.A.Engler, Pflanzenr., IV, 147, V: 92 (1912).
* *Cnemidostachys marginata* Mart., Nov. Gen. Sp. Pl. 1: 68 (1824). *Sebastiania marginata* (Mart.) Müll.Arg. in A.P.de Candolle, Prodr. 15(2): 1166 (1866).

Microstachys nummularifolia (Cordeiro) Esser, Kew Bull. 53: 959 (1998 publ. 1999).
Brazil (Minas Gerais). 84 BZL.
* *Sebastiania nummularifolia* Cordeiro, Bol. Bot. Univ. São Paulo 11: 77 (1989).

Microstachys revoluta (Ule) Esser, Kew Bull. 53: 959 (1998 publ. 1999).
Brazil (Bahia). 84 BZE. Hemicr.
* *Sebastiania revoluta* Ule, Bot. Jahrb. Syst. 42: 222 (1908).

Microstachys serrulata (Mart.) Müll.Arg., Linnaea 32: 90 (1863).
Brazil (Minas Gerais, São Paulo, Goiás), Paraguay. 84 BZC BZE 85 PAR. Nanophan.
* *Cnemidostachys serrulata* Mart., Nov. Gen. Sp. Pl. 1: 68 (1824). *Stillingia serrulata* (Mart.) Baill., Adansonia 5: 324 (1865). *Sebastiania serrulata* var. *genuina* Müll.Arg. in A.P.de Candolle, Prodr. 15(2): 1167 (1866), nom. inval. *Sebastiania serrulata* (Mart.) Müll.Arg. in A.P.de Candolle, Prodr. 15(2): 1167 (1866).
Microstachys serrulata var. *klotzschiana* Müll.Arg., Linnaea 32: 90 (1863). *Sebastiania serrulata* var. *klotzschiana* (Müll.Arg.) Müll.Arg. in A.P.de Candolle, Prodr. 15(2): 1167 (1866).
Microstachys serrulata var. *oncoblepharis* Müll.Arg., Linnaea 32: 90 (1863). *Sebastiania serrulata* var. *oncoblepharis* (Müll.Arg.) Müll.Arg. in A.P.de Candolle, Prodr. 15(2): 1167 (1866).
Sebastiania serrulata var. *hispida* Müll.Arg. in C.F.P.von Martius, Fl. Bras. 11(2): 549 (1874).
Sebastiania serrulata var. *oblongifolia* Müll.Arg. in C.F.P.von Martius, Fl. Bras. 11(2): 550 (1874).
Sebastiania serrulata var. *fastigiata* Pax & K.Hoffm. in H.G.A.Engler, Pflanzenr., IV, 147, V: 96 (1912).

Microstachys uleana (Pax & K.Hoffm.) Esser, Kew Bull. 53: 959 (1998 publ. 1999).
Brazil (Bahia). 84 BZE.
* *Sebastiania uleana* Pax & K.Hoffm. in H.G.A.Engler, Pflanzenr., IV, 147, V: 93 (1912).

Synonyms:
Microstachys bicornis A.Juss. === **Microstachys corniculata** (Vahl) Griseb.
Microstachys crotonoides Klotzsch ex Benth. === **Microstachys corniculata** (Vahl) Griseb.
Microstachys daphnoides var. *incana* Müll.Arg. === **Microstachys daphnoides** (Mart.) Müll.Arg.
Microstachys fruticulosa H.Karst. === **Microstachys corniculata** (Vahl) Griseb.
Microstachys guianensis Klotzsch === **Microstachys corniculata** (Vahl) Griseb.
Microstachys mercurialis (L.) Dalzell & Gibson === **Mircrococca mercurialis**
Microstachys micrantha Benth. === **Microstachys corniculata** (Vahl) Griseb.
Microstachys polymorpha Müll.Arg. === **Microstachys corniculata** (Vahl) Griseb.
Microstachys ramosissima A.St.-Hil. === **Sebastiania brasiliensis** Spreng.
Microstachys serrulata var. *klotzschiana* Müll.Arg. === **Microstachys serrulata** (Mart.) Müll.Arg.
Microstachys serrulata var. *oncoblepharis* Müll.Arg. === **Microstachys serrulata** (Mart.) Müll.Arg.
Microstachys sessilifolia Müll.Arg. === **Microstachys ditassoides** (Didr.) Esser
Microstachys sessilifolia f. *apiculata* Müll.Arg. === **Microstachys ditassoides** (Didr.) Esser
Microstachys sessilifolia var. *glabrata* Müll.Arg. === **Microstachys ditassoides** (Didr.) Esser
Microstachys sessilifolia f. *hastata* Müll.Arg. === **Microstachys ditassoides** (Didr.) Esser
Microstachys sessilifolia var. *ledifolia* Müll.Arg. === **Microstachys ditassoides** (Didr.) Esser

Microstachys sessilifolia var. *parvifolia* Müll.Arg. === **Microstachys ditassoides** (Didr.) Esser
Microstachys sessilifolia var. *vellerifolia* Müll.Arg. === **Microstachys ditassoides** (Didr.) Esser
Microstachys stipulacea (Müll.Arg.) Klotzsch ex Baill. === **Sebastiania stipulacea** (Müll.Arg.) Müll.Arg.
Microstachys vahlii A.Rich. === **Microstachys corniculata** (Vahl) Griseb.
Microstachys virgata Müll.Arg. === **Microstachys bidentata** (Mart. & Zucc.) Esser
Microstachys virgata var. *bidentata* (Mart. & Zucc.) Müll.Arg. === **Microstachys bidentata** (Mart. & Zucc.) Esser
Microstachys virgata var. *odontococca* Müll.Arg. === **Microstachys bidentata** (Mart. & Zucc.) Esser
Microstachys virgata var. *scoparia* (Mart.) Müll.Arg. === **Microstachys bidentata** (Mart. & Zucc.) Esser

Middelbergia

Synonyms:
Middelbergia Schinz ex Pax === **Clutia** Boerh. ex L.

Mildbraedia

3 species, tropical Africa (2 in W. and WC. Africa, 1 in E. Africa); shrubs or trees to 10 m, sometimes straggly or semi-scandent and with toothed or sometimes lobed leaves. The axillary inflorescences are rather more delicate than in *Paracroton* (formerly *Fahrenheitia*), with which they have been grouped in the most recent version of the Webster system (Webster, Synopsis, 1994). No modern revision has been published; any recent treatments are in standard floras (e.g. *Flore du Congo Belge*, 8(1)(1962); *Flora of Tropical East Africa*, Euphorbiaceae 1: 340-343 (1987)). Webster's inclusion of this genus (along with *Paracroton*) in his Crotoneae seems to be without a strong foundation. (Crotonoideae)

Pax, F. (1910). *Neojatropha*. In A. Engler (ed.), Das Pflanzenreich, IV 147 [I] (Euphorbiaceae-Jatropheae): 114-115. Berlin. (Heft 42.) La/Ge. — 2 species, E. Africa. [Now combined and included in *Mildbraedia*.]

Pax, F. (with K. Hoffmann) (1911). *Mildbraedia*. In A. Engler (ed.), Das Pflanzenreich, IV 147 III (Euphorbiaceae-Cluytieae): 11-12. Berlin. (Heft 47.) La/Ge. — 1 species, central Africa.

Pax, F. (with K. Hoffmann) (1914). *Mildbraedia*. In A. Engler (ed.), Das Pflanzenreich, IV 147 VII [Euphorbiaceae-Additamentum V]: 403. Berlin. (Heft 63.) La/Ge. — 4 species; revised key.

Mildbraedia Pax, Bot. Jahrb. Syst. 43: 319 (1909).
Trop. Africa. 22 23 25 26.
Neojatropha Pax in H.G.A.Engler, Pflanzenr., IV, 147, I: 114 (1910).

Mildbraedia carpinifolia (Pax) Hutch. in D.Oliver, Fl. Trop. Afr. 6(1): 801 (1912).
Kenya, Tanzania (incl. Zanzibar), Mozambique. 25 KEN TAN 26 MOZ. Nanophan. or phan.
* *Jatropha carpinifolia* Pax in H.G.A.Engler, Pflanzenw. Ost-Afrikas C: 240 (1895).
Neojatropha carpinifolia (Pax) Pax in H.G.A.Engler, Pflanzenr., IV, 147, I: 114 (1910).

var. **carpinifolia**
Kenya, Tanzania, Zanzibar, Mozambique. 25 KEN TAN 26 MOZ. Nanophan. or phan.
Jatropha fallax Pax, Bot. Jahrb. Syst. 33: 284 (1903). *Neojatropha fallax* (Pax) Pax in H.G.A.Engler, Pflanzenr., IV, 147, I: 115 (1910). *Mildbraedia fallax* (Pax) Hutch. in D.Oliver, Fl. Trop. Afr. 6(1): 800 (1912).

var. **strigosa** Radcl.-Sm., Kew Bull. 27: 507 (1972).
Tanzania, Mozambique. 25 TAN 26 MOZ. Nanophan. or phan.

Mildbraedia paniculata Pax
Artist: Not identified
Mildbraed, Wiss. Ergeb. Deutsch. Zentral-Afrika-Exped. 1907-1908, 2 (Botanik): pl. 58 (1912)
KEW ILLUSTRATIONS COLLECTION

Mildbraedia klaineana Hutch. in D.Oliver, Fl. Trop. Afr. 6(1): 799 (1912).
Gabon. 23 GAB. Nanophan.

Mildbraedia paniculata Pax, Bot. Jahrb. Syst. 43: 319 (1909). – FIGURE, p. 1194.
Liberia to Zaire. 22 GHA IVO LBR 23 ZAI. Phan.

subsp. **occidentalis** J.Léonard, Bull. Jard. Bot. État 31: 65 (1961).
W. Trop. Africa. 22 GHA IVO LBR. Phan.

subsp. **paniculata**
Zaire. 23 ZAI. Phan.
Croton cavalliensis Beille ex Hutch. in D.Oliver, Fl. Trop. Afr. 6(1): 800 (1912).

Synonyms:
Mildbraedia balboana Chiov. === **Croton alienus** Pax
Mildbraedia fallax (Pax) Hutch. === **Mildbraedia carpinifolia** (Pax) Hutch. var. **carpinifolia**

Minutalia

Synonyms:
Minutalia Fenzl === **Antidesma** L.

Mirabellia

Synonyms:
Mirabellia Bertol. ex Baill. === **Dysopsis** Baill.

Mischodon

1 species, Asia (Sri Lanka, S India); much-branched large shrubs or forest trees to 25 m with thick leaves in whorls of 3-4. The autochthony of the species in India has been questioned. Within the Webster system (Webster, Synopsis, 1994) it is considered most closely related to *Voatomalo* from Madagascar. (Oldfieldioideae)

Pax, F. & K. Hoffmann (1922). *Mischodon*. In A. Engler (ed.), Das Pflanzenreich, IV 147 XV (Euphorbiaceae-Phyllanthoideae-Phyllantheae): 292-293. Berlin. (Heft 81.) La/Ge. — 1 species; Asia (Sri Lanka, S India).

Raju, V. S. (1984). Notes on *Mischodon zeylanicus* Thwaites: a little-known euphorbiaceous plant from Sri Lanka and southern India. J. Econ. Taxon. Bot. 5: 165-167. En. — Rarely collected in S India and possibly not native there.

Mischodon Thwaites, Hooker's J. Bot. Kew Gard. Misc. 4: 299 (1854).
Sri Lanka, S. India. 40. Phan.

Mischodon zeylanicus Thwaites, Hooker's J. Bot. Kew Gard. Misc. 6: 300 (1854).
Sri Lanka, S. India ? 40 IND? SRI. Phan.

Moacroton

7 species, Middle America (Cuba); shrubs or small trees with narrow leaves, the surfaces in some contrasting. The genus is closely related to *Croton* and further study may show reduction to be in order; the main difference is in the presence of petals in pistillate flowers, these being reduced or absent in *Croton*. The least specialised species is said by Borhidi (1991) to be *M. gynopetalus*. In his revision, that author erected two sections, based in part on leaf thickness. In the latest version of the Webster system (Synopsis, 1994) the genus has been grouped with the African *Mildbraedia* and the Asian *Paracroton* in that author's Crotoneae but their inclusion therein was admittedly uncertain. (Crotonoideae)

Croizat, L. (1945). New or critical Euphorbiaceae from the Americas. J. Arnold Arbor. 26: 181-196. En. — Includes (pp. 189-191) a description of *Moacroton* and a synopsis (4 spp., Cuba); key to *Moacroton* and related genera but not to species of *Moacroton*.

• Borhidi, A. (1991). El género *Moacroton* Croiz. (Euphorbiaceae). Acta Bot. Hung. 36: 7-12, illus. Sp. — Synoptic treatment with key (this inclusive of distribution), conspectus of species and infraspecific taxa with types, description of one new species with documentation, list of references, and illustrations.

Moacroton Croizat, J. Arnold Arbor. 26: 189 (1945).
Cuba. 81.

Moacroton cristalensis (Urb.) Croizat, J. Arnold Arbor. 26: 191 (1945).
E. Cuba (Sierra del Cristal). 81 CUB. Nanophan.
* *Croton cristalensis* Urb., Symb. Antill. 9: 197 (1924).

Moacroton ekmanii (Urb.) Croizat, J. Arnold Arbor. 26: 191 (1945).
E. Cuba (Sierra Sagua Baracoa). 81 CUB. Nanophan.
* *Croton ekmanii* Leonard, J. Wash. Acad. Sci. 17: 69 (1927).

Moacroton gynopetalus Borhidi, Acta Bot. Acad. Sci. Hung. 36: 7 (1990).
Cuba. 81 CUB.

Moacroton lanceolatus Alain, Contr. Ocas. Mus. Hist. Nat. Colegio "De La Salle" 11: 4 (1952).
E. Cuba (Sierra de Moa). 81 CUB. Nanophan.
Moacroton lanceolatus var. *ellipticus* Borhidi & O.Muñiz, Acta Bot. Acad. Sci. Hung. 17: 10 (1971 publ. 1972).
Moacroton lanceolatus var. *varians* Borhidi, Acta Bot. Acad. Sci. Hung. 22: 306 (1976 publ. 1977).
Moacroton lanceolatus var. *longifolius* Borhidi, Acta Bot. Acad. Sci. Hung. 36: 10 (1991).

Moacroton leonis Croizat, J. Arnold Arbor. 26: 190 (1945).
E. Cuba (Sierra de Moa). 81 CUB. Nanophan.

Moacroton revolutus Alain, Contr. Ocas. Mus. Hist. Nat. Colegio "De La Salle" 11: 3 (1952).
Cuba (Matanzas). 81 CUB. Nanophan.

Moacroton tetramerus Borhidi & O.Muñiz, Acta Bot. Acad. Sci. Hung. 17: 10 (1971 publ. 1972).
Cuba. 81 CUB.

Synonyms:
Moacroton lanceolatus var. *ellipticus* Borhidi & O.Muñiz === **Moacroton lanceolatus** Alain
Moacroton lanceolatus var. *longifolius* Borhidi === **Moacroton lanceolatus** Alain
Moacroton lanceolatus var. *varians* Borhidi === **Moacroton lanceolatus** Alain
Moacroton trigonocarpus (Griseb.) Croizat === **Croton trigonocarpus** Griseb.

Moeroris

Synonyms:
Moeroris Raf. === **Phyllanthus** L.
Moeroris stipulata Raf. === **Phyllanthus stipulatus** (Raf.) G.L.Webster

Molina

Synonyms:
Molina Gay === **Dysopsis** Baill.

Monadenium

73 species, Africa; succulent (and sometimes caudiciform) geophytes, shrubs or small trees closely related to *Euphorbia* and perhaps a specialised offshoot from within that genus. Never revised for *Pflanzenreich* but subsequently treated in a fine monograph by Bally (1961). Subsequent contributions are listed in Malaisse et al. (1994) and a review of current knowledge was presented by Malaisse & Lecron (1994). Forster (1996) provided a lavishly illustrated checklist for enthusiasts, among whom the genus has long been popular. (Euphorbioideae (Euphorbieae))

Pax, F. & K. Hoffmann (1931). *Monadenium*. In A. Engler (ed.), Die natürlichen Pflanzenfamilien, 2. Aufl., 19c: 222. Leipzig. Ge. — Synopsis with description of genus; 25 species then known. [Superseded by Bally 1961 and later work.]

• Bally, P. R. O. (1961). The genus *Monadenium*. 111 pp., text-fig., 34 half-tones, maps, 32 col. pl. (loose in included portfolio). Bern: Benteli. En. — Well-illustrated standard monograph, covering 47 species with 'nearly a score' of additional infraspecific taxa; features keys (pp. 15-20), synonymy, descriptions, localities with exsiccatae, figures and distribution maps; album of coloured plates at end following general climatic and phytogeographical maps, the bibliography, the index and the half-tones. The general part includes an illustrated glossary including figures of growth-forms. [A good many additions have since been made.]

Jones, K. & J. B. Smith (1969). The chromosome identity of *Monadenium* Pax and *Synadenium* Pax (Euphorbiaceae). Kew Bull. 23: 491-498. En. — Karyological account; evidence for distinction of the genera.

Carter, S. (1987). New taxa and observations in *Monadenium* (Euphorbiaceae) in east Africa. Kew Bull. 42: 903-918. En. — Substantial suite of additions to Bally (1961); precursory to *Flora of Tropical East Africa* account.

Malaisse, F. & J. M. Lecron (1994). Problemes taxonomiques du genre *Monadenium* Pax (Euphorbiaceae). In J. H. Seyani & A. C. Chikuni (eds), *Proceedings of the 13th plenary meeting of AETFAT* (Zomba, Malawi, 2-11 April 1991), 1: 481-489, illus. Zomba, Malawi: National Herbarium and Botanic Gardens of Malawi. Fr. — Review of systematics in the genus (67 species by then accepted; additional ones known but not yet described, notably from Shaba, D.R. Congo). Many species are evidently quite local in distribution. A tabular checklist with dates of publication or recombination and key features appears on p. 484; representative life forms are illustrated on p. 485. No key is presented.

Malaisse, F., J. M. Lecron & M. Schaijes (1994 (1995)). Remarques à propos du genre *Monadenium* Pax (Euphorbiaceae), en particulier concernant les espèces de la région zambezienne. Bull. Acad. Roy. Sci. Outre-Mer 40: 389-418, illus. Fr. — Historical introduction, character review, and table of published species; review of genus in Shaba Province (D.R. Congo) with illustrations of habitat transects; examination of the geophytic *M. simplex* complex with descriptions of 4 new species and amplified description of *M. herbaceum* Pax; summary (importance of neoendemism and imperfect knowledge of geophytic species noted); literature (p. 408); many line drawings (with also 18 coloured photographs).

• Forster, P.I. (1996). A checklist of the genus *Monadenium* (Euphorbiaceae). Euphorbia J. 10: 142-161, col. illus. En. — Lavishly illustrated nomenclator with places of publication, basionyms and distribution; reduced names intercalated. [Accounts for all new taxa and other work published since 1961.]

Monadenium Pax, Bot. Jahrb. Syst. 19: 126 (1894).
Trop. & S. Africa. 23 24 25 26 27.
Lortia Rendle, J. Bot. 36: 29 (1898).
Stenadenium Pax, Bot. Jahrb. Syst. 30: 343 (1901).

Monadenium angolense Bally, Gen. Monaden.: 95 (1961).
Angola. 26 ANG.

Monadenium arborescens Bally, Candollea 17: 25 (1959).
Tanzania. 25 TAN. (Succ.) nanophan. or phan.

Monadenium bianoense Malaisse & Lecron, Bull. Séances Acad. Roy. Sci. Outre Mer 40: 400 (1995 publ. 1995).
Zaire. 23 ZAI.

Monadenium bodenghieriae Malaisse & Lecron, Bull. Séances Acad. Roy. Sci. Outre Mer 40: 402 (1994 publ. 1995).
Zaire. 23 ZAI.

Monadenium cannellii L.C.Leach, Garcia de Orta, Sér. Bot. 1: 35 (1973).
Angola. 26 ANG.

Monadenium capitatum Bally, Candollea 17: 26 (1959). – FIGURE, p. 1199.
Tanzania, Zambia. 25 TAN 26 ZAM. Tuber geophyte.

Monadenium catenatum S.Carter, Kew Bull. 42: 905 (1987).
S. Tanzania. 25 TAN. Tuber geophyte.

Monadenium chevalieri N.E.Br. in D.Oliver, Fl. Trop. Afr. 6(1): 1035 (1913).
Central African Rep., Zaire. 23 CAF ZAI.

Monadenium clarae Malaisse & Lecron, Bull. Jard. Bot. Belg. 59: 478 (1989).
Zaire. 23 ZAI.

Monadenium coccineum Pax, Bot. Jahrb. Syst. 19: 127 (1894).
Tanzania. 25 TAN. Succ. tuber cham.

Monadenium crenatum N.E.Br. in D.Oliver, Fl. Trop. Afr. 6(1): 461 (1911).
Zimbabwe, Mozambique. 26 MOZ ZIM.

Monadenium crispum N.E.Br. in D.Oliver, Fl. Trop. Afr. 6(1): 1034 (1913).
NE. Tanzania. 25 TAN. Tuber geophyte or cham.
Monadenium intermedium Bally, Candollea 17: 28 (1959).

Monadenium cupricola Malaisse & Lecron, Bull. Jard. Bot. Belg. 60: 301 (1990).
Zaire. 23 ZAI.

Monadenium depauperatum (Bally) S.Carter, Kew Bull. 42: 908 (1987).
Tanzania, Malawi, N. Zambia. 25 TAN 26 MLW ZAM. Tuber cham.
* *Monadenium laeve* f. *depauperata* Bally, Candollea 17: 35 (1959).

Monadenium descampsii Pax, Bull. Soc. Roy. Bot. Belgique 37: 108 (1898).
Zaire. 23 ZAI.

Monadenium dilunguense Malaisse & Lecron, Bull. Séances Acad. Roy. Sci. Outre Mer 40: 403 (1994 publ. 1995).
Zaire. 23 ZAI.

Monadenium discoideum Bally, Candollea 17: 26 (1959).
Zambia. 26 ZAM.

Monadenium echinulatum Stapf, Hooker's Icon. Pl. 27: t. 2666 (1901).
Zaire, Tanzania, Zambia. 23 ZAI 25 TAN 26 ZAM. Tuber cham.
Monadenium aculeolatum Pax, Bot. Jahrb. Syst. 43: 89 (1909).
Monadenium asperrimum Pax, Bot. Jahrb. Syst. 43: 90 (1909).
Monadenium echinulatum f. *glabrescens* Bally, Candollea 17: 35 (1959).

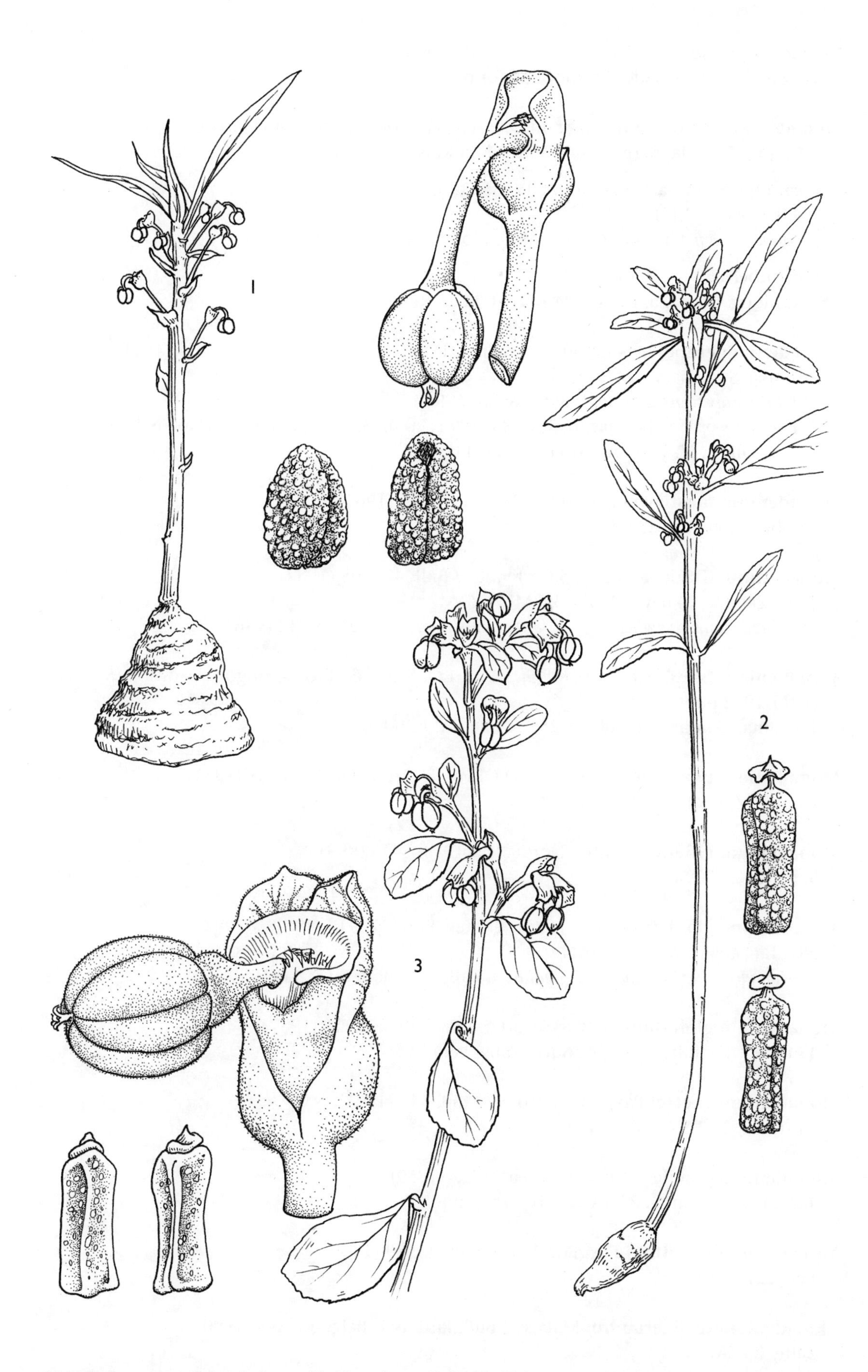

Monadenium pedunculatum S. Carter (1); *M. capitatum* Bally (2); *M. invenustum* N.E.Br. var *invenustum* (3)
Artist: Christine Grey-Wilson
Fl. Trop. East Africa, Euphorbiaceae 2: 543, fig. 102 (1988)
KEW ILLUSTRATIONS COLLECTION

Monadenium elegans S.Carter, Kew Bull. 42: 909 (1987).
Tanzania. 25 TAN. (Succ.) nanophan. or phan.

Monadenium ellenbeckii N.E.Br. in D.Oliver, Fl. Trop. Afr. 6(1): 454 (1911).
Ethiopia, Somalia, Kenya. 24 ETH SOM 25 KEN.

f. **caulopodium** Bally, Candollea 17: 35 (1959).
Ethiopia. 24 ETH.
Monadenium zavattarii Chiov., Miss. Biol. Borana 4: 103 (1939).

f. **ellenbeckii**
Ethiopia, Somalia, Kenya. 24 ETH SOM 25 KEN.

Monadenium erubescens (Rendle) N.E.Br. in D.Oliver, Fl. Trop. Afr. 6(1): 457 (1911).
Ethiopia, Somalia. 24 ETH SOM.
* *Lortia erubescens* Rendle, J. Bot. 36: 30 (1898).
Lortia major Pax, Bot. Jahrb. Syst. 33: 289 (1903). *Monadenium majus* (Pax) N.E.Br. in D.Oliver, Fl. Trop. Afr. 6(1): 457 (1911).

Monadenium fanshawei Bally, Gen. Monaden.: 26 (1961).
Zambia, Zimbabwe. 26 ZAM ZIM.

Monadenium filiforme (Bally) S.Carter, Kew Bull. 42: 905 (1987).
Zaire, Zambia, Malawi. 23 ZAI 26 MLW ZAM.
* *Monadenium chevalieri* var. *filiforme* Bally, Gen. Monaden.: 32 (1961).

Monadenium friesii N.E.Br. ex R.E.Fr., Wiss. Erg. Schwed. Rhod.-Kongo Exped. 1: 115 (1911-1912 publ. 1914).
Tanzania, Zambia, Zimbabwe. 25 TAN 26 ZAM ZIM.

Monadenium fwambense N.E.Br. in D.Oliver, Fl. Trop. Afr. 6(1): 461 (1911).
Zambia, Zimbabwe. 26 ZAM ZIM.

Monadenium gillettii S.Carter, Nordic J. Bot. 13: 541 (1993).
Somalia. 24 SOM.

Monadenium gladiatum (Bally) S.Carter, Kew Bull. 42: 913 (1987).
NC. Tanzania (Masai). 25 TAN.
* *Monadenium yattanum* var. *gladiatum* Bally, Candollea 17: 34 (1959).

Monadenium globosum Bally & S.Carter, Kew Bull. 42: 906 (1987).
Tanzania (Njombe). 25 TAN. Succ. Tuber geophyte.

Monadenium goetzei Pax, Bot. Jahrb. Syst. 30: 342 (1901).
Tanzania. 25 TAN. Tuber cham.

Monadenium gracile Bally, Candollea 17: 27 (1959).
Tanzania (Dodoma). 25 TAN. Tuber geophyte.

Monadenium guentheri Pax, Bot. Jahrb. Syst. 43: 89 (1909).
Kenya. 25 KEN.

Monadenium hedigerianum Malaisse, Bull. Jard. Bot. Belg. 60: 295 (1990).
Zaire. 23 ZAI.

Monadenium herbaceum Pax, Bot. Jahrb. Syst. 45: 241 (1910).
Zaire. 23 ZAI.

Monadenium heteropodum (Pax) N.E.Br. in D.Oliver, Fl. Trop. Afr. 6(1): 453 (1911).
Tanzania. 25 TAN. Succ. cham.
* *Euphorbia heteropoda* Pax, Bot. Jahrb. Syst. 34: 374 (1904).

var. **formosum** (Bally) S.Carter, Kew Bull. 42: 916 (1987).
Tanzania (Pare). 25 TAN. Succ. cham.
* *Monadenium schubei* var. *formosum* Bally, Candollea 17: 36 (1959).

var. **heteropodum**
Tanzania. 25 TAN. Succ. cham.
Monadenium guentheri var. *mamillare* Bally, Candollea: 17 (1959).

Monadenium hirsutum Bally, Candollea 17: 27 (1959).
Zambia, Malawi. 26 MLW ZAM.

Monadenium invenustum N.E.Br., Bull. Misc. Inform. Kew 1909: 329 (1909).
– FIGURE, p. 1199.
Kenya. 25 KEN. Succ. tuber cham.

var. **angustum** Bally, Candollea 17: 35 (1959).
Kenya. 25 KEN. Succ. tuber cham.

var. **invenustum**
Kenya. 25 KEN. Succ. tuber cham.

Monadenium kaessneri N.E.Br. in D.Oliver, Fl. Trop. Afr. 6(1): 459 (1911).
Zaire. 23 ZAI.

Monadenium kundelunguense Malaisse, Cact. Succ. J. (Los Angeles) 59: 204 (1987).
Zaire. 23 ZAI.

Monadenium laeve Stapf, Hooker's Icon. Pl. 27: t. 2666 (1901).
Tanzania, Malawi. 25 TAN 26 MLW. Tuber cham.

Monadenium letestuanum Denis, Bull. Mus. Natl. Hist. Nat. 28: 194 (1922).
Central African Rep. 23 CAF.

var. **letestuanum**
Central African Rep. 23 CAF.

var. **rotundifolium** Bally, Gen. Monaden.: 29 (1961).
Central African Rep. 23 CAF.

Monadenium letouzeyanum Malaisse, Bull. Mus. Natl. Hist. Nat., B, Adansonia 11: 337 (1989 publ.1990).
Zaire. 23 ZAI.

Monadenium lindenii S.Carter, Nordic J. Bot. 13: 542 (1993).
Somalia. 24 SOM.

Monadenium lugardiae N.E.Br., Bull. Misc. Inform. Kew 1909: 138 (1909).
SC. Trop. & S. Africa. 26 MLW MOZ ZIM 27 BOT NAT SWZ TVL. Succ. cham.

Monadenium mafingense Hargr., Cact. Succ. J. (Los Angeles) 53: 292 (1981).
Malawi, Zambia. 26 MLW ZAM.

Monadenium magnificum E.A.Bruce, Bull. Misc. Inform. Kew 1940: 51 (1940).
Tanzania. 25 TAN. (Succ.) tuber nanophan.

Monadenium mamfwense Malaisse & Lecron, Bull. Séances Acad. Roy. Sci. Outre Mer 40: 405 (1994 publ. 1995).
Zaire. 23 ZAI.

Monadenium montanum Bally, Candollea 17: 28 (1959).
Kenya, Tanzania. 25 KEN TAN.

Monadenium nervosum Bally, Candollea 17: 29 (1959).
Zambia, Tanzania, Zaire ? 23 ZAI? 25 TAN 26 ZAM. Tuber geophyte.

Monadenium nudicaule Bally, Candollea 17: 29 (1959).
Tanzania (Chunya). 25 TAN. Tuber geophyte or cham.

Monadenium orobanchoides Bally, Candollea 17: 30 (1959).
Zambia, Malawi, Tanzania. 25 TAN 26 MLW ZAM. Tuber geophyte.

Monadenium parviflorum N.E.Br. in D.Oliver, Fl. Trop. Afr. 6(1): 458 (1911).
Malawi. 26 MLW.

Monadenium pedunculatum S.Carter, Kew Bull. 42: 903 (1987). – FIGURE, p. 1199.
Zambia, Tanzania, Burundi, Zaire. 23 BUR ZAI 25 TAN 26 ZAM. Tuber geophyte.

Monadenium petiolatum Bally, Candollea 17: 30 (1959).
Tanzania (Dodoma). 25 TAN. Tuber geophyte or cham.

Monadenium pseudoracemosum Bally, Candollea 17: 30 (1959).
Tanzania, Zambia. 25 TAN 26 ZAM. Tuber geophyte.

var. **lorifolium** Bally, Gen. Monaden.: 49 (1961).
Zambia (Nmbulu). 26 ZAM. Tuber geophyte.

var. **pseudoracemosum**
Tanzania. 25 TAN. Tuber geophyte.

Monadenium pudibundum Bally, Candollea 17: 31 (1959). *Monadenium simplex* var. *pudibundum* (Bally) Bally, Gen. Monaden.: 40 (1961).
Zaire, Zambia. 23 ZAI 26 ZAM.

var. **pudibundum**
Zaire, Zambia. 23 ZAI 26 ZAM.

var. **rotundifolium** Malaisse & Lecron, Bull. Séances Acad. Roy. Sci. Outre Mer 40: 406 (1994 publ. 1995).
S. Zaire. 23 ZAI.

Monadenium reflexum Chiov. ex Chiarugi, Webbia 8: 235 (1951).
Kenya, Ethiopia. 24 ETH 25 KEN.

Monadenium renneyi S.Carter, Kew Bull. 42: 916 (1987).
Kenya. 25 KEN.

Monadenium rhizophorum Bally, Candollea 17: 32 (1959).
Kenya. 25 KEN.

Monadenium ritchiei Bally, Candollea 17: 32 (1959).
Kenya. 25 KEN. Succ. cham.

subsp. **marsabitense** S.Carter, Kew Bull. 42: 915 (1987).
NE. Kenya (Marsabit). 25 KEN. Succ. cham.

subsp. **nyambense** S.Carter, Kew Bull. 42: 915 (1987).
Kenya (Meru). 25 KEN. Succ. cham.

subsp. **ritchiei**
Kenya. 25 KEN. Succ. cham.

Monadenium rubellum (Bally) S.Carter, Kew Bull. 42: 913 (1987).
Kenya. 25 KEN.
* *Monadenium montanum* var. *rubellum* Bally, Candollea 17: 28 (1959).

Monadenium schaijesii Malaisse, Bull. Jard. Bot. Belg. 56: 487 (1986).
Zaire. 23 ZAI.

Monadenium schubei (Pax) N.E.Br. in D.Oliver, Fl. Trop. Afr. 6(1): 453 (1911).
Tanzania. 25 TAN.
* *Euphorbia schubei* Pax, Bot. Jahrb. Syst. 34: 373 (1904).

Monadenium shebeliense M.G.Gilbert, Bradleya 8: 45 (1990).
Ethiopia. 24 ETH.

Monadenium simplex Pax, Bull. Herb. Boissier 6: 743 (1898).
Angola. 26 ANG.

Monadenium spectabile S.Carter, Kew Bull. 42: 911 (1987).
Tanzania. 25 TAN. (Succ.) nanophan.

Monadenium spinescens (Pax) Bally, Candollea 17: 34 (1959).
Tanzania (Chunya). 25 TAN. (Succ.) phan.
* *Stenadenium spinescens* Pax, Bot. Jahrb. Syst. 30: 343 (1901).

Monadenium stapelioides Pax, Bot. Jahrb. Syst. 43: 89 (1909).
E. Trop. Africa. 25 KEN TAN UGA. Succ. cham.

var. **congestum** (Bally) S.Carter, Kew Bull. 42: 914 (1987).
Uganda, SW. Kenya, N. Tanzania. 25 KEN TAN UGA. Succ. cham.
* *Monadenium stapelioides* f. *congestum* Bally, Candollea 17: 36 (1959).

var. **stapelioides**
Kenya, Tanzania. 25 KEN TAN. Succ. cham.
Monadenium succulentum Schweick., Bull. Misc. Inform. Kew 1935: 274 (1935).

Monadenium stellatum Bally, Candollea 17: 33 (1959).
Somalia. 24 SOM.

Monadenium stoloniferum (Bally) S.Carter, Kew Bull. 42: 913 (1987).
Kenya. 25 KEN.
* *Monadenium rhizophorum* var. *stoloniferum* Bally, Gen. Monaden.: 55 (1961).

Monadenium torrei L.C.Leach, Garcia de Orta, Sér. Bot. 1: 37 (1973).
S. Tanzania, N. Mozambique. 25 TAN 26 MOZ. Succ. nanophan.

Monadenium trinerve Bally, Candollea 17: 33 (1959).
Kenya. 25 KEN.

Monadenium virgatum Bally, Gen. Monaden.: 65 (1961).
Kenya. 25 KEN.

Monadenium yattanum Bally, Candollea 17: 34 (1959).
Kenya. 25 KEN.

Synonyms:
Monadenium aculeolatum Pax === **Monadenium echinulatum** Stapf
Monadenium asperrimum Pax === **Monadenium echinulatum** Stapf
Monadenium chevalieri var. *filiforme* Bally === **Monadenium filiforme** (Bally) S.Carter
Monadenium echinulatum f. *glabrescens* Bally === **Monadenium echinulatum** Stapf
Monadenium gossweileri N.E.Br. === **Endadenium gossweileri** (N.E.Br.) L.C.Leach

Monadenium guentheri var. *mamillare* Bally === **Monadenium heteropodum** (Pax) N.E.Br. var. **heteropodum**
Monadenium intermedium Bally === **Monadenium crispum** N.E.Br.
Monadenium laeve f. *depauperata* Bally === **Monadenium depauperatum** (Bally) S.Carter
Monadenium lunulatum Chiov. === **Kleinia subulifolia** (Chiov.) P.Halliday (Asteraceae)
Monadenium majus (Pax) N.E.Br. === **Monadenium erubescens** (Rendle) N.E.Br.
Monadenium montanum var. *rubellum* Bally === **Monadenium rubellum** (Bally) S.Carter
Monadenium rhizophorum var. *stoloniferum* Bally === **Monadenium stoloniferum** (Bally) S.Carter
Monadenium schubei var. *formosum* Bally === **Monadenium heteropodum** var. **formosum** (Bally) S.Carter
Monadenium simplex var. *pudibundum* (Bally) Bally === **Monadenium pudibundum** Bally
Monadenium stapelioides f. *congestum* Bally === **Monadenium stapelioides** var. **congestum** (Bally) S.Carter
Monadenium subulifolium Chiov. === **Kleinia subulifolia** (Chiov.) P.Halliday (Asteraceae)
Monadenium succulentum Schweick. === **Monadenium stapelioides** Pax var. **stapelioides**
Monadenium yattanum var. *gladiatum* Bally === **Monadenium gladiatum** (Bally) S.Carter
Monadenium zavattarii Chiov. === **Monadenium ellenbeckii** f. **caulopodium** Bally

Monguia

Synonyms:
Monguia Chapel. ex Baill. === **Croton** L.

Monotaxis

11 species, Australia; microphyllous annuals, perennials or heath-like small shrubs. Its narrow cotyledons resulted it being placed in a subdivision known as 'Stenolobae' by Mueller and by Pax and Hoffmann. Like other Australian genera of that group, though, it appears instead to be specialised within one of the larger subfamilies, in this case the Acalyphoideae. Webster (1994) allied it to *Amperea*, also Australian (with the two genera forming tribe Ampereae); on the basis of pollen morphology, however, a possibility remains that the genus falls within Chrozophoreae. No modern revision of the genus is yet available, although a precursor related to *Flora of Australia* is possible. (Acalyphoideae)

Gruening, G. (1913). *Monotaxis*. In A. Engler (ed.), Das Pflanzenreich, IV 147 [Stenolobieae] (Euphorbiaceae-Porantheroideae et Ricinocarpoideae): 75-86. Berlin. (Heft 58.) La/Ge. — 9 species in 2 sections, Australia (many in W Australia).

Monotaxis Brongn. in L.I.Duperrey, Voy. Monde: 223 (1829).
Australia. 50.
Reissipa Steud. ex Klotzsch in J.G.C.Lehmann, Pl. Preiss. 2: 230 (1848).
Hippocrepandra Müll.Arg., Linnaea 34: 61 (1865).

Monotaxis gracilis (Müll.Arg.) Baill., Adansonia 6: 293 (1866).
Western Australia. 50 WAU. Cham.
* *Hippocrepandra gracilis* Müll.Arg., Linnaea 34: 62 (1865). *Monotaxis gracilis* var. *genuina* Grüning in H.G.A.Engler, Pflanzenr., IV, 147: 84 (1913), nom. inval.
Monotaxis gracilis var. *virgata* Grüning in H.G.A.Engler, Pflanzenr., IV, 147: 84 (1913).

Monotaxis grandiflora Endl. in S.L.Endlicher & al., Enum. Pl.: 19 (1837). *Monotaxis grandiflora* var. *typica* Grüning in H.G.A.Engler, Pflanzenr., IV, 147: 85 (1913), nom. inval.
Western Australia. 50 WAU. Cham.
Monotaxis ericoides Klotzsch in J.G.C.Lehmann, Pl. Preiss. 1: 177 (1845).
Monotaxis bracteata Nees ex Klotzsch in J.G.C.Lehmann, Pl. Preiss. 2: 230 (1848).

Hippocrepandra neesiana Müll.Arg., Linnaea 34: 62 (1865). *Monotaxis neesiana* (Müll.Arg.) Baill., Adansonia 6: 293 (1866).
Croton rosmarinifolius Drumm. ex Müll.Arg. in A.P.de Candolle, Prodr. 15(2): 208 (1866), nom. illeg.
Monotaxis grandiflora var. *obtusifolia* F.Muell. & Tate, Trans. Roy. Soc. South Australia 16: 341 (1892).
Monotaxis grandiflora var. *minor* Ewart, Proc. Roy. Soc. Victoria 22(1): 17 (1909).

Monotaxis linifolia Brongn. in L.I.Duperrey, Voy. Monde: 223 (1829). *Monotaxis linifolia* var. *genuina* Müll.Arg. in A.P.de Candolle, Prodr. 15(2): 212 (1866), nom. inval.
– FIGURE, p. 1206.
New South Wales, SE. Queensland. 50 NSW QLD. Cham.
Monotaxis tridentata Endl., Atakta Bot.: 8 (1833). *Monotaxis linifolia* var. *tridentata* (Endl.) Müll.Arg., Linnaea 34: 63 (1865).
Monotaxis simplex Brongn. ex Steud., Nomencl. Bot., ed. 2, 2: 158 (1841).
Monotaxis linifolia var. *cuneata* Grüning in H.G.A.Engler, Pflanzenr., IV, 147: 81 (1913).

Monotaxis lurida (Müll.Arg.) Benth., Fl. Austral. 6: 80 (1973).
Western Australia. 50 WAU. Cham.
* *Hippocrepandra lurida* Müll.Arg., Linnaea 34: 61 (1865).
Monotaxis oldfieldii Baill., Adansonia 6: 293 (1866).

Monotaxis luteiflora F.Muell., Fragm. 10: 51 (1876).
Western Australia, W. South Australia. 50 SOA WAU. Cham.

Monotaxis macrophylla Benth., Fl. Austral. 6: 79 (1873).
Queensland, New South Wales. 50 NSW QLD. Ther.

Monotaxis megacarpa F.Muell., Fragm. 4: 143 (1864).
Western Australia. 50 WAU. Cham.

Monotaxis occidentalis Endl. in S.L.Endlicher & al., Enum. Pl.: 19 (1837). *Monotaxis linifolia* var. *occidentalis* (Endl.) Müll.Arg., Linnaea 34: 63 (1865).
Western Australia. 50 WAU. Cham.
Monotaxis cuneifolia Klotzsch in J.G.C.Lehmann, Pl. Preiss. 2: 229 (1848).
Monotaxis porantheroides F.Muell. ex Benth., Fl. Austral. 6: 80 (1873).

Monotaxis paxii Grüning in H.G.A.Engler, Pflanzenr., IV, 147: 85 (1913).
Western Australia. 50 WAU. Cham.

Monotaxis stowardii S.Moore, J. Linn. Soc., Bot. 45: 192 (1920).
Western Australia. 50 WAU.

Monotaxis tenuis Airy Shaw, Muelleria 4: 239 (1980).
Northern Territory, Queensland. 50 NTA QLD.

Synonyms:
Monotaxis bracteata Nees ex Klotzsch === **Monotaxis grandiflora** Endl.
Monotaxis cuneifolia Klotzsch === **Monotaxis occidentalis** Endl.
Monotaxis ericoides Klotzsch === **Monotaxis grandiflora** Endl.
Monotaxis gracilis var. *genuina* Grüning === **Monotaxis gracilis** (Müll.Arg.) Baill.
Monotaxis gracilis var. *virgata* Grüning === **Monotaxis gracilis** (Müll.Arg.) Baill.
Monotaxis grandiflora var. *minor* Ewart === **Monotaxis grandiflora** Endl.
Monotaxis grandiflora var. *obtusifolia* F.Muell. & Tate === **Monotaxis grandiflora** Endl.
Monotaxis grandiflora var. *typica* Grüning === **Monotaxis grandiflora** Endl.

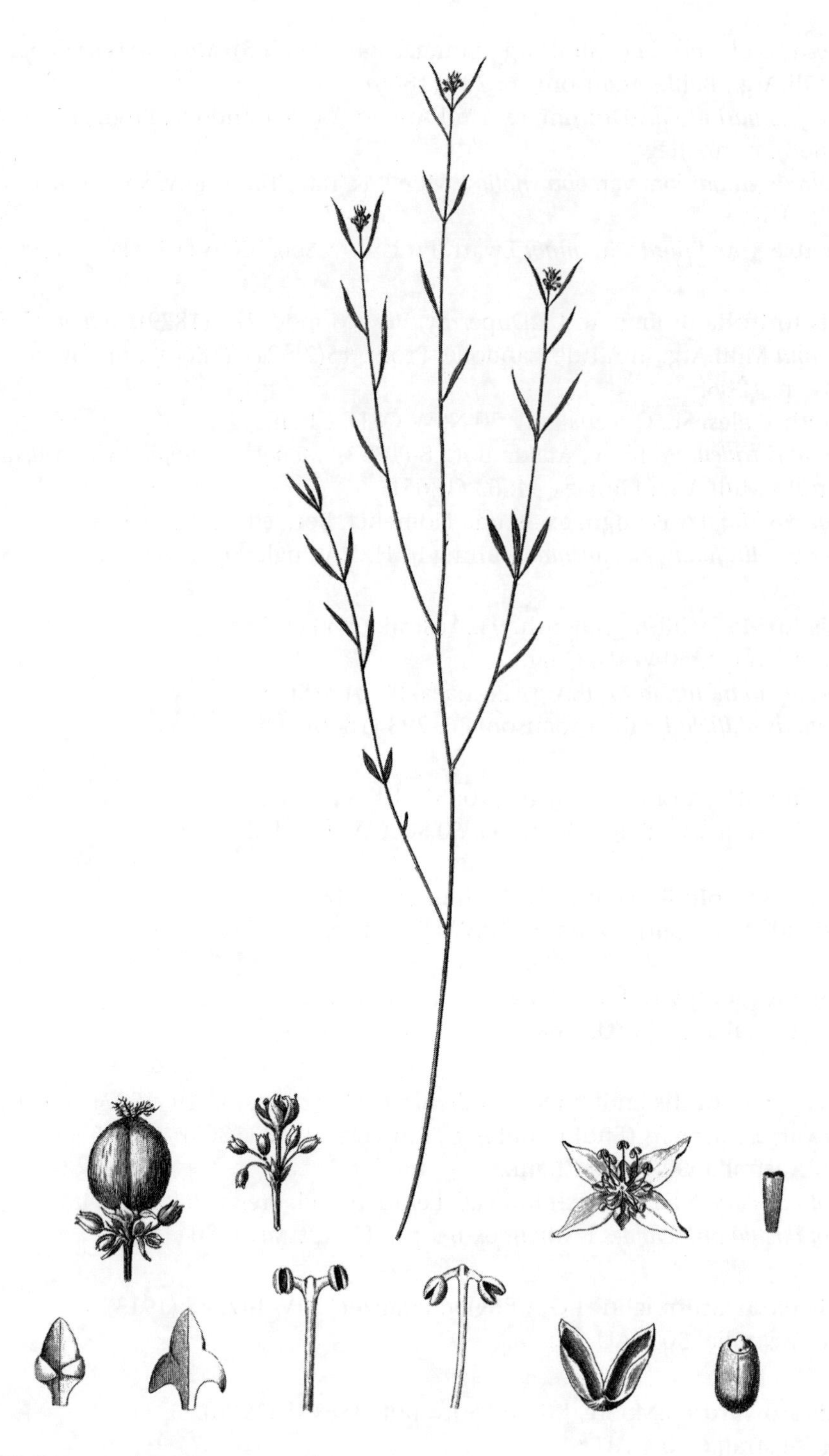

Monotaxis linifolia Brong. in L.I. Duperrey
Artist: Bessa
Duperrey, Voy. Monde, pl. 49, right [1834]
KEW ILLUSTRATIONS COLLECTION

Monotaxis linifolia var. *cuneata* Grüning === **Monotaxis linifolia** Brongn.
Monotaxis linifolia var. *genuina* Müll.Arg. === **Monotaxis linifolia** Brongn.
Monotaxis linifolia var. *occidentalis* (Endl.) Müll.Arg. === **Monotaxis occidentalis** Endl.
Monotaxis linifolia var. *tridentata* (Endl.) Müll.Arg. === **Monotaxis linifolia** Brongn.
Monotaxis neesiana (Müll.Arg.) Baill. === **Monotaxis grandiflora** Endl.
Monotaxis oldfieldii Baill. === **Monotaxis lurida** (Müll.Arg.) Benth.
Monotaxis porantheroides F.Muell. ex Benth. === **Monotaxis occidentalis** Endl.
Monotaxis simplex Brongn. ex Steud. === **Monotaxis linifolia** Brongn.
Monotaxis tridentata Endl. === **Monotaxis linifolia** Brongn.

Moultonianthus

1 species, Malesia (Sumatra, Borneo); shrubs to medium trees with opposite leaves. These plants are distinguished by persistent interpetiolar stipules between the leaves as well as dimorphic inflorescences. Most collections have been made in Borneo. Close relatives include *Erismanthus* and *Syndyophyllum*, as suggested by Webster (Synopsis, 1994) and van Welzen (1995). (Acalyphoideae)

Pax, F. & K. Hoffmann (1919). *Moultonianthus*. In A. Engler (ed.), Das Pflanzenreich, IV 147 XIV (Euphorbiaceae-Additamentum VI): 41-42. Berlin. (Heft 68.) La/Ge. — 1 species, Borneo. [Now known from Sumatra and Borneo.]

• Welzen, P. C. van (1995). Taxonomy and phylogeny of the Euphorbiaceae tribe Erismantheae G. L. Webster (*Erismanthus*, *Moultonianthus*, and *Syndyophyllum*). Blumea 40: 375-396, illus. En. — Treatment of *Moultonianthus*, pp. 384-388, with description, synonymy, references, citations, indication of distribution and habitat, and vernacular names. [Many more collections have been made in Borneo than Sumatra, but *M. leembruggianus*, described from the latter, has priority over the better-known *M. borneensis*.]

Moultonianthus Merr., Philipp. J. Sci., C 11: 70 (1916).
W. Malesia. 42.

Moultonianthus leembruggianus (Boerl. & Koord.) Steenis, Bull. Jard. Bot. Buitenzorg, III, 17: 405 (1948).
EC. Sumatera, Borneo. 42 BOR SUM. Phan.
* *Erismanthus leembruggianus* Boerl. & Koord. in A.Koorders-Schumacher, Syst. Verz. 2: 30 (1910).
Moultonianthus borneensis Merr., Philipp. J. Sci., C 11: 70 (1916).

Synonyms:
Moultonianthus borneensis Merr. === **Moultonianthus leembruggianus** (Boerl. & Koord.) Steenis

Mozinna

Synonyms:
Mozinna Ortega === **Jatropha** L.
Mozinna canescens Benth. === **Jatropha cinerea** (Ortega) Müll.Arg.
Mozinna cardiophylla Torr. === **Jatropha cardiophylla** (Torr.) Müll.Arg.
Mozinna cinerea Ortega === **Jatropha cinerea** (Ortega) Müll.Arg.
Mozinna cordata Ortega === **Jatropha cordata** (Ortega) Müll.Arg.
Mozinna pauciflora Rose === **Jatropha neopauciflora** Pax
Mozinna spathulata Ortega === **Jatropha dioica** Cerv.
Mozinna spathulata var. *sessiliflora* Hook. === **Jatropha dioica** Cerv.

Muricococcum

1 species, S. China; trees to 9 m in and outside of forest with alternate leaves and apparently terminal inflorescences, the staminate flowers in heads. The genus was referred by Kostermans (1961) to *Cephalomappa*, a move followed until a further review by Widuri & van Welzen (1998). It appears to be most closely allied with *Koilodepas* (Epiprininae), Kostermans having dismissed the original authors' suggestion that there was also a relationship with *Chaetocarpus*. (Acalyphoideae)

Chun, W.-Y. & How, F.-C. (1956). Species novarum arborum utilium Chinae meridionalis. Acta Phytotax. Sin. 5: 1-18, 7 pls. Ch. — Includes (pp. 14-16) protologue of the genus *Muricococcum* and description of the single species, *M. sinense*, from Guangxi (illus., pl. 6); assigned by the authors to subtribe Mercurialinae with a relationship to *Chaetocarpus* and *Koilodepas*.

Kostermans, A.J.G.H. (1961). The genus *Muricococcum* Chun & How (Euphorbiaceae). Reinwardtia 5: 413. En. — Reduction to *Cephalomappa*, based on comparison with *C. mallotricarpa*; close relationship with *Koilodepas* confirmed but *Chaetocarpus* not so.

Widuri, R. & P. van Welzen (1998). A revision of the genus *Cephalomappa* (Euphorbiaceae) in Malesia. Reinwardtia 11: 153-184, illus., maps. En. — Includes arguments for re-instatement of *Muricococcum* on grounds particularly of the position of the inflorescences (from the original description and illustration, no authentic material having been seen).

Muricococcum Chun & F.C.How, Acta Phytotax. Sin. 5: 14 (1956).
SE. China. 36.

Muricococcum sinense Chun & F.C.How, Acta Phytotax. Sin. 5: 15 (1956). *Cephalomappa sinensis* (Chun & F.C.How) Kosterm., Reinwardtia 5: 413 (1961).
SE. China. 36 CHS. Phan.

Murtekias

Synonyms:
Murtekias Raf. === **Euphorbia** L.

Myladenia

1 species, SE. Asia (Thailand); small tree to 7 m in riverine forests on sandstone, with distichously arranged, coarsely serrate leaves and tufts of flowers in their axils. [The genus was originally referred, with some doubt, to Phyllantheae subtribe Dissiliariinae sensu Pax & Hoffmann (1931), a group now, with some changes in circumscription, in Oldfieldioideae. It is, however, not accounted for by Webster in his Synopsis (1994).] (Oldfieldioideae)

Airy-Shaw, H. K. (1977). Additions and corrections to Euphorbiaceae of Siam. Kew Bull. 32: 69-83. En. — Includes (p. 79-80) protologue of *Myladenia* with description of one new species, *M. serrata* (described from staminate plants).

Myladenia Airy Shaw, Kew Bull. 32: 79 (1977).
Thailand. 41. Phan.

Myladenia serrata Airy Shaw, Kew Bull. 32: 79 (1977).
Thailand. 41 THA. Phan.

Myricanthe

1 species, New Caledonia; shrubs of maquis with small, narrow opposite leaves and terminal inflorescences. The genus is in the Webster system considered to be related to *Cocconerion*, also in New Caledonia, and the Australian *Bertya*, a former 'stenoloboid' genus. (Crotonoideae)

Airy-Shaw, H. K. (1980). Notes on Euphorbiaceae from Indomalesia, Australia and the Pacific, CCXXXV. *Myricanthe* Airy Shaw, a new genus from New Caledonia. Kew Bull. 35: 390-392. En. — Protologue of genus and description of *M. discolor* and a variety, *viridis*, from maquis in New Caledonia.

McPherson, G. & C. Tirel (1987). *Myricanthe*. Fl. Nouvelle-Calédonie, 14 (Euphorbiacées, I): 72-74. Paris. Fr. — Flora treatment (1 endemic species).

Myricanthe Airy Shaw, Kew Bull. 35: 390 (1980).
New Caledonia. 60.

Myricanthe discolor Airy Shaw, Kew Bull. 35: 390 (1980).
NW. New Caledonia. 60 NWC. Nanophan.
Myricanthe discolor var. *viridis* Airy Shaw, Kew Bull. 35: 391 (1980).

Synonyms:
Myricanthe discolor var. *viridis* Airy Shaw === **Myricanthe discolor** Airy Shaw

Myriogomphus

Synonyms:
Myriogomphus Didr. === **Croton** L.

Nageia

A later homonym of *Nageia* Gaertn. (Podocarpaceae).

Synonyms:
Nageia Roxb. === **Putranjiva** Wall.
Nageia putranjiva Roxb. === **Putranjiva roxburghii** Wall.

Nanopetalum

Synonyms:
Nanopetalum Hassk. === **Cleistanthus** Hook.f. ex Planch.
Nanopetalum myrianthum Hassk. === **Cleistanthus myrianthus** (Hassk.) Kurz

Nealchornea

2 species, Amazonian S. America (Colombia, Peru, Brazil); small to medium forest trees to 20 m or so with clear sap, the leaves spirally arranged and long-petioled. The genus was long considered monotypic, but an additional species was described in 1991. By Webster (Synopsis, 1994) grouped with the African *Hamilcoa* on account of similar pollen grains, with both placed in Pax & Hoffmann's subtribe Hamilcoinae; this was assigned to Stomatocalyceae which was separated in 1975 from Hippomaneae. To the writer of these lines the flushes are very reminiscent of the eastern Old World *Pimelodendron*, also in Stomatocalyceae but referred to its other subtribe, Stomatocalycinae; indeed, whether or not the two subtribes are really that distinct is open to question. (Euphorbioideae (except Euphorbieae))

Pax, F. & K. Hoffmann (1919). *Nealchornea.* In A. Engler (ed.), Das Pflanzenreich, IV 147 XIV (Euphorbiaceae-Additamentum VI): 51. Berlin. (Heft 68.) La/Ge. — 1 species, Amazonian S. America.

Nealchornea Huber, Bol. Mus. Goeldi Paraense Hist. Nat. Ethnogr. 7: 297 (1913).
S. Trop. America. 83 84.

Nealchornea stipitata B.Walln., Linzer Biol. Beitr. 23: 777 (1991).
Brazil (Amazonas). 84 BZN.

Nealchornea yapurensis Huber, Bol. Mus. Goeldi Paraense Hist. Nat. Ethnogr. 7: 298 (1913).
E. Colombia, Peru, Brazil (Amazonas). 83 CLM PER 84 BZN. Phan.

Necepsia

3 species, tropical Africa (3), Madagascar (1); includes *Neopalissya* (Bouchat & Léonard 1986). Shrubs or small trees to 12 m in forest understorey with more or less loosely spirally arranged leaves, these distinguished at the base by conspicuous stipules (their shape and relative persistence being characters of taxonomic value). Each of the recognised species is distinguished by one or more infraspecific taxa (two of these latter – one of them endemic – being in Madagascar), the whole forming a scattered set of 'stations' from west to east and south. The genus was referred by Webster (Synopsis, 1994) to Bernardieae. (Acalyphoideae)

Pax, F. (with K. Hoffmann) (1914). *Necepsia*. In A. Engler (ed.), Das Pflanzenreich, IV 147 VII (Euphorbiaceae-Acalypheae-Mercurialinae): 16-17. Berlin. (Heft 63.) La/Ge. — 1 species, W. and C. Africa.

Pax, F. (with K. Hoffmann) (1914). *Neopalissya*. In A. Engler (ed.), Das Pflanzenreich, IV 147 VII (Euphorbiaceae-Acalypheae-Mercurialinae): 16. Berlin. (Heft 63.) La/Ge. — 1 species, Madagascar. [Now combined with *Necepsia*.]

- Bouchat, A. & J. Léonard (1986). Révision du genre *Necepsia* Prain (Euphorbiacée africano-malgache). Bull. Jard. Bot. Natl. Belg. 56: 179-194, illus., map. Fr. — Detailed descriptive treatment (3 species and some additional infraspecific taxa) with key, synonymy, references and citations, types, observations, and indication of distribution (map, p. 182), vernacular names, uses, and localities with exsiccatae; illustration of *N. zairensis* (here newly described). *N. castaneifolia* subsp. *castaneifolia* and *capuronii* are the sole taxa in Madagascar. [The genus is widely scattered and disjunct, with for the most part vicariant taxa. A phylogenetic analysis, not essayed here, would be of interest.]

Necepsia Prain, Bull. Misc. Inform. Kew 1910: 343 (1910).
Trop. Africa, Madagascar. 22 23 25 26 29.
Palissya Baill., Étude Euphorb.: 502 (1858).
Neopalissya Pax in H.G.A.Engler, Pflanzenr., IV, 147, VII: 16 (1914).

Necepsia afzelii Prain, Bull. Misc. Inform. Kew 1910: 343 (1910).
W. & WC. Trop. Africa. 22 GHA IVO LBR SIE 23 CMN CON GAB. Nanophan. or phan.

subsp. **afzelii**
Sierra Leone, Liberia, Ghana, Ivory Coast. 22 GHA IVO LBR SIE. Nanophan. or phan.

var. **sitae** Bouchat & J.Léonard, Bull. Jard. Bot. Belg. 56: 186 (1986).
Congo. 23 CON.

subsp. **zenkeri** Bouchat & J.Léonard, Bull. Jard. Bot. Belg. 56: 186 (1986).
Cameroon, Gabon. 23 CMN GAB.

Necepsia castaneifolia (Baill.) Bouchat & J.Léonard, Bull. Jard. Bot. Belg. 56: 191 (1986).
Tanzania, Zimbabwe, Madagascar. 25 TAN 26 ZIM 29 MDG. Phan.
* *Palissya castaneifolia* Baill., Étude Euphorb.: 503 (1858). *Alchornea castaneifolia* (Baill.) Müll.Arg., Linnaea 34: 167 (1865), nom. illeg. *Neopalissya castaneifolia* (Baill.) Pax in H.G.A.Engler, Pflanzenr., IV, 147, VII: 16 (1911).

subsp. **capuronii** Bouchat & J.Léonard, Bull. Jard. Bot. Belg. 56: 194 (1986).
Madagascar. 29 MDG. Phan.

subsp. **castaneifolia**
Madagascar. 29 MDG. Phan.
Alchornea madagascariensis Müll.Arg. in A.P.de Candolle, Prodr. 15(2): 900 (1866).

subsp. **chirindica** (Radcl.-Sm.) Bouchat & J.Léonard, Bull. Jard. Bot. Belg. 56: 194 (1986).
Zimbabwe. 26 ZIM. Phan.
* *Neopalissya castaneifolia* subsp. *chirindica* Radcl.-Sm., Kew Bull. 39: 791 (1984).

subsp. **kimbozensis** (Radcl.-Sm.) Bouchat & J.Léonard, Bull. Jard. Bot. Belg. 56: 193 (1986).
Tanzania (Kimboza forest). 25 TAN. Phan.
* *Neopalissya castaneifolia* subsp. *kimbozensis* Radcl.-Sm., Kew Bull. 39: 792 (1984).

Necepsia zairensis Bouchat & J.Léonard, Bull. Jard. Bot. Belg. 56: 187 (1986).
Zaire. 23 ZAI. Phan.

var. **lujae** Bouchat & J.Léonard, Bull. Jard. Bot. Belg. 56: 190 (1986).
Zaire. 23 ZAI. Phan.

var. **zairensis**
Zaire. 23 ZAI. Phan.

Nellica

Synonyms:
Nellica Raf. === **Phyllanthus** L.

Neoboutonia

3 species, Africa, all variable and of relatively wide distribution. Small to medium trees of primary and secondary forest or regrowth or in open places to 20 m with palmately veined leaves, conspicuous persistent stipules and large axillary inflorescences; often near water or in wet ground. The genus is by Webster referred (along with *Benoistia*) to a unique subtribe, Neoboutonieae, in the Aleuritideae. A new revision would be desirable. [For illustrations, see Ic. Pl. 13: pls. 1298/1299 (1879).] (Crotonoideae)

Pax, F. (with K. Hoffmann) (1914). *Neoboutonia*. In A. Engler (ed.), Das Pflanzenreich, IV 147 VII (Euphorbiaceae-Acalypheae-Mercurialinae): 71-75. Berlin. (Heft 63.) La/Ge. — 3 species, Africa.

Neoboutonia Müll.Arg., J. Bot. 2: 336 (1864).
Trop. Africa. 22 23 24 25 26.

Neoboutonia macrocalyx Pax, Bot. Jahrb. Syst. 30: 339 (1901).
Zaire, Rwanda, Burundi, Uganda, Kenya, Tanzania, Malawi, Zambia, Zimbabwe. 23 BUR RWA ZAI 25 KEN TAN UGA 26 MLW ZAM ZIM. Phan.

Neoboutonia mannii Benth. & Hook.f., Gen. Pl. 3: 317 (1879). *Neoboutonia africana* var. *mannii* (Benth. & Hook.f.) Pax & K.Hoffm. in H.G.A.Engler, Pflanzenr., IV, 147, VII: 75 (1914).
Trop. Africa. 22 GUI LBR NGA 23 CMN EQG GGI ZAI 25 UGA 26 MLW MOZ ZAM. Phan.
Conceveiba africana Müll.Arg., Flora 33: 530 (1865). *Neoboutonia africana* (Müll.Arg.) Pax in H.G.A.Engler & K.A.E.Prantl, Nat. Pflanzenfam. 3(5): 57 (1890), nom. illeg.
Neoboutonia diaguissensis Beille, Bull. Soc. Bot. France 57(8): 125 (1910). *Neoboutonia africana* var. *diaguissensis* (Beille) Pax & K.Hoffm. in H.G.A.Engler, Pflanzenr., IV, 147, VII: 75 (1914).
Neoboutonia glabrescens Prain, Bull. Misc. Inform. Kew 1911: 265 (1911). *Neoboutonia africana* var. *glabrescens* (Prain) Pax & K.Hoffm. in H.G.A.Engler, Pflanzenr., IV, 147, VII: 75 (1914).

Neoboutonia melleri (Müll.Arg.) Prain, Bull. Misc. Inform. Kew 1911: 266 (1911).
– FIGURE, p. 1212.
Trop. Africa. 22 NGA 23 CMN GAB ZAI 24 SUD 25 KEN TAN UGA 26 ANG MLW MOZ ZAM ZIM? Phan.
Neoboutonia africana Müll.Arg., J. Bot. 2: 336 (1 Nov. 1864).
Croton niloticus Müll.Arg., Flora 47: 587 (1864).

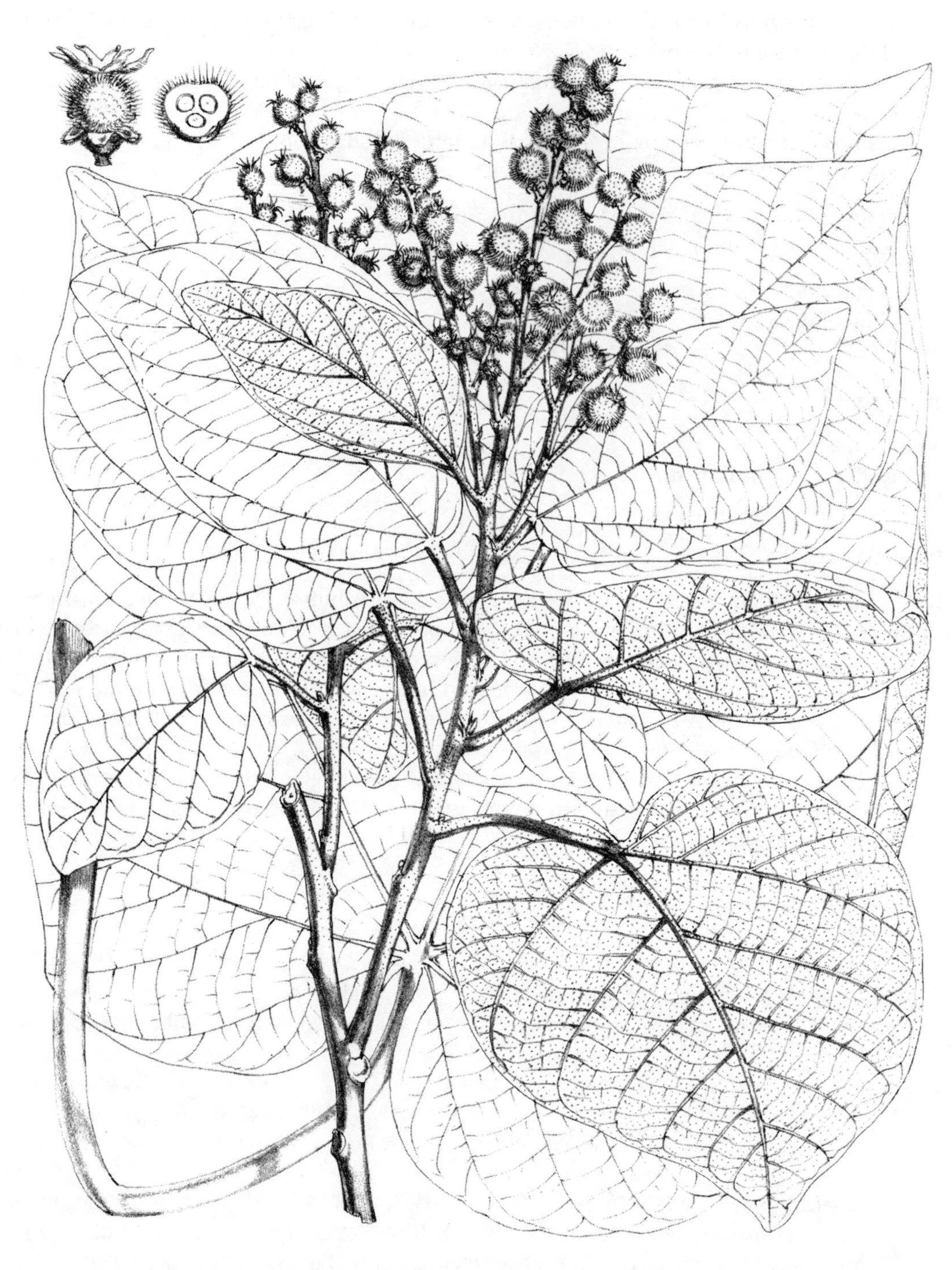

Neoboutonia melleri (Muell Arg.) Prain (as *Croton niloticus* Müll. Arg.)
Artist: W.H. Fitch
Trans. Linn. Soc. 29: pl. 95 (1873)
KEW ILLUSTRATIONS COLLECTION

* *Mallotus melleri* Müll.Arg., Flora 47: 468 (1864). *Neoboutonia melleri* var. *genuina* Pax & K.Hoffm. in H.G.A.Engler, Pflanzenr., IV, 147, VII: 74 (1914), nom. inval.

Neoboutonia canescens Pax, Bot. Jahrb. Syst. 19: 91 (1894). *Neoboutonia melleri* var. *canescens* (Pax) Pax in H.G.A.Engler, Pflanzenr., IV, 147, VII: 74 (1914).

Neoboutonia chevalieri Beille, Bull. Soc. Bot. France 55(8): 77 (1908).

Neoboutonia velutina Prain, Bull. Misc. Inform. Kew 1911: 266 (1911). *Neoboutonia melleri* var. *velutina* (Prain) Pax & K.Hoffm. in H.G.A.Engler, Pflanzenr., IV, 147, VII: 74 (1914).

Synonyms:
Neoboutonia africana (Müll.Arg.) Pax === **Neoboutonia mannii** Benth. & Hook.f.
Neoboutonia africana Müll.Arg. === **Neoboutonia melleri** (Müll.Arg.) Prain
Neoboutonia africana var. *diaguissensis* (Beille) Pax & K.Hoffm. === **Neoboutonia mannii** Benth. & Hook.f.
Neoboutonia africana var. *glabrescens* (Prain) Pax & K.Hoffm. === **Neoboutonia mannii** Benth. & Hook.f.
Neoboutonia africana var. *mannii* (Benth. & Hook.f.) Pax & K.Hoffm. === **Neoboutonia mannii** Benth. & Hook.f.
Neoboutonia canescens Pax === **Neoboutonia melleri** (Müll.Arg.) Prain
Neoboutonia chevalieri Beille === **Neoboutonia melleri** (Müll.Arg.) Prain
Neoboutonia diaguissensis Beille === **Neoboutonia mannii** Benth. & Hook.f.
Neoboutonia glabrescens Prain === **Neoboutonia mannii** Benth. & Hook.f.
Neoboutonia melleri var. *canescens* (Pax) Pax === **Neoboutonia melleri** (Müll.Arg.) Prain
Neoboutonia melleri var. *genuina* Pax & K.Hoffm. === **Neoboutonia melleri** (Müll.Arg.) Prain
Neoboutonia melleri var. *velutina* (Prain) Pax & K.Hoffm. === **Neoboutonia melleri** (Müll.Arg.) Prain
Neoboutonia musculiformis K.Schum. === **Neuburgia moluccana** (Boerl.) Leenh. (Strychnaceae)
Neoboutonia velutina Prain === **Neoboutonia melleri** (Müll.Arg.) Prain

Neochevaliera

Synonyms:
Neochevaliera A.Chev. & Beille === **Chaetocarpus** Thwaites

Neogoetzia

A Pax synonym of *Bridelia*.

Neoguillauminia

1 species, New Caledonia. A shrub or tree to 30 m of striking form in forest or (sometimes) maquis, evidently deserving of its specific epithet. The spirally arranged leaves are more or less in rosettes. It may well represent an 'outgroup' of *Euphorbia* s.l. In contrast to that genus and its immediate allies, the pseudopetals (petaloid appendages) are here a development of involucral bracts and do not arise from interbracteal glands; moreover, the cyathial glands are not on the rim of the tube. (Euphorbioideae (Euphorbieae))

Croizat, L. (1938). Notes on Euphorbiaceae, with a new genus and a new subtribe of the Euphorbieae. Philip. J. Sci. 64: 397-411, pl. 1. En. — Includes protologue of *Neoguillauminia* (and transfer therein of *Euphorbia cleopatra*) along with a diagnosis of subtribe Neoguillauminiinae (by the author also to include the African *Dichostemma*, an opinion not currently accepted). The rest of the paper is on the nature of the cyathium and, by extension, relationships within the Euphorbieae.

Croizat, L. (1942). Notes on the Euphorbiaceae, II. Bull. Bot. Gard. Buitenzorg, III, 17: 204-208. En. — Among the two contributions in this paper is one on the fruit and seeds of *Neoguillauminia*.

McPherson, G. & C. Tirel (1987). *Neoguillauminia*. Fl. Nouvelle-Calédonie, 14 (Euphorbiacées, I): 22-25. Paris. Fr. — Flora treatment (1 species, *N. cleopatra*, mainly in southern part of Grande-Terre; a shrub or tree to 30 m).

Neoguillauminia Croizat, Philipp. J. Sci. 64: 398 (1937 publ. 1938).
New Caledonia. 60.
Cleopatra Pancher ex Croizat, Philipp. J. Sci. 64: 398 (1937 publ. 1938).

Neoguillauminia cleopatra (Baill.) Croizat, Philipp. J. Sci. 64: 388 (1937 publ. 1938).
New Caledonia. 60 NWC. Nanophan. or phan.
* *Euphorbia cleopatra* Baill., Adansonia 2: 213 (1862).

Neoholstia

Recently reduced to *Tannodia* (Radcliffe-Smith 1998; see that genus).

Synonyms:
Neoholstia Rauschert === **Tannodia** Baill.
Neoholstia sessiliflora (Pax) Rauschert === **Tannodia tenuifolia** var. **glabrata** Prain
Neoholstia tenuifolia (Pax) Rauschert === **Tannodia tenuifolia** (Pax) Prain
Neoholstia tenuifolia var. *glabrata* (Prain) Radcl.-Sm. === **Tannodia tenuifolia** var. **glabrata** Prain

Neojatropha

Synonyms:
Neojatropha Pax === **Mildbraedia** Pax
Neojatropha carpinifolia (Pax) Pax === **Mildbraedia carpinifolia** (Pax) Hutch.
Neojatropha fallax (Pax) Pax === **Mildbraedia carpinifolia** (Pax) Hutch. var. **carpinifolia**

Neomanniophyton

Synonyms:
Neomanniophyton Pax & K.Hoffm. === **Crotonogyne** Müll.Arg.
Neomanniophyton chevalieri Beille === **Crotonogyne chevalieri** (Beille) Keay
Neomanniophyton ledermannianum Pax & K.Hoffm. === **Crotonogyne ledermanniana** (Pax & K.Hoffm.) Pax & K.Hoffm.
Neomanniophyton stenophyllum Pax === **Crotonogyne angustifolia** Pax
Neomanniophyton zenkeri f. *basicaudatum* Pax & K.Hoffm. === **Crotonogyne zenkeri** Pax
Neomanniophyton zenkeri f. *dasyanthum* Pax & K.Hoffm. === **Crotonogyne zenkeri** Pax
Neomanniophyton zenkeri f. *fallax* Pax & K.Hoffm. === **Crotonogyne zenkeri** Pax
Neomanniophyton zenkeri f. *glabratum* Pax & K.Hoffm. === **Crotonogyne zenkeri** Pax

Neomphalea

Synonyms:
Neomphalea Pax & K.Hoffm. === **Omphalea** L.

Neopalissya

Synonyms:
Neopalissya Pax === **Necepsia** Prain
Neopalissya castaneifolia (Baill.) Pax === **Necepsia castaneifolia** (Baill.) Bouchat & J.Léonard
Neopalissya castaneifolia subsp. *chirindica* Radcl.-Sm. === **Necepsia castaneifolia** subsp. **chirindica** (Radcl.-Sm.) Bouchat & J.Léonard
Neopalissya castaneifolia subsp. *kimbozensis* Radcl.-Sm. === **Necepsia castaneifolia** subsp. **kimbozensis** (Radcl.-Sm.) Bouchat & J.Léonard

Neopeltandra

Synonyms:
Neopeltandra Gamble === **Meineckia** Baill.

Neopycnocoma

Synonyms:
Neopycnocoma Pax === **Argomuellera** Pax
Neopycnocoma lancifolia Pax === **Argomuellera lancifolia** (Pax) Pax

Neoroepera

2 species, Australia (CE. and NE. Queensland: cf. map in Henderson, 1992); shrubs to 2 m or so in duneland, heathland, woodland or open forest with small, closely spaced, narrowly elliptic or obovate leaves. *N. buxifolia* may be locally dominant within its area of distribution. The genus (never revised for *Das Pflanzenreich*) is fully treated in the cited paper by Henderson, who relates it to *Micrantheum* (in Webster's system assigned to the same subtribe, Pseudanthinae, in Caletieae). However, the genus may be heterogeneous (Levin & Simpson 1994; see Oldfieldioideae). (Oldfieldioideae)

Airy-Shaw, H. K. (1980). *Neoroepera*. Kew Bull. 35: 658-659. (Euphorbiaceae-Platylobeae of Australia.) En. — Treatment of 2 species. [Succeeded by Henderson, 1992.]
• Henderson, R. J. F. (1992). Studies in Euphorbiaceae A. L. Juss. sens. lat., 2: A revision of *Neoroepera* Muell.-Arg. et F. Muell. (Oldfieldioideae Köhler et Webster, Caletieae Muell.-Arg.). Austrobaileya 3: 615-625, illus, map. En. — Full treatment (2 species) with key, descriptions, synonymy, references, types, indication of distribution and habitat, localities with exsiccatae, and commentary along with figures and a map. [The two species are mutually disjunct in southern-central and northern Queensland. The genus is apparently related to *Micrantheum* as first recognised by F. von Mueller in 1871.]

Neoroepera Müll.Arg. in A.P.de Candolle, Prodr. 15(2): 488 (1866).
Queensland. 50. Phan.

Neoroepera banksii Benth., Fl. Austral. 6: 117 (1873).
Queensland (Cook). 50 QLD. Phan.
Phyllanthus banksii A.Cunn. ex Benth., Fl. Austral. 6: 117 (1873).

Neoroepera buxifolia Müll.Arg. in A.P.de Candolle, Prodr. 15(2): 489 (1866).
Queensland (Port Curtis). 50 QLD. Phan.
Securinega muelleriana Baill., Adansonia 6: 333 (1866).

Neoscortechinia

6 species, Nicobar Is. and SE. Asia to the Philippines and Solomons (van Welzen 1994); includes *Alcinaeanthus*. Small to large trees of well-drained (or swamp) forest with coarse foliage, sometimes reaching to as much as 35 m. The genus is most closely allied to *Cheilosa* which occurs within its range; the two form tribe Cheiloseae in the Webster system. [Formerly *Scortechinia* Hook. f., non Sacc.; its type species, the former *Scortechinia kingii*, is illustrated in Ic. Pl. 18: pl. 1706 (1887).] (Acalyphoideae)

Pax, F. (with K. Hoffmann) (1914). *Alcineanthus*. In A. Engler (ed.), Das Pflanzenreich, IV 147 VII [Euphorbiaceae-Additamentum V]: 415. Berlin. (Heft 63.) La/Ge. — 1 species, Philippines. (Now in *Neoscortechinia*).
Airy-Shaw, H. K. (1963). Notes on Malaysian and other Asiatic Euphorbiaceae, XLV. A synopsis of the genus *Neoscortechinia* Pax and K. Hoffm. Kew Bull. 16: 368-371. En. — Synoptic treatment with key, synonymy, localities with exsiccatae, and commentary. [Now superseded.]
Airy-Shaw, H. K. (1966). Notes on Malaysian and other Asiatic Euphorbiaceae, LXXX. *Neoscortechinia* Pax et K. Hoffm. in New Guinea. Kew Bull. 20: 413-414. En. — Extension of range.

Airy-Shaw, H. K. (1972). Notes on Malesian and other Asiatic Euphorbiaceae, CLXVIII: Note on *Neoscortechinia forbesii*. Kew Bull. 27: 93. En. — Revised authority.

- Welzen, P.C. van (1994). Taxonomy, phylogeny and geography of *Neoscortechinia* Hook. f. ex Pax (Euphorbiaceae). Blumea 39: 301-318, illus., maps. En. — Introduction; character and phylogenetic analysis (the latter 'rather difficult'); biogeography; treatment of 6 species (including one new combination) with key, descriptions, synonymy, literature, indication of distribution, habitat, biology, uses and vernacular names, illustrations, maps and notes; list of exsiccatae at end.

Neoscortechinia Pax in H.G.A.Engler & K.A.E.Prantl, Nat. Pflanzenfam., Nachtr. 1: 213 (1897).
Trop. Asia to Solomon Is. 41 42 60.
* *Scortechinia* Hook.f., Fl. Brit. India 5: 366 (1887), nom. illeg.
Alcinaeanthus Merr., Philipp. J. Sci., C 7: 379 (1912 publ. 1913).

Neoscortechinia angustifolia (Airy Shaw) Welzen, Blumea 39: 309 (1994).
Borneo (Sabah, Kalimantan). 42 BOR. Phan.
* *Neoscortechinia sumatrensis* var. *angustifolia* Airy Shaw, Kew Bull. 16: 368 (1963).

Neoscortechinia forbesii (Hook.f.) S.Moore, J. Bot. 62: 54 (1924).
New Guinea (incl. Admiralty Is.), Bismarck Archip., Solomon Is. 42 BIS NWG 60 SOL. Phan.
* *Scortechinia forbesii* Hook.f., Hooker's Icon. Pl. 18: t. 1706 (1887).

Neoscortechinia kingii (Hook.f.) Pax & K.Hoffm. in H.G.A.Engler, Pflanzenr., IV, 147, XIV: 52 (1919).
Pen. Malaysia, Sumatera, Borneo (Sarawak, Brunei, E. Kalimantan). 42 BOR MLY SUM. Phan.
* *Scortechinia kingii* Hook.f., Hooker's Icon. Pl. 18: t. 1706 (1887).
Neoscortechinia kingii var. *pedicellata* Airy Shaw, Kew Bull. 16: 371 (1963).

Neoscortechinia nicobarica (Hook.f.) Pax & K.Hoffm. in H.G.A.Engler, Pflanzenr., IV, 147, XIV: 53 (1919).
Nicobar Is., Pen. Malaysia, Sumatera, Borneo, Jawa, Philippines (Palawan), Sulawesi, Irian Jaya. 41 NCB 42 BOR JAW MLY NWG PHI SUL SUM. Phan.
* *Scortechinia nicobarica* Hook.f., Hooker's Icon. Pl. 18: t. 1706 (1887).
Alchornea arborea Elmer, Leafl. Philipp. Bot. 4: 1274 (1911). *Neoscortechinia arborea* (Elmer) Pax & K.Hoffm. in H.G.A.Engler, Pflanzenr., IV, 147, XIV: 52 (1919).

Neoscortechinia philippinensis (Merr.) Welzen, Blumea 39: 316 (1994).
Burma, W. & C. Malesia. 41 BMA 42 BOR MLY PHI SUM. Phan.
* *Alcinaeanthus philippinensis* Merr., Philipp. J. Sci., C 7: 380 (1912 publ. 1913).
Alcinaeanthus parvifolius Merr., Philipp. J. Sci., C 9: 461 (1914 publ. 1915).
Neoscortechinia arborea var. *parvifolia* (Merr.) Pax & K.Hoffm. in H.G.A.Engler, Pflanzenr., IV, 147, XIV: 52 (1919). *Neoscortechinia parvifolia* (Merr.) Merr., Enum. Philipp. Fl. Pl. 2: 456 (1923).
Neoscortechinia coriacea Merr., Univ. Calif. Publ. Bot. 15: 164 (1929).

Neoscortechinia sumatrensis S.Moore, J. Bot. 63(Suppl.): 99 (1925).
S. Pen. Malaysia, N. Sumatera (incl. Simeuluë), Borneo. 42 BOR SUM. Phan.

Synonyms:

Neoscortechinia arborea (Elmer) Pax & K.Hoffm. === **Neoscortechinia nicobarica** (Hook.f.) Pax & K.Hoffm.
Neoscortechinia arborea var. *parvifolia* (Merr.) Pax & K.Hoffm. === **Neoscortechinia philippinensis** (Merr.) Welzen
Neoscortechinia coriacea Merr. === **Neoscortechinia philippinensis** (Merr.) Welzen
Neoscortechinia kingii var. *pedicellata* Airy Shaw === **Neoscortechinia kingii** (Hook.f.) Pax & K.Hoffm.

Neoscortechinia parvifolia (Merr.) Merr. === **Neoscortechinia philippinensis** (Merr.) Welzen
Neoscortechinia sumatrensis var. *angustifolia* Airy Shaw === **Neoscortechinia angustifolia** (Airy Shaw) Welzen

Neoshirakia

1 species, C. & S. China, Korea (incl. Cheju Do), Nansei-shoto and C. & S. Japan; formerly in *Sapium* sect. *Parasapium*. Deciduous shrubs to 5 m with slender branches and leaf-opposed inflorescences. The generic name replaces the illegitimate *Shirakia* Hurus. Current thinking (Esser et al., 1997; see **Euphorbioideae (except Euphorbieae)**) suggests a relationship with *Shirakiopsis* and *Gymnanthes* within the Hippomaneae. (Euphorbioideae (except Euphorbieae))

Pax, F. (with K. Hoffmann) (1912). *Sapium*. In A. Engler (ed.), Das Pflanzenreich, IV 147 V (Euphorbiaceae-Hippomaneae): 199-258. Berlin. (Heft 52.) La/Ge. — No. 89 (p. 252), *S. japonicum*, is the type of *Neoshirakia* (=*Sapium* sect. *Parasapium* s.s.).

Esser, H.-J. (1998). *Neoshirakia*, a new name for *Shirakia* Hurus. (Euphorbiaceae). Blumea 43: 129-130. En. — Hurusawa's name antedated by *Shirakia* S. Kawasaki (for a fern fossil); one new combination, *N. japonica*. [Other species formerly in *Shirakia* would be transferred to a new genus, *Shirakiopsis*, and to *Triadica* Lour.]

Neoshirakia Esser, Blumea 43: 129 (1998).
C. & S. China, Temp. E. Asia. 36 38. Phan.
* *Shirakia* Hurus., J. Fac. Sci. Univ. Tokyo, Sect. 3, Bot. 6: 317 (1954), nom. illeg.

Neoshirakia japonica (Siebold & Zucc.) Esser, Blumea 43: 129 (1998).
C. & S. Japan, Korea (incl. Cheju Do), Nansei-shoto, C. & S. China. 36 CHC CHN CHS 38 JAP KOR NNS. Phan.
Croton sirakii Siebold & Zucc., Fl. Jap. Fam. Nat. 1: 36 (1845), nom. nud.
* *Stillingia japonica* Siebold & Zucc., Abh. Math.-Phys. Cl. Königl. Bayer. Akad. Wiss. 4(2): 145 (1846). *Triadica japonica* (Siebold & Zucc.) Baill., Étude Euphorb.: 512 (1858). *Excoecaria japonica* (Siebold & Zucc.) Müll.Arg., Linnaea 32: 123 (1863). *Sapium japonicum* (Siebold & Zucc.) Pax & K.Hoffm. in H.G.A.Engler, Pflanzenr., IV, 147, V: 252 (1912). *Shirakia japonica* (Siebold & Zucc.) Hurus., J. Fac. Sci. Univ. Tokyo, Sect. 3, Bot. 6: 318 (1954).

Neotrewia

1 species, Philippines and Sulawesi; trees to 15 m with coarse mallotoid foliage most usually found along or near rivers and sometimes reported as common. Regarded by Webster (Synopsis, 1994) as being closely related to the geographically vicarious *Trewia* and perhaps not separable from it. (Acalyphoideae)

Pax, F. (with K. Hoffmann) (1914). *Neotrewia*. In A.Engler (ed.), Das Pflanzenreich, IV 147 VII (Euphorbiaceae-Acalypheae-Mercurialinae): 211-212. Berlin. (Heft 63.) La/Ge. — Monotypic, Philippines.

Neotrewia Pax & K.Hoffm. in H.G.A.Engler, Pflanzenr., IV, 147, VII: 211 (1914).
Philippines, Sulawesi. 42.

Neotrewia cumingii (Müll.Arg.) Pax & K.Hoffm. in H.G.A.Engler, Pflanzenr., IV, 147, VII: 212 (1914).
Philippines, Sulawesi, 42 PHI SUL. Phan.
* *Mallotus cumingii* Müll.Arg., Linnaea 34: 195 (1865).
Trewia ambigua Merr., Philipp. J. Sci. 1 (Suppl.): 79 (1906).

Neotrigonostemon

Synonyms:
Neotrigonostemon Pax & K.Hoffm. === **Trigonostemon** Blume

Neowawraea

This distinctive Hawaiian tree was transferred to *Flueggea* by Hayden (1987; see that genus).

Synonyms:
Neowawraea Rock === **Flueggea** Willd.
Neowawraea phyllanthoides Rock === **Flueggea neowawraea** W.J.Hayden

Nepenthandra

Synonyms:
Nepenthandra S.Moore === **Trigonostemon** Blume
Nepenthandra lanceolata S.Moore === **Trigonostemon lanceolatus** (S.Moore) Pax

Nephrostylus

Synonyms:
Nephrostylus Gagnep. === **Koilodepas** Hassk.

Niedenzua

Synonyms:
Niedenzua Pax === **Adenochlaena** Boivin ex Baill.

Niruri

Usually regarded as an infrageneric taxon of *Phyllanthus*.

Synonyms:
Niruri Adans. === **Phyllanthus** L.

Niruris

Synonyms:
Niruris Raf. === **Phyllanthus** L.

Nisomenes

Synonyms:
Nisomenes Raf. === **Euphorbia** L.

Nymania

Currently treated as an infrageneric taxon within *Phyllanthus*.

Synonyms:
Nymania K.Schum. === **Phyllanthus** L.
Nymania insignis K.Schum. === **Phyllanthus clamboides** (F.Muell.) Diels

Nymphanthus

Synonyms:
Nymphanthus Lour. === **Phyllanthus** L.
Nymphanthus chinensis Lour. === **Phyllanthus villosus** Poir.
Nymphanthus niruri (L.) Lour. === **Phyllanthus niruri** L.
Nymphanthus pilosus Lour. === **Glochidion pilosum** (Lour.) Merr.
Nymphanthus ruber Lour. === **Phyllanthus ruber** (Lour.) Spreng.
Nymphanthus squamifolia Lour. === **Phyllanthus squamifolius** (Lour.) Stokes

Ocalia

Synonyms:
Ocalia Klotzsch === **Croton** L.
Ocalia sellowiana Klotzsch === **Croton antisyphiliticus** Mart.

Octospermum

1 species, Malesia (New Guinea); trees of primary or secondary forest to 36 m with pubescent mallotoid foliage and terminal panicles. The single species has a scattered but largely northern distribution from Kepala Burung to the Milne Bay Province of Papua New Guinea. Related to *Mallotus* (particularly sects. *Rottlera* (no. 6 in Pax & Hoffmann) and *Rottleropsis* (nos. 1-2 in Pax & Hoffmann)), but differs in a 8-9-locular gynoecium and fleshy, indehiscent, baccate fruit. The genus is grouped with *Mallotus, Neotrewia* and *Trewia* in subtribe Rottlerinae (tribe Acalypheae) in the Webster system. (Acalyphoideae)

Airy-Shaw, H. K. (1965). Notes on Malaysian and other Asiatic Euphorbiaceae, LIII. A new mallotoid genus from New Guinea. Kew Bull. 19: 311-313. En. — Protologue of *Octospermum* and transfer thereto of *Mallotus pleiogynus* Pax and K. Hoffm., first described from the mid-Sepik basin in Papua New Guinea; formerly comprised *Mallotus* sect. 7, *Pleiogyni*, of Pax (1914: 187; q.v.).
Airy-Shaw, H. K. (1974). *Octospermum pleiogynium* (Pax & Hoffm.) Airy Shaw. Ic. Pl. 38: pl. 3716. En. — Plant portrait with descriptive text.

Octospermum Airy Shaw, Kew Bull. 19: 311 (1965).
New Guinea. 42.

Octospermum pleiogynum (Pax & K.Hoffm.) Airy Shaw, Kew Bull. 19: 312 (1965).
– FIGURE, p. 1220.
New Guinea. 42 NWG. Phan.
* *Mallotus pleiogynus* Pax & K.Hoffm. in H.G.A.Engler, Pflanzenr., IV, 147, VII: 187 (1914).

Odonteilema

Synonyms:
Odonteilema Turcz. === **Acalypha** L.
Odonteilema claussenii Turcz. === **Acalypha claussenii** (Turcz.) Müll.Arg.

Odotalon

Synonyms:
Odotalon Raf. === **Argythamnia** P.Browne

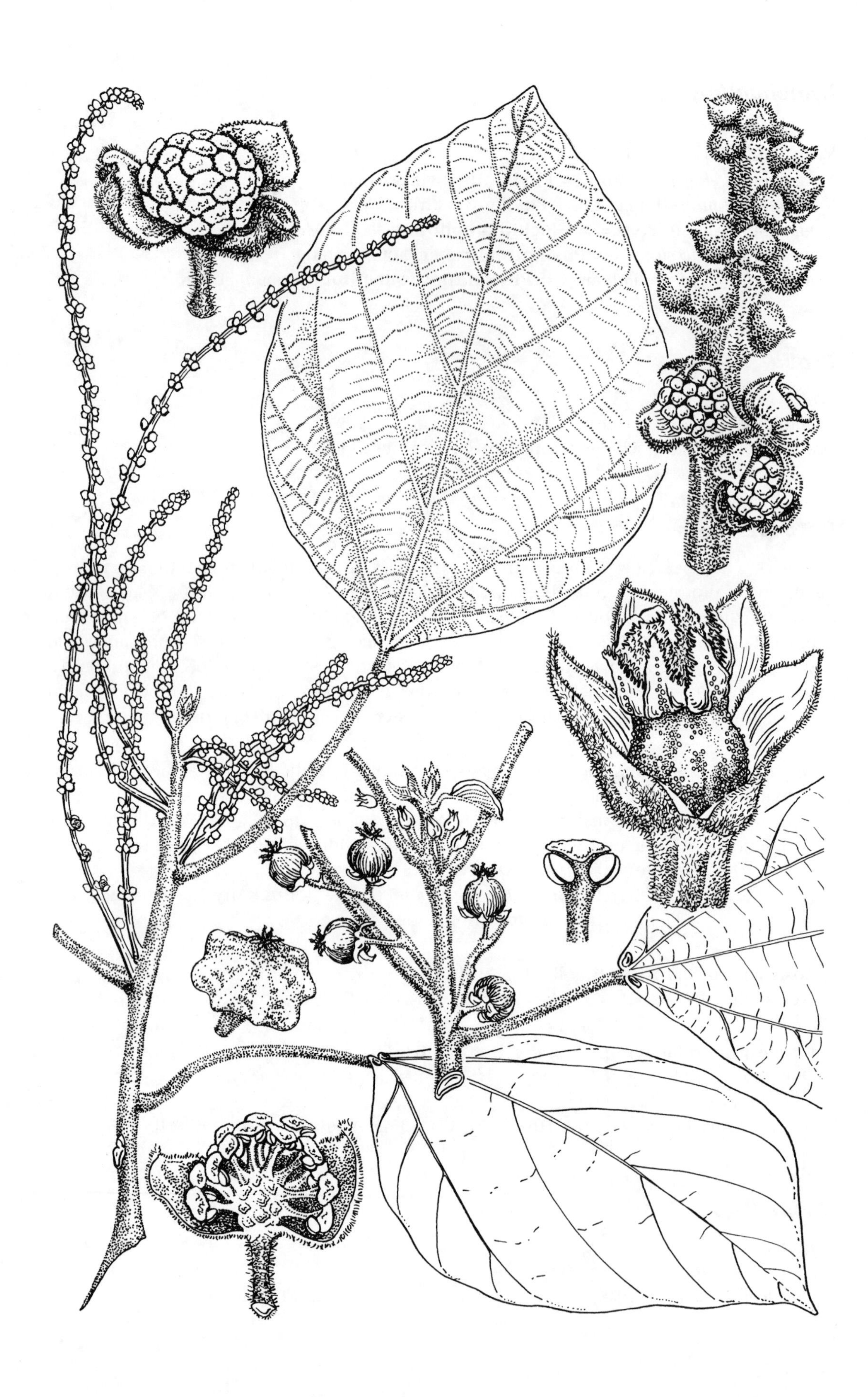

Octospermum pleiogynum (Pax & K. Hoffm.) Airy Shaw
Artist: Mary Grierson
Ic. Pl. 38: pl. 3716 (1974)
KEW ILLUSTRATIONS COLLECTION

Oldfieldia

4 species, tropical Africa (including *Paivaeusa*), the species partly disjunct in distribution. Much-branched, sometimes seasonally deciduous trees of woodland and dry forest and on inselbergs but also (*O. africana* and *macrocarpa*) up to 40 m in forest, the leaves spirally arranged or (more usually) opposite or whorled and trifoliolate to digitately compound. The flushes are reminiscent of *Vitex* (Labiatae). Related to the S. American *Piranhea* (Radcliffe-Smith in *Flora of Tropical East Africa*); Webster, however, assigns *Oldfieldia* to its own subtribe (Paivaeusinae) and *Piranhea* to the all-American Picrodendrinae (both, however, in Picrodendreae). Stuppy (1995; see **Phyllanthoideae**) has called for exclusion of the genus from Euphorbiaceae, with Meliaceae proposed as an alternative. (Oldfieldioideae)

Pax, F. & K. Hoffmann (1922). *Oldfieldia*. In A. Engler (ed.), Das Pflanzenreich, IV 147 XV (Euphorbiaceae-Phyllanthoideae-Phyllantheae): 297. Berlin. (Heft 81.) La/Ge. — 1 species; Africa (W Africa).

Pax, F. & K. Hoffmann (1922). *Paivaeusa*. In A. Engler (ed.), Das Pflanzenreich, IV 147 XV (Euphorbiaceae-Phyllanthoideae-Phyllantheae): 296-297. Berlin. (Heft 81.) La/Ge. — 1 species; Africa. [Genus now reduced to *Oldfieldia*.]

• Léonard, J. (1956). Notulae systematicae XXI. Observations sur les genres *Oldfieldia*, *Paivaeusa* et *Cecchia* (Euphorbiaceae africanae). Bull. Jard. Bot. État 26: 335-343. Fr. — Main revision of *Oldfieldia* with synonymy, types, selected exsiccatae, commentary, and description of *O. macrocarpa* (but no key); 4 species accounted for. *Paivaeusa* and *Cecchia* are reduced.

Oldfieldia Benth. & Hook.f., Hooker's J. Bot. Kew Gard. Misc. 2: 184 (1850).
Trop. Africa. 22 23 24 25 26.
Paivaeusa Welw. ex Benth. in G.Bentham & J.D.Hooker, Gen. Pl. 1: 993 (1867).
Cecchia Chiov., Fl. Somala 2: 397 (1932).

Oldfieldia africana Benth. & Hook.f., Hooker's J. Bot. Kew Gard. Misc. 2: 185 (1850).
W. Trop. Africa. 22 CMN IVO LBR SIE. Nanophan. or phan.

Oldfieldia dactylophylla (Welw. ex Oliv.) J.Léonard, Bull. Jard. Bot. État 26: 340 (1956).
– FIGURE, p. 1222.
Tanzania, Angola, Zambia, Zaïre. 23 ZAI 25 TAN 26 ANG ZAM. Phan.
* *Paivaeusa dactylophylla* Welw. ex Oliv., Fl. Trop. Afr. 1: 328 (1868).

Oldfieldia macrocarpa J.Léonard, Bull. Jard. Bot. État 26: 341 (1956).
Zaire. 23 ZAI. Phan.

Oldfieldia somalensis (Chiov.) Milne-Redh., Kew Bull. 3: 456 (1948 publ. 1949).
Kenya, Tanzania, Somalia, Mozambique. 24 SOM 25 KEN TAN 26 MOZ. Phan.
* *Cecchia somalensis* Chiov., Fl. Somala 2: 397 (1932).
Paivaeusa orientalis Mildbr., Notizbl. Bot. Gart. Berlin-Dahlem 12: 710 (1935).

Oligoceras

1 species, SE. Asia (Vietnam); glabrous trees to 20 m with thick branches and ovate-deltoid leaves similar to *Aleurites*. Now considered related to *Deutzianthus*, following Webster and, in turn, Pax & Hoffman (1931); it is, however, monoecious rather than dioecious. Thin (1995; see **Asia**) groups these two genera along with *Aleurities* and *Vernicia* as national representatives of Aleuritideae. (Crotonoideae)

Gagnepain, F. (1925). Quelques genres nouveaux d'Euphorbiacées. Bul. Soc. Bot. France 71: 864-879. Fr. — Includes (pp. 872-873) generico-specific protologue of *Oligoceras eberhardtii* (described from central Vietnam).

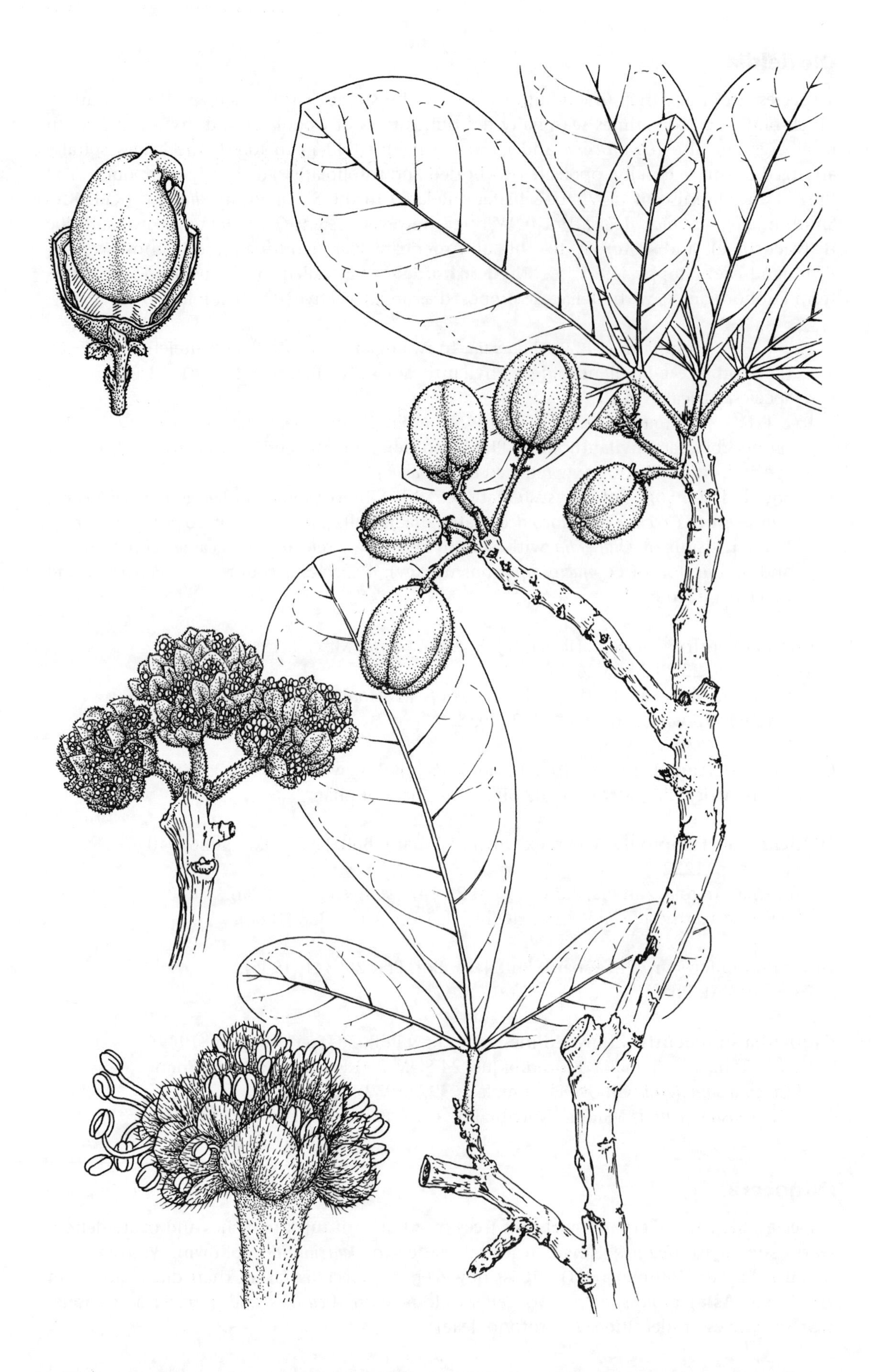

Oldfieldia dactylophylla (Welw. ex Oliv.) J. Léonard
Artist: Christine Grey-Wilson
Fl. Trop. East Africa, Euphorbiaceae: 116, fig. 21 (1987)
KEW ILLUSTRATIONS COLLECTION

Airy-Shaw, H. K. (1960). Notes on Malaysian Euphorbiaceae, XII. The taxonomic position of *Oligoceras* Gagnep. Kew Bull. 14: 392. En. —*Oligoceras* referred to Pax's tribe Cluytieae. [The author, however, also noticed a *habit* similarity with *Deutzianthus*, an opinion earlier expressed by Pax & Hoffmann (1931) and supported in the Webster system.]

Oligoceras Gagnep., Bull. Soc. Bot. France 81: 872 (1924 publ. 1925).
Vietnam. 41.

Oligoceras eberhardtii Gagnep., Bull. Soc. Bot. France 71: 872 (1924 publ. 1925).
S. Vietnam. 41 VIE. Phan.

Omalanthus

The original (and orthographically correct) spelling of *Homalanthus* (q.v.) as used by Adrien de Jussieu and, most recently, by Esser; with conservation of the commonly used orthography accepted, however, *Homalanthus* has been adopted for this *World Checklist.*

Synonyms:
Omalanthus A.Juss. === **Homalanthus** A.Juss.

Omphalandria

Rejected against *Omphalea* L.

Synonyms:
Omphalandria P.Browne === **Omphalea** L.
Omphalandria linearibracteata Millsp. === **Omphalea trichotoma** Müll.Arg.

Omphalea

22 species, widely scattered in the tropics (for the greater part in the Americas, but also in E. Africa (1), Madagascar (4), Asia, Malesia, Australia and the Solomon Islands); includes *Hecatea* and *Neomphalea*. Shrubs, small or medium trees (in *O. ekmanii* to 10 m and *O. oppositifolia* to 15 m) or (in 6 species) large tendrilloid canopy lianas – some at least light-demanding – with clear but scant reddish latex, simple leaves (sometimes also trilobed, more so in juveniles) and more or less large terminal inflorescences with foliaceous bracts. Some species are widespread (but may not be common), while the majority have limited ranges. A revision of the genus by Gillespie (1988), recognising 17 species, has yet to appear in full although some novelties have been published (e.g. Gillespie, 1997). The large seeds of *O. queenslandiae* (*O. papuana*) are in New Guinea threaded into necklaces along with those of *Coix lachryma-jobi* (*Coode s.n.*, K). The position of the genus has been controversial but general support has been lent to Croizat's view of a a relationship with Plukenetieae. Affinities with Stomatocalyceae (Euphorbioideae) have, however, also been proposed (Webster, Synopsis, 1994) and, more recently, molecular and biological studies suggest Crotonoideae, albeit probably as an early offshoot thereof (Gillespie, 1997). Rudall (1994; see **Special**) also believes that the genus is misplaced in Acalyphoideae. The name is conserved against *Omphalandria* P. Br. (Acalyphoideae)

Pax, F. (with K. Hoffmann) (1912). *Omphalea.* In A. Engler (ed.), Das Pflanzenreich, IV 147 V (Euphorbiaceae-Hippomaneae): 14-22. Berlin. (Heft 52.) La/Ge. — 15 species (12 in Americas) in 2 sections (based on leaf venation). [For additions, see *idem*, IV 147 VII (Additamentum V): 419 (1914) and *idem*, IV 147 XIV (Additamentum VI): 54 (1919).]

Pax, F. & K. Hoffmann (1919). *Neomphalea.* In A. Engler (ed.), Das Pflanzenreich, IV 147 XIV (Euphorbiaceae-Additamentum VI): 54. Berlin. (Heft 68.) La/Ge. — Protologue of genus and transfer thereto of *O. papuana*. [The genus is no longer regarded as distinct from *Omphalea*.]

Croizat, L. (1942). Notes on the Euphorbiaceae, II. Bull. Bot. Gard. Buitenzorg, III, 17: 204-208. En. — Among the two contributions in this paper is one on the systematic position of *Omphalea*, regarded as near *Dalechampia* and *Tragia*.

Airy-Shaw, H. K. (1966). Notes on Malaysian and other Asiatic Euphorbiaceae, LXXXI. New records for *Omphalea* L. Kew Bull. 20: 414. En. — Additional records for two species.

Airy-Shaw, H. K. (1966). Notes on Malaysian and other Asiatic Euphorbiaceae, LXXXII. The genus *Neomphalea* Pax et K. Hoffm. in Papua. Kew Bull. 20: 414-415. En. — Range extension of *N. papuana* to the mainland (previously known from New Ireland). [Genus since reduced to *Omphalea*, hinted at here in the discussion.]

Jablonski, E. (1967). *Omphalea*. Euphorbiaceae, Guayana Highland (Mem. New York Bot. Gard. 17(1)): 162-163. New York. En. — 1 species recorded.

Airy-Shaw, H. K. (1969). Notes on Malesian and other Asiatic Euphorbiaceae, CXIX. Notes on *Omphalea* L. and *Neomphalea* Pax et Hoffm. Kew Bull. 23: 130-131. En. — Notes on 2 species, respectively in Australia and New Guinea, in the latter involving reduction of *Neomphalea gageana* to *Omphalea* (subsequently itself referred to *O. queenslandiae*).

Airy-Shaw, H. K. (1971). Notes on Malesian and other Asiatic Euphorbiaceae, CXLVI. *Omphalea* L. in the Solomon Islands. Kew Bull. 25: 550-551. En. — Report of *O. queenslandiae* from Guadalcanal.

Alain H. Liogier (1971). Novitates antillanae, IV. Mem. New York Bot. Gard. 21(2): 107-157, illus., map. En. — Pp. 124-131 comprise a revision of Hispaniolan *Omphalea* (4 species) with key, description of *O. ekmanii*, commentary, and map. There is also a key to 3 Cuban species (exclusive of *OO. commutata* and *diandra*).

Airy-Shaw, H. K. (1978). Notes on Malesian and other Asiatic Euphorbiaceae, CCXXI. *Omphalea* L. Kew Bull. 33: 77. En. — The flowers of *O. bracteata* var. *pedicellaris* (Borneo) are those of an *Erycibe* (Convolvulaceae), making the type collection a mixed gathering.

• Gillespie, L. J. (1988). A revision and phylogenetic analysis of *Omphalea* (Euphorbiaceae). 297 pp., 53 fig. Davis, Calif. (Unpubl. Ph.D. dissertation, University of California, Davis.) En. — Includes a full revision of the genus. [Not seen; cited from Gillespie & Armbruster 1997 (see **Acalyphoideae**).]

Forster, P. I. (1995). *Omphalea celata*, a new species of Euphorbiaceae from central Queensland. Austrobaileya 4: 381-385, illus. En. — Novelty (*O. celata*); key to Australian *Omphalea*.

Gillespie, L. J. (1997). *Omphalea* (Euphorbiaceae) in Madagascar: a new species and a new combination. Novon 7: 127-136, illus., map. En. — In fact a revision of the genus in Madagascar (4 species, one new, one transferred from *Hecatea*) with key, synonymy, references, types, vernacular names, localities with exsiccatae, indication of distribution and habitat, and commentary; all species mapped. A general review of the genus is also presented.

Omphalea L., Syst. Nat. ed. 10, 2: 1264 (1759).
Trop. 25 29 41 42 60 80 81 82 83 84.
Omphalandria P.Browne, Civ. Nat. Hist. Jamaica: 335 (1756).
Duchola Adans., Fam. Pl. 2: 357 (1763).
Ronnowia Buc'hoz, Pl. Nouv. Découv.: 6 (1779).
Hecatea Thouars, Hist. Vég. îles France: 27 (1804).
Hecaterium Kunze ex Rchb., Handb. Nat. Pfl.-Syst.: 281 (1837).
Hebecocca Beurl., Kongl. Vetensk. Acad. Handl. 1854: 146 (1856).
Neomphalea Pax & K.Hoffm. in H.G.A.Engler, Pflanzenr., IV, 147, XIV: 54 (1919).

Omphalea ankaranensis L.J.Gillespie, Novon 7: 128 (1997).
N. Madagascar. 29 MDG. Nanophan. or phan.

Omphalea bracteata (Blanco) Merr., Sp. Blancoan.: 230 (1918).
S. Burma, Pen. Malaysia, Laos, Borneo (Sabah, E. Kalimantan), Philippines, Sulawesi, Flores. 41 BMA LAO 42 BOR LSI MLY PHI SUL. Cl. nanophan. or phan.

* *Tragia bracteata* Blanco, Fl. Filip., ed. 2: 481 (1845).
Omphalea philippinensis Merr., Philipp. J. Sci., C 3: 236 (1908).
Omphalea sargentii Merr., Philipp. J. Sci. 16: 574 (1920).

Omphalea brasiliensis Müll.Arg., Linnaea 32: 86 (1863).
Brazil (Bahia). 84 BZE. Phan.

Omphalea cardiophylla Hemsl., Pharm. J. Trans.: 301 (1882).
El Salvador. 80 ELS. Phan.

Omphalea celata P.I.Forst., Austrobaileya 4: 381 (1995).
Queensland. 50 QLD.

Omphalea commutata Müll.Arg., Linnaea 32: 86 (1863).
Dominican Rep. 81 DOM. Phan.
Omphalea triandra Tussac, Fl. Antill. 4: 18 (1827), nom. illeg.

Omphalea diandra L., Syst. Nat. ed. 10, 2: 1264 (1759). *Omphalea diandra* var. *genuina* Müll.Arg. in A.P.de Candolle, Prodr. 15(2): 1135 (1866), nom. inval.
Trop. America. 80 COS ELS HON PAN 81 CUB DOM JAM LEE TRT 82 FRG GUY SUR VEN 83 CLM PER 84 BZL BZN. Cl. phan.
Omphalea cordata Sw., Prodr.: 95 (1788).
Omphalea diandra var. *panamensis* Klotzsch in B.Seemann, Bot. Voy. Herald: 101 (1853). *Hebecocca panamensis* (Klotzsch) Beurl., Kongl. Vetensk. Acad. Handl. 1854: 146 (1856). *Omphalea panamensis* (Klotzsch) I.M.Johnst., Sargentia 8: 177 (1949).
Omphalea diandra var. *paraensis* Baill., Adansonia 5: 335 (1865).
Omphalea guyanensis Klotzsch ex Lanj., Euphorb. Surinam: 163 (1931), nom. illeg.

Omphalea ekmanii Alain, Mem. New York Bot. Gard. 21: 128 (1971).
Dominican Rep. 81 DOM. Phan.

Omphalea elaeophoroides Steyerm., Publ. Field Mus. Nat. Hist., Bot. Ser. 22: 152 (1940).
Brazil (Amazonas). 84 BZN.

Omphalea frondosa Juss. ex Baill., Étude Euphorb.: 529 (1858).
Jamaica. 81 JAM. Phan.

Omphalea grandifolia Merr., Philipp. J. Sci. 16: 574 (1920).
Philippines. 42 PHI.

Omphalea hypoleuca Griseb., Nachr. Königl. Ges. Wiss. Georg-Augusts-Univ. 1: 177 (1865).
W. Cuba. 81 CUB.

Omphalea malayana Merr., Philipp. J. Sci., C 11: 71 (1916).
Pen. Malaysia (Pulao Tioman), Borneo (Sarawak), Philippines (Luzon). 42 BOR MLY PHI. Nanophan. or phan.
Trigonostemon arboreus Ridl., Bull. Misc. Inform. Kew 1928: 75 (1928).

Omphalea mansfeldiana Mildbr., Notizbl. Bot. Gart. Berlin-Dahlem 13: 276 (1936).
– FIGURE, p. 1226.
Tanzania. 25 TAN. Cl. nanophan.

Omphalea megacarpa Hemsl., Hooker's Icon. Pl. 26: t. 2537 (1897).
Trinidad, Tobago, Grenada. 81 TRT WIN. Cl. nanophan.

Omphalea mansfeldiana Mildbr.
Artist: Christine Grey-Wilson
Fl. Trop. East Africa, Euphorbiaceae 1: 372, fig. 69 (1987)
KEW ILLUSTRATIONS COLLECTION

Omphalea occidentalis Leandri, Bull. Soc. Bot. France 85: 529 (1938 publ. 1939).
WC. & NW. Madagascar. 29 MDG. Nanophan. or phan.

Omphalea oleifera Hemsl., Pharm. J. Trans.: 301 (1882).
El Salvador. 80 ELS. Phan.

Omphalea oppositifolia (Willd.) L.J.Gillespie, Novon 7: 133 (1997).
E. Madagascar. 29 MDG. Nanophan. or phan.
Hecatea alternifolia Willd., Sp. Pl. 4: 514 (1805). *Omphalea alternifolia* (Willd.) Baill., Étude Euphorb.: 529 (1858).
* *Hecatea oppositifolia* Willd., Sp. Pl.: 513 (1805).
Hecatea biglandulosa Pers., Syn. Pl. 2: 588 (1807). *Omphalea biglandulosa* (Pers.) Baill., Étude Euphorb.: 529 (1858).

Omphalea palmata Leandri, Bull. Soc. Bot. France 85: 530 (1938 publ. 1939).
C. Western Madagascar. 29 MDG. Nanophan. or phan.

Omphalea queenslandiae F.M.Bailey, Rep. Exped. Bellenden-Ker: 58 (1889).
Papua New Guinea, Bismarck Archip., Queensland (Cook), Solomon Is. 42 BIS NWG 50 QLD 60 SOL. Cl. nanophan.
Omphalea papuana Pax & K.Hoffm. in H.G.A.Engler, Pflanzenr., IV, 147, VII: 419 (1914).
Omphalea papuana Gage, Nova Guinea 12: 485 (1917), nom. illeg. *Omphalea gageana* Airy Shaw, Kew Bull. 23: 131 (1969).

Omphalea triandra L., Syst. Nat. ed. 10, 2: 1264 (1759). *Omphalea triandra* var. *genuina* Pax & K.Hoffm. in H.G.A.Engler, Pflanzenr., IV, 147, V: 16 (1912), nom. inval.
Jamaica, Haiti. 81 JAM HAI. Phan.
Omphalea nucifera Sw., Prodr.: 95 (1788).
Omphalea laevigata Desf., Tabl. École Bot., ed. 3: 411 (1829).
Omphalea triandra var. *robusta* Pax & K.Hoffm. in H.G.A.Engler, Pflanzenr., IV, 147, V: 16 (1912).
Omphalea trinitatis Pax & K.Hoffm. in H.G.A.Engler, Pflanzenr., IV, 147, V: 16 (1912).

Omphalea trichotoma Müll.Arg., Linnaea 32: 86 (1863).
Cuba. 81 CUB. Nanophan. or phan.
Omphalandria linearibracteata Millsp., Publ. Field Columbian Mus., Bot. Ser. 2: 59 (1900). *Omphalea linearibracteata* (Millsp.) Pax in H.G.A.Engler, Pflanzenr., IV, 147, V: 20 (1912).

Synonyms:
Omphalea alternifolia (Willd.) Baill. === **Omphalea oppositifolia** (Willd.) L.J.Gillespie
Omphalea axillaris Sw. === **Phyllanthus axillaris** (Sw.) Müll.Arg.
Omphalea biglandulosa (Pers.) Baill. === **Omphalea oppositifolia** (Willd.) L.J.Gillespie
Omphalea bracteata var. *pedicellaris* Airy Shaw === **Erycibe** sp. (Convolvulaceae)
Omphalea cauliflora Sw. === **Phyllanthus cauliflorus** (Sw.) Griseb.
Omphalea cordata Sw. === **Omphalea diandra** L.
Omphalea diandra Vell. === **Sebastiania sp.**
Omphalea diandra var. *genuina* Müll.Arg. === **Omphalea diandra** L.
Omphalea diandra var. *panamensis* Klotzsch === **Omphalea diandra** L.
Omphalea diandra var. *paraensis* Baill. === **Omphalea diandra** L.
Omphalea eglandulata Vell. === **Sebastiania eglandulata** (Vell.) Pax
Omphalea epistylium Poir. === **Phyllanthus axillaris** (Sw.) Müll.Arg.
Omphalea gageana Airy Shaw === **Omphalea queenslandiae** F.M.Bailey
Omphalea glandulata Vell. === **Sapium glandulosum** (L.) Morong
Omphalea guyanensis Klotzsch ex Lanj. === **Omphalea diandra** L.
Omphalea lactescens Vell. === **Mabea piriri** Aubl.

Omphalea laevigata Desf. === **Omphalea triandra** L.
Omphalea linearibracteata (Millsp.) Pax === **Omphalea trichotoma** Müll.Arg.
Omphalea nucifera Sw. === **Omphalea triandra** L.
Omphalea panamensis (Klotzsch) I.M.Johnst. === **Omphalea diandra** L.
Omphalea papuana Gage === **Omphalea queenslandiae** F.M.Bailey
Omphalea papuana Pax & K.Hoffm. === **Omphalea queenslandiae** F.M.Bailey
Omphalea philippinensis Merr. === **Omphalea bracteata** (Blanco) Merr.
Omphalea sargentii Merr. === **Omphalea bracteata** (Blanco) Merr.
Omphalea triandra Tussac === **Omphalea commutata** Müll.Arg.
Omphalea triandra var. *genuina* Pax & K.Hoffm. === **Omphalea triandra** L.
Omphalea triandra var. *robusta* Pax & K.Hoffm. === **Omphalea triandra** L.
Omphalea trinitatis Pax & K.Hoffm. === **Omphalea triandra** L.
Omphalea verticillata Vell. === **Senefeldera verticillata** (Vell.) Croizat

Ophellantha

2 species, Mexico (Oaxaca, Chiapas) and C. America (Guatemala, El Salvador); small trees of thickets to 6 m with thin, small, somewhat rhombic leaves and short, sometimes obscure stipular spines. The genus is very close to *Acidocroton* and Webster (Synopsis, 1994) recommends merger. Both are in tribe Codiaeae. (Crotonoideae)

Standley, P. C. (1924). New species of plants from Salvador, III. J. Washington Acad. Sci. 14: 93-99. En. — Includes (pp. 97-98) description of *Ophellantha* and *O. spinosa* from El Salvador. [Also treated under that genus in *Flora of Guatemala*.]

Ophellantha Standl., J. Wash. Acad. Sci. 14: 97 (1924).
C. America. 80.

Ophellantha spinosa Standl., J. Wash. Acad. Sci. 14: 98 (1924). *Acidocroton spinosus* (Standl.) G.L.Webster, Ann. Missouri Bot. Gard. 81: 107 (1994).
El Salvador. 80 ALS. Nanophan.

Ophellantha steyermarkii Standl., Publ. Field Mus. Nat. Hist., Bot. Ser. 23: 123 (1944). *Acidocroton steyermarkii* (Standl.) G.L.Webster, Ann. Missouri Bot. Gard. 81: 107 (1994).
S. Mexico, Guatemala. 79 MXS 80 GUA. Nanophan.

Ophthalmoblapton

4 species, S America (Brazil); forest trees with relatively narrow, finely-veined, laurel-like leaves and very poisonous latex. The potential action of the latter gave rise to the generic name. Related to *Hura* according to Webster (Synopsis, 1994) but this is contested by Esser (1994: 20; see **Euphorbiaceae (except Euphorbieae**). (Euphorbioideae (except Euphorbieae))

Pax, F. (with K. Hoffmann) (1912). *Ophthalmoblapton*. In A. Engler (ed.), Das Pflanzenreich, IV 147 V (Euphorbiaceae-Hippomaneae): 278-281. Berlin. (Heft 52.) La/Ge. — 3 species in two sections, E and SE Brazil; *O. macrophyllum* of distinctive appearance.
Emmerich, M. (1981). Controbuição ao estudo das Euphorbiaceae brasileiras, I. Duas espécies novas. Bol. Mus. Nac. Rio de Janeiro, Bot. 62: 1-7, illus. Pt. — Includes description of *Ophthalmoblapton parviflorum* along with a key to 4 species (in 3 sections). [The author's new sect. *Monantha*, erected for the new species, may be invalid.]

Ophthalmoblapton Allemão, Pl. Novas Brasil: 4 (1849).
Brazil. 84.

Ophthalmoblapton crassipes Müll.Arg. in C.F.P.von Martius, Fl. Bras. 11(2): 532 (1874).
Brazil (São Paulo, Santa Catarina). 84 BZL BZS. Phan.

Ophthalmoblapton macrophyllum Allemão, Pl. Novas Brasil: 4 (1849).
Brazil (Rio de Janeiro). 84 BZL. Phan.
Ophthalmoblapton brasiliense Walp., Ann. Bot. Syst. 3: 362 (1852).
Ophthalmoblapton megalophyllum Müll.Arg. in C.F.P.von Martius, Fl. Bras. 11(2): 707 (1874).

Ophthalmoblapton parviflorum Emmerich, Bol. Mus. Nac. Rio de Janeiro, Bot. 62: 1 (1981).
Brazil (Bahia). 84 BZE. Phan.

Ophthalmoblapton pedunculare Müll.Arg. in C.F.P.von Martius, Fl. Bras. 11(2): 533 (1874).
Brazil (Minas Gerais, São Paulo, Rio de Janeiro). 84 BZL. Phan.

Synonyms:
Ophthalmoblapton brasiliense Walp. === **Ophthalmoblapton macrophyllum** Allemão
Ophthalmoblapton megalophyllum Müll.Arg. === **Ophthalmoblapton macrophyllum** Allemão

Orbicularia

Synonyms:
Orbicularia Baill. === **Phyllanthus** L.
Orbicularia foveolata Britton === **Phyllanthus myrtilloides** subsp. **erythrinus** (Müll.Arg.) G.L.Webster
Orbicularia orbicularis (Kunth) Moldenke === **Phyllanthus orbicularis** Kunth
Orbicularia phyllanthoides Baill. === **Phyllanthus orbicularis** Kunth
Orbicularia scopulorum Britton === **Phyllanthus scopulorum** (Britton) Urb.

Oreoporanthera

2 species, New Zealand and Tasmania. The New Zealand species is a small 'alpine' shrub of South Island, while that in Tasmania, described only in the 1980s, is similarly 'alpine'. Like its near relative *Poranthera,* the presence of narrow cotyledons resulted in placement in the 'Stenolobae' by Mueller and by Pax and Hoffmann. As with other Australasian genera of that group, however, the genus appears instead to be a specialised offshoot from the *Antidesma* line (Webster, Synopsis, 1994). In contrast to *Poranthera* it is dioecious and lacks petals. [Additional coverage of *O. alpina* may be sought in standard works on the New Zealand flora and its 'alpine' element.] (Phyllanthoideae)

Hooker, J. D. (1881). *Poranthera alpina.* Ic. Pl. 14: pl. 1366. En. — Plant portrait with descriptive text (the effective protologue of this species).
Gruening, G. (1913). *Poranthera.* In A. Engler (ed.), Das Pflanzenreich, IV 147 [Stenolobieae] (Euphorbiaceae-Porantheroideae et Ricinocarpoideae): 13-21. Berlin. (Heft 58.) La/Ge. — Of the 7 species, the last, *P. alpina* of New Zealand, the sole species of subgen. *Oreoporanthera,* was raised to generic rank by Hutchinson in his 1969 system of the family, a move since upheld.

Oreoporanthera (Grüning) Hutch., Amer. J. Bot. 56: 747 (1969).
Tasmania, New Zealand. 50 51. Cham.

Oreoporanthera alpina (Cheeseman ex Hook.f.) Hutch., Amer. J. Bot. 56: 747 (1969).
New Zealand. 51 NZS. Cham.
* *Poranthera alpina* Cheeseman ex Hook.f., Hooker's Icon. Pl. 14: t. 1366 (1881).

Oreoporanthera petalifera Orchard & J.B.Davies, Pap. & Proc. Roy. Soc. Tasmania 119: 62 (1985).
Tasmania. 50 TAS. Cham.

Orfilea

4 species, Madagascar (3) and Mauritius; includes *Lautembergia*. Forest shrubs or small trees with spirally arranged leaves and scurfy terminal inflorescences. The two sections recognised by Pax (1914) were distinguished on the basis of presence or absence of a hypogynous disk. Along with *Bossera*, the genus is related to *Alchornea* (Webster, Synopsis, 1994). No recent revision is available. (Acalyphoideae)

Pax, F. (with K. Hoffmann) (1914). *Lautembergia*. In A. Engler (ed.), Das Pflanzenreich, IV 147 VII (Euphorbiaceae-Acalypheae-Mercurialinae): 253-254. Berlin. (Heft 63.) La/Ge. — 3 species in 2 sections, Madagascar.

Orfilea Baill., Étude Euphorb.: 452 (1858).
Madagascar, Mauritius. 29.
Lautembergia Baill., Étude Euphorb.: 451 (1858).
Diderotia Baill., Adansonia 1: 274 (1861).

Orfilea ankafinensis (Baill.) Radcl.-Sm. & Govaerts, Kew Bull. 52: 480 (1997).
W. Madagascar. 29 MDG.
* *Macaranga ankafinensis* Baill., Bull. Mens. Soc. Linn. Paris 2: 992 (1892).

Orfilea coriacea Baill., Étude Euphorb.: 453 (1858). *Alchornea coriacea* (Baill.) Müll.Arg., Linnaea 34: 168 (1865).
Madagascar. 29 MDG.

Orfilea multispicata (Baill.) G.L.Webster, Ann. Missouri Bot. Gard. 81: 81 (1994).
Madagascar. 29 MDG.
* *Lautembergia multispicata* Baill., Étude Euphorb.: 452 (1858). *Alchornea multispicata* (Baill.) Müll.Arg., Linnaea 34: 168 (1865).

Orfilea neraudiana (Baill.) G.L.Webster, Ann. Missouri Bot. Gard. 81: 81 (1994).
Mauritius. 29 MAU.
* *Claoxylon neraudianum* Baill., Adansonia 1: 280 (1861).
Cleidion cafcaf Croizat, Trop. Woods 77: 16 (1944).

Ostodes

2 species, Asia and Malesia (S. China, Vietnam and India to Sumatra); small to medium trees with large paniculate inflorescences (borne below the terminal flush of leaves). The genus is closely related to *Dimorphocalyx* and *Paracroton*, although in the Webster system the latter is included in Crotoneae on account of having stellate indumentum. (Crotonoideae)

Pax, F. (with K. Hoffmann) (1911). *Ostodes*. In A. Engler (ed.), Das Pflanzenreich, IV 147 III (Euphorbiaceae-Cluytieae): 17-22. Berlin. (Heft 47.) La/Ge. — 9 species in 2 sections; a further species is of uncertain position. [Several additions in later Pflanzenreich installments; however, for the most part these have been referred to other genera or even families.]

Airy-Shaw, H. K. (1966). Notes on Malaysian and other Asiatic Euphorbiaceae, LXXIX. Realignments in the *Ostodes-Dimorphocalyx*-complex. Kew Bull. 20: 409-413. En. — Key to genera; new combinations with synonymy and indication of distribution; list of *Ostodes* names with their disposition. [Some taxa are now in other genera.]

Chakrabarty, T. & N. P. Balakrishnan (1985(1987)). A note on the genus *Ostodes* (Euphorbiaceae). Bull. Bot. Surv. India 27: 259-260. En. — Reduction of *OO. katherinae* and *thyrsantha* to *O. paniculata* at varietal rank; key and additional records not previously documented in the literature along with a historical review.

Ostodes Blume, Bijdr.: 619 (1826).
S. China, Trop. Asia. 36 40 41 42.

Ostodes kuangii Y.T.Chang, Acta Phytotax. Sin. 20: 224 (1982).
S. China. 36 CHS.

Ostodes paniculata Blume, Bijdr.: 620 (1826).
E. Nepal to W. Malesia. 36 CHC CHH CHT 40 ASS EHM NEP 41 BMA THA VIE 42 JAW MLY SUM. Phan.

var. **katharinae** (Pax) Chakrab. & N.P.Balakr., Bull. Bot. Surv. India 27: 260 (1985 publ. 1987).
SE. Tibet, China (S. Yunnan), NW. Thailand. 36 CHC CHT 41 THA. Phan.
* *Ostodes katharinae* Pax in H.G.A.Engler, Pflanzenr., IV, 147, III: 19 (1911).

var. **paniculata**
China (Yunnan, Hainan), E. Nepal, Sikkim, Assam, Burma, Thailand, Pen. Malaysia, Vietnam, Sumatera, Jawa. 36 CHC CHH 40 ASS EHM NEP 41 BMA THA VIE 42 JAW MLY SUM. Phan.
Ostodes corniculata Baill., Étude Euphorb.: 391 (1858).
Ostodes kerrii Craib, Bull. Misc. Inform. Kew 1911: 464 (1911).
Ostodes thyrsantha Pax in H.G.A.Engler, Pflanzenr., IV, 147, II: 18 (1911). *Ostodes paniculata* var. *thyrsantha* (Pax) Chakrab. & N.P.Balakr., Bull. Bot. Surv. India 27: 260 (1985 publ. 1987).
Ostodes prainii Gand., Bull. Soc. Bot. France 66: 287 (1919 publ. 1920).

Synonyms:
Ostodes angustifolia Merr. === **Dimorphocalyx angustifolius** (Merr.) Airy Shaw
Ostodes appendiculata Hook.f. === **Lepisanthes tetraphylla** (Vahl) Radlk. (Sapindaceae)
Ostodes collina (Rchb.f. & Zoll. ex Müll.Arg.) Pax === **Paracroton pendulus** (Hassk.) Miq. subsp. **pendulus**
Ostodes corniculata Baill. === **Ostodes paniculata** Blume var. **paniculata**
Ostodes helferi Müll.Arg. === **Popowia pauciflora** Maing. ex Hook.f. & Thoms. (Annonaceae)
Ostodes integrifolia Airy Shaw === **Paracroton integrifolius** (Airy Shaw) N.P.Balakr. & Chakrab.
Ostodes ixoroides C.B.Rob. === **Dimorphocalyx ixoroides** (C.B.Rob.) Airy Shaw
Ostodes katharinae Pax === **Ostodes paniculata** var. **katharinae** (Pax) Chakrab. & N.P.Balakr.
Ostodes kerrii Craib === **Ostodes paniculata** Blume var. **paniculata**
Ostodes macrophylla (Müll.Arg.) Benth. & Hook.f. === **Paracroton pendulus** (Hassk.) Miq. subsp. **pendulus**
Ostodes minor (Thwaites) Müll.Arg. === **Paracroton zeylanicus** (Müll.Arg.) N.P.Balakr. & Chakrab.
Ostodes muricata Hook.f. === **Dimorphocalyx muricatus** (Hook.f.) Airy Shaw
Ostodes muricata var. *genuinus* Pax === **Dimorphocalyx muricatus** (Hook.f.) Airy Shaw
Ostodes muricata var. *minor* Hook.f. === **Dimorphocalyx muricatus** (Hook.f.) Airy Shaw
Ostodes paniculata var. *thyrsantha* (Pax) Chakrab. & N.P.Balakr. === **Ostodes paniculata** Blume var. **paniculata**
Ostodes pauciflora Merr. === **Dimorphocalyx denticulatus** Merr.
Ostodes pendula (Hassk.) A.Meeuse === **Paracroton pendulus** (Hassk.) Miq.
Ostodes prainii Gand. === **Ostodes paniculata** Blume var. **paniculata**
Ostodes serratocrenata Merr. === **Paracroton pendulus** (Hassk.) Miq. subsp. **pendulus**
Ostodes thyrsantha Pax === **Ostodes paniculata** Blume var. **paniculata**
Ostodes villamilii Merr. === **Tapoides villamilii** (Merr.) Airy Shaw

Ostodes zeylanica (Thwaites) Müll.Arg. === **Paracroton pendulus** subsp. **zeylanicus** (Thwaites) N.P.Balakr. & Chakrab.
Ostodes zeylanica var. *minor* (Thwaites) Bedd. === **Paracroton zeylanicus** (Müll.Arg.) N.P.Balakr. & Chakrab.

Owataria

Synonyms:
Owataria Matsum. === **Suregada** Roxb. ex Rottl.

Oxalistylis

Synonyms:
Oxalistylis Baill. === **Phyllanthus** L.

Oxydectes

Synonyms:
Oxydectes L. ex Kuntze === **Croton** L.
Oxydectes hauthalii Kuntze === **Croton hauthalii** (Kuntze) K.Schum.